ReCombinatorics

I dedicate this book
to the memory of my mother
Irma Gusfield,
and to the future of my daughters
Talia and Shira

MIT Press books may be purchased at special quantity discounts for business or sales promotional use. For information, please email special_sales@mitpress.mit.edu.

This book was set in Times Roman using LATEX by TEXnology Inc. and was printed and bound in the United States of America.

Library of Congress Cataloging-in-Publication Data

Gusfield, Dan.
ReCombinatorics : the algorithmics of ancestral recombination graphs and explicit phylogenetic networks / Dan Gusfield.
        p. cm
Includes bibliographical references and index.
ISBN 978-0-262-02752-6 (hardcover : alk. paper)
1. Genetic recombination. 2. Evolution (Biology) I. Title. II. Title: Re combinatorics.
QH443.G87 2014
572.8'77dc23
2013043434

10   9   8   7   6   5   4   3   2   1

Dan Gusfield

with contributions from

Charles H. Langley
Yun S. Song and
Yufeng Wu

ReCombinatorics

The Algorithmics of Ancestral Recombination Graphs and Explicit Phylogenetic Networks

The MIT Press
Cambridge, Massachusetts   London, England

Contents

Preface

Where This Book Came From

Early history: It started with haplotyping

About fourteen years ago, Chuck Langley (population geneticist in the Department of Evolution and Ecology at UC Davis) asked me to look at a ten-year-old paper [71] on an algorithm to solve the *Haplotyping* problem (discussed in chapter 12). That seminal paper, written by Andrew Clark in 1990, was ahead of its time because the genetic data that the algorithm was designed to analyze was not yet plentiful. But by the year 2000, with the introduction of widespread, high-throughput genotyping, and later sequencing, the data flood had begun, and computational haplotyping was becoming a central issue in analyzing genotypic data for population genetics. Digging into the 1990 paper, and then developing new ideas based on it, I began a focus on computational problems in *population genetics*, and also began a continuing collaboration and conversation with Chuck. After we obtained a joint NSF grant, additional graduate students, postdocs, and visitors joined our group at UC Davis, for differing periods of time.[1]

Next came haplotype blocks and haplotyping in blocks

The announcement in 2001 of the identification of *haplotype blocks* in humans (discussed in chapter 12), motivated the next question: Finding haplotyping solutions that conform to

---

1 In chronological order of participation, the other people involved at UC Davis on topics in this book were: John Kececioglu (as a postdoc before this history, and later on sabbatical), R. H. Chung, Yelena Frid, Satish Eddhu, Vladimir Filkov, Zhihong Ding, Dean Hickerson, Yun S. Song, Yufeng Wu, Dan Brown (visiting from Waterloo), Fumei Lam, Simone Linz, Rob Gysel, Kristian Stevens, Balaji Venkatachalam, and Michael Coulombe. Collaborators from outside of UC Davis were V. Bafna, V. Bansal, S. Hannenhalli, G. Lancia, S. Orzack, and S. Yooseph.

the *perfect-phylogeny* **tree** model. That problem is called the *perfect-phylogeny haplotyping (PPH) problem*, and it is also discussed in chapter 12. The PPH problem is a beautiful algorithmic and combinatorial problem that at first sounds artificial but actually models biological reality in the appropriate contexts. Its solution partly uses deep combinatorial mathematics that goes back to the 1930s. Rarely does one encounter such a well-structured combinatorial optimization problem, with beautiful, efficient algorithmic solutions, where the problem captures a real application without the need for artificial assumptions. But still, the perfect-phylogeny tree model, which involves no recombination, needed to be extended for even greater genomic application. Somehow, recombination had to be added into the model.

Which led to constrained recombination and galled trees

In a fortuitous development, a seminal paper [426] had appeared a couple of years earlier, introducing an extension of the perfect-phylogeny tree model to phylogenetic *networks* with constrained recombination. That network model, later called the "galled tree" model (discussed in chapter 8), seemed appropriate for a *"modest"* level of recombination. Chuck was particularly interested in this model, explaining that we want to find regions in the genome where some amount of recombination has occurred, but not a huge amount. Recombination causes variation in genomes, which can be exploited to connect what appears in the genome (*genotypes*) to important genetic traits (*phenotypes*) in an individual. This allows one to identify genomic loci that influence genetic traits, and to better understand details (and maybe the mechanisms) of those genetic influences. So, some recombination is needed, but too much recombination obscures the signal, making the analysis too hard. Therefore, an algorithm that finds regions of *modest*, but not zero, recombination is of interest. Hence, with graduate student S. Eddhu, we dug into the network paper [426], leading to new ideas and methods on galled trees. From that introduction to networks with recombination, my focus shifted sharply to problems of phylogenetic *networks with recombination*, of the type that arise in population genetics.

The next development: Simon Myers's thesis, and Yun and Jotun's constructions

The next turning point was the publication of the paper [302] by S. Myers and R. Griffiths in 2003 (based on the Oxford University dissertation of Myers [299]), on computing *lower bounds* on the number of recombination events needed in an *ancestral recombination graph (ARG)* (formally defined in chapter 3), to create a given set of binary sequences. This paper was the first real advance on that question since 1985, and it pointed to several different research directions. The lower bounds in Myers's thesis will be discussed in chapters 5 and 10. At about the same time as Myers and Griffiths were working on lower bounds, Yun Song and Jotun Hein (also at Oxford) were developing algorithms to explicitly *build* networks (ancestral recombination graphs) that *construct* a given set of binary sequences, minimizing the number of recombination events in the network. Those methods continued the earlier work of Jotun's a decade before. That work will be partly discussed in chapter 9.

Shortly thereafter, Yun joined our group in Davis as a postdoctoral student (supervised jointly by myself and Chuck), and Yufeng Wu and Zhihong Ding joined as Ph.D. students. This core group (along with some of the participants mentioned earlier), focused almost exclusively on problems involving networks and recombination, motivated by issues in population genomics. In our first summer together, we organized a readings seminar on coalescent theory, reading papers and the newly published book on gene genealogies and coalescent theory [168] by Hein, Schierup, and Wiuf. From that seminar, we wrote out a series of open questions that moved from the most specific to the most general, where an answer to any question would also answer the prior ones. Additional nonserial questions were also enumerated.

Middle history: The growth of phylogenetic networks

Before continuing the story of how this book came about, I digress to say something of the history of the field of phylogenetic networks. In my view, the field of phylogenetic networks has three early sources.[2] It was initially identified with the seminal work on *median networks* and *splits decomposition* and associated ideas, developed by Hans J. Bandelt, and Andreas Dress and their students, starting in the early 1990s. Those networks are now often called "data display" networks, because they represent patterns of incompatibility in data but do not try to tell an explicit story of the evolution of the data. The second source of the field actually began before the first one, but was only later considered part of the field of phylogenetic networks. That second source is the work of R. Hudson, P. Marjoram, and R. Griffiths, who defined models and networks (ancestral recombination graphs) building on coalescent theory, to explicitly represent the evolution of binary sequences through mutation and recombination. The third source of the field is the work of Jotun Hein, starting in 1990, exploring algorithms "reconstructing evolution of sequences subject to recombination." (Ancestral recombination graphs and the related networks of Jotun are now called "explicit" phylogenetic networks.)

The three sources that started in the late 1980s and early 1990s gave rise to the growth and the broadening of the field of phylogenetic networks about fifteen years later. "Who Is Who in Phylogenetic Networks" [121] shows the rapid growth of the field. The number of papers published in the years 2001 through 2005, were 10, 8, 12, 28, and 42,

---

2 Of course, there were other early contributors to the field, and I apologize to all those whose names I have omitted.

respectively, and 55 in 2012. More impressive, there has been a broadening of the lines of inquiry, and models, beyond the three sources. Many new problems and models have been examined (for example, *galled trees, level-k networks, cluster networks, hardwired and softwired reticulation networks, normal networks, tree-child networks*, and more). In the last decade there have also been several dissertations on phylogenetic networks, and two books published.

Additionally, there were (at least) six international meetings (that I was privileged to attend) that helped propel the field: The 2004 meeting on phylogenetic combinatorics at Uppsala University; the 2005 meeting on the mathematics of evolution and phylogeny, at the Henri Poincaré Institute in Paris; two meetings on mathematical methods in phylogenetics, sponsored by the Isaac Newton Institute at Cambridge University in 2007 and 2011; the 2009 meeting on algorithmics in human population genomics, at the Dimacs Center at Rutgers University; and most recently, a 2012 meeting sponsored by the Lorentz Institute at the University of Leiden, on the future of phylogenetic networks. These meetings brought people together from different parts of the field and from different parts of the world, and helped to define the broader field of phylogenetic networks and to create a more coherent community of phylogenetic network researchers.

Back to the story of the book

During that middle period, as the field blossomed, the work of our group at UC Davis also blossomed. In the four years after we wrote our list of questions, all but one of them were answered (by us and others). And a week before the book went to the publisher, a proposed answer to the last question was explicitly verified (see section 14.6.2).

Late history: The emerging book

With the maturation of the field, and with the graduation of Zhihong (now working at Adobe) and of Yufeng (now tenured at U Connecticut) and the departure of Yun (now tenured at UC Berkeley), it seemed that it was time to revisit the whole area, to write a book for a broad audience of computer scientists, mathematicians, and biologists, with several goals in mind.

The most concrete goal was to give a more scholarly, integrated, and unified exposition of computational issues involving ARGs; standardizing notation, adding many illustrations and examples, and simplifying, completing, correcting, and generalizing various proofs and algorithms. A book would allow a deeper treatment of various topics, and yet at a more leisurely pace than is possible in a journal publication, and it would allow new full expositions, and integration, of difficult material that had been developed by other researchers. The next goal was to tell the story of the series of results (ours and others) in which an increasingly general understanding of combinatorics and algorithmics of ARGs was developed; but to tell it *backward*—that is, to first develop the most abstract and general results (in chapters 6 and 7), and then use that machinery to explain and prove more specific results (for example on galled trees), which chronologically had been developed before, and had led to, the more general results. The next goal was to identify, and make explicit, common ideas (mostly inspired by the coalescent theory *viewpoint* but not by any actual coalescent theory) that underlie many disparate methods to construct ARGs. This is done in the first sections of chapter 9. This goal also led to the unification of constructive and destructive methods (explained in chapter 9) and to the connection of ARG construction methods to the history lower-bound method (in chapter 10). An additional goal was to explain the use of ARGs in a variety of applications, for example in association mapping (discussed in chapters 4 and 13), and in the logic behind association mapping methods that *don't use* ARGs. Another goal was to show how ARGs fit into the larger field of phylogenetic networks, relating ARG problems and models to phylogenetic network problems and models that seem at first unrelated to ARGs.

Finally, the most general goal was to widen the biological focus beyond problems involving humans, and widen the methodological focus beyond population genomics. In particular, to explicitly (and in several ways) make the point that although the biological contexts of population genetics and phylogenetics are very different, there are mathematical and algorithmic ideas that are common to both fields, where the biological differences do not matter. In fact, there are formal ways to *transform* certain problems and results in one domain to problems and results in the other. This point is made throughout the book (perhaps ad nauseam), but most explicitly in section 3.2.3.3 and in chapter 13.

In addition to these goals, the envisioned book would illustrate the importance of recombination in solving biological problems (what I sometimes call the bio-*logical* importance of recombination), in addition to the biological importance of recombination. This is done most explicitly in chapter 4. And the book would be a vehicle to introduce and explain the utility and versatility of computational techniques (most notably, integer linear programming and dynamic programming) to a broad audience, some of whose members may not have been exposed to those techniques. Most importantly, the envisioned book would help shape the research and education agendas of the computational biology community, and enable and encourage people outside the community to enter the field.

So, having decided that there should be a book on the combinatorial structure and algorithmics of problems defined on ARGs, and how ARGs relate to other phylogenetic networks, I took a yearlong sabbatical starting July 2008, sure that everything would be finished by its end, in September 2009. Well, four years later, in October 2013 (after putting in over 2,800 "billable hours" and typing over 1.7 million keystrokes) the book is almost done.[3]

The Title: ReCombinatorics

It is a *portmanteau* word[4] derived from the single-crossover recombination of the words "recombination" and "combinatorics":

r	e	c	o	m	b	i	n	a	t	i	o	n	
-	-	-	-	-	-	-	-	-	-				
		c	o	m	b	i	n	a	t	o	r	i	c	s
										-	-	-	-	-

Independently, other word-playful people [202] hit on "Recombinatorics" as the natural word for this field.

This Is Not a How-To Book

This book is about *ideas, models, and methods*. It is not a how-to book. I am positive that most of the general ideas exposed in this book will have productive uses or productive progeny, long after current software can no longer be compiled, let alone be executed. Moreover, I also believe that understanding the ideas and models that underly methods is helpful, if not essential, to applying the methods most effectively.

A Comment on Empirical Testing and Software

In several sections of the book, specific programs are discussed and some empirical results are mentioned, obtained from simulations or from executions on biological data, using those programs. The main purpose in mentioning programs and empirical results is to

---

3 Perhaps like giving birth to a child, it's good we quickly forget how hard it is or we would never do it again.

4 Lewis Carroll in Through the Looking Glass introduced this term: "Well, 'slithy' means 'lithe' and 'slimy' ... You see it's like a portmanteau – there are two meanings packed up into one word."

establish that *ideas* discussed in the book have been implemented, and that some of them work (some better than others). The empirical results provide very *crude* indications of the practicality of the methods. I am generally very skeptical and uninterested in detailed empirical results on program times, and to some extent accuracy. I use empirical results to answer in a broad-brush way, whether or not a method underlying a program is ballpark practical for some likely data of interest. So in this book, I make no effort to provide an in-depth comparison of software speed and reliability. If you have data, and there are choices for the appropriate programs, try them out. However, you can find links to the programs mentioned in this book at www.cs.ucdavis.edu/~gusfield/recsoftware.

A Note about Citations

There are expositions of established material that are new in this book, but regardless of the origin of the exposition, the book will cite the original source of any result that is not new to the book. For such results, please be sure to cite the *original* authors and the *original* publications, rather than citing only this book. If the material is original to this book, or you wish to direct a reader to this book for a new exposition of established material, I welcome that, but please cite the original papers as well. It is good scholarship to do so, and it shows proper respect for the original authors.

Acknowledgments

There are so many people to acknowledge and to thank. First are all of my research collaborators mentioned earlier. Foremost among those are Chuck Langley, Yun Song, and Yufeng Wu. In addition to their critical, central contributions in the development of many of the results that are re-exposed in this book, they watched over me during the writing of the book, providing wisdom and encouragement. I am particularly indebted to Yufeng, who provided several new ideas that are discussed in the book. Of course, any mistakes in the book are my own.

Additionally, I want to thank (in completely random order) Jane Gitscher, Jotun Hein, John Wakeley, Ladan Doroud, Laxmi Parida, Steven Kelk, Leo van Iersel, Celine Scornavacca, Chris Whidden, Charles Semple, Mike Steel, all the students in the fall 2011 UC Davis Computer Science course 224, Julia Matsieva, Tandy Warnow, Katherine St. John, Luay Nakhleh, Gabriel Valiente, Simon Myers, Bob Griffiths, Ken Burtis, Nick Shepard, Iain Mathieson, Mike Waterman, Eleazar Eskin, Eran Halperin, Dick Karp, Richard Durbin, Richard Mott, Sorin Istrail, Daniel Huson, David Morrison, Andy Clark, Ron Shamir, David Fernández-Baca, Martin Tompa, all the students in the 2012 International Winter School on Methods in Bioinformatics in Tarragona, Spain, Earl Barr, Katharina Huber, Vincent Moulton, Philippe Gambette, Sylvain Guillemot, Shawnie Briggs (who made the tree sculpture), and my neighbors Rob and Lacey Thayer (who own it).

I also want to thank the National Science Foundation for continuous support of our research on the combinatorics and algorithmics of phylogenetic networks with recombination. This was made possible through grants: SEI-BIO 0513910, CCF-0515378, IIS-0803564, and CCF-1017580. In particular, I thank the NSF program officers, Sylvia Spengler, Ding-Zhu Du, and Mitra Basu, for their trust, support, and encouragement. I thank the Simons Institute for the Theory of Computation which partially supported my efforts in the final phase of completing the book.

I thank Bob Prior at MIT Press for his trust in this project and persistence in bringing it to MIT. I thank the seven anonymous reviewers who made many helpful comments, and Virginia Crossman and Amy Hendrickson for their contributions to editing and layout.

Finally, I thank my wife, Carrie, for her love and support — although words do not suffice to thank her for understanding and tolerating the enormous amount of time I spent locked up in my upstairs office — rather than attending to house and family affairs. Thank you Carrie — I will never do it again (or maybe just a little book next time).

for example by *lateral gene transfer or hybrid speciation*. The central algorithmic problems are to reconstruct *plausible* histories, with mutations, treelike events, and nontreelike events that generate a given set of extant, observed genomic *sequences*; to determine the *minimum* number of such biological events needed to derive the sequences; to enumerate a range of plausible histories, and assess their biological fidelity; and to characterize properties of plausible or optimal histories.

This book primarily concerns combinatorial and algorithmic issues involved in reconstructing the evolutionary history of extant sequences observed in populations of diploid organisms (such as humans), where the sequences are generated by mutations and recombinations. However, many of the combinatorial and algorithmic results apply equally well at the *phylogenetic* level, namely to reticulate evolution of *species*, rather than populations, and we will point these out when they arise. Indeed, one of the goals of this book is to expose common mathematical and algorithmic structure[3] that occurs both in populations and species, despite the differences in biological origin, and differences in the biological communities that study the two areas.

The book is aimed broadly at computer scientists, mathematicians, and biologists. We will explain the various biological phenomena; the mathematical, population genetic, and phylogenetic models that capture the essential elements of those phenomena; the resulting combinatorial and algorithmic problems that derive from those models, and from biological questions that are formulated in terms of those models; the theoretical results (both combinatorial and algorithmic) that have been obtained; related software that has been developed; and the results of empirical testing of that software on simulated and real biological data. In addition, we will explain some needed combinatorial and algorithmic background for those readers who might not be familiar with particular existing results or techniques. We begin with some essential definitions.

Definition A *chromosome* is a single linear molecule consisting of double-stranded DNA. An individual's genes are arranged on their chromosomes.

Definition A *locus* refers to a discrete, specifiable interval of sites or positions in a chromosome. The set of *sequences* that can occur at a particular locus specifies the set of *states* or *alleles* of the locus. The plural of *locus* is *loci*.

The word "specifiable" is part of the definition to emphasize the fact that we don't always know where the locus is. It has a specifiable location, but it might not be known.

---

3 "Poetry is the art of giving different names to the same thing; mathematics is the art of giving the same name to different things." This misquotation is attributed to Henri Poincaré.

An *allele* is one of a number of alternative forms of the same gene or same genetic locus. (from Wikipedia)

Definition A *diploid* organism (such as a human) has two (not necessarily identical) "copies" of each chromosome. The two copies are called *homologs* or *homologous chromosomes*, and they form a *homologous pair*.

Homologous chromosomes are similar but not identical. Each carries the same genes in the same order, but the alleles at each site might not be the same. (from Wikipedia, again; somewhat modified)

For example, humans have 22 homologous pairs of (autosomal) chromosomes, and one pair of sex chromosomes (X, Y in males, and X, X in females).

Definition The corresponding sequences on two homologous chromosomes are called *homologous sequences*.

1.2.1   Recombination and Genealogical Networks

Meiotic recombination

The best-known biological event that creates variation in genomes is a *point mutation* where a single nucleotide changes state, say from A to one of the other three states $T, C,$ or G. However, mutations (that don't quickly die out) are relatively rare events, and in short time periods (even thousands of years in humans) mutations are not the primary cause of variation in genomes. Instead, **meiotic recombination** during meiosis is the key biological event that creates high-variation genomes over relatively short time periods in human (and other diploid) *populations* (i.e., individuals in a single species).

Meiosis is the process in which a *gamete* (egg or sperm), containing one copy of each chromosome, is created from a cell that has a homologous pair of each chromosome. In meiosis, recombination uses a pair of homologous chromosomes to create two *recombinant* chromosomes consisting of alternating segments (usually a small number) of the two homologs (see figure 1.1). Any child of that individual then inherits *one* of the resulting recombinant chromosomes. Similarly, recombination between two homologous chromosomes in the other parent creates two recombinant chromosomes, one of which is passed down to the child. Hence, the child receives one chromosome from their mother and one from their father, and each chromosome is a recombinant chromosome created from two homologous chromosomes of one parent.

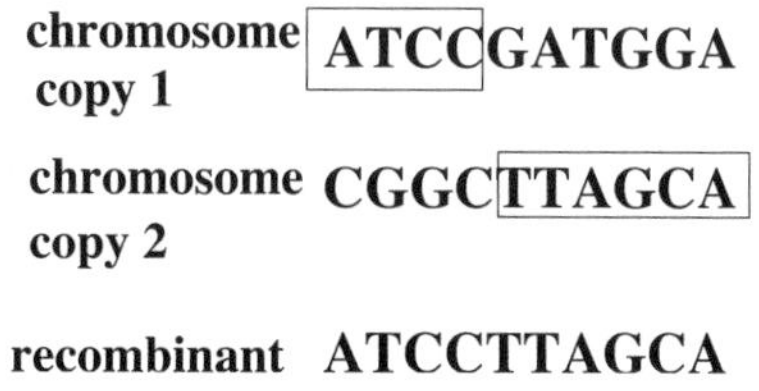

chromosome ATCC|GATGGA
copy 1

chromosome CGGC|TTAGCA|
copy 2

recombinant ATCCTTAGCA

Figure 1.1 Meiotic recombination of two sequences creates a third sequence, called a *recombinant* sequence. The recombinant sequence is created from the boxed segments of the two parental sequences. This example illustrates *single-crossover* recombination. Double and multiple-crossover recombination will be introduced later.

The key observation

Because of recombination in all the prior generations, it follows that the genome that any individual inherits is a mixture and a reflection of the DNA of *all* of the individual's ancestors. In this way, meiotic recombination allows the rapid creation of *chimeric* chromosomes even without mutations. This ability to create new chromosome sequences allows species to rapidly respond to changes in the environment, and to drive out deleterious mutations. Recombination is therefore an important adaptive property that occurs (along with sexual reproduction that enables it) in almost all eukaryotic species. The existence of such rich combinatorial genomes compels the study of genomic variation in populations to discover relationships between genome content and genetically influenced traits of interest.

To a computer scientist, it is almost irresistible to view combinatorial genomes, and the associated observed traits (phenotypes), as nature's way of implementing a kind of *binary search* (or a similar *divide and conquer* method) to identify the locations of important genomic features. In that view, nature has already done the experiments, posing and answering most of the required search queries. Now, in a kind of *Genomic Jeopardy* game, we have to find the right *questions* to match and exploit nature's *answers*.

1.2.2 Why Networks?

We are interested in reconstructing plausible histories of mutations and recombinations that might have derived chromosomal sequences observed in current populations. Such histories are not in the form of trees, but rather in the form of *networks*. To explain the need for networks in describing the history of chromosome sequences, we consider first the related, but simpler, issue of family *pedigrees*.

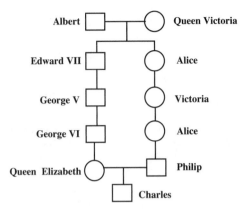

Figure 1.2   A partial pedigree of Prince Charles, who was married to Princess Diana, and is the father of Princes William and Harry (not shown). Note that the three females (Alice, Victoria, and Alice) drawn opposite to Edward and the two Georges, are not their wives. Also, we only show the ancestors of Charles who are descendants of Albert and Queen Victoria. Prince Charles had two parents (Elizabeth and Philip), who each had two parents, who each had two parents, etc. Following five generations back in time, Charles would have $2^5 = 16$ ancestors in that generation, if they were all distinct. But in fact, Charles does not have sixteen ancestors in the fifth generation back from him. In the fifth generation back from Charles, there are two *convergence* events: Edward and Alice converge at a common father, Albert, and they converge at a common mother, Victoria. That is, Edward and Alice are full siblings with the same parents. This part of the pedigree forms a *cycle* or *loop*. Note that the pedigree would also have had a cycle if Edward and Alice had only been half-siblings, i.e., if they had only shared a single parent. No parents of Albert or Queen Victoria are shown, and hence they are the *founders* of this partial pedigree.

1.2.2.1   Pedigrees

If we trace the ancestry of an individual *backward* in time, their two parental lines will expand into multiple lines (as parents expand to grandparents and great-grandparents, etc.). But some lines will eventually "converge," meaning that two distinct ancestors of the individual will have one or two common parents.[4] See figure 1.2 showing a partial pedigree of recent English royalty. It follows that the full genealogical history, or *pedigree*, of a set of individuals will contain *cycles* (often called "loops" in the genetics literature) and therefore *cannot* be represented by a *tree*; instead, the representation requires a **network**. We will develop precise definitions later in this chapter and in chapter 3.

---

4 A convergence event in a pedigree is sometimes called a *coalescent* event, but we will reserve that term for an event involving sequences.

1.2.2.2 Back to Sequences

Above, we considered family pedigrees in order to introduce the notions of *convergence*, *cycles*, and *networks*, and in order to distinguish pedigrees from *genealogical networks* which we will introduce shortly. But our main interest is the history of DNA *sequences*, and not the history of families or the individuals who carry those sequences. So, we now shift attention back to sequences on chromosomes. We first consider the case where there is *no* recombination and *no* mutation.

Recall that a diploid individual receives one copy of a particular chromosome (say chromosome 21) from the individual's mother and receives one copy from the individual's father. Moreover, without recombination or mutation, the copy received from the individual's mother is identical to one of the mother's homologs, and the same is true for the copy received from their father. For example, in figure 1.3, two four-character sequences are shown; these are sequences from the same location in a homologous pair of chromosomes.

Tracing the transmission history of sequences

Now consider an individual in a population and just *one* of the homologs of a homologous pair of some chromosome (say from chromosome 21). The sequence on that homolog was transmitted to the individual from just one parent, and since there is no recombination, that sequence was passed down from exactly one grandparent, etc. It follows that the transmission history of the chromosome sequence forms a *path* through ancestors in the individual's pedigree. That path begins at a founder sequence and descends to the individual. Further, since we have assumed that there is no mutation, each of the individual's ancestors on the path possesses and transmits an identical copy of the sequence.

Definition The path through sequences (overlayed on a pedigree), showing the transmission of a sequence from an ancestor of an individual, to that individual, is called a *sequence-transmission path*. Note that the elements on the path are sequences, rather than the people who possess those sequences.

For example, in figure 1.3, Charles's CCCC *must have* been transmitted from his father, Philip, even though his mother, Queen Elizabeth, also has CCCC. We deduce this because Charles also has TAAT, which he could only have gotten from his mother. Next, we deduce that Philip must have received CCCC from his mother, Alice. We deduce this because Alice must either have passed CCCC or GCTA to Philip, but he does not have sequence GCTA. Continuing with this logic, we deduce that Charles received his CCCC from Philip through a right-hand path, originating with Albert, his great-great-great-grandfather. The path tracing the transmission of CCCC from Albert to Charles is a *sequence-transmission*

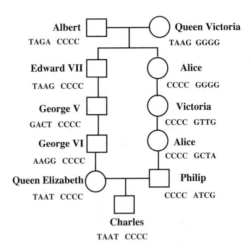

Figure 1.3 Each individual is shown with two fictitious sequences on two homologous chromosomes (say, from the two homologs of chromosome 21). For example, Charles's two sequences are TAAT and CCCC. Each individual receives one sequence from their mother and the other from their father. The order of the two sequences displayed for an individual does not indicate which came from the mother and which from the father. That information must be deduced, if possible. For example, we deduce that Elizabeth's copy of CCCC descended from a copy of CCCC possessed by Albert, via a sequence-transmission path through copies of CCCC possessed by Edward, and George V. Similarly, we deduce that Charles received CCCC from a copy possessed by Albert, via a sequence-transmission path containing sequences possessed by Alice, Victoria, Alice, and Philip. We see from this that the copy of CCCC possessed by Elizabeth is *identical by descent* with the copy of CCCC possessed by Philip. In fact, all of the copies of CCCC (other than Albert's) are identical by descent, from Albert. Looking backward in time, we say that the two *sequence-transmission* paths of the sequence CCCC possessed by Elizabeth and by Philip *coalesce* at the copy of CCCC possessed by Albert.

path. We can also deduce that Queen Elizabeth received her copy of CCCC along the left-hand path, a sequence-transmission path, also originating with Albert. Note that a sequence-transmission path traverses *sequences, not people.*

Definition If a sequence s is on a sequence-transmission path that leads (forward in time) to a sequence s' (where s might be identical to s'), then s is an *ancestral sequence* of s'.

Now, consider the *two* sequence-transmission paths of an *identical* sequence possessed by two individuals who share a common ancestor. For example, in figure 1.3, consider the sequence-transmission paths of the sequence CCCC, possessed by Elizabeth and Philip.

Definition If, traversing the sequence-transmission paths backward in time, two sequence-transmission paths intersect (at some *sequence*), then we say that the paths *coalesce* at that intersection point. This is also called a *coalescent event*.

Coalescence versus convergence

The two types of events are related, but a *coalescence event* (defined on sequences) implies a *convergence event* (defined on a pedigree), while a convergence event does not always imply a coalescence event. For example, in figure 1.4*a*, Edward VII and Alice (the upper one) have the same (two) parents (Albert and Queen Victoria), and that represents two convergence events in the pedigree. However, there is only one coalescence event. That event is when the two sequence-transmission paths that contain CCCC coalesce at the copy of CCCC possessed by Albert. Even though Edward and Alice have the same mother, Queen Victoria, they did not receive the same sequence from her (Edward received TAAG, and Alice received GGGG). So there is a convergence at Victoria, but not a coalescence.

Note that without mutations, all the sequences on two sequence-transmission paths that coalesce at some sequence s, must be identical to s.

Definition Without mutation, the sequences on two sequence-transmission paths that coalesce, are called *identical by descent*. See figure 1.3.

A history of sequences

To emphasize that our interest is in the history of *sequences* and sequence *transmissions*, and *not* in the history of the *people* through whom the sequences flow, in figure 1.4*a*, we redraw the sequence-transmission paths from figure 1.3, removing any sequence that is not on a deduced sequence-transmission path. However, we maintain the names of the people for ease of reference.

The underlying trees

When there is no recombination, the key feature of a set of sequence-transmission paths is that *no* sequence has two edges directed into it. There can be two sequences possessed by the same individual, where each sequence has a directed edge into it, but those two edges are directed to *different* sequences. For example, in figure 1.4 there are two edges directed into the sequences possessed by Edward, but one is directed to TAAG and one is directed to CCCC. If, in a set of sequence-transmission paths, we replace each sequence with a node, the resulting graph will not have any node with two edges directed into it. Hence, the graph will consist of a set of rooted *trees* that *partition* the sequences (each sequence is in one and only one tree). See figure 1.4*b*.

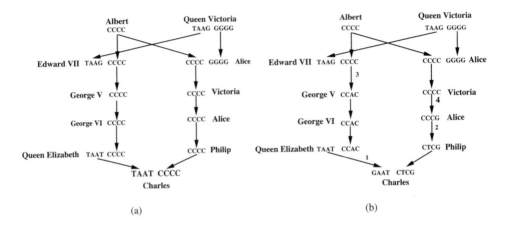

(a)   (b)

Figure 1.4 (*a*) The deduced sequence-transmission paths from figure 1.3. The sequence-transmission paths form four disjoint trees that partition the sequences. One tree consists of the two copies of TAAT; a second tree consists of the two copies of TAAG; a third tree consists of two copies of GGGG; and the fourth tree consists of all of the copies of CCCC.

(*b*) Now we allow sequences to mutate, but there is still no recombination. The number written on an edge identifies the site that mutates on that edge; the actual mutation is seen by comparing the nucleotide for that site at the head and tail of the edge. For example, on the edge with label 3, the nucleotide at site 3 changes from C to A.

Finding Adam and Eve in sequences

If we sample *sequences* in the current population that were transmitted without recombination, and we could trace their transmission paths back in time, we should eventually reach a point where all of the transmission paths coalesce (not simultaneously) at one common ancestor. That is, all the sampled sequences descend from one ancestral sequence, and the transmission paths jointly form a tree. Of course, we can't trace back in time, but we can *estimate* that tree. This was recently done for human patrilineal lineages, using the Y chromosome [340]. Only males have a Y chromosome, and only a small segment of the Y chromosome recombines (with a segment of the X chromosome). Hence, sequences from most regions on the Y chromosome have descended without recombination. The most recent common ancestor of those sequences is called the "Y-chromosome Adam." Similarly, using *mitochondrial* DNA, which is mostly (if not exclusively) inherited from one's mother, a matrilineal tree was estimated. The most recent common ancestor of the sampled mitochondrial sequences is called the "mitochondrial Eve."

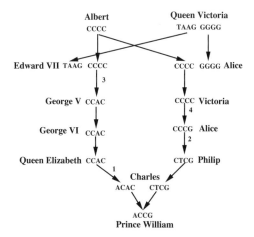

Figure 1.5 A history with recombination. The sequence, ACCG, that William receives from his father, Charles, is created by a recombination of the two sequences, ACAC and CTCG, transmitted to Charles from Elizabeth and Philip respectively. Note that sequence ACAC, transmitted to Charles from Elizabeth, is derived from Elizabeth's paternal sequence (rather than from her maternal sequence, as in figure 1.4). Also, it contains one mutated site (site 1), and hence is not identical to either of the sequences that Elizabeth possesses. The recombinant sequence ACCG contains the first two letters of ACAC and the last two letters of CTCG. William also receives a sequence from his mother, Diana, who is not shown in this partial pedigree.

1.2.2.3   *Adding in Mutation and Recombination*

Returning to the royals, we next add in mutation. Mutations do not change any sequence-transmission paths, but do change the sequences. So, it is no longer true that all the sequences on a sequence-transmission path are identical. For example, see figure 1.4*b*.

Finally, we add in recombination between two sequences. Then, when a sequence is transmitted to an individual from one of their parents, say their father, it might be a recombinant sequence created from the father's two sequences during meiosis. In that case, it could be different from both of the copies of the sequence that the father received. For example, in figure 1.5 we include one of Charles's sons, William, and see that the sequence William receives from Charles is a recombinant sequence, different from either of Charles's sequences.

Recombination creates cycles
In contrast to the case when there is no recombination, if a set of sequence-transmission paths contains a sequence with two incoming edges (due to recombination), then the

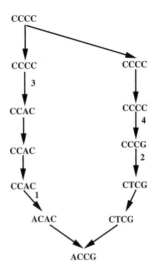

Figure 1.6   A recombination cycle. The two sequence-transmission paths in figure 1.5 form a *recombination cycle* with coalescent (or root) sequence CCCC and recombinant sequence ACCG.

transmission history is *not* a set of disjoint rooted trees. Moreover, as in figure 1.6, two sequence-transmission paths that meet at a recombinant sequence x might also coalesce at an ancestral sequence of x, say s. In that case, the two paths define a *recombination cycle* with *coalescent sequence s* (also called a *root* sequence), and *recombinant* sequence x. Further, if all the sequence-transmission paths ultimately coalesce at a *single* root sequence, then if there is any recombination, there must be a recombination cycle. For example, start at a recombinant sequence and trace back along two paths; since all paths eventually coalesce, the two traced paths must eventually coalesce, at which point the two paths specify a recombination cycle. A directed graph with a recombination cycle is not a directed tree; it is a **network**.

1.2.2.4   *Transmission Paths Form (Parts of) a Genealogical Network*

The network that represents the transmission history of chromosome sequences, shaped by mutation and recombination, is often called a *"phylogenetic network"* in the computer science literature, although the term *"genealogical network"* is more appropriate. So, the network shown in figure 1.4*b* is part of a genealogical network, as are the networks shown in figures 1.5 and 1.6. A first, informal definition of a genealogical network will be given in section 1.6.2, after some concepts from graph theory have been introduced, and a more complete, formal definition will be given in chapter 3. When some additional restrictions apply (explained in chapter 3) a genealogical network is called an *"ancestral recombination graph"* (ARG). ARGs will be the networks of greatest importance in this book.

1.2.2.5   *Genealogical Networks Relate Sequences, Not People*

It is easy to confuse pedigrees and genealogical networks, so it is critical to note that genealogical networks represent relationships between *sequences*, rather than relationships between *people*, as in a pedigree. In fact, in most cases, we do not know the pedigree of the individuals who possess the sequences of interest. It is true that a genealogical network is constrained by that (unknown) pedigree, but a genealogical network displays information not contained in a pedigree, and the pedigree contains information not contained in the network. Moreover,

> Even though each individual possesses a pair of homologous chromosomes, the two homologous chromosomal sequences are represented *independently* in the genealogical network.

That is, the genealogical network (representing the history of sequences) *decouples* the two homologous sequences of each individual. In figure 1.6, the two sequences that Charles possesses, which recombine to create William's recombinant sequence, are drawn further apart than they are in figure 1.5, in order to emphasize that point.

> I may be a twin, but I am one of a kind.
> — Anonymous

Hence, the genealogical network represents a set of sequences in a history, without indicating which pairs of sequences (if any) were possessed by the same individual. The only way we can deduce that two sequences in a genealogical network were possessed by the same individual is if they recombine. Two sequences that recombine during meiosis must both be possessed by a single individual. As an example, figure 1.13 (page 31) shows a genealogical network which is unrelated to any shown pedigree. We can deduce that the sequences shown at nodes u and v must have been possessed by a single individual.

We want the historically correct genealogical network
Knowing the true, historically correct genealogical network that explicitly reveals the origin and derivation of the sequences in a current population, and shows the locations of all

the mutations and recombinations (both in the genome and in time), would tremendously facilitate the use of genomic data to address many basic biological questions, and be of use in biotechnology.

The rub

Unfortunately, we cannot directly examine the past so we cannot know (for sure) the historically correct genealogy of the extant sequences. However, a robust literature on *algorithms* that construct plausible genealogical networks, or deduce well-defined aspects of a genealogy, has developed, particularly in the last several years. Related questions about *hybridization networks*, which are similar to genealogical networks but do not generally involve explicit sequences, have also been addressed. This algorithmic research has been encouraged by a growing appreciation by biologists that many evolutionary and population genetic phenomena must be represented by networks rather than by trees.

1.3   The Central Thesis of the Book

Even though we can never know for sure that an algorithm has deduced the correct genealogical network (or features of it), we will detail in this book that applications of these algorithms have correctly answered certain biological questions, suggesting that important parts of true genealogies are captured in, or reflected by, the computations. These applications go to the heart of the book's central thesis.

> **Central Thesis:** *Explicit* genealogical networks representing a derivation of extant sequences in a population can be effectively computed, and even if those networks do not perfectly capture the true transmission history of the sequences, they can reveal parts of the history and give significant insight into basic biological phenomena, and be used to address applied problems in biotechnology.

1.4   Fundamental Definitions

The atomic objects of concern in this book are *individuals* in the context of population genetics, or *species* in the context of phylogenetics, or *molecular sequences* in the context of molecular evolution. Sometimes the particular biological context affects the mathematical models and the algorithmic problems that are defined on that model. However, many of the mathematical and algorithmic results we discuss in this book apply to all the biological models. We want to be as general as possible and so we will use a generic term for the objects of interest.

Definition We interchangeably use the terms *taxa* or *individuals* for the objects of interest, and *taxon* or *individual* for a single object.

This is intentional

There is a huge difference in the biological contexts in which the terms "taxon" and "individual" are typically used. The first term is used in phylogenetics, i.e., in the classification of *species, genus, family*, etc., and the second term refers to a single organism at the lowest level of the classification. The International Code of Zoological Nomenclature [320] defines a taxon as

> ... a population, or group of populations of organisms which are usually inferred to be phylogenetically related and which have characters in common which differentiate the unit (e.g., a geographic population, a genus, a family, an order) from other such units. A taxon encompasses all included taxa of lower rank and individual organisms.

The differences in size and time scales between taxa and individuals are huge. So why are we mixing the terms here? In particular, why use the term "taxon" at all, when the primary biological motivation for topics in this book comes from population genetics?

The answer is partly historical: the term "taxon" has become the default term in the phylogenetic networks literature for any unit of interest, while "individual" is an essential term when talking about individuals (duh) in populations. But perhaps a better reason for mixing the terms is that one of the key points we make in this book (repeatedly), is that many problems arising in vastly different biological contexts have the same (or related) mathematical and algorithmic structure.[5] That point is not well known, and barriers between the communities are very strong. So, the mixing is both due to laziness (following a crowd), and to be intentionally provocative—reflecting the effort to identify and exploit common structure. Similar justifications (or apologies) apply to other mixing of biological terms.

Definition A *character* or *trait* is a discrete property or characteristic of a taxon. By "discrete," we mean that there is a finite number of *states* (*alleles* in more biological vernacular) that a character can take on.[6]

For example, if the taxon of interest is a human, the gender (male or female) of the taxon is a *binary character* taking on one of *two* possible states. As another example, the

---

5 For example, in the way that multiple-crossover recombination can be used to model phylogenetic phenomena, even though recombination is not a phylogenetic phenomenon.

6 There are also continuous characters where the states are not discrete, but these are not of concern in this book.

nucleotide ($A, T, C,$ or G) present at a particular site of a DNA sequence is a *four-state* character. The character would be binary if we only record whether the nucleotide at that site is a purine (A or G) or a pyrimidine (C or T). Another example of a binary character results from the phenomenon of DNA *inversions* where the orientation of a whole segment of DNA is inverted. Hence, the character is "segment orientation," which takes on two possible states. Similar, large-scale modifications of DNA (insertions, deletions, duplications, etc.) can give rise to additional binary characters, if those sequence modifications are independent, identifiable events [357].

Note that the meaning of the word "character" in the context of evolutionary biology is different from its colloquial meaning. In normal use, a character is a letter or a symbol in an alphabet, but in evolutionary biology a character is a trait of an individual. To confuse matters even more, a site in a DNA sequence can be considered to be a character in the sense of evolutionary biology, but the four possible states of that character are characters in the colloquial sense of the word, i.e., letters in the four-letter DNA alphabet.

1.4.1   Mutation, Infinite Sites, Perfect Characters, and Binary Sequences

Which mutations are modeled?
We introduced the concept of a *point mutation* in section 1.2.1, illustrated by a change in the state at a single DNA site (i.e., change of the nucleotide at that site). However, there are other changes that are biologically quite different from a single nucleotide change, but have characteristics that allow them to be modeled in the same way we model a nucleotide change at a single site. Hence, we want a more general definition to handle such changes.

Definition A *locus mutation* is a change of state at that locus, which is *independent* of changes at any other locus.

A point mutation is certainly a locus mutation. However, events such as DNA inversions, or DNA insertions and deletions, are locus mutations but not point mutations. Note that a change of state due to a locus mutation is distinct from a change of state due to recombination. Further, for a locus mutation to be a useful binary character, it must be short enough that it is unlikely to be split by a recombination event.

A notational conversion
In this book we are concerned with locus mutations that can be modeled as *binary characters*, that is, that have two permitted states, and are unlikely to be interrupted by recombination. As discussed before, there are several common locus mutations (e.g., point

mutations, inversions, insertions, deletions, duplications, etc.) that, under certain conditions, serve as useful binary characters. Most of the algorithmic and mathematical theory developed in this book is unaffected by which particular biological events give rise to the binary characters. Hence, we will often use the simpler term "mutation" in place of "locus mutation" or "point mutation," but will use the latter terms when the distinction is necessary. Similarly, we will generally use the term "site" instead of "locus," even though the underlying mutational event may involve an interval of sites.

A mutation model is needed

Genealogical networks represent the derivation of extant sequences which change due to mutation and recombination. Unrestricted, mutation events alone (without recombination) can derive any set of sequences, but that derivation would not likely reflect biological reality. Therefore, we need a *model* of the mutations that are permitted in sequence data.

1.4.1.1   The Infinite-Sites Model

> Essentially, all models are wrong, but some are useful.
> — George Box

The most commonly used mutation model in *population genetics* is the *infinite-sites model* where any locus (in the sample) mutates *at most once* in the entire history of the sequences. The infinite-sites model was introduced in the context of *point mutations*, but also has a clear meaning, and biological reality, for particular, more complex, locus mutations.

The infinite-sites model is justified when the probability of a mutation at any given locus is so low that the possibility of multiple mutations there can be ignored. For example, the infinite-sites model is justified when the history of a sample covers a *relatively short* time period, and mutations occur at random sites, so that a mutation at any given site is a low frequency event. Hence, the probability that a point mutation occurs twice at a site is extremely low.[7] The infinite-sites model is not appropriate for all evolutionary phenomena, but it is widely assumed in the population genetics literature, particularly for short time periods, and it is the basic mutational model for most of the results discussed in this book. See [168, 421] for more in-depth justifications of this model.

---

7 The origin of the term "infinite sites" comes from the view of a genome as having a huge (essentially infinite) number of sites so that each successive mutation, occurring at a random position in the genome, occurs at a site where no mutation has occurred before. It follows that a mutation at any given site can occur at most once, and that the character is binary.

The number two ... is the geneticist's favorite number.
— A. Knudson, quoted in [296]

The infinite-sites model implies that each character in any of the studied sequences can take on only one of *two* possible states: the *ancestral* and the *derived* states. Hence the sequences we observe in a population can be considered, and recoded as *binary* sequences. Note that the assumption of binary sequences does *not* come from assuming that the DNA (or other) alphabet has been reduced from four letters to two. The DNA alphabet still contains four letters, but in the set of DNA sequences found in a population, it is rare to observe more than two different ones (above a low frequency) at any given site. That is, if we fix a particular site and look at the individuals in the population to see what letters occur at that site, we rarely see more than two different letters (and generally, we only see one). A DNA site where we see *two* states in the population (above some minimum frequency) is called a *single-nucleotide polymorphism* (SNP, pronounced "snip") site.

Binary recoding

With the assumption that there will be at most two different states at any site, we can code each site as a binary character, using the alphabet $\{0, 1\}$ to represent the character states at the site, and so each sequence observed in the population becomes a binary sequence. It is common to use 0 for the most frequent of the two letters at that site (assumed to be the ancestral state), and 1 for the least frequent of the two letters (the assumed derived state). See figure 2.2 (page 38).

1.4.1.2 SNPs

The strongest empirical validation for the infinite-sites model comes from DNA sequence data in populations, and the *single-nucleotide polymorphism* (SNP) sites that are observed in the data. At a SNP site, only two of the four possible nucleotides appear in the population (with a frequency above some minimum threshold). In humans, and other well-studied organisms [116], millions of SNP sites have been found and cataloged, most prominently by the International HapMap Project [198, 199], the Human Genome Diversity Project [251], and the One Thousand Genomes Project [1, 2]. See also [62, 176] for additional empirical, population-genetic validation for the infinite-sites model.

1.4.1.3 Perfect Characters and Homoplasy

The infinite-sites model comes from population genetics. In contrast, phylogenetics concerns the history of *species*, rather than *populations*, and the term "infinite-sites" is not used in that literature. However, phylogenetics does have a similar model. In phylogenetics a character that mutates only once in the history of the sample, has been called a *perfect character* [174, 351] or an *ideal* character [432]. Since it mutates only once, all descendants of the species where the character state changes, must also have that character state, and no other species will. Hence, at an abstract level, a perfect character in phylogenetics behaves the same as a binary character in population genetics that obeys the infinite-sites model.

> Taxonomists intuitively select character states which they postulate to define monophyletic[8] sets of species. The ideal character contains some state that both uniquely defines a set of species and has not been reversed in evolution, so that all existing species which possess this state can be said to have descended from one species in the past that evolved the state. For every such character state that can be identified, a branch in the phylogenetic tree can be added. [432]

Although the term "perfect character" originated in the phylogenetics community, we will use the term to refer to any character that obeys the infinite-sites model, whether it is a character used to study populations, or to study species.

Morphological support for perfect characters

The perfect character model in phylogenetics is supported by certain morphological data. A *morphological* character (such as having horns, or hair, or a tail) may be a trait that required many successive mutations. In that case, the probability is low that the trait will have evolved twice, independently in different species. Instead, a complex morphological trait that is common to several species is generally thought to have arisen only once, in a species that is ancestral to all of the species containing that trait. Further, once acquired, many complex traits will be retained in all descendant's species. This makes complex morphological traits ideal candidates for perfect characters in phylogenetics.

However, not all morphological traits are believed to be perfect (or near-perfect) characters, and it also believed that *convergent* or *parallel* independent evolution of highly valuable traits, such as flight or vision, has occurred. See [267] for a recently studied, exceptional case of convergent evolution. In phylogenetics, "homoplasy" is the term for violations of the perfect character model [364], due to recurrent or back mutation:

> Homoplasy is similarity that is the result not of simple ancestry, but of either reversal to an ancestral trait in a lineage or of independent evolution. [420]

---

8 A monophyletic set of species is a set that possess some character(s) in common that are not possessed by any individual outside the set. The term "clade" is a synonym.

Note that in phylogenetics, recombination is not considered a cause of homoplasy since recombination is not a phylogenetic-scale event. However, we will relax the original meaning of "homoplasy" and say that it is any violation of the perfect character model, regardless of the biological phenomena that causes it.

The phylogenetics community also uses the term *incongruence* for phylogenetic trees that don't agree with each other. Two characters are incongruent if they cannot be derived together on a single phylogeny without back or recurrent mutation.

Other perfect characters

Point mutations are the best-known and most widely studied mutations in DNA, but many other kinds of mutations also occur. The most common of these are *insertions*, *deletions*, *inversions*, or *duplications* of whole segments of DNA. When these kinds of locus mutations occur at low frequency, they generally obey the infinite-sites model, and can be considered perfect characters. For example, if a segment of DNA is inserted into some location in a chromosome of an individual, and the descendants of the individual inherit the augmented chromosome, and the same segment is never again inserted at the same location in another individual, and the segment is not changed by recombination, then the existence or non-existence of the segment at that location, is a binary character that obeys the infinite-sites assumption. Equivalently, it is a perfect character.

Specific kinds of mutations, other than point mutations, that might be used as perfect characters are the insertion of *micro-RNA* sequences [164, 201]; or insertions of mobile elements such as *SINEs* or *LINEs* [174, 314, 351]; or increases in the number of *tandem repeats*; or whole *gene duplications* in place; or duplications on a different part of the chromosome, leading to the existence of *pseudo-genes*; or inversions of segment of DNA. All of these kinds of mutations have been suggested to be perfect characters in the right biological contexts, and this list is certainly not exhaustive. Other suggestions for perfect or near-perfect characters include *ultra-conserved* elements and their flanking DNA [81]. For a general discussion of perfect characters, see [357].

> Because the probability that a SINE/LINE will be lost once it has been inserted into the genome is extremely small, and the probability that the same SINE/LINE will be inserted independently into an identical region in the genomes of two different taxa is also very small, the probability that homoplasy will obscure phylogenetic relationships is, for all practical purposes, zero. [314]

> ... microRNAs ... are highly conserved, non-coding genes that can be treated in datasets as presence/absence, like most phenotypic characters, ... with the additional advantage of being rarely lost in evolution. [201]

1.4.1.4 Perfection Is an Abstraction

> Have no fear of perfection—you'll never reach it.
> — Salvador Dali

We do not claim that every binary biological character is a perfect character, or even that the specific ones discussed here are always perfect (there is continuing debate about some of these [174, 351]). And, in fact, most characters (particularly in phylogenetics) are not perfect [364, 420]. But we claim that perfect characters are either plentiful (as in the case of SNPs) or can be found (as in micro-RNA, LINEs, SINEs, and complex morphological traits). So, there are important contexts where the infinite-sites, perfect-character models hold sufficiently well to justify their use, whether they occur in the context of population genetics, or in phylogenetics, or in other contexts such as in linguistics [306].

Further, when a site shows evidence of homoplasy, many studies will often remove that site from the data [112]. This is justified when the primary focus is on the history of the taxa, not the history of the characters, and when it is believed that the true history of the taxa, stripped of characters, is a tree with one leaf for each taxon. In that view, one should seek perfect characters (which are necessarily binary) that are sufficient to construct a history of the taxa in the form of a tree T. Characters that are not perfect (show evidence of homoplasy), are unneeded in order to define T, and can be ignored. In that view, those characters are noise imposed on top of the historically correct tree, definable with perfect characters. Similarly, in the context of SNP sites, any DNA site that has more than two variants (above some specified frequency) in a population can be removed.

> Recombination, parallel or back mutations, gene-conversion or genotypic misclassification can cause the perfect phylogeny condition to be violated. Such data may be pruned using an algorithm that deletes haplotypes or SNPs or a combination of both to give a reduced set of data consistent with a gene tree.[9] [75]

A further justification for the perfect character model is that often a clean, ideal model can be used in the *core* of a practical method that handles messier data not completely conforming to the model.[10] In that approach, the core is executed repeatedly by wrapper software that iteratively modifies the data. In the context of perfect characters, some examples of this approach are found in [105, 106, 158, 240, 363, 368, 369, 392].

---

9 "Gene tree" is a term for a perfect phylogeny in [75].

10 This is one of the foundational principles of theoretical, algorithmic computer science—one of the tribes I belong to.

Additional support for the perfect character model comes from the discussion (on the relationship of the perfect-phylogeny model to coalescent theory) in section 12.3.1.

1.4.1.5   And, We Can Often Incorporate Homoplasy

Above, we gave the argument that perfect characters (or nearly perfect characters) do exist in real applications. That is our biological justification for perfect characters. Still, from a *biological* perspective, the assumption of infinite-sites, and of perfect characters, is limiting. However, from a *mathematical, modeling* perspective, the limitations are much less severe. This is because a back or a recurrent mutation, or parallel evolution, *can be modeled* as two-crossover recombination, with no modification to the infinite-sites assumption. This will be detailed in section 3.2.3.3. So, in this book, *any* result that holds for two-crossover recombination (and of course for multiple-crossover recombination), holds for trees or genealogical networks *with homoplasy* allowed. It is only the results that depend on single-crossover recombination that do not automatically apply when homoplasy is allowed.

We will sometimes explicitly point out that a result holds with homoplasy, but we will usually leave it to the reader to realize this, since our primary expositional model is the ARG, which does not *explicitly* allow homoplasy.

1.5   The Observed Data

In this book, the data for most of the problems of interest is represented by a set of taxa and a set of binary characters, together with information on the state of each character for each taxon. This data is usually presented in the form of an n by m *matrix* M whose n rows represent taxa, and whose m columns represent characters. Each cell (f, c) of M specifies the state of character c for taxon f. For example, see figure 2.1 (page 36) showing a matrix M with five taxa and five characters. Except for discussions of empirical data, we will generally not care what biological phenomena produced M, as long as each character is a perfect character (or near-perfect character).

When talking about the matrix M we will often use the terms "taxon" and "row" interchangeably, and will often use the terms "character," "column," "locus," and "site" interchangeably, choosing whichever term is most informative for the context. Further, the ordered entries in a row f of M can be considered to form a *sequence*, and so we have the following:

Definition The *sequence s_f for taxon f*, or the *sequence for f*, is the ordered sequence formed from the entries in row f of matrix M.

Given this definition, we will also consider M to be a *set* of sequences, as well as a *matrix* representing that set of sequences. Context will often determine whether M is a set or a matrix.

1.6 A Few Graph Definitions

The principle combinatorial objects that we deal with in this book are *graphs*, and so we state a few basic definitions and facts about the kinds of graphs we will encounter.

Definition An *undirected graph* $G = (V, E)$ is a combinatorial object consisting of a set of *nodes* (also called *vertices*) V, and a set of *edges* E. Each edge in E is specified by an *unordered* pair of nodes (u, v) from V, where $u \neq v$.

Definition For an edge $e = (u, v)$ in E, nodes u and v are called the *endpoints* of edge e.

For example, the undirected graph in figure 1.7a (page 25) has node set $V = \{a, b, c, d, e\}$ and edge set $E = \{(a, b), (a, c), (a, e), (b, d), (c, d), (c, e)\}$.

The definition of an undirected graph given here does not allow an edge whose two endpoints are the same, and it does not allow multiple copies of the same edge. Alternate definitions of an undirected graph do allow such self-loops and parallel edges, but they will never appear in the graphs considered in this book.

Definition An undirected graph $G = (V, E)$ is called *bipartite* if the nodes of V can be *partitioned* into two subsets V_1, V_2, so that for every edge in E, one of the endpoints of the edge is in V_1, and the other endpoint is in V_2. See figure 1.8.

Definition A *directed graph* $G = (V, E)$ is defined by a set of nodes V, and a set of *directed edges* E, where each directed edge is specified by an *ordered* pair of nodes (u, v), where $u \neq v$. By convention, the directed edge $e = (u, v)$ is directed *from* the first node, u, in the ordered pair *to* the second node, v, in the ordered pair. The first node is called the *tail* of e and the second node is called the *head* of e, so e is directed from its tail to its head. See figure 1.7d.

Definition A *subgraph* $G' = (V', E')$ of a graph $G = (V, E)$ is a graph where $V' \subseteq V$ and $E' \subseteq E$, and for every edge $(u, v) \in E'$, both u and v are in V'. An *induced* subgraph $G' = (V', E')$ of G is a graph where $V' \subseteq V$, and E' consists of *every* edge $(u, v) \in E$ such that both u and v are in E'. See figures 1.7b and c.

Definition If G is a directed graph, or a graph with some directed and some undirected edges, the *underlying undirected graph* of G is the graph formed by ignoring the directions on the edges of G. That is, each ordered pair of nodes that defines an ordered edge in G is now considered as an unordered pair of nodes.

Definition For any node v in an undirected graph, the *degree* of v is the number of edges that touch v, that is, the number of edges where v is one of the endpoints. For a node v in a directed graph, the *in-degree* of v is the number of edges directed into v; the *out-degree* of v is the number of edges directed out of v. See figures 1.7a and d.

Definition An *undirected path* from a node v_1 to a node v_k in an undirected graph $G = (V, E)$ is specified by an ordered list of nodes $v_1, v_2, \ldots, v_k$, such that for every i from 1 to $k - 1$, the node pair (v_i, v_{i+1}) is an edge in E.

Definition A *directed path* from a node v_1 to a node v_k in a directed graph $G = (V, E)$ is specified by an ordered list of nodes $v_1, v_2, \ldots, v_k$, such that for every i from 1 to $k - 1$, the ordered node pair (v_i, v_{i+1}) is an edge in E, that is, an edge directed from v_i to v_{i+1}.

Definition An undirected graph G is *connected* if for ever pair of nodes u, v in G, there is a path between u and v in G. It is *biconnected* if for every pair of nodes u, v there are at least *two* paths between u and v that share no nodes other than u and v.

Definition A *cut edge* in a connected graph is an edge whose removal disconnects the graph.

Definition A directed graph G is connected (sometimes called "weakly connected") if the underlying undirected graph of G is connected. It is *strongly connected* if for every pair of nodes u, v, there is a directed path from u to v, and also a directed path from v to u. Those two paths may share nodes in addition to u and v.

Definition An *undirected cycle* in an undirected graph G is an undirected path that starts and ends at the same node. A *directed cycle* in an directed graph G is an directed path which starts and ends at the same node.

Definition A graph G (possibly with some directed edges) is called a *network* if the underlying undirected graph of G contains an undirected cycle.

For example, when we consider the pedigree in figure 1.2 as an undirected graph, then it contains an undirected cycle. The directed graph in figure 1.6 does not contain a directed cycle, as defined before, but its underlying undirected graph does contain an undirected cycle. That cycle is what we called a *recombination cycle*.

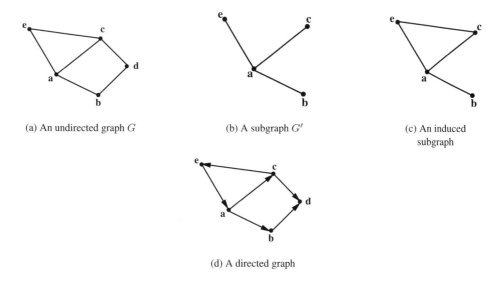

(a) An undirected graph G (b) A subgraph G' (c) An induced subgraph

(d) A directed graph

Figure 1.7   (a) The undirected graph G contains several cycles; one is $\{a,b,d,c\}$. The degree of node a is three. (b) A subgraph G' of G containing nodes $\{a, b, c, e\}$ and edges $\{(a, b), (a, c), (a, e)\}$. (c) The *induced* subgraph of G, induced by the node set $\{a, b, c, e\}$. (d) The directed graph contains the directed cycle $\{a,c,e\}$. It is connected but not strongly connected. Node a has in-degree one and out-degree two. Node a is the tail of edge (a, c), and the head edge (e, a).

Figure 1.8   A bipartite graph. Every edge in the graph has one endpoint in V_1 and one in V_2.

Definition The *node-contraction* of a node v of degree two removes v and merges the two edges incident with v into a single edge. The new edge is labeled by the union of the characters that labeled the two merged edges. See figure 1.9.

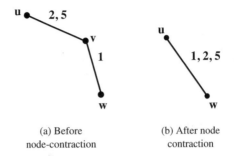

(a) Before
node-contraction

(b) After node
contraction

Figure 1.9   (a) Two edges incident with node v of degree two.(b) The result of contracting node v.

Definition An *edge-contraction* of an edge $e = (u, v)$ in a graph (either directed or undirected) is the operation that superimposes nodes u and v, removing the edge e between them.

1.6.1   Trees and DAGs: The Most Central Graphs in This Book

Now we can define the most important types of graphs needed in this book, and some properties of those graphs. We have already used some of these terms, relying on the reader's general background and intuition, but now we present formal definitions.

1.6.1.1   Trees and Subtrees

Definition An undirected graph is called an *unrooted tree* or an *undirected tree* if it is connected and contains no undirected cycles. See figure $1.10a$.

When it is clear by context that G is undirected, we will use "tree" in place of "unrooted tree" or "undirected tree."

Definition In an undirected tree T, a node of degree one is called a *leaf* of T, and every other node is called an *internal* or *interior* node.

The following is one of the most basic facts about trees.

Lemma 1.6.1 *Every undirected tree with n nodes has exactly $n - 1$ edges. Conversely, any connected, undirected graph with n nodes and $n - 1$ edges is a tree.*

Definition A directed graph G is called a *rooted tree* with a root node r, if the underlying undirected graph of G is a tree, and if every node in G is reachable from r via some directed path. See figure 1.10b. Equivalently, a *rooted tree* is a directed tree with one node designated as the root, where all edges are directed *away from* the root.

Definition In a rooted tree T, a node with out-degree zero is called a *leaf*; the root is the unique node with in-degree zero. Every other node is called an *internal* or *interior* node.

Definition A *subtree* T' of a tree T is a tree contained in T, where T' has fewer nodes and edges than T. Equivalently, T' is a connected subgraph of T. A subgraph of tree T that is not connected is called a *subforest* of T.

Often we will be concerned with the subtree of a special form.

Definition A subtree of a directed tree T, consisting of all the nodes and edges reachable from a particular node v in T, is called *the subtree of T rooted at node v*. This is usually denoted T_v. See figure 1.11.

Definition A rooted tree where every non-leaf node has out-degree two is called a *binary tree*.

The following is one of the most basic theorems concerning binary trees. It is easy to prove by induction on the number of leaves.

Theorem 1.6.1 *Any binary tree with n leaves has exactly $n-1$ non-leaf nodes, and $2n-2$ edges.*

1.6.1.2 *Directed Acyclic Graphs (DAGs)*

Definition A directed graph that contains no *directed* cycles is called a *directed acyclic graph*, or DAG for short. See figure 1.12a.

Note the underlying undirected graph of a DAG G may contain undirected cycles, even though G does not contain any directed cycles. Further, a DAG may contain a recombination cycle, since a recombination cycle is not a directed cycle.

Definition Two directed paths in a DAG that have the same start and end nodes, but otherwise share no nodes, form a *recombination cycle*. The start node of the paths is called the *coalescent* node of the cycle, and the end node of the paths is called the *recombination* node. In figure 1.5, the coalescent node is labeled Albert and the recombination node is labeled Prince William.

Now we establish some simple properties of DAGs.

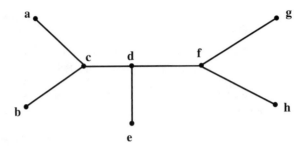

(a) An undirected tree

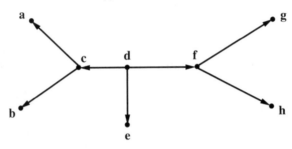

(b) A rooted tree T with root node d

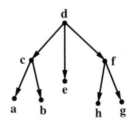

(c) Tree T drawn with
edges directed down

Figure 1.10   (a) An undirected tree. By definition, it has no root; its leaves are $\{a,b,e,g,h\}$.
(b) A rooted tree T, rooted at node d.
(c) Tree T drawn so that the root is at the top and all edges are directed downward. This is the standard way that directed trees are drawn in this book.

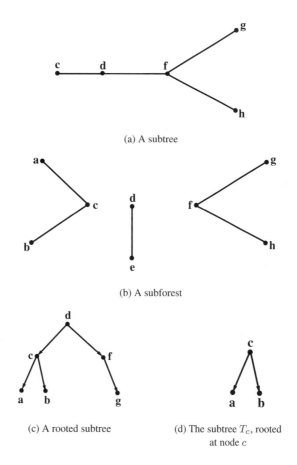

(a) A subtree

(b) A subforest

(c) A rooted subtree   (d) The subtree T_c, rooted
                                             at node c

Figure 1.11   (a) A subtree of the tree shown in figure 1.10*a*.
(b) A subforest of the tree in figure 1.10.
(c) A subtree of the rooted tree T in figure 1.10*c*.
(d) The subtree T_c, rooted at node c, of the rooted tree T in Figure 1.10*c*.

Lemma 1.6.2 *Every DAG G has at least one node with out-degree zero, and at least one node with in-degree zero.*

Proof: Arbitrarily pick a node u in G and find a directed path starting at u, by following any directed edge out of each successive node on the path. Since G does not contain any directed cycles, no node on this path is included twice, and since the number of nodes in G

is finite, ultimately the path must reach a node v that has out-degree zero. So the first part of the lemma is proved.

To prove that every DAG G has at least one node with in-degree zero, reverse the directions on every edge in G. This graph will also be a DAG. By what was proved above, this graph has a node u with out-degree zero, meaning that u is a node with in-degree zero in G.   ■

Note that a DAG might have more than one node of in-degree zero, or more than one node of out-degree zero.

Lemma 1.6.3 *The nodes of a DAG G can be partitioned into layers, and the layers can be ordered so that for any node v, all the nodes that can reach v via a directed path in G are in layers before the layer containing v.*

Proof: The first layer consists of all the nodes of in-degree zero. Remove all of those nodes from G. Then the second layer consists of all the nodes of in-degree zero in the resulting graph. The successive layers are obtained by repeating this process.   ■

Definition A total order, Π, of the nodes of a DAG G that is consistent with the ordering of the layers defined in lemma 1.6.3 is called a *topological sort* of the nodes of G. See figure 1.12*b*.

In Π, a node u must appear before a node v, if v is reachable via a directed path in G, from node u. Lemma 1.6.3 implies that a topological sort is always possible for any DAG.

1.6.2   Genealogical Networks and ARGs: First Definitions

Genealogical networks and *ancestral recombination graphs* are the DAGs that most centrally model the biological phenomena of interest in this book. Indeed, most of the book concerns these models. Here we give somewhat *informal* definitions for these graphs. More complete, formal definitions will be given in chapter 3. Also, in chapter 14, we will introduce another DAG, called a *reticulation network*.

A *genealogical* network $\mathcal{N}$, generating a set of sequences M, each of length m, is a directed acyclic graph containing *one root* node with in-degree zero, and n leaves, each with in-degree one and out-degree zero. Every other node has in-degree one (tree nodes) or two (recombination nodes). Each site in M is assigned to a set of edges in $\mathcal{N}$, but none is assigned to any edge entering a recombination node. Each node v in $\mathcal{N}$ is labeled by an m-length binary sequence, denoted s_v. The label for any non-recombination node v is obtained from the label of v's parent $p(v)$, by changing the state of any sites labeling the

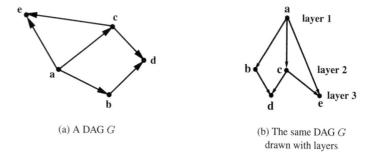

(a) A DAG G

(b) The same DAG G drawn with layers

Figure 1.12 (a) The directed graph G contains no *directed* cycles and so is a DAG. The underlying, undirected graph does contain a cycle.

(b) In this drawing of graph G, the layers defined in lemma 1.6.3 are drawn top down. Thus, the first layer consists of the single node a; the second layer consists of $\{b, c\}$; and the third consists of $\{d, e\}$. The DAG has four different topological sorts, for example (a, b, c, d, e) and (a, c, b, e, d). If we view node a as the root node, the DAG has two recombination cycles: one with coalescent node a and recombination node d; and one with coalescent node a and recombination node e.

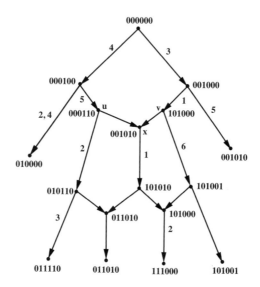

Figure 1.13 A genealogical network. In the genealogical network, sites $\{1, 2, 3, 4, 5\}$ each mutate more than once. The sequence 001010 that labels recombination node x, could have taken 00 from the parental sequence at node u, then 10 from the parental sequence at v, and then the last 10 again from the parental sequence at u. Note that the parental sequences are identical at sites 2 and 6, so those are the only sites where there is a choice for which parental sequence contributes the state. For all other sites, the choice is forced.

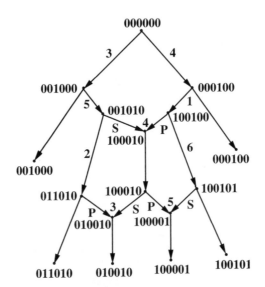

Figure 1.14 In any ARG, each site mutates exactly once, and in this ARG, each recombination event is a single-crossover recombination. The labels "P" and "S" on the edges into a recombination node indicate which parental sequence contributes the prefix and which contributes the suffix; the number over the recombination node indicates where the recombinant sequence switches from prefix to suffix contributions.

edge $(p(v), v)$. The label on a recombination node x can be any binary sequence s such that for each site c, the state of s at c equals the state at c in at least one of x's parental sequences. The sequences labeling the leaves of $\mathcal{N}$ must define the sequences in M. figure 1.13 shows a genealogical network. In the taxonomy of phylogenetic networks shown in [195] (page 69 of that book), these networks are also called *recombination networks*. A modified version of that taxonomy is shown in figure 3.5.

> An ARG is a directed graph that simultaneously describes vertical and nonvertical evolutionary events. [37]

We informally define an *ancestral recombination graph* (ARG) for M, as a genealogical network $\mathcal{N}$ for M, where each site labels *exactly one* edge in $\mathcal{N}$. Hence, the main distinction between a genealogical network and an ARG is that the mutations on an ARG obey the infinite-sites model.

Further, genealogical networks always allow multiple-crossover recombination, while ARGs can be required to only allow single-crossover recombination. When an ARG is constrained in that way, the recombinant sequence s_x labeling a recombination node x, must be formed from a prefix of one of x's parental sequences, followed by a suffix of the other parental sequence. Figure 1.14 shows an ARG with the same DAG used for the genealogical network in figure 1.13, but deriving a different set of sequences, since each site mutates only once. Relating this definition to the quote above, mutations are *vertical* events, and recombinations are *nonvertical* events.

1.7   The Book

This book discusses algorithmic and mathematical results, mostly obtained in the last decade, concerning combinatorial structure of genealogical networks, particularly ARGs, (sometimes extending to other phylogenetic networks). The algorithms exploit the structure, and are used to deduce information (sometimes only partial) about the networks, or are used to explicitly construct networks that generate observed sequences through the biological events of mutation and recombination (and sometimes other events). The networks serve as hypotheses for the true genealogical history of the extant sequences, and help to address fundamental biological questions, or are used in practical problems such as association mapping, location of recombination hotspots, computing local recombination rates, phasing genotypic data, and identification of SNP sites (all of which will be discussed in depth). Moreover, algorithms that create explicit networks form a complement, or an alternative, to methods based on the more commonly used numerical, statistical, measures that less directly reflect the underlying genealogy.

Problems of constructing genealogical networks from sequence data, or deducing features of such networks, are significantly more complex than for the analogous problems in trees, and the field of network reconstruction is much less developed than the field of tree reconstruction. The book develops ideas and methods for reconstructing or obtaining information about genealogical networks, particularly ARGs. It concentrates on ideas and methods that we believe are *fundamental*, and on applications that illustrate the utility of these ideas and methods.

2   Trees First

Master your twos and trees. Half of computer science exploits the *powers of twos*, and half exploits the *powers of trees*.[1]

Our main interest is in genealogical and phylogenetic *networks*, which by definition are *not* trees. However, many of the network models derive from tree models, and many of the tools that address networks rely critically on tools for trees. Further, if we are to understand when and why recombination is needed, we need to understand when and why recombination is *not* needed. Therefore we must first understand some models of treelike evolution and some combinatorial and algorithmic results about evolutionary *trees*. The main tree-based model of evolution that we use is called the (rooted, binary character) *perfect-phylogeny* model.

2.1   The Rooted Perfect-Phylogeny Problem

Definition Let M be an n by m matrix representing n taxa in terms of m characters or traits that describe the taxa. Each character takes on one of two possible *states*, 0 or 1; a cell (f, c) of M has a value of one if and only if the state of character c is 1 for taxon f. Thus the characters of M are *binary-characters* and M is called a *binary matrix*.

When a taxon f has state 1 for a binary character c, we also say that "f possesses (or contains or has) character c."

---

1   A bad pun I often use in my algorithms classes.

	c_1	c_2	c_3	c_4	c_5
r_1	1	1	0	0	0
r_2	0	0	1	0	0
r_3	1	1	0	0	1
r_4	0	0	1	1	0
r_5	0	1	0	0	0

Table 2.1 Matrix M has a perfect phylogeny T shown in figure 2.1.

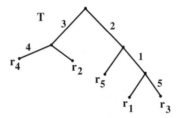

Figure 2.1 Perfect phylogeny T for matrix M shown in table 2.1. The root is shown at the top, and the (unshown) direction of the edges is away from the root, toward the leaves.

Definition Given an n by m binary-character matrix M for n taxa, a *perfect phylogeny for* M is a *rooted* (directed) tree T with exactly n leaves, obeying the following properties:

1. Each of the n taxa labels exactly one leaf of T.

2. Each of the m characters labels *exactly one* edge of T.

3. For any taxon f, the characters that label the edges along the unique path from the root to the leaf labeled f, specify all of the characters that taxon f possesses (i.e., whose state is 1).

When needed, we will also assume that if a leaf of T is labeled by a taxon f, then it is labeled by the characters that f possesses, or equivalently, is labeled by the binary sequence defined by row f of M.

A perfect phylogeny exists for some binary matrices, but not for all. The binary matrix M shown in table 2.1 has a perfect phylogeny, shown in figure 2.1.

In the definition given for a perfect phylogeny T, it is required that T be rooted and directed, and that the characters be binary. This is the *default* case, the technically simplest case and most common case that will be discussed in this book. However, we will sometimes relax these assumptions; when we do, we will always explicitly state the alternative assumptions being used. So, when we use "perfect phylogeny" with no additional modifiers, we are referring to a *rooted, binary-character* perfect phylogeny.

The interpretation of a perfect phylogeny T for M is that it gives an estimate of the rooted evolutionary history of the taxa (in terms of branching pattern, but not time), based on the following biological and technical assumptions:

1. The taxa in M are generally taxa whose states have been observed and are known.

2. There exists a taxon r (possibly unknown) that is *ancestral* to all the taxa in M. Taxon r is called the *root*. The character-state sequence for the root is called the *ancestral sequence* or the *root sequence*, and is denoted s_r. For technical and expositional convenience, we assume first that the state of r for each character is zero, so s_r is the all-zero sequence. We will see later how to relax that assumption.

3. In the evolutionary history of the taxa, each of the characters mutates from the zero state to the one state *exactly* once, and never from the one state to the zero state. Hence every character c labels exactly one edge e in a perfect phylogeny T for M, indicating the unique point in the evolutionary history of the taxa when character c mutates. It follows that any taxon that labels a leaf below $e = (u, v)$ (or in more graph-theoretic terminology, in the subtree T_v, rooted at v) must possess character c. As stated in chapter 1, a character that satisfies these assumptions is called a *perfect character* [174, 351].

The key combinatorial constraint

The key biological and combinatorial feature of the perfect-phylogeny model is that each character mutates *exactly once* in the evolutionary history of the taxa. That is, each character is a *perfect character*. This assumption is principally motivated by the *infinite-sites* model from population genetics and widely collected SNP data (discussed in section 1.4.1). The infinite-sites assumption is not always valid, but is appropriate in the biological settings that are the major focus of this book. Hudson and Kaplan [183] summarize the population genetics viewpoint as follows:

> For the infinite-site model the mutation rate for any site is infinitesimal; therefore, at most one mutation event can occur in the history of the sample at that site. Thus, for any two sites there are at most four gametic types in the population. Furthermore, since the model does not allow for back mutation and recurrent mutation, the only way for all four gametic

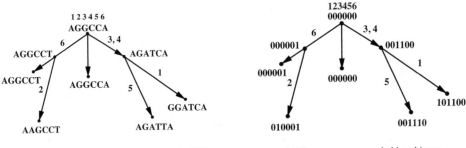

(a) Perfect phylogeny of five sequences at six SNP sites.

(b) The sequences recoded into binary.

Figure 2.2   The perfect phylogeny developed in [297] for five sequences at six SNP sites in the Human DTNBP1 gene (Dysbindin) on chromosome 6.   It has been suggested that variations in these sequences are associated with schizophrenia, although the evidence is said to be contradictory. There are two DNA variants at each SNP site (for example A and G at site 1, and C, T at site 5), and so the SNPs can be recoded into binary, using 0 for the ancestral state and 1 for the derived state. The six sites labeled 1 through 6 here are actually SNPs 2,3,5,6,8,11 in [297].

types to be in the sample is for at least one recombination event to have occurred in the history of the sample between the two sites.

The last statement in this quotation will be formally established in theorem 5.2.1, and extended in lemma 5.2.3 (page 131) in section 5.2.2.

Further motivation

The assumption of only one mutation per character is also motivated by the biological basis of *complex characters* in species as discussed in section 1.4.1; and by low-frequency insertions or deletions or of inversions of whole DNA segments; and by the use of other rare genetic features that allow one to build phylogenies extending far into the past [81, 164, 174, 351, 357]. Additional discussion of the perfect-phylogeny model (in relation to coalescent theory) is presented in section 12.3.1. Further, the perfect-phylogeny model is justified by the success of methods that assume the perfect-phylogeny model, in order to solve applied problems in genetics (for example in chapter 13).

Figure 2.2 shows the phylogeny of six SNP sites, published in [297]. The SNP data fits the infinite-sites model and the phylogeny is a perfect phylogeny.

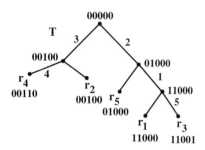

Figure 2.3   Perfect phylogeny T with nodes labeled by the derived binary sequences.

2.1.1   Alternative Definitions of a Perfect Phylogeny

There are two alternative, but equivalent, definitions of a perfect phylogeny that are often technically helpful, and generalize nicely to *nonbinary* and to *undirected* or *unrooted* perfect-phylogeny problems, which we will discuss later. Recall that a node in a rooted tree T is called an *internal node* if it is neither a leaf node nor the root of T.

A second definition of a perfect phylogeny A perfect phylogeny for M is a rooted tree T with n leaves, where each leaf is labeled by a distinct taxon of M, and where the root of T and each internal node of T, is labeled by an m-length binary sequence specifying a state for each of the m characters, such that:

> For every character c and each state i (either 0 or 1) of c, the nodes labeled with state i for character c form a *connected* subtree of T, denoted $T_c(i)$.

This property is called the *convexity requirement*. Clearly, for any character c and states $i \neq j$, the subtrees $T_c(i)$ and $T_c(j)$ of perfect phylogeny T are node disjoint. For example, the perfect phylogeny from figure 2.1, with its node labels, is shown in figure 2.3.

Note that in this second definition of a perfect phylogeny there is no mention of mutations, or labels on edges, or the requirement that each character mutates only once. However, it is easy to see that the two definitions of a perfect phylogeny are equivalent. We leave the proof to the reader.

There is a third related, and equivalent, view of a perfect phylogeny that is also often used.

A third definition of a perfect phylogeny Given M, let T be a rooted tree with n leaves, where each leaf is labeled by a distinct taxon of M, where the nonleaf nodes of T are *unlabeled*.

Definition For a state i of character c, let $S_c(i)$ denote the *smallest* subtree of T connecting all the leaves of T that are labeled with state i for character c.

> Then T is called a *perfect phylogeny* if and only if for each c, the subtrees $S_c(0)$ and $S_c(1)$ are node disjoint.

We again leave it to the reader to convince themselves that this definition is equivalent to the two prior definitions.

2.1.2   The Perfect-Phylogeny Problem and Solution

The perfect-phylogeny problem: Given an n by m binary matrix M, determine whether there is a perfect phylogeny for M, and if so, build one.

We will solve the perfect-phylogeny problem with a simple $O(nm)$-time algorithm where each comparison operation and each reference to M takes one time unit.

For the algorithm and its proof of correctness, it will be helpful to first sort the columns of M by the *number* of ones they contain, largest first, breaking ties arbitrarily. Let $\overline{M}$ denote the sorted matrix M. For an example, see table 2.2. Certainly, M has a perfect phylogeny if and only if $\overline{M}$ does, and the perfect phylogeny for $\overline{M}$ differs from the perfect phylogeny for M only by a change in edge labels corresponding to the sorting of the columns of M. For example, see figure 2.4.

In the rest of this chapter, the name of any character will be the same as the column that it occupies in $\overline{M}$, rather than in M. For example, the character at the left of $\overline{M}$ will be called character one. Therefore, for two characters c and d, with $c < d$, character c must be to the left of character d in $\overline{M}$. Similarly, if character c is to the left of character d in $\overline{M}$, then it must be that $c < d$. We also assume that the m rows (taxa) of $\overline{M}$ are labeled $r_1, r_2, ..., r_m$, but we also use f, g, h as variables, to refer to rows in general.

Theorem 2.1.1 The perfect-phylogeny theorem *Matrix $\overline{M}$ (or M) has a perfect phylogeny (with all-zero ancestral sequence) if and only if no pair of columns c, d contains the three binary pairs 0,1; 1,0; and 1,1.*

Proof: First, suppose that T is a perfect phylogeny for $\overline{M}$ and consider two characters c and d. Let e_c be the edge of T on which character c changes from state 0 to state 1, and let e_d be the similar edge for character d. Note that all of the taxa that possess character c

original column	2	1	3	5	4
	c_1	c_2	c_3	c_4	c_5
r_1	1	1	0	0	0
r_2	0	0	1	0	0
r_3	1	1	0	1	0
r_4	0	0	1	0	1
r_5	1	0	0	0	0

Table 2.2   Matrix $\overline{M}$ resulting from sorting the columns of the matrix M shown in table 2.1. The first row of numbers above $\overline{M}$ indicates the original column of each character in M. The second row of numbers gives the new name for each character.

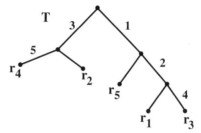

Figure 2.4   Perfect phylogeny for the sorted matrix $\overline{M}$ shown in table 2.2.

(or d) are found at the leaves of T below edge e_c (or edge e_d), and one of four cases must hold: Either (1) $e_c = e_d$, or (2) e_c is on the path from the root of T to e_d, or (3) e_d is on the path from the root of T to e_c, or (4) the paths to e_c and e_d diverge before reaching either of those edges.

In case (1), there cannot be a taxon with (ordered) states 0,1 or 1,0 for character pair c, d. In case (2) there cannot be a taxon with (ordered) states 0,1 for c, d; similarly in case (3) there cannot be a taxon with (ordered) states 1,0 for c, d. In case (4) there cannot be a taxon with states 1,1 for c, d. This proves the "only if" direction of the theorem.

We now consider the "if" direction. The proof will be constructive, and establishes the *perfect-phylogeny algorithm*. We can assume that no pair of columns, c, d, are identical to each other; if they were identical we could remove one, column c say, and then if T is a perfect phylogeny for the remaining characters, we could add character c to the label of the edge labeled by character d. The resulting tree would be a perfect phylogeny for all characters. Hence, it suffices to prove this direction of the theorem assuming that every column is distinct.

By definition, in *any* perfect phylogeny T for $\overline{M}$ (assuming one exists), and for any taxon f, the characters that appear on the path from the root of T to the leaf labeled f, are exactly the characters that have state 1 for taxon f. Moreover, those characters must appear in exactly the same order that they appear (left to right) in row f of $\overline{M}$. To see this, suppose taxon f possesses both characters c and d, and that $c < d$. As above, characters c and d are both on the path from the root of T to leaf f. Since the columns of $\overline{M}$ are sorted by the number of ones they contain, there are more taxa with state one for character c then there are with state one for character d, and so the edge e_c labeled by character c occurs above the edge labeled by character d. This is because all the taxa that label the leaves below e_c must possess character c. Hence, in a perfect phylogeny T (if one exists) the characters on the path from the root of T to a leaf f must be in exactly the same order that they are in $\overline{M}$. So, assuming there is a perfect phylogeny T for $\overline{M}$, for any taxon f, the set of characters and the order that those characters appear on the path from the root of T to leaf f, is precisely and *uniquely* determined. It then follows that there is a perfect phylogeny for $\overline{M}$ if and only if those n separate, forced paths, can be assembled into a single tree, i.e., where each character labels exactly one edge.

We will show constructively how to assemble those n forced paths into a perfect phylogeny for $\overline{M}$, under the stated premise of the theorem, that no pair of columns in $\overline{M}$ contains all the binary pairs $0, 1$; $1, 0$ and $1, 1$. To that end, we first develop a property that $\overline{M}$ must have when no pair of columns contains all those three binary pairs.

The shared-prefix property

> For two taxa f and g, let d be the largest index character (rightmost in $\overline{M}$) that taxa f and g both possess (i.e., where both have state 1). Then, assuming no pair of columns contain all three binary pairs $0, 1$; $1, 0$; and $1, 1$, rows f and g in $\overline{M}$ must be *identical* from column one (at the left end of $\overline{M}$) to column d.

For example, in table 2.2, character c_3 is the largest index character that taxa r_2 and r_4 both possess; as required, the rows for r_2 and r_4 are identical (containing $0, 0, 1$) from columns 1 to 3 of $\overline{M}$.

To establish the shared-prefix property, suppose taxon f possesses a character $c < d$ in $\overline{M}$, so that columns c, d contain the binary pair $1, 1$. Since the columns of $\overline{M}$ are distinct and are sorted by the number of ones they contain, columns c, d must also contain the ordered binary pair $1, 0$. Therefore (by the premise of the theorem), columns c, d cannot contain the ordered pair $0, 1$, and hence taxon g must also possess character c. The choice of taxon f in this argument was arbitrary, so the conclusion holds for taxon g also, and hence if either f or g possess a character $c < d$, then both f and g possess character c. It follows that rows f and g are identical in $\overline{M}$ from column one to column d. This establishes the *shared-prefix property* for $\overline{M}$.

Constructing a perfect phylogeny The shared prefix property allows a simple algorithm to construct a perfect phylogeny. The algorithm builds up the perfect phylogeny T for $\overline{M}$ by processing the rows of $\overline{M}$ in order. It first creates a root node for T and adds to it a single path from the root to a leaf labeled by taxon r_1. If taxon r_1 possesses t characters, that path will contain t edges successively labeled by one character possessed by taxon r_1, in the order that those characters appear in row 1 of $\overline{M}$, followed by a single unlabeled edge leading to a leaf labeled by taxon r_1. Note that this single path is a perfect phylogeny for the first taxon of $\overline{M}$.

Let T_f denote the intermediate tree that contains all the paths for the taxa from r_1 to f. We assume inductively that T_f is a perfect phylogeny for the first f taxa of $\overline{M}$. Then tree T_{f+1} is constructed from T_f as follows: Starting at the root of T_f, examine the characters that taxon $f + 1$ possesses (from left to right in $\overline{M}$) and in parallel, walk from the root of T_f down the (unique) path in the tree, as long as the successive characters on the path match the successive characters that taxon $f + 1$ possesses. For example, see figure 2.5. The path is unique because no character appears more than once anywhere in the perfect phylogeny T_f, and in particular, no character appears more than once on the edges leading out of any node. The walk ends at a node, denoted v_{f+1}, where no label on any edge out of v_{f+1} matches the next character that taxon $f + 1$ possesses, or where all the characters that taxon $f + 1$ possesses have been matched. Let c denote the last matched character on the walk. Then create a new path out of v_{f+1} containing all the characters to the right of c that taxon $f + 1$ possesses (in the order they appear in $\overline{M}$), followed by an unlabeled edge to a leaf labeled $f + 1$. The result is a tree T_{f+1}.

We claim that T_{f+1} is a perfect phylogeny for the first $f + 1$ taxa of $\overline{M}$. Clearly, each path to a leaf $h \leq f + 1$ in T_{f+1} contains exactly the characters that taxon h possesses. Also, since T_f is a perfect phylogeny, no character on the path to node v_{f+1} is anywhere else in T_f. So to prove that T_{f+1} is a perfect phylogeny for the first $f + 1$ taxa, we only need to prove that none of the characters on the new path out of v_{f+1} are in T_f. Let d be the rightmost character (in $\overline{M}$) that taxon $f + 1$ possesses, such that d labels some edge in T_f. Let e_d denote that edge in T_f. Any taxon h labeling a leaf below e_d in T_f possesses character d, and by the shared prefix property, rows h and $f + 1$ of $\overline{M}$ are identical from column 1 to column d. Hence, the walk to v_{f+1} is a walk toward the leaf labeled by taxon h. Moreover, by the choice of character d, and the fact that all characters that taxon h

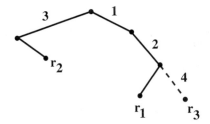

Figure 2.5   The extension from T_{r_2} to T_{r_3} in the creation of the perfect phylogeny T shown in figure 2.4. Here, $f = r_2$, and the walk from the root follows edges labeled with characters 1 and 2. Then, a new edge to a leaf labeled r_3 is added, and that edge is labeled with character 4.

possess are in T_f, taxa h and $f + 1$ do not possess any common characters to the right of d. Hence, the characters on the walk from the root of T_f to node v_{f+1} exactly match all the characters that taxon $f + 1$ possesses, from the left end of $\overline{M}$ to character d. Therefore, by the choice of d, none of the characters on the new path out of v_{f+1} are in T_f, and hence T_{f+1} is a perfect phylogeny for the first $f + 1$ taxa of $\overline{M}$.

When all taxa have been processed, the resulting tree T is perfect phylogeny for $\overline{M}$, and for M. This proves the *if* direction, and finishes the proof of theorem 2.1.1.   ∎

Note that in this construction, each internal node is labeled. If the resulting perfect phylogeny contains a node v with degree two, other than the root node, we can contract node v, merging the two edges that were incident with v, possibly creating an edge labeled by more than a single character. Note also that the only unlabeled edges are edges into leaves of the tree.

The proof of theorem 2.1.1 shows that *any* perfect phylogeny for $\overline{M}$ (assuming no duplicate columns) *must* be the superposition of n forced paths. This establishes the following:

Corollary 2.1.1 *If there is a perfect phylogeny for M, and every column in M is distinct, there is only one, unique, perfect phylogeny for M where every internal node has degree at least three.*

That is, after contracting all the unlabeled edges that are not directed into a leaf, two perfect phylogenies for M must be identical. Corollary 2.1.1 applies when all the columns of M are distinct, but it is easy to remove this assumption. As discussed in the proof of theorem 2.1.1, we can always remove duplicate characters and still construct a perfect phylogeny (if one exists) for all of the data. It then follows from corollary 2.1.1 that a perfect phylogeny is always unique if we consider that the reinserted duplicate characters label their common edge as a *set* of characters. It is also possible to build a perfect phylogeny where duplicate characters label different edges. Then the edges labeled by those duplicate characters must be in a consecutive path and the intermediate nodes (nodes not at the two ends of the path) on that path must each have degree two. In that case, the perfect phylogeny is unique except for the order of the characters on that path. We leave the details to the reader. We summarize these facts as:

Theorem 2.1.2 *Suppose M contains some columns which are duplicated and there is a perfect phylogeny for M. If we require that no internal node of a perfect phylogeny has out-degree one, then all the perfect phylogenies for M have the same leaf-labeled topology (i.e., the leaf-labeled trees are identical after removal of all edge labels). Similarly, if we require that every edge be labeled with at most one character, but allow internal nodes with out-degree one, then all the perfect phylogenies for M have the same leaf-labeled topology.*

The proof of theorem 2.1.1 not only establishes the theorem—it also gives a constructive method to build a perfect phylogeny for M, when one exists. We now examine the running time for this method.

Theorem 2.1.3 *If there is a perfect phylogeny T for M, it can be constructed in $O(nm)$ operations. And if each row f of M is presented as the set of characters that taxon f possesses, then T can be constructed in time proportional to the number of ones in M.*

Proof: First, to create $\overline{M}$ from M we count the number of ones in each column of M, in $O(nm)$ operations. Those counts range from 0 to n, so we can sort the numbers, in decreasing order, and sort the associated columns, in $O(m)$ operations by standard bucket sort or counting sort [80]. The number of operations needed to create T_{f+1} from T_f is $O(m)$, since no character appears twice in T_f, implying that T_f has at most m labeled edges. Thus the total number of operations needed to build the perfect phylogeny for M is $O(nm)$.

For the second part of the theorem, let q denote the number of ones in M; when the characters possessed by each taxon are presented as a set, the value of q is the sum of the sizes of these n sets. By scanning the sets, we can create a linked list $Q(c)$, for each character c, containing the taxa that possess c. These linked lists can be built, and the size of each list can be determined, in $O(q)$ operations. The size of the list for character c indicates the number of taxa which possess character c; as before, those numbers (and their associated characters) can be sorted in $O(m)$ operations. Let L be the ordered list of characters, based on this sort. Next, we want to reorder the set of characters that each taxon f possesses, to agree with the order of the characters in L. Let $S(f)$ denote the desired ordered set of characters that f possesses. These ordered sets are obtained by processing the characters in L in order: when a character c is processed, we put c at the end of the growing ordered set $S(f)$, for every taxon f in $Q(c)$. The n ordered sets are thus built in $O(q)$ total operations, and essentially describe the matrix $\overline{M}$.

Next we use the $S(f)$ sets to build up the perfect phylogeny T as described in the proof of theorem 2.1.1. The only added detail needed is that when building up T, we create a pointer indexed by c, to the (unique) location of each character c in the tree, at the time that character c is on a new path that is added to the tree. Then when inserting the path for any taxon $f + 1$, we process the set $S(f + 1)$ in order (simulating the left to right scan of row $f + 1$ in $\overline{M}$), and when a character c is the next character in $S(f)$, we use the pointer indexed by c to determine if c appears in T_f, and if so, where it appears. In this way, the number of operations needed to build T_{f+1} from T_f is proportional to the size of $S(f+1)$, and hence the perfect phylogeny T can be built in $O(q)$ operations in total.   ■

The first proved $O(nm)$-time perfect-phylogeny algorithm was given in [136], with a method that is different from the one given here. An alternate version appears in [137]. An earlier perfect-phylogeny algorithm, developed in [285] and discussed in [373], can also be shown to run in $O(nm)$ time, but no time bound was established in [285]. The $O(q)$ method for building a perfect phylogeny from sets was first established in [4]. Note that a straightforward implementation of theorem 2.1.1 would give an $\Omega(nm^2)$-time algorithm to determine whether M has a perfect phylogeny, and would not construct one.

2.1.2.1   *An Alternate Statement of the Perfect-Phylogeny Theorem*

There is an alternate statement of the perfect-phylogeny theorem that is often used. For emphasis, we remind the reader that a perfect phylogeny is a *rooted* tree and that the characters are binary.

Definition For any column c of $\overline{M}$, let O_c be the set of taxa that possess character c.

Theorem 2.1.4 *There is a perfect phylogeny for M if and only if for every two characters c and d, either $O_c \cap O_d = \emptyset$, or one set is contained in the other.*

We leave the justification of this to the reader. A direct proof of theorem 2.1.4 appears in [137]. Other proofs of theorems 2.1.1 and 2.1.4 and of corollary 2.1.1 appear in a number of places, for example [107, 108, 109], and in somewhat different language in [373].

2.2 The Case of a Known, Non-Zero Ancestral Sequence

We will now examine some ways that the basic assumptions in the definition of a perfect phylogeny can be relaxed, and how the modified perfect-phylogeny problem is solved in those cases.

We have assumed that the root of a perfect phylogeny is labeled with the all-zero sequence, corresponding to the assumption that the ancestral taxon does not possess any of the characters in M. This is a convenient *technical* assumption, but it is not necessary. Suppose that the binary characters in M are such that the known ancestral taxon r *does* possess some of the characters in M. Therefore, the binary sequence labeling the root should not be the all-zero sequence, but rather a binary sequence where values of one indicate the characters that r possesses. What is the modified model for a perfect phylogeny in this case, and how can we solve the problem of determining if there is such a perfect phylogeny?

The root-known perfect phylogeny
We start with the following

Definition Given an n by m binary-character matrix M for n taxa, and a given binary sequence s_r, whose ones indicate the characters that the ancestral taxon r possesses, an s_r-*perfect phylogeny for* M is a *rooted* tree T with exactly n leaves, obeying the following properties:

1. The root of T is labeled by the ancestral sequence s_r.

2. Each of the n taxa labels exactly one leaf of T.

3. Each of the m characters labels *exactly one* edge of T.

4. For any taxon f, let C_f be the set of characters labeling edges on the path in T from the root to the leaf labeled f. Then, the binary sequence for f (i.e., the row in M for f) and the binary sequence s_r differ at *exactly* the characters in C_f.

Another way to visualize this definition is that each leaf f is labeled with a binary sequence (the row in M for f) that is derived from the ancestral sequence s_r by walking from the root to leaf f, changing the state of character c when an edge labeled by c is encountered.

Note that a perfect phylogeny is simply an s_r-perfect phylogeny in the case that the specified ancestral sequence, s_r, is the all-zero sequence. Thus, the concept of a root-known perfect phylogeny generalizes a perfect phylogeny, with a small change in the definition of a perfect phylogeny given earlier.

	c_1	c_2	c_3	c_4	c_5
s_r	1	1	0	0	1
r_1	1	1	0	0	0
r_2	0	0	1	0	0
r_3	1	1	0	0	1
r_4	0	0	1	1	0
r_5	0	1	0	0	0

Table 2.3   When the matrix above is transformed based on s_r, the result is the matrix M shown in table 2.1. Matrix M has a perfect phylogeny (with the all-zero ancestral sequence), so there is an s_r-perfect phylogeny for the above matrix, and it has the same leaf-labeled topology as the perfect phylogeny for M.

The root-known perfect-phylogeny problem: Given M and m-length binary sequence s_r, determine if there is an s_r-perfect phylogeny for M.

We efficiently solve this problem by *reducing* it back to the original (all-zero root) perfect-phylogeny problem.

Theorem 2.2.1 *Let M' be the binary matrix obtained from M by interchanging all the 0 and 1 entries in each column c of M, where $s_r(c) = 1$. Then there is an s_r-perfect phylogeny T for M if and only if there is a perfect phylogeny T' (with all-zero ancestral sequence) for M'. Moreover, T and T' have the same leaf-labeled topology.*

As an example, see table 2.3. Note that no changes are made in any column c of M where $s_r(c) = 0$. Hence, the same transformation (changing any ones to zeros) applied to s_r, creates the all-zero sequence.

Theorem 2.2.1 gives an efficient, constructive way to determine whether there is a perfect phylogeny with a specified ancestral sequence s_r. The proof of theorem 2.2.1 is simple and is left to the reader as an exercise.

The reduction of the root-known perfect-phylogeny problem to the original perfect-phylogeny problem, combined with the perfect-phylogeny theorem (theorem 2.1.1), yield the following:

The root-known perfect-phylogeny theorem

Theorem 2.2.2 *Binary matrix M has an s_r-perfect phylogeny if and only if no pair of columns c, d in M contains the three binary pairs that differ from the (ordered) binary* pair in positions c, d of s_r. *Moreover, if all the columns of M are distinct, and there is an s_r-perfect phylogeny for M, there is one, unique s_r-perfect phylogeny for M, where all the internal nodes have degree at least three.*

Proof: By case analysis, we see that a pair of columns c, d in M contain the three distinct binary pairs (in rows f, g, h say) that differ from the ordered binary pair in positions c, d of s_r, if and only if the same rows f, g, h in M' contain the distinct ordered binary pairs that differ from the transformed s_r at c, d. But the transformed s_r is the all-zero sequence, so columns c, d in M' contain all the binary pairs $0, 1; 1, 0; 1, 1$, if and only if columns c, d in M contain all the binary pairs that differ from the binary pair in positions c, d of s_r. Now, by theorem 2.1.1 there is a perfect phylogeny for M' if and only if no pair of columns in M' contain all three binary pairs $0, 1; 1, 0; 1, 1$. The first part of the theorem then follows by application of theorem 2.2.1.

To show uniqueness, let T_1 be an s_r-perfect phylogeny for M, and let T_1' be the perfect phylogeny for M'. By theorem 2.2.1, T_1 and T_1' have the same leaf-labeled topology. Now if there exists a different s_r-perfect phylogeny T_2 for M, then there exists a perfect phylogeny T_2' for M that differs from T_1', which violates the statement of uniqueness in theorem 2.1.1. ∎

Theorem 2.2.2 can be extended to handle the case of duplicate characters, in the same way that theorem 2.1.2 extends corollary 2.1.1. We leave this as an exercise for the reader.

2.3   The Root-Unknown Perfect-Phylogeny Problem

We now consider the most relaxed perfect-phylogeny model, when *no* ancestral sequence is specified as part of the problem instance.

Definition Given a binary matrix M, but *no* specified ancestral sequence, the *root-unknown perfect-phylogeny problem* is to determine whether there exists some binary sequence s_r, such that there is an s_r-perfect phylogeny. Moreover, an ancestral sequence s_r should be explicitly identified if one exists.

Definition An *undirected perfect phylogeny* for M is a tree that is obtained from a rooted perfect phylogeny T for M, after removing all the directions on the edges of T, and successively contracting nodes of degree two. See figure 2.6.

The contraction of nodes of degree two is done in order to address the case of duplicate characters, and the fact that the root of an s_r-perfect phylogeny might have degree two even if there are no duplicate characters. Note that in the definition of an undirected perfect phylogeny T, all internal nodes of T have degree greater than two.

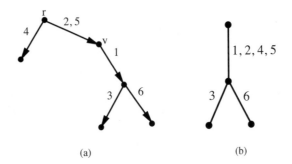

Figure 2.6   (a) A rooted, directed tree with two nodes, r and v of degree two.
(b) The unrooted tree obtained by removing the directions on all edges, and successively contracting the nodes r and v of degree two.

The root-unknown perfect-phylogeny problem is equivalent to the problem of determining if there is an undirected perfect phylogeny for M, so the root-unknown perfect-phylogeny problem is also called the *undirected* or *unrooted* perfect-phylogeny problem. The term "root-unknown perfect phylogeny" reminds us that a perfect phylogeny is a *rooted* tree. However, when no ancestral sequence is known (and this is often the case), it may be more consistent with the known information if a perfect phylogeny T is converted to an undirected perfect phylogeny, so we will often use that term.

To solve the root-unknown perfect-phylogeny problem, we again use reduction, in this case to the root-known perfect-phylogeny problem. To see the reduction, suppose there is a perfect phylogeny T for M with some ancestral sequence s_r. Each taxon f in M defines the sequence labeling a leaf of T, and hence there is a directed path in T from the root to leaf f. Now imagine grabbing the leaf f and raising it above the root of T, making T hang down from f. We could then consider f as the new root of the resulting directed tree T_f, where all of the sequences in M are derived from f. In particular, the sequence for any taxon g is obtained from the sequence for f by changing the state of the characters on the path from f to g. Tree T_f is almost an f-perfect phylogeny for M, but it does not strictly obey the definition of a perfect phylogeny because f does not label a leaf. To rectify that, we simply add an unlabeled edge from the root of T_f to a new leaf labeled f. This establishes the following:

Theorem 2.3.1 *Given a binary matrix M and no specified ancestral sequence, if there is an undirected perfect phylogeny for M, then there is an f-perfect phylogeny for M, where f is any taxon in M.*

Note that in the transformation from T to T', no edge labels change, and the leaf-labeled topology also does not change, except for the addition of the edge to the new leaf labeled f.

Theorem 2.3.1 leads to an efficient algorithm for the root-unknown perfect-phylogeny problem: declare the sequence of some (any) taxon f in M to be the ancestral sequence s_r, and then solve the root-known perfect-phylogeny problem. If a (rooted) s_r-perfect phylogeny T is obtained, but an unrooted tree is more biologically valid, remove the directions on all edges of T.

2.3.1 The Four-Gametes Theorem

Theorem 2.3.1 not only leads to an efficient algorithm, it leads to the classic necessary and sufficient condition for the existence of a *root-unknown* perfect phylogeny for binary matrix M:

Theorem 2.3.2 The four-gametes theorem   *When no ancestral sequence is known, matrix M has an undirected perfect phylogeny if and only if no pair of columns contains all four binary pairs 0,0; 0,1; 1,0; and 1,1.*

Proof: If there is an undirected perfect phylogeny for M, then by theorem 2.3.1 there is an f-perfect phylogeny for any taxon f in M. Therefore, by theorem 2.2.2 no pair of columns c, d in M can have all three binary pairs that differ from the binary pair in columns c, d for sequence f. Then, since f is also in M, no pair of columns in M can have all four binary pairs.

Conversely, for any taxon f, if no pair of columns in M contain all the four binary pairs, then no pair of columns of M contain all three binary pairs that differ from the binary pair that sequence f has in those columns. Therefore, by theorem 2.2.2, M has an f-perfect phylogeny T, and so M has the undirected perfect phylogeny obtained from T.   ∎

The condition in theorem 2.3.2 is called the *four-gametes condition*, or *four-gametes test* in the population genetics literature [168, 183], and is called the *compatibility condition* or *consistency condition* in the phylogenetics literature [112, 373, 432].[2]

---

2 A discussion in [112] sorts out most of the origin of theorem 2.3.2. But more detail is possible. The early papers [432, 248] each establish the *necessary* direction of theorem 2.3.2, but do not state the (more interesting and more difficult) sufficient direction, although vague language in [248] might hint at it. The first paper that states and proves the sufficient direction is [54], although it does not explicitly state the necessary direction. Both the necessary and sufficient conditions were stated in [109] and a proof was provided, although in such a highly mathematical style that I have not verified

By analogy, the condition in theorem 2.1.1, the perfect-phylogeny theorem, could be called the *three-gametes* condition, although that phrase is not generally used.

Back to a rooted problem

Through the reductions used in the proofs, we have seen a very close relationship between unrooted, undirected versions of problems and results and root-known versions. As another reflection, we can now give a different, but equivalent, way to discuss theorem 2.2.2, as follows:

Definition Given an n by m binary matrix M and a fixed binary sequence s of length m (which need not be in M), let $M \cup \{s\}$ be the binary matrix created by adding s to M.

Corollary 2.3.1 *Given a binary matrix M, and an ancestral sequence s_r (which need not be in M), M has an s_r-perfect phylogeny, if and only if no pair of sites in $M \cup \{s_r\}$ contains all four gametes.*

2.3.1.1   An Alternate Reduction

There is an alternative reduction that can solve the undirected perfect-phylogeny problem, based on an interesting fact detailed in theorem 2.3.3 below.

Definition For any character c, if strictly more than half of the taxa have state i for c, then i is called the *majority state* of c. A sequence s_m is said to be a *plurality sequence* if s_m has the majority state for every character c that has a majority state. If character c does not have a majority state, then the value of s_m at position c is permitted to be either 0 or 1.

Theorem 2.3.3 *Given a binary matrix M and no specified ancestral sequence, if there is an undirected perfect phylogeny for M, then there is an s_m-perfect phylogeny for M, where s_m is any plurality sequence for M.*

We leave the proof of theorem 2.3.3 as a simple exercise for the reader. There are applications where the use of theorem 2.3.3 is preferred to the use of theorem 2.3.1. We will see one in chapter 12.

---

it. The term "four-gametes test" was introduced in [183], where the necessary direction of theorem 2.3.2 is stated (and is all that is required for the purposes of that paper), without mention of the sufficient direction.

We also note the undirected analog to theorem 2.1.4. Recall that O_c and O_d are the sets of taxa that possess characters c and d respectively. Let $\overline{O_c}$ and $\overline{O_d}$ be the sets of taxa that *do not* possess characters c and d respectively.

Theorem 2.3.4 *There is an undirected perfect phylogeny for M if and only if for every two characters c and d, one of the sets $O_c \cap O_d$, $\overline{O_c} \cap O_d$, $O_c \cap \overline{O_d}$, $\overline{O_c} \cap \overline{O_d}$ is empty.*

2.3.2   Uniqueness of an Undirected Perfect Phylogeny

The four-gametes theorem establishes when there is an undirected perfect phylogeny for a binary matrix, but it does not address the question of uniqueness. We address that now.

Theorem 2.3.5 *Suppose M has an undirected perfect phylogeny. Then there is one, unique, undirected perfect phylogeny for M, where all of the internal nodes have degree at least three. If an edge can only have one label, but an internal node can have degree two, then all of the undirected perfect phylogenies for M have the same leaf-labeled topology.*

Proof: Suppose all the columns of M are distinct and that there are two different undirected perfect phylogenies, T and T', for M. Let f be a taxon in M, so there is a leaf labeled f in both T and T'. By theorem 2.3.1 there are two f-perfect phylogenies T_f and T'_f. These trees must be different because T and T' are different, and the transformations of T and T' to T_f and T'_f respectively add a new edge from the root to the new leaf f, but preserve all the prior edges and edge labels. However, by theorem 2.2.2, when all columns of M are distinct, the f-perfect phylogeny for M is unique, a contradiction. Extension to the case of duplicate columns is immediate.

The proof of the second claim in the theorem is immediate from the first claim.   ∎

A note on the importance of uniqueness and improbability

Theorems 2.2.2 and 2.3.5, and corollary 2.3.1, all establish that when there is a perfect phylogeny (under different models), there is only one *unique* perfect phylogeny. Uniqueness has arguably been a critical factor in making the perfect phylogeny model, under different names in different areas of biology, important. Uniqueness implies that *if or when* the underlying biological reality fits the mathematical assumptions of the perfect-phylogeny model, the obtained tree will be *the* biologically and historically correct tree. It is not just one tree among many that display some aspects of the data—it is the tree that explains *the* true evolutionary history of the sequences. In contrast, even if a model perfectly captures the underlying biological reality, when a large set of trees fit that model, one does not know how much biologically correct information can be extracted from any *one* of those trees, or which of the trees is the most biologically informative.

So, the uniqueness of the perfect phylogeny is a very significant feature. Of course, the caveat is that this feature is of greatest utility *when* the biological reality fits, or nearly fits, the mathematical assumptions of the model. We will see in chapter 8 that uniqueness, or essential-uniqueness, is also one of the primary attractions of a generalization of perfect phylogeny, to networks called *galled trees*.

In addition to uniqueness, one of the arguments in favor of the perfect-phylogeny model, is that a set of random binary sequences is very unlikely to be derivable on a perfect phylogeny. Therefore, when one has a set of binary sequences that can be derived on a perfect phylogeny (or can be after some small modification of the data), this is strong evidence that the history of the sequences does conform to the perfect-phylogeny model.

2.4 The Splits-Equivalence Theorem: The Fundamental Theorem of Trees

In this book we have chosen to first expose the combinatorial structure of evolutionary trees through the viewpoint of the perfect-phylogeny model and variants of it. However, a different approach is also common [373], using the viewpoint of *splits* and the *splits-equivalence theorem*. In this section we will develop the splits-equivalence theorem and show that it is essentially the same as theorem 2.3.5 developed for the root-unknown perfect phylogeny problem. Thus, although the two viewpoints may at first seem different, they are really addressing the same combinatorial phenomena. But despite the equivalence, the language and viewpoint of splits will be very helpful in several parts of the book.

Let T be an undirected tree whose leaves have distinct labels. The removal of any edge e from T creates exactly two connected subtrees. An associated *bipartition* of the leaves of T is created by those two subtrees, i.e., two leaves are in the same class of the bipartition if and only if they are in the same subtree.

Definition We define the *split for* e as the bipartition of the leaves (equivalently, the leaf labels) defined by the two undirected subtrees resulting from the removal of edge e from T. Given a tree T with m edges, we define the *splits of* T as the set of m splits, one split for each edge in T. Note that if T has a node of degree two, then there will be two adjacent edges that define exactly the same split. This will not happen if each non-leaf node has degree at least three.

For example, there are eight splits in the tree shown in figure 2.4 (page 41). Those splits are:

$$\{r_4\}, \{r_1, r_2, r_3, r_5\}; \quad \{r_2\}, \{r_1, r_3, r_4, r_5\}; \quad \{r_2, r_4\}, \{r_1, r_3, r_5\};$$

$$\{r_2, r_4\}, \{r_1, r_3, r_5\}; \quad \{r_5\}, \{r_1, r_2, r_3, r_4\};$$

$$\{r_1, r_3\}, \{r_2, r_4, r_5\}; \quad \{r_1\}, \{r_2, r_3, r_4, r_5\}; \quad \{r_3\}, \{r_1, r_2, r_4, r_5\}.$$

Note that the two edges labeled with sites 1 and 3 define the same split. The splits of a tree are very informative, as shown next.

Theorem 2.4.1 (The splits-equivalence theorem) *For any undirected tree T with distinct leaf labels, the splits of T uniquely define T.*

Stated differently, for any (distinctly) leaf-labeled, undirected tree T, there is no other undirected tree with the same set of splits as T. Therefore, we can *uniquely* reconstruct T if we know the splits of T. This is one of the most fundamental and useful facts about the combinatorial structure of trees.[3] We will prove the splits-equivalence theorem by relating it to the root-unknown perfect phylogeny problem and theorem 2.3.5.

Proof (Of theorem 2.4.1): We assume that the splits of T are distinct, and leave the case when they are not distinct to the reader. We represent the splits of T in a binary matrix SP where each leaf of T is represented by a row of SP and each split of T (equivalently, each edge in T) is represented by a column of SP. The zeros in the column for split e identify the leaves on one side of the split, and the ones in the column identify the leaves on the other side of the split. Note that we could interchange all the zeros and ones in any column and still define exactly the same bipartition; the zeros and ones only serve to specify the bipartition and have no other meaning.

Now consider matrix SP as an input matrix to the *root-unknown* perfect phylogeny problem. We claim that there is an undirected perfect phylogeny for SP with the same leaf-labeled topology as T. To show this in detail, we must exhibit an s_r-perfect phylogeny for SP which becomes T when all edge directions are removed and every node of degree two is contracted. To create the desired s_r-perfect phylogeny, label each edge e in T by the column in SP associated with e (i.e., the column that describes the split for e in T), and label each leaf in T by the row in SP associated with that leaf. Next, choose any leaf f in T to be the root and set the ancestral sequence s_r to the sequence for f in SP. Finally, add a new edge from the root to a new leaf labeled f and direct all the edges away from the

---

3 Lior Pachter has suggested it should be called "the fundamental theorem of trees."

root. The result is an s_r-perfect phylogeny for SP that establishes that T is an undirected perfect phylogeny for SP.

Since all the splits in T are distinct, so all of the columns of SP are distinct, theorem 2.3.5 implies that T is the *unique* undirected perfect phylogeny for matrix SP. Further, if we interchange the zeros and ones in any column of SP, creating matrix SP', tree T will also be the unique undirected perfect phylogeny for SP'. The needed s'_r-perfect phylogeny for SP' is obtained from the s_r-perfect phylogeny for SP by interchanging the zeros and ones in s_r at every position where the values in SP were interchanged to create SP'. Hence, no matter how the splits of T are encoded in SP, tree T is the unique undirected perfect phylogeny for SP.

Now suppose there is another tree T' which is different from T but has exactly the same splits as T, encoded in a matrix SP'. By the argument in the prior paragraphs, T' is an undirected perfect phylogeny for SP'. But no matter how the splits of T' are encoded, matrix SP' also describes the splits of T, so T is the *unique* undirected perfect phylogeny for SP', contradicting the assumption that T and T' are different. ∎

The proof of the splits-equivalence theorem assumes that all splits are distinct, but it can be easily extended to the case when there are duplicate splits. In that case, all undirected perfect phylogenies for the matrix SP (encoding the splits of T) will have the same leaf-labeled perfect phylogenies, and this can be used to prove that the splits of T uniquely define T, even if some of the splits are not distinct. We leave the details to the reader. We note that T can have a duplicate split only if it has an internal node of degree two.

2.4.1   The Existence Problem

The splits-equivalence theorem says that the splits of an existing undirected tree T uniquely define T. But often we are given a set of splits, and need to determine *whether* they come from an undirected tree T. A little reflection shows that this existence problem has already been solved.

We showed above that when the splits come from a tree T and are encoded in a binary matrix SP, tree T is the unique undirected perfect phylogeny for SP. So to solve the existence problem, given a set of splits (bipartitions) of a set χ, we encode the splits in a binary matrix M and apply the four-gametes theorem. However, the splits literature uses somewhat different terminology, as follows.

Definition When no tree T is known, a set of *splits* of a set χ is a set of *bipartitions* of χ.

Note that the word "split" is used in two different ways. One is when a tree T is known and each split corresponds to an edge in T; the other is when a split just refers to a bipartition of a set χ, represented in a matrix M. The context should make clear which use is intended. The two uses of "split" are related because we represent the splits defined by edges in a tree by bipartitions represented in a matrix.

Definition A pair of columns in a binary matrix is called *incompatible* if they contain all four binary pairs 0,0; 0,1; 1,0; and 1,1. Otherwise the pair is called a *compatible pair*, or just *compatible*. A split can be represented as a column in a binary matrix M. We say that two splits in M are compatible if and only if their associated columns in M are compatible.

Using this terminology, the four-gametes theorem becomes:

Theorem 2.4.2 *Let M be a binary matrix defining a set of splits of a set χ. Then there exists an undirected tree T whose leaves are labeled by χ, and whose splits, defined by the edges of T, contain the splits of M, if and only if every pair of columns in M is compatible.*

Definition An edge $e = (u, v)$ in a tree is called an *internal edge* if neither u nor v is a leaf node.

We can strengthen theorem 2.4.2, requiring that every internal edge of T define a *distinct* split of M, by successively identifying and contracting any edge that defines a split defined by another remaining edge of T. We call such a tree a "reduced tree." Note that in a reduced tree, the split defined by an edge that touches a leaf need not be a split in M. However, every tree whose leaves are labeled by χ will contain this set of $|\chi|$ splits defined by an edge touching a leaf. Then by theorem 2.4.1, it follows that:

Theorem 2.4.3 *If there is an undirected tree T whose splits contain the splits of M, then there is a unique reduced tree whose splits contain the splits of M.*

Although compatibility and incompatibility have been defined as properties of *pairs* of columns (sites, characters), we will sometimes need to focus on individual characters, and will somewhat abuse the definitions as follows.

Definition An individual character (site) c in M is called *fully compatible* if c is not incompatible with any character in M.

A rooted version of the splits-equivalence theorem

Suppose T is a *rooted* tree with n leaves and m edges, where each leaf has a distinct label. A split for an edge e again creates a *bipartition* of the leaves of T, but now the two sides of the split can be distinguished by noting which side contains the root of T. These *rooted splits* can be represented by an n by m matrix SP where each row represents a leaf and each column represents an edge of T; a cell $SP(i, e)$ has the value of 0 if leaf i is in the subtree of $T - e$ containing the root of T and has value 1 otherwise. Note that the set of leaves with value 1 in the column for edge e precisely specifies all of the leaves reachable from e in T.

Definition In a rooted tree T, the set of leaves reachable from an edge e in T is called the *cluster* defined by e.

Hence, there is a cluster for each rooted split in T. In some parts of the phylogenetics literature, a cluster is called a *clade*.

Observe that $SP(i, e) = 1$ if and only if edge e is on the path from the root of T to leaf i. That is, each row in SP precisely specifies the edges in the path from the root of T to leaf i. Hence we can interpret SP as input M to the perfect phylogeny problem, and interpret T as a perfect phylogeny for M. Moreover, any other tree T' that has exactly the same splits as given in SP, is also a perfect phylogeny for M. By the perfect-phylogeny theorem, when a binary matrix M can be represented by a (rooted) perfect phylogeny (with all-zero ancestral sequence), the perfect phylogeny for M is unique. This implies the following:

Theorem 2.4.4 *The set of clusters of a rooted tree T uniquely determine T, including its root and the direction of each edge.*

Incompatibility is defined for undirected problems. There is also a notion of incompatibility that is used for rooted or directed problems.

Definition Given an n by m binary matrix M and a binary sequence s of length m, two sites c and d in M are said to *conflict relative to s* if c and d are incompatible in $M \cup \{s\}$.

With this definition, we can restate theorem 2.2.2 as follows:

Theorem 2.4.5 *A binary matrix M has an s_r-perfect phylogeny if and only if no pair of sites in M conflict relative to s_r.*

Sometimes we can assume that s_r is part of M, in which case we will only need results concerning incompatibility, rather than results explicitly about conflict.

2.5 General References on Phylogenetic Trees

Compared to *phylogenetic networks*, the field of *phylogenetic trees* is extremely well developed, although it is still an area of continuing, vigorous research. There are several thousand methodological papers on phylogenetic trees, and many more times that on applications to specific datasets. Whole journals are devoted to phylogenetic trees, and there are several excellent books. We only mention four: At the most biological, practical end, the classic book is *Molecular Systematics* [172], by David Hillis and Craig Moritz, now available in a second edition [173]; for a biological, statistical, philosophical discussion and overview of many topics in phylogenetic trees, the essential text (almost an encyclopedia) is *Inferring Phylogenies* [112], by Joseph Felsenstein; for a crisp, deterministic mathematical treatment of phylogenetic trees, the invaluable book is *Phylogenetics* [373], by Charles Semple and Mike Steel; for a deeper treatment of combinatorial issues in phylogenetic trees, the comprehensive book is *Basic Phylogenetic Combinatorics* [96], by Andreas Dress, Katharina Huber, Jacobus Koolen, and Vincent Moulton (don't let the word "Basic" in the title fool you). What is lacking in this field is a book on statistical aspects of phylogenetic trees with the rigor and depth of the books [96, 373], which only treat the deterministic and combinatorial aspects of phylogenetics.

3   A Deeper Introduction to Recombination and Networks

In the last chapter we discussed necessary and sufficient conditions for binary sequences to be representable by a perfect phylogeny. When there is a perfect phylogeny (i.e., when all sites are pairwise compatible), it serves as a hypothesis for the actual evolutionary history of the sequences. However, a perfect phylogeny does not exist for most sets of binary sequences encountered in populations, because some pairs of sites are incompatible. The principal biological reason, in the context of populations (over a relatively short historical time period), is that *meiotic recombination* creates new mosaic sequences in each generation. Changes due to recombination are in addition to any changes due to locus mutations. Thus, the main focus of this book concerns algorithmic and combinatorial questions about the evolution of sequences in populations, when *both* recombination and mutation shape the sequences. However, we will also see that some of the combinatorial structure and algorithmic results apply to other models of phylogenetic networks and reticulate evolution.

> If we understood something just one way, we would not understand it at all.
> — Marvin Minsky

In this chapter we discuss recombination in a biological context and also in an algorithmic/mathematical context. Building on the general introduction to recombination given in chapter 1, we give more formal and complete definitions of many central terms and models.

3.1   The Biological and Physical Context of Recombination

Repeating the quote from Watson from chapter 1:

> All DNA is recombinant DNA. ... [The] natural process of recombination and mutation have acted throughout evolution. ... Genetic exchange works constantly to blend and rearrange chromosomes, most obviously during meiosis.

Recombination means that every diploid genome has pieces from many long-dead ancestors, reflecting many different histories:

> Modern chromosomes are essentially stitched together, by recombination, from fragments of ancestral chromosomes that have different coalescent histories. [380]

Recombination is central to many diverse biological phenomena, at the molecular, population, and evolutionary levels. For example:

> Understanding the determinants of recombination is ... crucial for the study of genome evolution. [79]

And yet:

> Little is known about the rules that govern the distribution of recombination events, although age, sex, DNA sequence, chromatin structure, chromosomal location, and chromosome sizes have been shown to be important. [365]

Meiotic recombination is a principal force creating sequence variation in populations, and has been observed to be involved with, and often central to, many other biological phenomena. However, there are many unresolved questions about the role of recombination in those phenomena, and there are many basic questions about recombination itself.

3.1.1   Crossing Over

The best-known form of recombination, occurring during every meiosis, is *single-crossover recombination*, also called *crossing over*.

Definition In the context of DNA sequences, single-crossover recombination takes two equal-length sequences and produce a third *recombinant* sequence of the same length, consisting of a *prefix* of one of the sequences, followed by a *suffix* of the other sequence. The point of change is the *breakpoint*.

Definition Two sequences s and s' that recombine to create the recombinant sequence s_x are called the *parental* sequences of s_x.

Note that in the phrase "parental sequence," the word "parental" does not refer to a parent in a pedigree. It only refers to the two sequences which contribute to the sequence s_x. What makes this additionally confusing is that in the ARG model, both parental sequences s and

$$\underline{101}00 \quad \text{parent 1}$$

$$010\underline{11} \quad \text{parent 2}$$

$$10111 \quad \text{recombinant}$$

Figure 3.1 A single-crossover recombination, and a recombinant sequence. The prefix (underlined) contributed by parental sequence 1 consists of the first three characters of sequence 1. The suffix (underlined) contributed by parental sequence 2 consists of the last two characters of sequence 2.

s' are possessed together by only *one* of the parents (in the pedigree sense) of the individual who *receives* the recombinant sequence s_x. To make matters worse, we sometimes write "parent of sequence s_x" instead of explicitly writing "the parental sequence of s_x." This should not cause confusion, because we will rarely refer to parents in the pedigree sense.

Lining up

Since the two parental sequences, and the resulting recombinant sequence, are all of the same length, we can view the operation of crossing over as first "lining up" the parental sequences end-to-end, then cutting both sequences at the *same* location in the aligned sequences, and then exchanging strands to create two recombinant sequences, only one of which is kept. See figure 3.1. In fact, this is a very high-level description of crossing over in *meiosis*. Recombination where the parental sequences line up *end-to-end* is sometimes [188] called *allelic recombination.*[1] Recombination where the two recombining sequences are very *similar* around the recombination breakpoint is called *homologous recombination* and is common even in organisms, such as bacteria, that do not replicate by meiosis. Generally, allelic recombination is also homologous recombination. So we will assume that the term "allelic recombination" implies "homologous recombination," but not the converse.

As we have discussed, crossing over in meiosis is one of the major forces shaping genetic variation within a eukaryotic species. It allows the mixing of genes from two homologous chromosomes, creating two new chimeric chromosomes, one of which is passed to a child.

---

1 Actually, the term "allelic recombination" is used infrequently in comparison with the term "non-allelic recombination," which is used very heavily. If you Google "allelic recombination," you will be swamped with links to "non-allelic recombination." From this we deduce that "allelic" recombination is "non-non-allelic" recombination. "Non-allelic recombination" simply means that the sequences involved in crossing over, are not alleles, i.e., do not come from the same region of a homologous pair of chromosomes.

$\underline{10}10\underline{0}$ parent 1

$01\underline{01}1$ parent 2

10010 recombinant

Figure 3.2   This figure illustrates both double-crossover recombination and gene-conversion, although those are biochemically very different phenomena, and the figures are illustrated wildly out of scale. The prefix and suffix, underlined, are contributed by parental sequence 1, and consist of the first two characters, and the last character of sequence 1, respectively. The internal segment, underlined, is contributed by parental sequence 2, and consists of the third and fourth characters of sequence 2. In gene-conversion, the internal segment is called the *conversion tract*.

Hence, it allows the rapid creation of hybrid chromosomes even without mutations. In addition to its role in fundamental biological processes and questions, meiotic crossing over is central to several critical applied problems. Three such problems are discussed in chapter 4.

3.1.2   Double-Crossover, Gene-Conversion, and Multiple-Crossover Recombination

Double-crossover
Another form of meiotic recombination, called *double-crossover* recombination,   creates a recombinant sequence from a prefix of one sequence, followed by an internal segment of a second sequence, followed by a suffix of the first sequence. Both parental sequences and the recombinant sequence are of the same length, and the two parental sequences align end-to-end before the recombination. See figure 3.2.

Gene-conversion
There is another common form of recombination called *gene-conversion*, in which a short segment of DNA is copied, in place, from one homolog of a chromosome to the other homologous copy of that chromosome. See figure 3.2. Gene-conversion can sometimes be *modeled* as a double-crossover, although biochemically it is a *single* event, rather than two events: "In a modeling perspective it can be thought of as two very close crossover points even though this rarely is the mechanical way it occurs" [168]. In gene-conversion, the copied segment is called a *conversion tract*, and it is generally short, between 50 and 500 base pairs.

Gene-conversion is believed to be more common than single-crossover recombination. It is known to play an important role in successful meioses, and in repair of DNA damage throughout the development of an organism, and in all tissues. Until recently, it has

chromosome copy 1   ATCC GATGGA TAAGGG CAT

chromosome copy 2   CGGC TTAGCA GGTATT AAC

recombinant   ATCCTTAGCA TAAGGGAAC

Figure 3.3   Multiple-crossover recombination. The recombinant sequence is created from the boxed segments of the two parental sequence. This example is wildly out of scale, as the four segments are very short. In humans, the number of segments in a single recombinant chromosome is generally under six, and so true segments are much longer than in this example.

been hard to study gene-conversion in populations, partly because of the lack of analytical tools and the lack of fine-scale data. For example, little is known about the distribution of tract lengths in humans, except for its range (50 to 500 bases). Gene-conversion events that are mistaken for single-crossover recombination may also cause problems in association mapping and in other efforts to deduce information about recombination [168]. Eventually complete genomic resequencing will allow quantification of the fundamental parameters of gene-conversion, and the contribution of gene-conversion to the overall patterns of sequence variations in populations. Recent detailed studies of gene-conversion and its frequency compared to crossing over, in *Arabidopsis thaliana*, *Saccharomyces cerevisiae* (brewer's yeast), and *Drosophila melanogaster* appear in [78, 269, 347, 402]. The role of gene-conversion in creating genomic diversity is explored in [77].

Multiple-crossover recombination

In *multiple-crossover* recombination, a recombinant sequence is created from two equal-length sequences by *more than* two crossovers. As before, the recombinant sequence is the same length as the two parental sequences, so the two parental sequences alternate contributing segments. Each segment starts at the site just after the site where the previous segment ends. See figure 3.3.

Multiple-crossover recombination occurs on a chromosomal scale, and will generally only be seen when looking at very long intervals of a chromosome. The number of crossovers that occur between two sites on a chromosome is central to the concept of the *genetic distance* between those points. Genetic distance was the primary information about the arrangement of genes that was obtainable before DNA sequencing methods became available, and it continues to remain important. This is discussed more in section 4.2.

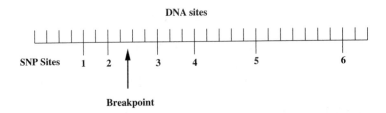

Figure 3.4 A cartoon representation of six SNP sites occurring on a longer sequence of DNA sites. The DNA sites represented by short, equally spaced vertical lines; the SNP sites are represented by the longer vertical lines. The arrow indicates that the breakpoint of the recombination event is physically between two DNA sites that are not SNP sites. In the extracted SNP sequence $(1, 2, 3, 4, 5, 6)$, the breakpoint is considered to occur between the two adjacent SNP sites 2 and 3. Therefore, the crossover index (defined in section 3.2.3) is 3.

3.1.3   The Physical Context of SNP Sequences

As discussed in section 1.4.1.2, there are several types of biological phenomena (point mutation, locus mutation, and complex morphological traits) that give rise to perfect characters and binary sequences. However, most of the data we are concerned with are from SNP sites (introduced in section 1.4.1.2), and so for concreteness, we will focus on SNP sequences (i.e., sequences of state values at a set of linearly ordered SNP sites). The reader can extend the discussion to the other cases of binary data.

Adjacent positions in a SNP sequence correspond to DNA sites, on the same chromosome, that could be physically very far apart, separated by a large and varying number of nucleotides (in the hundreds or thousands). We abstractly think of a breakpoint of a recombination as occurring between two *adjacent* SNP sites (or before or after the first/last SNP site) in a SNP sequence. But *physically* the breakpoint of the recombination can be anywhere in the *interval* on the chromosome between the two adjacent SNP sites. So, the physical breakpoint will likely be between two DNA sites that are not part of the SNP sequence. Since a SNP sequence does not represent the DNA that lies between or around SNP sites, we only represent the location of a breakpoint *relative* to the SNP sites (see figure 3.4). It will sometimes be important to keep this physical reality in mind, for example when discussing the accuracy of methods to find the location of mutations, given SNP data. We will return to this issue in section 3.2.3 when we more formally define a *genealogical network* and the crossover index of a recombination.

3.2 The Algorithmic and Mathematical Context of Recombination

In our treatment of recombination, we abstract away the biological and physical detail, and focus on recombination as an *operation* on *binary* sequences. In this context, the key distinction is the number of crossovers allowed at a recombination event. We distinguish the cases of a *single-crossover* event, of a *double-crossover* event, and of a *multiple-crossover* event. Some algorithmic results apply only to single-crossover recombination, while some apply to single and double-crossover recombination, and some apply to multiple-crossover recombination.

3.2.1 Representing a History of Recombinations and Mutations

As we saw in chapter 2, the evolutionary history of a set of sequences that derive from a single ancestral sequence, and are modified only by successive mutations, can be represented by a directed tree where each node represents a sequence, and each mutation is represented as a label on an edge. Tree representations work because mutation is an operation that creates a new sequence from *one* existing sequence. Often, but not always, a tree is an adequate representation of evolution at the *species* level. But a tree is generally not adequate to represent the history of a population inside a sexually reproducing species. As J. Felsenstein writes: "Once we are inside a species, we have lost by genetic recombination the single branching genealogy that exists between species" [112].

Recombination is an operation that creates a new sequence from *two* sequences and so the historically correct derivation of a set of sequences created by both mutations and recombinations cannot be represented by a tree. Moreover, if the sequences contain a pair of sites that are incompatible, no perfect phylogeny (even a historically incorrect one) can derive the sequences. Instead, we represent a recombination event by two directed edges entering a node (see figure 3.6), and we represent the derivation of a set of sequences as a *directed acyclic network* or a *directed acyclic graph* (DAG). See figures 3.7 and 3.8.

Terminological confusion

Depending on the underlying biological context in which the sequences are derived, and on the research community, the DAGs that are used to represent evolution have been called *phylogenetic networks*, *cluster networks*, *reticulate networks*, *recombination networks*, *genealogical networks*, *ancestral recombination graphs (ARGs)*, *hybridization networks*, and other terms. Specialized terminology for restricted classes of networks such as *galled trees* (to be discussed in chapter 8), *galled networks* and *level-k networks*, and so on, have also been used. Terminology for related *undirected* networks includes *median networks*, *quasi-median networks*, and *splits-trees*. The terminology has been evolving and is confusing and sometimes contradictory.

"Phylogenetic network" confusion

The biggest source of confusion is that the term "phylogenetic network" has been defined differently in different literatures, and in some literatures it has often been the only term used.[2] The confusion caused by the overuse of the term "phylogenetic network" motivates us now to use terms that make more precise distinctions between different kinds of networks and the different biological contexts in which they arise. We follow the definition in [195, 196]:

> A *phylogenetic network* is any graph used to represent evolutionary relationships (either abstractly or explicitly) between a set of taxa that labels some of its nodes (usually the leaves).

Under this definition, all the networks discussed in this book are phylogenetic networks. See [191, 195, 196] for a taxonomy of many different biological networks that are called "phylogenetic networks." That taxonomy is partly replicated and extended in figure 3.5. In the taxonomy in [195], the main networks considered in this book are called *recombination networks*, but we prefer the term *genealogical networks*. See [294] for a more biological introduction to a large variety of networks that have been called "phylogenetic networks." See also the March 3, 2013 blog post on the website *The World of Genealogical Phylogenetic Networks* [414] for a review of different kinds of networks that have been proposed as models of reticulate biological evolution.[3] See [121] for a nearly exhaustive bibliography of papers and programs on phylogenetic networks. See [10] for description of a website that takes in aligned sequences and determines which of four explicit phylogenetic networks can construct those sequences.

We focus in this book on a particular subset of phylogenetic networks that are called *explicit phylogenetic networks* in the taxonomy in [195]. Those are networks that *explicitly*

---

2 In fact, in much of our own research papers, we used the term "phylogenetic network" for what we now refer to as a "genealogical network" or "ancestral recombination graph."

3 That website also contains many other interesting posts on phylogenetic trees and networks, including photographs of people with phylogenetic tattoos, historical artistic rendering of networks, and a cartoon of a phylogenetic network, summarizing the evolution of humans as explained on the television show *South Park*, series 10, episode 12. See also [120] for more information on the cartoon, and other interesting material on phylogenetic networks.

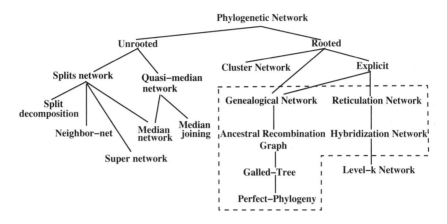

Figure 3.5 This taxonomy of phylogenetic networks is a modification of the taxonomy appearing in [195]. The boxed networks are the ones discussed in this book.

relate, display, embed, or generate the input data. That is in contrast to phylogenetic networks such as *median networks* or *splits-networks* where the network *reflects* the data, but does not represent an explicit hypothesis about its evolutionary history. Of greatest importance in the book are explicit phylogenetic networks that generate a set of given *binary sequences*. In the last chapter of the book, we also look at *reticulation networks*, which *explicitly* display a given set of *trees* or *clusters* derived from a set of trees. To distinguish those two types of explicit networks, we will use the following:

Definition A *genealogical network* is a phylogenetic network that models the derivation of *sequences* by both mutation and recombination events.

Note that the term "genealogical network" does not specify the number of allowed crossovers at a recombination event, nor the number of times that a site can mutate.

Although the focus is on genealogical networks and sequences, some of the structural and algorithmic results on genealogical networks apply, or can be extended, to another type of widely studied phylogenetic network that we call a "reticulation network." Reticulation networks, which are sometimes called "hybridization networks," are used to model biological phenomena such as *species hybridization*, and *lateral gene transfer*. A precise definition of a reticulation network will be presented in chapter 14.

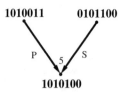

Figure 3.6   A graphical representation of a *single-crossover recombination* event. The contribution of the prefix is indicated by the character P, and the contribution of the suffix is indicated by the character S. The *crossover index*, where the recombinant sequence begins to take characters from the suffix, is written above the recombination node. Remember that the sequences recombine during meiosis, and hence both parental sequences are possessed by the *same* individual, who transmits the recombinant sequence to a child.

3.2.2   Ancestral Recombination Graphs (ARGs)

Now we define the central combinatorial object of this book.

Definition An "ancestral recombination graph," abbreviated "ARG," is a genealogical network that obeys the additional restriction that *each site mutates at most once;* hence, each site labels at most one edge of the network. That is, each character is a *perfect character*.

    The assumption that each site mutates only once in an ARG is again a consequence of the infinite-sites assumption. ARGs will be defined more fully and formally in the next section.

3.2.3   Formal Definitions for a Genealogical Network and an Ancestral Recombination Graph

We begin with a formal definition of a *genealogical network*, and then specialize it to an *ancestral recombination graph*. The reader may be aided by referring to the informal definitions given in section 1.6.2, and the genealogical network shown in figure 1.13 (page 31).

    There are *four* components needed to specify a genealogical network for a given set of binary sequences M: the underlying graph; the edge labels; the node labels; and the observed sequences. We next discuss these four components in detail.

1. The underlying graph Given a set of n binary sequences M, each of length m, a *genealogical network* $\mathcal{N}$ for M is built on a *directed acyclic graph* (DAG) containing exactly one node, the *root* with no incoming edges, a set of *internal* nodes that have *both* incoming and outgoing edges, and exactly n nodes, the *leaves*, each with exactly one incoming edge and no outgoing edge. Each node other than the root has either *one* or *two* incoming edges. An internal node with a single incoming edge is called a *tree node*; a node with two incoming edges is called a *recombination* node; a node with exactly one *incoming* edge and no outgoing edges is a *leaf node*. An edge into a recombination node is called a *recombination edge*; an edge into a tree node is called a *tree edge*; and an edge into a leaf is called a *leaf edge*.

The root node and any internal node can have any number of outgoing edges, representing the process of replication. Figure 3.7 shows a genealogical network illustrating the definitions given here.

2. The edge labels Each edge can be labeled with a *set* of integers from 1 to m, but may also be unlabeled. No integer label is given to any recombination edge. (This assumption is for technical purposes, and does not constrain any of the results in this book.) Note that in a genealogical network, the same integer might label different edges. The integer labels on an edge represent sites in M where a mutation occurs in the time interval represented by the edge.

3. The node labels Each node in $\mathcal{N}$ is labeled by one m-length binary sequence, starting with the root node, which is labeled with some sequence s_r, called the "ancestral sequence" or the "root sequence." Since $\mathcal{N}$ is a DAG, by lemma 1.6.3 (page 30), the nodes in $\mathcal{N}$ can be (topologically) sorted into a list, so that every node occurs in the list only *after* its parent node(s)). Using that list, we can constructively label the non-root nodes with well-defined sequences in the order of their appearance in the list, as follows:

3a. The tree-node labels For a tree node v, let $e = (u, v)$ be the unique edge directed into v. The sequence labeling v is obtained from the sequence labeling v's parent node u by changing the state (from 0 to 1, or from 1 to 0) of site c, for every integer c that labels edge e.

3b. The recombination-node labels For a recombination node x, let s and s' denote the two m-length parental sequences of node x. Then the "recombinant sequence" s_x labeling node x can be any m-length sequence, provided that at every site c in s_x, the state (0 or 1) is equal to the state of site c in (at least) one of the sequences s or s'.

The creation of sequence s_x from s and s' at a recombination node is called a "recombination event," and models *multiple-crossover* recombination. To fully specify the recombination event, we must specify, for every site c in s_x, whether the state of c in s_x equals the state of c in s or in s'. The specification is forced when the states in s and s' of site c are different. When they are the same, a choice must be specified.

For a recombination event at node x in $\mathcal{N}$, we indicate which parental sequence contributes the *prefix* of the recombinant sequence, by labeling the edge from that parent node with the letter P. That parent is called the *P-parent*. We say that a *crossover* occurs *between* sites c and $c+1$ if the states in s_x of sites c and $c+1$ come from different parental sequences, and we set a *crossover index* equal to $c+1$. So, there is one crossover index for each crossover. The recombinant sequence can be deduced from the parental sequences, the designated P-parent, and the crossover indices. When drawing network $\mathcal{N}$ (as in figures 3.7 and 3.8), we display the crossover indexes above the recombination node x. Sometimes there is a choice for crossover index, and one can determine a range of possible crossover indices, but in the definition for a genealogical network, only a single crossover index is specified for each crossover.

Also, we will sometimes need to determine the *minimum* number of crossovers needed to create sequence s_x by a recombination of parental sequences s and s'. That problem has an easy solution using a *greedy* method, algorithm *min-crossover*, that we will discuss in section 9.1.3.3.

As discussed earlier, in the case of *single-crossover* recombination, between sites c and $c + 1$, *the crossover index of x is denoted b_x and set to $c + 1$ (see figure 3.4). In some treatments of ARGs, b_x is a real number between 0 and 1, but here b_x is always an integer.

With single-crossover recombination, the recombinant sequence s_x is formed from a *prefix* of one of its parental sequences (s or s') followed by a *suffix* of the other parental sequence. The parent contributing the suffix is called the *S-parent*.

4. The observed sequences The sequences labeling the leaves of $\mathcal{N}$ are the *observed* sequences, i.e., the sequences in M. Hence, for every row f in M, there is exactly one leaf in $\mathcal{N}$ that is labeled by the sequence s_f specified by row f of M.

These four components completely define a genealogical network.

Definition We say that a genealogical network $\mathcal{N}$ *derives (or explains or generates)* a set of n sequences M if and only if each sequence in M labels one of the leaves of $\mathcal{N}$.

It can also happen that an internal node will be labeled with a sequence $s \in M$, but by definition, there will be a leaf that is also labeled with sequence s.

Definition A node in a genealogical network $\mathcal{N}$ for M that is labeled with a sequence *not* in M is called a *Steiner* node.

Breakpoint versus crossover index

As was discussed in section 3.1.3, a "breakpoint" refers to a physical location in a chromosome, while a "crossover index" is an integer that is part of the specification of an genealogical network. If we know the true breakpoint for a recombination event, and it

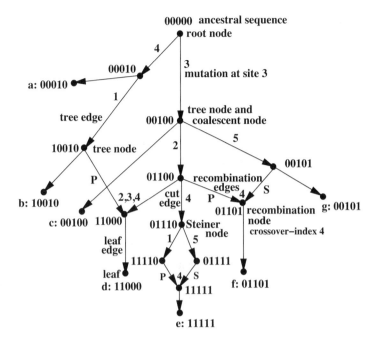

Figure 3.7 A genealogical network $\mathcal{N}$ generating a set of sequences M, labeling the leaves. This figure illustrates many of the elements of the definition of a genealogical network. Two of the recombination nodes show single-crossover recombination events, and one shows a triple-crossover event. At the triple-crossover event, three crossover indexes are shown (2,3,4), but only one recombination edge is labeled, with P, since the last segment of the recombinant sequence might also come from the same parental sequence as the prefix.

occurs between sites c and $c + 1$ in the sequences M, then the crossover index for the recombination event is $c + 1$, even though the physical breakpoint could be very far from site $c + 1$.

When we don't know the exact (physical) breakpoint for a recombination event associated with a recombination node x, but do know that it must be between sites c and $d > c+1$ of M, we say that the crossover index b_x is in the interval $(c, ..., d]$. Note that this interval is *open* on the left and *closed* on the right, meaning that the crossover index for node x must be specified by an integer b_x that is *strictly* larger than c and less than or equal to d. The asymmetry of the interval (open on the left and closed on the right) comes from the convention that a crossover index b_x indicates that the choice of which parental sequence contributes to the recombinant sequence, changes *at* site b_x.

Since each of the n sequences in M has the same length, m, we will often consider the sequences arranged in an n by m matrix with one sequence per row, and refer to that matrix as M. In that case, we sometimes refer to a site or character or locus as a "column."

3.2.3.1   *Where Are the Meioses?*

It is natural to ask: Where in the genealogical network does meiosis occur? In the interpretation of a genealogical network given in this book, meiosis occurs somewhere in the *ether*, between the nodes representing the two parental sequences and the recombination node representing a child who *receives* the recombinant sequence. An alternative interpretation is that meiosis occurs *at* the recombination node, which corresponds to an individual. In that interpretation, the two parental sequences of a recombination node x, are the sequences separately passed to the individual at x, by their mother and father. Then, the recombinant sequence written at x is the sequence of a *gamete* produced there. In this book, we follow the first interpretation, but either interpretation leads to the same mathematical model, and the same algorithmic questions and results.

3.2.3.2   *Specializing Genealogical Networks to Ancestral Recombination Graphs*

Definition Given a set of binary sequences M, an *ancestral recombination graph* (ARG) for M is a genealogical network $\mathcal{N}$ that generates M, where each integer (site) from 1 to m labels *exactly* one edge in $\mathcal{N}$. See figure 3.8.

The assumption that each integer labels only a *single* edge reflects the *infinite-sites model*, and the *perfect character* concept discussed in chapters 1 and 2.

Note that in the definitions of a genealogical network (and hence of ARG), there is *no* bound on the number of crossovers that are allowed at a recombination event (other than the number of sites minus one). Allowing an unbounded number of multiple crossovers is a convenient mathematical assumption that will allow us to model a wide variety of biological phenomena. In particular, it will be a way that we can apply results about genealogical networks to problems concerning *reticulation and hybridization* networks, discussed in chapter 14.

However, as a biological reality, in *meiotic recombination*, the number of crossovers is typically small. Multiple-crossover recombination occurs on a chromosomal scale, and in humans the number of crossovers on a single chromosome is typically under five. Until recently, only data from relatively short segments of chromosomes where available, so

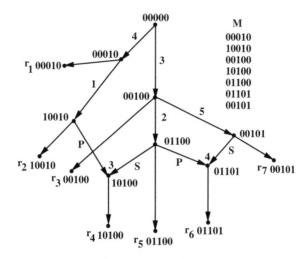

Figure 3.8 An ARG $\mathcal{N}$ with two recombination nodes, each representing a single-crossover recombination event. The matrix of sequences M that are derived by $\mathcal{N}$ is shown at the right. Note that the node with sequence label 01100 is sequence S for the left recombination node, and is sequence P for the right recombination node. In this example, every label of an internal node also labels a leaf, hence $\mathcal{N}$ has no Steiner nodes, but that is not a general property of all ARGs.

it was rare to see more than a single crossover in a segment. Therefore, the algorithmic/mathematical literature motivated by meiotic recombination has mostly assumed that only a single-crossover recombination is allowed at a recombination node.

Default crossover assumption

For many of the mathematical and algorithmic results in this book, it does not matter which type of recombination is allowed. Therefore, in our treatment of ARGs, the *default* assumption is that a recombination event in an ARG is a *single-crossover* recombination, although the definition of an ARG does not require that. When the distinction matters, we will explicitly state whether single-crossover or multiple-crossover recombination is intended.

When a recombination event, with crossover index b_x, is a *single-crossover* recombination, the recombinant sequence consists of the prefix of P-parent's sequence, from site 1 through $b_x - 1$, followed by the suffix of S-parent's sequence, from sites b_x to m. See figures 1.1 and 3.6.

ARGs, coalescent theory, and phylogenetic networks

The term "ancestral recombination graph" arose in the population genetics literature [130, 131]. More precisely, an ARG is the *graphical representation* of the genealogical relations generated by the *stochastic process* called the "*coalescent-with-recombination*" model [130, 180]. The distinction between an ARG and the coalescent-with-recombination is made in [286] as follows:

> We are using the term "ARG" to mean the data structure for representing genealogical histories. The distribution of these under the Wright-Fisher model with recombination is described by the stochastic process called the "coalescent-with-recombination" model.

A similar point is made in [37]:

> The phylogenetic network community also deals extensively with nonvertical evolution. Recognizing that ARGs are graph-theoretic objects distinct from the coalescent framework, this computational biology community has abstracted the definition of an ARG into a phylogenetic network. Using definitions suggested by Huson and Bryant [190], ARGs simply represent a special type of "explicit" or "reticulate" network.

With these definitions, a perfect phylogeny is just is an ARG with *no* recombination nodes.

Although coalescent theory (with recombination) addresses the distribution of ARGs in the context of stochastic models, our focus in this book is on *structural, combinatorial, nonstochastic* features of ARGs. A critical distinction between our treatment of an ARG as a DAG, and the distribution of ARGs generated by the coalescent-with-recombination, is that in the latter, the edges of an ARG have *lengths* representing the passage of time. In our treatment of ARGs, edges have no lengths and the only temporal information in an ARG is the relative order of events implied by node and edge reachability relations in the ARG. Additionally, in this book, we are not concerned with the stochastic distribution of ARG *topologies* generated by the coalescent-with-recombination. Instead, the combinatorial and algorithmic results in this book apply to all ARGs, or to subclasses of ARGs that we define. Still, coalescent theory and the coalescent-with-recombination provide some of the insight underlying certain combinatorial methods and results discussed in this book, and so it is helpful to understand a bit about the coalescent approach and viewpoint. See [168, 274, 315, 316, 421] for modern treatments of coalescent theory and applications. See [130, 131, 180, 225] for the founding papers, and see chapter 1 of [421] for a more comprehensive discussion of the origin of the field.

3.2.3.3 *Double-Crossover Recombination Models Back and Recurrent Mutation*

Back mutation occurs when the state of a site mutates back to its ancestral state. Recurrent mutation occurs when the state of a site is permitted to mutate from its ancestral state more than once in an evolutionary history. The infinite-sites model prohibits both of these kinds of mutations, but there are many models of phylogenetic trees where they are not only allowed, but are a key feature of interest. So how can these events be modeled? Do we have to discard the infinite-sites model?

The answer is that each back or recurrent mutation can be *modeled* as a *two-crossover recombination* in an ARG, *without* violating the infinite-sites model. For example, a single back mutation at site i in a sequence s can be modeled by a two-crossover recombination of the ancestral sequence s_r and sequence s, where the prefix and suffix come from s, and only site i comes from s_r. That is, the first crossover index is i, and the second crossover index is $i + 1$. See figure 3.9. Similarly, a recurrent mutation can be modeled by a two-crossover recombination. See figure 3.10. Note that each back or recurrent mutation is modeled by a *single* two-crossover recombination event.

Hence, although this book explicitly discusses most results through the language of recombination and ARGs, many results that allow double-crossover recombination implicitly apply to back and recurrent mutation. For example, all *lower bounds* that are valid with multiple-crossover recombination are also valid with back and recurrent mutation. Similarly, many results apply to lateral gene transfer and hybrid speciation.

More generally
From a mathematical, technical standpoint, one of the most significant differences between problems that arise in phylogenetics—where the characters are, for example, morphological traits—and problems involving meiotic recombination in populations, is that the linear arrangement of the characters (sites, columns) in a phylogenetic problem is arbitrary, but the linear arrangement of sites on a chromosome is *fixed*. And problems involving single-crossover recombination can be strongly affected by the particular arrangement. In many mathematical and algorithmic problems, that distinction is critical to the success of the result. However, when *multiple-crossover* recombination is allowed, the distinction disappears. As we have illustrated here, multiple-crossover recombination can *simulate* any linear ordering of characters. So all mathematical, algorithm results involving reticulation in phylogenetic networks translate to results involving recombination, provided that multiple-crossovers are allowed (and sometimes, as in the case of back and recurrent mutation, double-crossovers suffice). This is one of several approaches to breaking down the technical (but not biological) barriers between the study of networks in population genetics, and networks in phylogenetics.

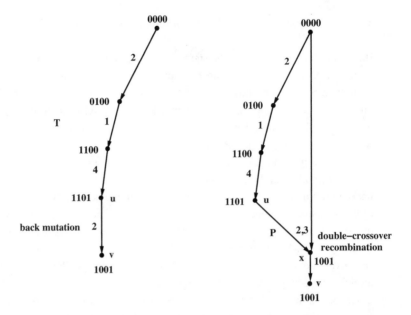

Figure 3.9   (*a*) A back mutation at site 2, on edge (u, v) in (part of) a phylogenetic *tree* T. The back mutation changes the state of site 2 from 1 back to 0.
(*b*) The back mutation can be modeled as a double-crossover recombination in an ARG. The parental sequences of the recombination come from node u (the tail of the edge with the back mutation), and the root of T. The crossover indexes are 2 and 3, and the prefix (just site 1, in this example) comes from node u.

3.2.3.4   Additional Helpful, but Not Limiting, Assumptions

For ease of exposition, there are several additional assumptions that we make about M and about any ARG for M. None of these assumptions limits the results obtained.

Definition A site (character, column) c is called *polymorphic* if at least one taxon has state 0 for site c, and at least one taxon has state 1 for site c. A site that is not polymorphic is called *uninformative*.

Polymorphic assumption
Generally, and without the need to state this each time, we assume that all sites in M are polymorphic. This is not a limiting assumption. To see this, suppose there is a site c in M that is not polymorphic, so that all the sequences have the same state at c (assume state

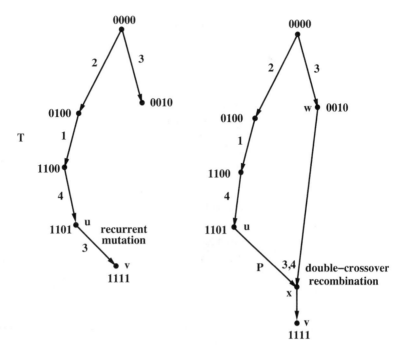

Figure 3.10 (*a*) A recurrent mutation at site 3, on edge (u, v) in (part of) a phylogenetic *tree T*. (*b*) The recurrent mutation can be modeled as a double-crossover recombination in an ARG. The parental sequences of the recombination come from node u (the tail of the edge with the recurrent mutation at site 3), and node w (the head of the edge with the first mutation at site 3). The crossover indexes are 3 and 4, and the prefix comes from node u.

one), and suppose $\mathcal{N}$ is any ARG for all the sites in M other than c. Then, if we insert site c with state one into all of the sequences labeling nodes in $\mathcal{N}$, we obtain an ARG for M.

Root degree assumption
We also assume that the root node of any ARG has out-degree *at least* two. To see that this is not a limiting assumption, note first that by definition, the root node of an ARG is not a leaf node, and every sequence in M must label a leaf. Hence, if the ancestral sequence s_r is in M, then the root must have at least two outgoing edges, one to the leaf labeled s_r, and one to the rest of the ARG. It follows that if the root has out-degree one, then the ancestral sequence is not in M. So, suppose that the root r has out-degree one, and let $\mathcal{P}$ be the unique path from r to the first node v with out-degree more than one. If any edge on $\mathcal{P}$ is

labeled with a site c, then all sequences in M will have the same state of c, contradicting the assumption that all sites are polymorphic. If no edges on $\mathcal{P}$ are labeled with a site, then all the edges between r and v can be contracted, making v the new root node, with the same ancestral sequence as before.

Internal node degree assumption

We further assume (unless otherwise stated) that every internal node of an ARG has degree at least three, and that each recombination node has degree exactly three (two in-edges and one out-edge). The latter assumption is valid because if a recombination node x has more than one out-edge, then we can insert a single, unlabeled out-edge from x to the tail node of each of the edges currently out of x. To see that any other internal node v can be assumed to have degree at least three, note first that an internal node v with degree two has one incoming edge and one outgoing edge (v, w). If v is not the tail of a recombination edge (i.e., w is not a recombination node), then node v can be contracted without changing the set of sequences generated on the ARG. Similarly, node v can be contracted if w is a recombination node, but there are no mutations on the edge into node v. However, if w is a recombination node, and there is a mutation at some site c that labels the edge into node v, then the situation is more involved (see figure 3.11).

Without loss of generality, assume that c mutates from state zero to state one. The state of c must be one at the recombination node w, for if it returned to zero (through recombination), site c would have state zero in every sequence in M, which contradicts the assumption that every site is polymorphic. So, the state of c is one at both v and w. But since the only edge out of v goes to w, we could remove c from the unique edge into v and use it to label the unique edge out of w. Exactly the same sequences would be generated, and the edge into v would have one fewer labels. Through this logic, all of the labels on the edge into v can be moved to the edge out of w, so that the edge into v becomes unlabeled. At that point we can contract node v. Hence, we can assume that every internal node has degree at least three.

Duplicate rows and sites assumptions

Also, we assume that M does not contain any duplicate rows (sequences), but note that M might contain duplicate columns (sites). In some special conditions, a duplicate *site can* be removed without limiting any of the obtained results. The simplest case is when the two duplicate sites, c_1 and c_2 are *adjacent* in M, so $c_2 = c_1 + 1$. In that case, we can remove c_1 from M, creating M', find an ARG $\mathcal{N}'$ for M', and then modify $\mathcal{N}'$ to create an ARG $\mathcal{N}$ for M. The critical detail is to change any crossover index in $\mathcal{N}'$ that is equal to c_2, to c_1. Then, sites c_1 and c_2 will be treated identically in $\mathcal{N}$, and $\mathcal{N}$ will generate M using the

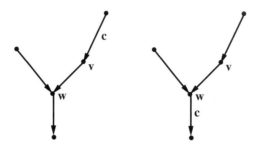

Figure 3.11 Move of the mutation c from the edge into node v, to the edge out of the recombination node w.

same number of recombination nodes as in $\mathcal{N}'$. So, when two *adjacent* sites are identical, we can remove one without affecting the generality of the solution(s) obtained.

A related condition that allows a site removal is if two adjacent sites *differ* in every position (i.e., they are perfect complements of each other), and the root sequence is *not* specified. We leave the justification to the reader. These two kinds of permitted site removals, and the removal of duplicated rows, can have a noticeable, practical impact, through reducing the size of a problem instance. We can assume that such removals have been made, but this will generally not be a critical assumption for the results in the book.

The above two approaches do not work when two identical, or perfectly complementary, sites are *not* adjacent, and in general we *cannot* remove duplicate sites without affecting the generality of the solution(s). Still, for some results, we will need to assume that M does *not* contain any duplicate sites, knowing that the result might not hold when M contains non-adjacent duplicate sites. Later in the book, we will see some other special conditions that allow duplicate site removals, for example in section 8.8.1, in the context of *galled trees*.

3.2.4   There Is No ARG-Feasibility Problem

We saw in chapter 2 that not every set of sequences M can be derived on a perfect phylogeny with all-zero ancestral sequence, or on a perfect phylogeny where the ancestral sequence is not specified. Therefore, the feasibility question of whether there is a perfect phylogeny for M is of interest. However, the feasibility question for ARGs is *not* interesting because the answer is always "yes," as we show next.

Theorem 3.2.1 *For any set of binary sequences M, and any sequence s_r, there is an ARG $\mathcal{N}$ with ancestral sequence s_r that derives M, and $\mathcal{N}$ has at most nm recombination nodes.*

Proof: We prove this constructively. Using the one allowed mutation per site, create the sequence s'_r where every site c has the (binary) state that is the opposite of the state for c in s_r. The result is that at every site c, one of the two sequences has state 0, and the other has state 1. Now let s be an arbitrary binary string. At any site c, the state of c in s is identical to the state of c in s_r or in s'_r, so s can be created from s_r and s'_r, using at most $m - 1$ single-crossover recombinations. Therefore all the sequences in M can be generated using at most nm recombination nodes. Since in the definition of an ARG, there is no bound on the number of allowed recombination nodes, any set of sequences M can be created in this way. ∎

Theorem 3.2.1 was improved in [441] to show that for any M, there is an ARG $\mathcal{N}$ for M that has at most $\frac{2nm}{\log_2 n}$ recombination nodes. We will detail that result in section 5.3.

Which ARGs are meaningful?

Because of theorem 3.2.1, unlike the case of perfect phylogeny, the mere *existence* of an ARG $\mathcal{N}$ that generates M does not provide evidence that $\mathcal{N}$ has captured significant features of the true historical evolution of M, or even adds evidence in support of the infinite-sites model of mutation. To obtain biologically plausible ARGs, we need to constrain the set of ARGs we produce, to focus on significant *properties* that such an ARG should possess. The most striking property of realistic ARGs is that they contain a *relatively* small number of recombination nodes, leading to the core algorithmic problem in ARG construction.

3.3   The Core Algorithmic Problem: Recombination Minimization

We now introduce one of the key computational problems that has been formulated in order to reconstruct plausible genealogical networks, and to study the extent of historical recombination in populations. The problem is:

> Determine or estimate the *minimum* number of recombination events needed to generate an observed set of binary sequences M that derived from an ancestral sequence (which might not be known), when the observed sequences were generated by both mutations and recombinations.

To make this problem precise, we must specify a *model* for permitted mutations. As discussed in chapters 1 and 2, the most common mutation model in population genetics is the *infinite-sites* model, that implies that any site (in the study) has mutated at most once in the entire history of the sequences. As observed earlier, this also implies that each site in any of the sequences in M has only two states. With that mutation model, we can now formalize the *core algorithmic problem*.

Definition Given a set of binary sequences M, we let $Rmin(M)$ denote the *minimum* number of *single-crossover* recombination events needed to generate the sequences M from any (unspecified) ancestral sequence, allowing *only one* mutation per site, over the entire history of the sequences.

An alternative, equivalent definition is:

Definition $Rmin(M)$ is the minimum number of recombination nodes that appear in any ARG $\mathcal{N}$ that derives M, when only single-crossover recombinations are allowed.

Sometimes we will explicitly emphasize that only single-crossover recombination events are allowed, and in that case we will use the notation "$R^1min(M)$" in place of "$Rmin(M)$."

Clearly, $Rmin(M)$ is the tightest possible *lower bound* on the true number of recombinations that occurred in the derivation of M, assuming that the true history conformed to the infinite-sites model, and that only single-crossover recombinations occurred. That is one of motivations for studying $Rmin(M)$, since $Rmin(M)$ can (in principle) be computed from M, but the true number of historical recombinations cannot be known.

To handle the case of multiple-crossover recombinations, we have the following:

Definition $R^mmin(M)$ is the minimum number of recombination nodes that appear in any ARG $\mathcal{N}$ that derives M, when a *multiple-crossover* recombination is allowed at any recombination node.

From the discussion in section 1.4.1.5, $R^mmin(M)$ can be interpreted as the minimum number of *homoplasy-causing* events (recombination, back mutation, or recurrent mutation) in any genealogical network that derives M. It follows that $R^mmin(M)$ is a *lower bound* on the minimum number of homoplasy-causing events in a phylogenetic *tree* for M.

When the ancestral sequence is known
Sometimes the ancestral sequence is known and specified, and we need definitions that reflect that case.

Definition $Rmin_s(M)$ is the minimum number of recombination nodes that appear in any ARG $\mathcal{N}$ that derives M, where $\mathcal{N}$ has ancestral sequence s.

Clearly, $Rmin_s(M) \geq Rmin(M)$ for any particular s. A common case is when the ancestral sequence is the all-zero sequence, so we have:

Definition $Rmin_0(M)$ is the minimum number of recombination nodes that appear in any ARG $\mathcal{N}$ that derives M, where $\mathcal{N}$ has the all-zero ancestral sequence.

Definition The number of recombination nodes in an ARG $\mathcal{N}$ is denoted $R(\mathcal{N})$.

Definition An ARG $\mathcal{N}$ that derives a set of binary sequences M, where $R(\mathcal{N}) = Rmin(M)$, is called a *MinARG*; the problem of finding a MinARG for M is called the *MinARG Problem*.

Unable to resist the play on words,[4] we also offer the following more formal definition:

A MinARG for M is an element of ArgMin$\{R(\mathcal{N}) : \mathcal{N}$ is an ARG for $M\}$.

The problem of computing $Rmin(M)$ (or computing closely related values) is NP-hard [44, 45, 426], and hence the problem of constructing a MinARG is also NP-hard. Of course, $Rmin(M)$ is zero and the MinARG is a perfect phylogeny, if and only if there is no incompatible pair of sites in M. Note that knowing a MinARG for M reveals $Rmin(M)$, but it is conceivable that we can determine $Rmin(M)$ without knowing any MinARG for M.

3.4   Why Do We Care about $Rmin(M)$ and MinARGs?

In this book, and in the research that has led up to it, considerable attention is given to the problem of computing $Rmin(M)$, or computing information about $Rmin(M)$. Similarly, considerable attention is given to problems of finding a MinARG, and the problem of finding "near-optimal" ARGs. The MinARG problem, and the problem of computing $Rmin(M)$, are motivated by the general utility of *parsimony* in biological problems, and because most evolutionary histories are thought to contain a small number of "detectable" recombinations. Moreover, of all the statistics that we would like to determine concerning the history of recombinations, $Rmin(M)$ is one that can be concretely defined and, in principle, computed.

---

4 Don't worry if you don't understand this. It is an attempt at a math joke.

Even *lower bounds* on $Rmin(M)$ have been shown to be useful. They can be used to answer questions about recombination, such as finding potential recombination *hotspots* in genomes [19, 111, 447], in estimating the *recombination rate* in observed sequences [422, 423]; and in determining various *population genetic parameters* [183]. Lower bounds on $Rmin(M)$ are also important in several *branch-and-bound* algorithms that compute more complex information about recombination. Similarly, explicitly computing a MinARG or a near-Min ARG has been useful in addressing biological problems such as gene finding via *association mapping* [286, 403, 439, 440] or determining *haplotypes* (defined in section 4.1), or *genotype* data [447], or *locating SNPs* in low-coverage sequence data [246], or *distinguishing gene-conversion* from single-crossover recombination [289, 387], or detecting *gene-flow* in yeast [209], or detecting coevolution in *fungi* [65], or identifying genomic regions that are *identical by descent*. These applications will all be discussed throughout this book.

MinARGs impose dependence

More technically, a MinARG (or an ARG with a relatively small number of recombinations) imposes shared patterns and correlations between the sequences it generates. So, for the inverse problem of deducing information about the shared history of a set of sequences, the use of MinARGs leads to a reduction in the *degrees-of-freedom* permitted in the solution. The constraining nature of ARGs and the utility of reducing the degrees-of-freedom will be reflected in many of the topics discussed in this book.

Minimizing recombination minimizes size

The focus on minimizing the number of recombination *nodes* is further motivated by the fact that for any ARG $\mathcal{N}$ deriving a set of n distinct sequences, M, if $\mathcal{N}$ contains $R(\mathcal{N})$ recombination nodes and t tree nodes, then $t \leq n + R(\mathcal{N}) - 2$, and the total number of nodes and edges are at most $2n + 2R(\mathcal{N}) - 1$ and $2n + 3R(\mathcal{N}) - 2$, respectively. That fact requires (as previously assumed) that every internal node has degree at least three, and that every sequence in M labels a leaf (although it can also label a non-leaf node). Also, since every sequence in M labels a leaf and we have assumed that every site is polymorphic, there can be no mutations on a leaf edge, so every sequence in M labels a leaf and its unique parent. Then, it follows that the number of Steiner nodes in $\mathcal{N}$ is at most $2R(\mathcal{N}) - 1$. We can be more precise in the following important case:

Theorem 3.4.1 *Suppose every internal node has degree exactly three, and the root has degree two (which are biologically sensible assumptions under the coalescent-with-recombination model), and every sequence in M labels a leaf. Then, $t = n + R(\mathcal{N}) - 2$, and* *the total numbers of nodes and edges in $\mathcal{N}$ are exactly $2n+2R(\mathcal{N})-1$ and $2n+3R(\mathcal{N})-2$, respectively.*

Proof: At each recombination node x in $\mathcal{N}$, arbitrarily remove one of the recombination edges (v, x) into x, reducing the degrees of both nodes v and x to two. Next, perform a node-contraction at v and x, creating a binary tree T with exactly n leaves. By theorem 1.6.1, T will have exactly $n - 1$ non-leaf nodes and $2n - 2$ edges. Putting back the removed recombination edges, and the contracted nodes adds $2R(\mathcal{N})$ nodes and $3R(\mathcal{N})$ edges, so $\mathcal{N}$ must have $2R(\mathcal{N}) + 2n - 1$ total nodes, including the root node, and must have $2n + 3R(\mathcal{N}) - 2$ edges. Subtracting the recombination nodes, the root, and the leaves from the the total number of nodes, $\mathcal{N}$ must have $t = n + R(\mathcal{N}) - 2$ tree nodes.   ∎

So among such ARGs that derive M, the *size* of an ARG is captured by the *single* parameter $R(\mathcal{N})$, the number of recombination nodes in ARG $\mathcal{N}$. In that case, the goal of minimizing the number of *recombination nodes* in an ARG is equivalent to the goal of minimizing the *total size* of an ARG (i.e., the number of nodes, and/or the number of edges).

3.4.1 A Robust Literature

Although we cannot know for sure the history of mutations and detectable recombinations that has created a given set of extant sequences, a robust literature has developed on *algorithms* to construct *plausible, biologically informative* genealogical networks, MinARGs, and near-MinARGs; or to study the history of recombinations; or to deduce well-defined aspects of a genealogy. This literature has grown particularly in the last twenty years, and includes [13, 17, 18, 19, 59, 91, 102, 103, 135, 133, 134, 142, 144, 145, 146, 147, 150, 165, 166, 183, 189, 191, 192, 193, 195, 196, 202, 203, 209, 218, 222, 235, 241, 246, 259, 265, 277, 286, 289, 290, 299, 302, 305, 310, 324, 326, 325, 327, 328, 329, 372, 387, 388, 389, 390, 393, 394, 403, 405, 412, 413, 416, 426, 434, 439, 440, 441, 445, 447, 448, 449, 454, 461], although this list certainly misses many papers. Similar questions about reticulation networks, and related networks (discussed in section 14.3), have also been addressed. That literature is very large, and the following list is only a sample [7, 23, 26, 27, 44, 45, 47, 96, 189, 190, 192, 193, 195, 194, 288, 294, 305, 306, 307, 308, 372]. In particular, see t he book *Phylogenetic Networks* by Huson, Rupp, and Scornavacca [195] for an overview of many types of phylogenetic networks, and an in-depth treatment of *splits-networks*, *cluster networks*, and related topics.

3.4.2 Networks Replacing Trees

> It's more a network than a tree.
> — Paleogeneticist Carles Lalueza, in discussing the history of interbreeding in early human groups. [333]

The need for networks in place of trees has long been understood in population genetics, where it is formalized by the *coalescent-with-recombination*, and the *ancestral recombination graph*. This understanding has been more recently matched by a wider range of biologists who have presented the view that many evolutionary phenomena must be represented by networks, rather than by trees [11, 12, 25, 37, 82, 93, 94, 126, 160, 171, 232, 276, 292, 293, 294, 338, 339, 342, 346, 371]. Of most importance, see the book *Introduction to Phylogenetic Networks* [294] (and its precursor [293]) by David Morrison, for a biologically oriented introduction to a wide range of phylogenetic networks and an extensive discussion of the causes of reticulation. See [277, 435, 453] for surveys and evaluations of several computer programs that *detect* recombination in sequences, and see [11] for a discussion of programs that *simulate* sequence generation with recombination. See [73] for an excellent review of methods to *quantify* the structure of recombination in humans.

3.5 Non-Meiotic Recombination, and an Extension of the Model beyond Animals and Plants

So far, we have discussed recombination that occurs in meiosis, and built an ARG model based on allelic recombination that (essentially) involves the end-to-end aligning of a pair of homologous chromosomes. This is well motivated for diploid eukaryotic species. But the ARG model can also be used more generally for non-eukaryotic organisms where homologous recombination occurs, but not through meiosis, and where it can be non-allelic. In those cases, it is not true that the homologs align end-to-end—instead, homologous *patches* of the chromosome align to enable recombination.

> Recombination events in bacteria are localized to a small fraction of the genome. Segments transduced or transformed in the laboratory are frequently less than several kilobases in length. Surveys of sequences in nature show that recombination in nature is likewise highly localized within the chromosome. [76]

Homologous recombination, particularly in the form of gene-conversion, occurs frequently in organisms (such as bacteria) where the recombining sequences have very different lengths, or are only fragments of a full sequence, and may not originate in the same location of their genomes. In fact, recombination in bacteria is very common:

> In contrast to the case for animals and plants, recombination in bacteria is promiscuous. Whereas animal groups typically lose the ability to exchange genes entirely by the time their mitochondrial DNA sequences are 3% divergent, bacteria can undergo homologous recombination with organisms at least as divergent as 25% in DNA sequence. [76]

So even though the main biological *motivation* in this book is from allelic, particularly meiotic, recombination, it is still possible to use ARGs to address recombination in organisms that do not undergo meiosis. For example, it is possible to use ARGs to address problems of recombination in bacteria. The key is to identify homologous regions in the bacteria that have experienced recombination. The sequences in those regions can be aligned, and those aligned sequences can be used for input to problems concerning ARGs. This is done in [37]. See also [250, 339] for a general discussion of recombination in bacteria, viruses, and human mitochondria; and for a survey on methods to detect recombination in aligned sequences, and to estimate recombination rates. See [404] for an example of aligning virus sequences in order to deduce the history of recombinations that generated the observed viruses. It is also possible to use ARGs to study recombination in organisms that do have meiosis, but are very different from plants and animals, such as fungi [37, 65].

3.6   Mind the Gap

Much of this book concerns the study of MinARGs, so before going further, we address a concern about their biological fidelity.

3.6.1   The Gap Between Total Recombinations and $Rmin(M)$

Several simulation studies and analytical results [19, 183, 247, 250, 394, 397, 435, 436] have established that $Rmin(M)$ is generally much lower than the *total* number of recombinations that occur in the ARG that generates M. The simulation and analytical results are obtained under the *coalescent-with-recombination* model, and while the issue has not been fully examined under other models of sequence evolution, we expect that it is common to many models.

  In the first and best-known paper to study the question [183], Hudson and Kaplan showed through simulations and analytical results that a particular *lower bound* on $Rmin(M)$ (called the *HK bound*, which will be discussed in detail in section 5.2.2) can be much lower than the total number of recombinations that are used in the generation of M. The *HK bound* is based on the pattern of incompatibilities of sites in M.

Assuming the coalescent-with-recombination model, the expected size of the gap is influenced by the number of distinct sequences, the mutation rate, and the recombination rate. The expected size of the gap decreases as the number of sequences increases; as the mutation rate increases; and as the recombination rate decreases. When the mutation rate is infinite, the expected size of the gap goes to zero as the number of sequences increases, but the convergence is very slow. For example, with eleven sequences, the probability of detecting a given recombination (using a method based on the *HK bound*) is about one-third, and with five hundred sequences the probability only increases to about 0.67. Moreover, although the analytical results were obtained for the *HK bound*, the results hold for any method that detects recombinations by examining the pattern of incompatible sites (as the method based on the *HK bound* does), rather than by using information such as the frequencies of sequences in a population, or correlations between states at two or more sites.

For non-infinite mutation rates, the observed gaps reported in [183] are even larger than those obtained by analytical results. In simulations with mutation and recombination rates reflecting human rates (but with fewer than twenty-five sequences) the mean number of recombination events was as large as two hundred times the mean *HK lower bound*, and no smaller than six times as large. Later reexamination of this issue in [394] and [19] showed less severe gaps—the mean ratios were at most twenty, which is still large.

Some of the gap size is due to problems in the *HK bound*. Development of better lower bound methods (discussed in chapters 5, 7, and 10), and methods to compute $Rmin(M)$ exactly (discussed in chapter 9) have shown that the *HK bound* is a weak bound, and in many tests $Rmin(M)$ or lower bounds on $Rmin(M)$ can be three to four times the size of the *HK bound* (for example, see [19, 391, 394]). Further, a result in [183] suggests that the problem of detecting a given recombination is less severe when there are many[5] SNP sites on each side of the recombination breakpoint, a condition that is generally true in the long SNP sequences currently being obtained. And, there are methods that use more than the pattern of incompatibilities to detect recombination, in which case there may be even fewer missed recombinations.

These considerations reduce but do not eliminate the significance of the gaps. In fact, now that we can compute $Rmin(M)$ exactly for some data and we observe that some of the better lower bounds are often close to $Rmin(M)$, it is clear that the gap is not just

---

5 Around two hundred in the specific data examined in [183].

between the lower bounds and the total number of recombinations used in the generation of M—the gap is also between $Rmin(M)$ and the total number of recombinations.

3.6.2 Undetectable Recombination

Here we introduce the analytical results in [183], which are more completely developed in [168, 435], through the concept of an *undetectable* recombination. For simplicity of exposition, we will assume that the ancestral sequence is known and is the all-zero sequence. It is straightforward to extend this discussion to the case when no ancestral sequence is known.

Definition Let $\mathcal{N}$ be an ARG and $\overline{\mathcal{N}}$ be $\mathcal{N}$ after removal of all mutations. A recombination event with crossover-index b_x at a node x in ARG $\mathcal{N}$ is called *detectable* if it is possible to place the sites onto $\overline{\mathcal{N}}$ so that sites that bracket b_x (i.e., closest on either side of b_x) are incompatible. A recombination even that is not detectable is called *undetectable*.[6]

For example, the recombination in figure 3.12a is undetectable, but the recombination in figure 3.12b is detectable, as evidenced by the placement of mutations shown in figure 3.13a. Although the definition involves the placement of sites, the question of whether a recombination at node x is detectable depends only on the topology of $\mathcal{N}$ and the crossover-index b_x. If a recombination is undetectable, then methods that are based on the incompatibilities of sites in M will not be influenced by that recombination. However, being detectable is only a necessary condition, and whether a recombination will be recognized by an algorithm is influenced by the placement of mutations on the ARG.

In [183] and [168, 435], analytical results are developed to estimate the probability that a given recombination is undetectable. The analysis is done under the coalescent-with-recombination model for the generation of the topology of ARG $\mathcal{N}$. Recombination events are divided into types 1, 2, and 3 in [168, 435], based on how the recombination affects the topology of the ARG. Only type 1 recombinations are said to be undetectable in [168], but in our interpretation both type 1 and 2 recombinations are undetectable. A table is shown in [168] that gives the probabilities, as a function of the number of sequences, that a recombination is of type i ($i = 1, 2, 3$). In that table, the probabilities that a recombination is of type 3 are consistent with the probabilities stated in [183] that a recombination will be detected by using the method based on the *HK lower bound*, under the assumption of

---

6 The term used in [168] is "invisible", but we avoid that term here to reduce confusion with the term "visible" used later in the book.

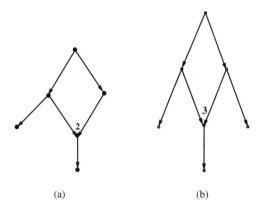

(a)   (b)

Figure 3.12   The recombination in (a) is undetectable, but the recombination in (b) is detectable.

an infinite mutation rate.[7] Later in the book (in section 9.2.2.1), after additional needed concepts have been developed, we will return to the issue of undetectable recombinations and be more precise about the approach in [168, 435].

The pointed question
Given the gap problem, the question arises: How biologically valid is a MinARG for M?

We have discussed in section 3.4 several ways that the deduction of MinARGs or near-min ARGs have addressed and solved biological problems. In arguing for the biological fidelity of the ARGs produced by the program *Margarita* (to be discussed in chapter 9), the authors state: The "justification is that the inferred ARGs work for disease mapping [286]." The fact that ARG-based approaches "work" is their strongest validation. Ultimately, a full answer to the *pointed question* will be empirical. So far, the answer looks good for MinARGs.

A better question
The question of how well a MinARG captures the true *number* of recombinations is of less importance than the question:

---

7 An infinite mutation rate effectively means that only the topology of the ARG and the crossover indices affect the ability of the method to detect a given recombination, since mutations throughout the sequence will appear on every edge of the ARG.

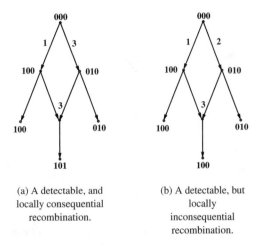

(a) A detectable, and
locally consequential
recombination.

(b) A detectable, but
locally
inconsequential
recombination.

Figure 3.13 The same ARG topology and crossover index as in figure 3.12*b*, but with placement of mutations. In figure (a), the recombination creates the new sequence 101, and so does affect the set of observed sequences. But in the placement of mutations in figure (b), the recombination creates an existing sequence, 100. So that recombination is locally inconsequential—it does not contribute to the *set* of observed sequences although it does affect the frequency distribution of the sequences. The recombination is of type 3 in the schema in [168, 435].

Are the functions that map M to $Rmin(M)$, or to ARGs with few recombinations, sufficiently sensitive to biologically important changes in M, and to changes in the structure of the ARGs that generate M?

For example, if $Rmin(M)$ is strongly correlated with the true number of recombinations, then $Rmin(M)$ is sensitive to changes in the true number and will work well as a surrogate for the true number. In fact, with some mathematical and genetic assumptions, the total number of recombinations, and local recombination rates, can be *estimated* from computable features of M, such as lower bounds on $Rmin(M)$. This idea has been developed and heavily exploited in [247, 422]. In a similar way, the computed *HK bound* on $Rmin(M)$ was used in [183] to estimate the effective population size of *Drosophila melanogaster*. Additional analytic results connecting undetectable recombinations and the total number of recombinations are derived or reviewed in [435]. Consistent with these results, it is empirically shown in [19] that better (higher) lower bounds are more sensitive to the changes in the true total number of recombinations, than are weaker bounds.

> The reason we need a model to estimate recombination rates is because we are not able to directly count the total number of recombination events. [250]

Useful models relate unmeasurable or undetectable phenomena of interest to well-defined constructs that are measurable or computable. $Rmin(M)$ is a useful part of models of recombination, because it is a well-defined and computable (in principle) construct.

Structure

More important, if the *structure* of the MinARGs reflects the recombinations that shape genetic variation (a vague concept), then the structure of those ARGs will reveal important biological information. And that is part of the central thesis of this book, substantiated by the successful uses of computed ARGs. Further evidence for this view comes from simulations in [348, 394], where the structurally informative *marginal trees* (which we will discuss in section 9.2.1) obtained from computed MinARGs were very similar to the marginal trees in the simulated ARGs that created the data. This similarity of marginal trees occurred even though the true ARG for M had many more (undetectable) recombinations than the computed MinARG for M. More strongly stated, these observations suggest that if undetectable recombinations are removed from the true ARG, the resulting ARG will be similar to a MinARG. Shortly, we will discuss what it means to remove a recombination.

3.6.3   Back to the Gap. What's in It?

The gap between $Rmin(M)$ and the actual number of recombinations used to generate M may be large, but many of the recombinations have no effect on which sequences are observed at the leaves, although they might affect the frequencies of the sequences.

> We are not able to directly count the total number of recombination events due to the fact that most leave no trace of their occurrence. [250]

For example, a recombination between two *identical* sequences produces another identical copy of those sequences, so that recombination event can have no effect on the set of derived sequences observed later. Similarly, a recombination between two sequences that differ at exactly one site produces a recombinant sequence that is identical to one of the parental sequences. Those recombinations "leave no trace of their occurrence". No method based on the *set* of observed sequences (ignoring their frequencies) could deduce that such a recombination occurred. That is a problem if we want to reconstruct or count the actual recombinations that occurred in the generation of the sequences. But for other applications, such as finding recombination hotspots that create *sequence diversity*, or for finding mutations contributing to genetically-influenced traits, many recombinations will be of no consequence since they have no genetic effect. Writing about methods to deduce historical recombination, Wakeley states:

> Their aim is to capture recombination events that have had an observable impact on the data ... we would like to count recombination events that are important in shaping patterns of genetic variation, but this is surprisingly difficult ... the effect of recombination on genetic variation is rather indirect. [421]

And it is difficult to count the number of recombinations because "Some fraction of the events ... perhaps even the vast majority, will leave no observable trace" [421].

So, there are suggestions in the literature that the gap between $Rmin(M)$ and the total number of recombinations contains a large number of events that have no genetic effect on the observed sequences. That suggests that a MinARG captures the "essential" recombinations in the generation of M. We would like to say definitively that most recombinations contributing to the gap do not affect the observed sequences, i.e., "leave no trace", but the question of which recombinations contribute to "shaping patterns of genetic variation" is subtle, and is not yet settled.

3.6.4 A Modest Proposal: Locally Inconsequential Recombination

The concept of an undetectable recombination does not resolve the issue of which recombinations leave a trace, because there are undetectable recombinations that affect the observed sequences (see figure 3.15); and there are detectable recombinations that do not affect the observed sequences, depending on where particular mutations occur (see figure 3.13).

Given the difficulty of determining, or even defining, which recombinations shape patterns of genetic variation, and the deficiencies of the concept of undetectable recombinations, we propose here a different concept called a *locally inconsequential recombination*. A locally inconsequential recombination takes into consideration all of the information in an ARG, i.e., the topology of ARG; the distribution of mutations on the ARG; the recombination indices; and the sequences generated by the ARG. This approach has been preceded by a very related one in a seminal paper by Parida et al. [329] (see also [326]), who show that most of the recombinations in ARGs generated the same way they were in [183] can be removed without changing the sequences generated in the ARG.

Definition A recombination event at a recombinant node x in ARG $\mathcal{N}$ is *locally inconsequential* if one of the two recombination edges *into* x can be removed (along with any newly formed nodes of degree one), and some mutations possibly moved to the edge out of x, without changing the set of sequences observed at the leaves of $\mathcal{N}$. Otherwise, the recombination event is called *locally consequential*.

So, a recombination is locally inconsequential in $\mathcal{N}$ if small, local changes allow it to be removed. Even if every other feature of the true ARG is deducible, no algorithm based on the *set* of observed sequences can distinguish the two scenarios at a locally inconsequential recombination node, although the use of frequency data and site/state correlations can help identify the most likely scenario. See figure 3.14 for an example of a locally inconsequential recombination that is more interesting than the case where the two parental sequences differ in at most one site.

Clearly, no recombination event in a MinARG for M is locally inconsequential, so we have:

Theorem 3.6.1 $Rmin(M)$ *is the tightest lower bound on the number of locally consequential recombinations that occurred in the historically correct ARG that generated M.*

The concept of a locally inconsequential recombination is meant to be more inclusive than that of an undetectable recombination, and to include recombinations that actually do "leave a trace." The recombination leaves a trace because it does affect the observed sequences. But it doesn't contribute to the pattern of site incompatibilities, and it is easily removed, so it does not seem essential to the generation of M. If the gap is mostly populated by locally inconsequential recombinations, as suggested by the results in [329], then the significance of the gap is greatly reduced. More empirical work is needed to further clarify this issue.

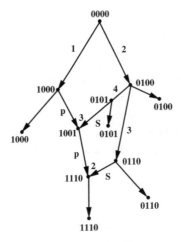

Figure 3.14   The first recombination, at node labeled 1001 is undetectable, but the second recombination is detectable. With the existing mutations, the recombination labeled 1001 is also locally inconsequential. The *S*-edge into that recombination node can be removed from the ARG without changing the sequences labeling the leaves (although other sequences change). Note that the two parental sequences are different, and each is different from 1001.

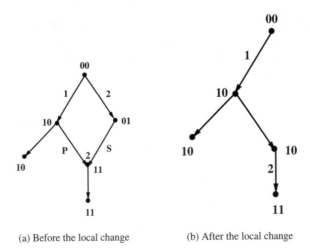

(a) Before the local change   (b) After the local change

Figure 3.15   The recombination in figure (*a*) is undetectable because no matter where mutations occur in the ARG, sites 1 and 2 will remain compatible. With the specific mutations shown in figure (*a*), the recombination is locally inconsequential. If the *S*-edge into the recombination node is removed, along with the newly created node of degree one, and mutation 2 is moved to the edge out of the recombination node, then the sequences created remain the same. Even though the recombination is undetectable, and is locally inconsequential, it does affect the observed sequences.

4　Exploiting Recombination

> We know very little, ... yet ... so little knowledge can give us so much power.
> — Bertrand Russell (a bit out of context)

Three problems and solutions where recombination is central

To further motivate the importance of understanding patterns of recombination in populations, we discuss three high-value practical problems whose solutions exploit properties of meiotic recombination. These three illustrations are highly simplified, with the intent of showing the role of recombination in the *logic* of the solutions,[1] particularly for readers who may not have had any prior exposure to these problems or solutions.

The first illustration, *genetic mapping by linkage analysis*, is the oldest one, devised more than one hundred years ago, well before any molecular understanding of genes or DNA. Building linkage maps, following the basic outline of the first success, was a central focus of genetics[2] until the introduction of more powerful molecular methods in the 1980s. At that point, new methods became available (although the classic methods are still in use), but the underlying logic and the importance of linkage maps continues today. The next two illustrations, *identifying signatures of recent positive selection* and *locating causal loci (usually genes) by association mapping*, are more recent exploitations of recombination, and are ongoing, high-visibility applications. The discussion of association mapping also illustrates the utility of computationally reconstructing explicit plausible ARGs.

This chapter can be skipped by the reader whose interests are only the algorithmic and mathematical topics in the book. However, the concepts of *haplotypes*, *genotype*, *Mendelian trait*, and *genome-wide association mapping* are introduced in this chapter, and

---

1　They are bio-*logical* illustrations.

2　Thousands of linkage maps were constructed for differing organisms.

will be needed in later chapters. The reader who skips this chapter can, when reading later chapters, use the index to find the needed definitions presented in this chapter.

4.1   Haplotypes and Genotypes

The concepts of *haplotypes* and *genotypes* will be needed in the three problems discussed in this chapter, and in many other parts of the book. So, it is worthwhile to clearly define and illustrate these concepts here.

Definition A *haplotype* is a sequence in an individual's genome that occurs together on one of the two homologous copies of a particular chromosome. An individual inherits a haplotype as a single unit from one of their parents. See figure 4.1.

In *diploid* species, each individual has two haplotypes in any specified region of the genome. See figure 4.2. Often, for an individual, the two haplotypes in a region are not known, but the two alleles that appear at the same site on the two homologous chromosomes are known.

Definition The two alleles at the same site c on two homologous chromosomes form the *genotype* at site c. If the states of the alleles at site c are P and Q, then we use "$P|Q$" to denote the genotype at site c.

For example, individual 1 in figure 4.2 has haplotypes

$$A, G, G, C, C, A$$

and

$$A, A, G, C, C, T$$

so has genotypes $A|A$, $G|A$, $G|G$, $C|C$, $C|C$, $A|T$ at the six sites. Because of the second and sixth sites, it is not possible to uniquely deduce the two haplotypes of individual 1, given only the genotypes. This is because, in addition to the above pair of haplotypes, the haplotype pair

$$A, A, G, C, C, A$$

and

$$A, G, G, C, C, T$$

is also consistent with the genotype data. We will discuss the computational problem of deducing haplotypic information from genotype data in chapter 12. But you can look ahead to figure 12.1 (page 390) now.

	1	2	3	4	5	6
42%	A	G	G	C	C	A
6%	A	A	G	C	C	T
33%	A	G	G	C	C	T
8%	A	G	A	T	T	A
11%	G	G	A	T	C	A

Figure 4.1 The five distinct haplotypes in the *Human Dysbindin* gene, and their reported frequencies in the sampled population [297].

	1	2	3	4	5	6
individual 1: haplotype 1	A	G	G	C	C	A
individual 1: haplotype 2	A	A	G	C	C	T
individual 2: haplotype 1	A	G	G	C	C	T
individual 2: haplotype 2	A	G	G	C	C	A
individual 3: haplotype 1	A	G	A	T	T	A
individual 3: haplotype 2	A	G	A	T	T	A
individual 4: haplotype 1	G	G	A	T	C	A
individual 4: haplotype 2	A	G	A	T	T	A

Figure 4.2 Four hypothetical pairs of haplotypes in four individuals. These haplotypes, but not the pairings, are the haplotypes in the *Human Dysbindin* gene on chromosome six, reported in [297]. Each individual has two homologous copies of chromosome six, and so has two haplotypes in this region. Notice that there are only five distinct haplotypes among the eight haplotypes possessed by the four individuals. These are shown in figure 4.1. The reported phylogenetic history of these haplotypes is shown in figure 2.2 (page 38).

Definition A site c is called *homozygous* if the genotype at site c consists of two *identical* alleles. The site is called *heterozygous* if the genotype at site c consists of two different alleles. For example, in figure 4.2, sites 1, 3, 4, and 5 of individual 1 are homozygous, but sites 2 and 6 of individual 1 are heterozygous.

4.2 Problem/Solution 1: Genetic Mapping by Linkage

We now discuss the first problem where the exploitation of recombination is central in the logic of the solution.

Genetic mapping is based on the perhaps counterintuitive notion that it is possible to find *where* a gene is without knowing *what* it is. [244]

Before DNA sequencing, or any other molecular interrogation of the genome, was possible, geneticists exploited recombination to learn the relative order and rough location of "factors" (genes) contributing to observable traits in an organism,[3] even when they had no idea what those factors were or how they worked. The first example of such genetic mapping was done through *linkage analysis* or *linkage mapping*. We will illustrate the idea of linkage mapping by explaining the way it was first devised more than one hundred years ago, by Alfred Sturtevant in a study of *Drosophila* [401], when Sturtevant was an undergraduate student at Columbia.

The basic idea of linkage mapping in diploid organisms is simple, although implementations in specific organisms can be very complex. The key is to exploit the following:

> **First recombinant fact**: During meiosis, the *probability* of a recombination between two sites on a single chromosome, *increases* with the physical distance between the sites.

More precise facts are known, but we do not need them in order to explain the basic idea of linkage mapping.

4.2.1 Recombinant Gametes

Definition A *Mendelian trait* is a trait that is genetically influenced by only a single locus (e.g., a single gene or a single site).

Consider two Mendelian traits influenced by two genes at locations c_1 and c_2 on the same chromosome (for example, chromosome eight). We assume that each gene has two possible states (alleles), denoted A, a and B, b, respectively. Now let I be an individual with alleles A and B on one homolog of the chromosome containing c_1 and c_2; and with alleles a and b on the other homolog of the chromosome. That is, individual I has haplotypes A, B and a, b, so I is *heterozygous* at both sites c_1 and c_2. See figure 4.3a.

After meiosis, a gamete produced by I will have one of the four possible haplotypes (at sites c_1 and c_2): A, B; a, b; A, b; or a, B. The first two gametes are *non-recombinant*, meaning that each has a haplotype that was present in individual I. However, the last two

---

3 Observable traits are technically called "phenotypes."

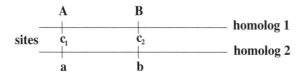

(a) Individual I has two heterozygous sites c_1, c_2 with haplotypes
A, B and a, b before meiosis.

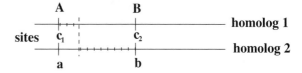

(b) A single-crossover recombination during meiosis.

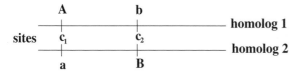

(c) Two recombinant gametes created during meiosis.

Figure 4.3 (a) Individual I is heterozygous at both sites c_1 and c_2, with haplotypes A, B and a, b.
(b) A single-crossover recombination between c_1 and c_2 creates two recombinant gametes with haplotypes A, b
and a, B, shown in panel (c). Only one of the two gametes is transmitted to an offspring, via an egg (if individual
I is female) or a sperm (if individual I is male).

gametes are *recombinant*, since each gamete has a haplotype that is *not* present on either
homolog that individual I has.

To create either of the recombinant gametes, a recombination during meiosis must have
occurred, with breakpoint between c_1 and c_2. In fact, a recombinant gamete will be created by any (multiple-crossover) recombination event with an *odd* number of crossovers
between c_1 and c_2; but if there is an *even* number of crossovers, the resulting gamete will be
counted as a non-recombinant. See figure 4.4a. Linkage mapping using three sites simultaneously can sometimes avoid this problem of incorrect counting, as shown in figures 4.4b
and 4.4c.

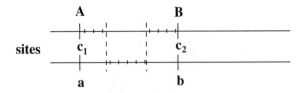

(a) A double-crossover recombination creates non-recombinant
gametes with haplotypes A, B and a, b at c_1, c_2.

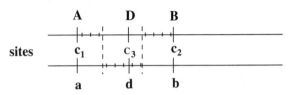

(b) Double-crossover recombination with three observed sites.

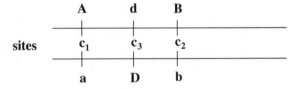

(c) Double-crossover recombination creates two gametes, with
haplotypes A, d, B and a, D, b at c_1, c_3, c_2.

Figure 4.4   (a) When only two sites are observed, a double-crossover recombination, or gene-conversion, between them produces *non*-recombinant gametes with haplotypes A, B and a, b. In general, a recombination event with an *odd* number of crossovers produces recombinant gametes, and a recombination event with an *even* number of crossovers produces a non-recombinant gamete, with respect to two sites c_1 and c_2.

(b) However, if three sites are observed, and the heterozygous site c_3 is between two breakpoints of a double-crossover, then two crossovers would be deduced from either of the resulting gametes with haplotypes A, d, B or a, D, b. This is shown in panel (c).

4.2.1.1   Recombination Fractions

Now suppose that there are many individuals with the same pair of haplotypes as individual I, at sites c_1, c_2, and assume that each one produces one gamete. From the *first recombinant fact* (that the probability of a recombination between two sites increases with physical distance), it follows that the fraction of the produced gametes that are recombinant, called the *recombination fraction*, should increase with physical distance between two sites. Therefore, the recombination fraction can be used as a rough *measure* of the distance between two sites on a chromosome. It is known now that this measure is *not* proportional to physical distance, and is not perfectly additive. Moreover, the relationship between recombination fraction and physical distance varies throughout the genome, and is different for males and females in some species (for example, humans). But despite these difficulties, measurements of the recombination fraction *can* be used to reliably deduce the *linear order* of a set of sites on a chromosome and to get some rough estimate of the relative distances between the sites, particularly if they are close.

4.2.1.2 *Deducing the Linear Order and Rough Locations*

Suppose now that we have three Mendelian traits, influenced respectively by three genes at sites c_1, c_2, c_3 on the same chromosome. Let $\Theta(c_1, c_2), \Theta(c_2, c_3)$, and $\Theta(c_1, c_3)$ be the recombination fractions observed for pair (c_1, c_2), pair (c_2, c_3), and pair (c_1, c_3), respectively. Exploiting the *first recombinant fact*, we can use those three numbers to deduce the linear order of sites c_1, c_2, c_3 as follows:

> If $\Theta(c_i, c_j)$ (for $i, j \in \{1, 2, 3\}$) is the largest of the three recombination fractions, then the linear order of the sites is deduced to be c_i, c_k, c_j, or its reverse c_j, c_k, c_i, where k is the index $\{1, 2, 3\} - \{i, j\}$.

Moreover, the ratio $\Theta(c_i, c_k)/[\Theta(c_i, c_k) + \Theta(c_k, c_j)]$ can be used as a rough, initial estimate of where c_k should be placed in the interval between c_i and c_j. That is, c_k should be placed at a fraction $\Theta(c_i, c_k)/[\Theta(c_i, c_k) + \Theta(c_k, c_j)]$ of the way between c_i and c_j.

Extending this idea, if there are n sites, then the $\binom{n}{2}$ recombination fractions (or certain subsets of them) can be used to deduce a linkage map showing the linear order of the n sites, and to make a rough estimate of the relative distances between the sites. The deduced linear order of sites in a linkage map is generally more reliable than estimates of distances between sites, because the recombination fraction is *not proportional* to physical distance and is not perfectly additive.

Definition A placement of the sites showing their linear order and an estimate of their distances is called a *linkage map*.

Sturtevant's original linkage map of *Drosophila* contained six sites, and correctly deduced the relative order of those sites. See figure 4.5 and figure 4.6.

Site Pair	recombination fraction
B, C	1.2
B, P	32.2
B, O	0.5
C, P	30.0
C, R	34.6
P, R	3.0
P, M	26.9
C, M	47.2

Figure 4.5 Some of the recombination fractions reported in Sturtevant's original paper [401]. From these values, the relative order of the six sites can be determined. For example, consider sites B, C, and P, and recombination fractions for B, C; B, P; and C, P, of 1.2, 32.2, and 30, respectively. These values support the order of B, C, P, with C significantly closer to B than to P. These numbers also support placing site O between B and C, but the full data was more ambiguous, suggesting that sites C and O are indistinguishably close to each other.

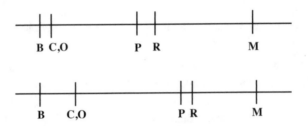

Figure 4.6 The top drawing shows the relative order and the linkage map of the six sites deduced by Sturtevant in 1911. The bottom drawing shows the true physical positions of those six sites, drawn at the same scale, determined from the *Drosophila* genome, which was sequenced in the year 2000. Sturtevant correctly determined the relative order, and many gross features of the true distances between sites. The physical distance between the beginning of site B and the end of site M is about seventeen million nucleotides.

4.2.1.3 Complications

There are, of course, many complications in implementing the abstract linkage mapping method, and its implementation differs in different organisms. But before discussing any specific implementation, we note one additional biological-mathematical point. Because of the possibility of *multiple* crossovers during a meiosis in individual I, if the number of crossovers is *even*, the gamete produced will have an allele combination that *is* present on one of I's chromosome homologs. Therefore, that gamete will be counted incorrectly as a *non-recombinant*. It follows that the counted number of recombinant gametes, $\Theta(c_1, c_2)$, is actually the fraction of meioses that have a recombination event with an *odd* number of crossovers between sites c_1 and c_2. An extension of the *first recombinant fact* adds more detail:

> **Second recombinant fact**: The expected number of crossovers in a meiosis increases with physical distance, and as the expected number increases, the probability of an odd number of crossovers converges to the probability of an even number of crossovers.

An immediate consequence is that $\Theta(c_1, c_2)$ can never be larger than *one-half*, because when c_1 and c_2 are physically very far apart on a chromosome, the probability of a gamete being counted as a non-recombinant converges to the probability of it being counted as a recombinant. This also means that when $\Theta(c_1, c_2)$ is close to 0.5, we cannot distinguish the case that sites c_1 and c_2 are on the same chromosome, but very far apart, from the case that they are on two different chromosomes.

4.2.1.4 *How Could Sturtevant Recognize Recombinant Gametes?*

In the preceding abstract description of linkage mapping, we assumed that we could determine whether a gamete is recombinant or not, and so determine the recombination fraction for two sites on a chromosome. But it was not possible to directly examine or interrogate DNA in a gamete until half a century after linkage mapping was first introduced. In the early 1900s, when Morgan and Sturtevant were already computing recombination fractions in *Drosophila*, it was not even universally accepted that chromosomes carried the "factors of heredity," and if so, whether each factor (gene) occupied a discrete location on the chromosome. In fact, the term "gene" had only been introduced a few years before Sturtevant's work. So how could he recognize that a gamete was recombinant?

The general answer, for both Sturtevant's initial work and for the huge number of successive linkage maps, was to devise a series of matings (*genetic crosses*) that produce the required heterozygous individuals, i.e., with haplotypes A, B and a, b. Then, from those heterozygous individuals, produce descendants whose observable traits indicate whether the heterozygote produced a recombinant gamete during meiosis. The details of the mating system depend on the genetics of the particular organism, and sometimes require considerable cleverness to exploit special features (or oddities) of the organism's genetics.

4.2.1.5 *Sturtevant's Success*

The method that Sturtevant used relies on morphological features (i.e., phenotypes) that are observable in the fly itself, and the nature (recessive or dominant) of the alleles that determine the features.[4] For illustration (and in the original Sturtevant paper), we consider two sites that can have states (alleles) denoted P or p and M or m, respectively. We will need the following facts:

1. By 1911 (when Sturtevant began his work), it had already been deduced that both of the sites of interest reside on the X chromosome.

2. Female flies have two X chromosomes, and male flies have one. So for any region of the X chromosome, female flies have two haplotypes, while male flies have only one.

3. A female fly receives one X chromosome from its mother and one from its father. A male fly receives its single X chromosome from its mother.

4. A female fly with the genotype $p|p$ at the first site will have *vermilion-colored* eyes, and otherwise will have *red* eyes.

5. A female fly with genotype $m|m$ at the second site will have *rudimentary* wings (whatever they are), and otherwise will have *long* wings.

6. A male fly with state m at the first site will have rudimentary wings, and otherwise will have long wings; a male fly with state p at the second site will have vermilion-colored eyes, and otherwise will have red eyes.

Restating facts 4 and 5, the two characters are "eye color" and "wing type," and in the female fly, both of the phenotypes "vermilion" and "rudimentary" are determined by *recessive* alleles. So, one can be sure that any female fly with vermilion-colored eyes has the genotype $p|p$ at the first site, and similarly any female fly with rudimentary wings has genotype $m|m$ at the second site. Thus, the fly is homozygous at both sites. See figure 4.7.

---

4 A *recessive* allele is one that determines the phenotype only when both homologs of a chromosome have that allele. A *dominant* allele is one that determines the phenotype, even if only one of the copies has that allele.

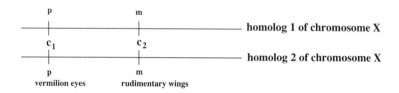

Figure 4.7 When a female fly has *vermilion eyes* and *rudimentary wings*, we know it must be homozygous at the two sites, with genotype $p|p$ at site c_1, and $m|m$ at site c_2.

The method in detail

The first step in the linkage mapping method that Sturtevant could have used[5] was to breed a large population of flies, and then select a set of females with vermilion-colored eyes and rudimentary wings; and also select a set of males with red eyes and long wings. These selected flies form the *parent generation*. See figure 4.8. So, from the observed traits, one can be sure of the genotypes of the flies in the parent generation,[6] even without the ability to directly observe the DNA (or even knowing its role in heredity).

The second step in the method is to cross (mate) males and females in the parent generation. Their progeny are called the F_1 generation. Clearly, by fact 3, each F_1 female will have haplotypes P, M and p, m, and each F_1 male will have haplotype m, p. Now we have the conditions in the female that were assumed at the start of abstract discussion of linkage mapping. That is, all the females will have heterozygous genotypes, so any non-recombinant gamete produced by the females (i.e., an egg) will have haplotype P, M or p, m, and any recombinant gametes will have haplotype P, m or p, M. If those haplotypes could be observed at this point in the experiment, then the recombination fraction could be immediately computed. However, the gametes could not be directly seen or interrogated by Sturtevant, so the method required another step.

The third step is to cross the F_1 females with the F_1 males, creating the F_2 generation. See figure 4.9. Recalling, from fact 3, that a male's single X chromosome is inherited

---

5 We have modified Sturtevant's actual method a bit for simplicity, but it illustrates the same essential logic.

6 This is not true in the actual method introduced by Sturtevant in [401], where the female parents are specified to have haplotypes p, M and p, M (and so are vermilion-long). The male parents have the haplotype P, m (and so are red-rudimentary). It is hard to understand how the female parents were obtained, since a female with haplotypes p, M and p, m would also be vermilion-long. Also, in contrast with the method we describe here, with Sturtevant's choice of parents, only the F_2 males can be used to recognize recombinant gametes.

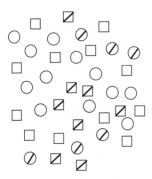

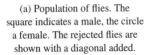

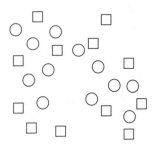

(a) Population of flies. The
square indicates a male, the circle
a female. The rejected flies are
shown with a diagonal added.

(b) Select the parents:
vermilion-rudimentary females
and red-long males.

Figure 4.8   The first step is to breed a population of flies, and then select the vermilion-rudimentary females
and the red-long males. Every selected female has haplotypes $p, m; p, m$ and every selected male has haplotype
P, M on the X chromosome. The selected flies form the *parent generation*.

from the fly's mother, the X chromosome in an F_2 male will be identical to the gamete produced by its mother during meiosis. Hence, a male's haplotype is either $p, m; P, M; P, m$
or p, M, depending on the gamete produced by its mother. These haplotypes cause the
male to be rudimentary-vermilion, long-red, rudimentary-red, or long-vermilion, respectively. Since those phenotype combinations are all different, the males that inherited a
recombinant chromosome are trivially identified. Similarly, each F_2 female will have haplotype pairs $p, m; p, m$, or $P, M; p, m$, or $P, m; p, m$, or $p, M; p, m$, which give rise to
four distinct phenotype combinations: vermilion-rudimentary, red-long, red-rudimentary,
and vermilion-long. The latter two phenotype combinations identify a recombinant gamete
(egg) produced by the female fly's F_1 mother. Hence we can identify all of the F_1 females
that produce a recombinant gamete, and so the recombination fraction can be calculated.
See figure 4.9.

4.2.1.6   Linkage Mapping: In Summary

Genetic maps were a triumph of abstract mathematical reasoning: Sturtevant was able
to chart the location of mutations affecting fly development—even though he understood
neither the biochemical basis of the defects nor even that genes were made of DNA! [244]

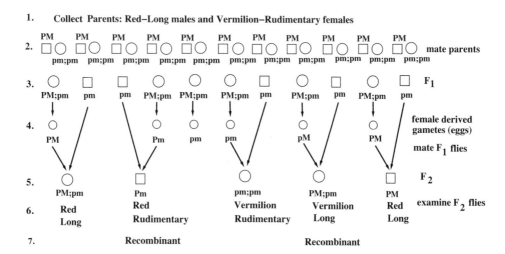

Figure 4.9   The full flow of logic of Sturtevant's method (modified as described in the text). In line 1, the parent generation is selected. All the females in the parent generation are heterozygous, with haplotypes P, M and p, m. The males each have haplotype p, m. The haplotypes are written in the figure without commas between alleles, to reduce space. In line 2, pairs from the parent generation are mated, generating the F_1 generation shown in line 3. All the females in the F_1 generation have haplotype pair $P, M; p, m$ and all the males have haplotype p, m. If recombination (with an odd number of crossovers) occurs when an F_1 female produces a gamete (egg), shown on line 4, then when that female is mated to an F_1 male, the offspring, shown on line 5, will either be a female with haplotypes $P, m; p, m$, or a female with haplotypes $p, M; p, m$, or a male with haplotype P, m, or a male with haplotype p, M, and these haplotypes do not occur otherwise. So F_1 females who produce a recombinant gamete can be identified by their F_2 offspring, and the number of recombinant gametes can be counted. In line 4, the eggs produced by the F_1 females are shown; some of these gametes are recombinant, but we cannot directly examine them to determine how many. Line 5 shows the result of the mating of F_1 females and males, producing the F_2 generation. Note that one of the F_1 females is not mated, and two of the five F_2 flies (one male and one female) are recombinant. In line 5, the haplotypes of the F_2 generation are shown. The phenotypes of the F_2 generation are shown on line 6. The phenotypes uniquely identify recombinant and non-recombinant individuals in the F_2 generation. Line 7 identifies the recombinant haplotypes in the F_2 generation. These can be recognized by the particular observable traits that they have, shown in line 6. The recombination fraction in the figure is 0.4. The recombination fraction, between sites P and M, that Sturtevant reported was 0.24 (after correcting an arithmetic error discovered years after publication). We can explain this deviation by the fact that our sample is very small (only five flies) and, of course, that the example is completely fictional.

To understand the true triumph of this abstract reasoning, it is necessary to remember how little was known about genes and chromosomes in 1911. Not only was it still debated whether chromosomes carry the "factors of heredity," but it was also not completely accepted that chromosomes are *linear* objects. In fact, the *relative* consistency (near addi-tivity) of the observed recombination fractions shown in Sturtevant's map gave support to the conjecture of chromosomal inheritance and linearity. As Sturtevant put it:

> These results ... form a new argument in favor of the chromosome view of inheritance, since they strongly indicate that the factors investigated are arranged in a linear series, at least mathematically. [401]

Moreover, the near additivity of the fractions gave support to the conjecture that the factors are *discrete* objects at the same *fixed* locations on a chromosome in all of the individuals:

> The idea that individual genes occupy regular positions on chromosomes was one of the great insights of early genetics. [354]

Currently

Once DNA sequencing, and earlier molecular interrogation methods, became available (starting in the 1970s), observable *molecular* markers replaced organismal or morpholog-ical traits as the phenotypes used in linkage mapping. The major advance was initiated in a paper by David Botstein (and coauthors) [46] who suggested using *restriction fragment-length polymorphisms* (RFLPs) as markers. Today, many other directly observ-able molecular markers are used, and RFLPs have themselves been passé for some time. However, the general idea of linkage mapping remains the same, and even with the current availability of complete DNA sequencing, linkage mapping is still used to map traits of interest. Additionally, linkage mapping remains important for a number of other biological problems. For example, *sequence assembly* in *de novo shotgun sequencing* of a genome is greatly facilitated by having a dense linkage map for that genome.

4.3   Problem/Solution 2: Locating Signatures of Recent Positive Selection

Here we discuss a second problem where the exploitation of recombination is central in the *logic* of the solution, using a highly idealized version that still illustrates the key ideas.

With the recent and increasing availability of data on genomic variation in humans and other species, population geneticists and evolutionary biologists have developed several methods to detect signatures of *positive selection* [185, 284, 312, 335, 360, 361, 362, 418]. In this section, we introduce the logic of one of the more sophisticated methods, originally developed by Hudson, Saez, and Ayala [185], to study selection in *Drosophila*. More recently, it has become known as the "Extended Haplotype Homozygosity" method [360], which we simply call the "long-haplotype" method. The method, and variants of it, have been used to identify chromosomal regions containing likely causal mutations for recent positively selected traits, or to suggest that specific traits of interest were recently positively selected, or to provide evidence that known causal mutations for known traits are recent and that the trait was positively selected. Most notably, it has been used to detect positive selection in humans that occurred within the last 30,000 years. Similar methods have been used to determine the relative ages of different alleles, i.e., estimates of when particular mutations occurred. Recombination is at the heart of the logic of the long-haplotype and related methods.

4.3.1 Positive Selection

Positive selection refers to a process through which a *beneficial* genetic trait becomes more frequent in a population. When a *heritable* genetic trait first appears in a population, for example through a mutation in a germ-line cell (egg or sperm), it appears in a single individual called a *founder*. Through purely stochastic effects of uneven numbers of offspring (i.e., *random drift*), the trait can go extinct in the population, or with a much lower probability, the *frequency* of the trait can increase over time. But even if random drift causes the frequency of the trait to increase, the increase will be very slow and it should occur in low frequency in the population for a long time. However, if the (heritable) trait is beneficial, that is, it contributes to an individual having more viable offspring than do individuals without the trait, the frequency of individuals who have the trait will likely increase in successive generations, and the rate of increase can be very rapid. Such a trait is said to be *under positive selection* or just *under selection* and to *sweep* the population (or sweep away the variation in the population). Ultimately, if (nearly) everyone in the population comes to have that trait, the trait is said to be *fixed* in the population. The kind of sweep described here is more precisely called a *hard sweep* [341], where a new beneficial mutation sweeps the population, and linked sites also sweep through *hitchhiking* [184, 280]. There are also *soft sweeps*, where (possibly multiple) preexisting variants (also called *standing polymorphisms* jointly sweep the population [287, 341], perhaps after a change in the environment. The relative prevalence of hard versus soft sweeps is a topic of ongoing research, and the consensus seems to be tilting toward the importance of soft sweeps. This issue is not central to our discussion; but we have based the exposition on the concept of a hard sweep, for simplicity.

The speed by which a beneficial trait under positive selection can sweep the population, compared to random drift, can be very dramatic. It is known [161] that *if* a trait becomes fixed in a population due to random drift alone, the expected number of generations until fixation is proportional to the size of the "effective population" [64] (which crudely can be thought of as the size of a *virtual* subpopulation in which there is *true random* mating). In humans, the effective population size is believed to be somewhere between three and ten thousand. So, even if a human trait becomes fixed due to random drift (and there is a much higher probability that it will go extinct), the *expected time to fixation* is proportional to thousands of generations. In contrast, there are beneficial traits such as changes in the color, size, and patterns of spots that act as protective camouflage in fish, that have been *observed in nature* to become fixed in a population in a handful of generations [104]. It has also been suggested that adaptive amino acid substitutions in *Drosophila* can reach fixation in a population in roughly ten to fifty years [287]. The generation time for *Drosophila* is about ten days, so this corresponds to somewhere between 360 to 2000 generations, even though the effective population size of *Drosophila melanogaster* is estimated to be over one million [64].

Another well-studied rapid sweep is antibiotic resistance in populations of bacteria, which is a beneficial trait to the bacteria if not to us. In particular, the evolution of drug-resistant tuberculosis is studied in [61], and the role of recombination in the rapid evolution of antibiotic resistance by *S. pneumoniae* is studied in [68]. A similar well-studied sweep is the development of *insecticide resistance* [117] in populations of insects. Most recently [14, 337], the development of artemisinin-resistant malaria (rapidly sweeping parts of southeast Asia only a few decades after the introduction of artemisinin-based antimalarial drugs) has been found to be strongly associated with SNP mutations in a single gene in the malaria parasite *Plasmodium falciparum*. This is a dramatic demonstration of strong positive selection at a single locus. Earlier evidence of drug-resistance selection in *Plasmodium falciparum* appears in [295].

It is of great interest to identify traits that have swept a population due to recent positive selection. The issue discussed in this section is how positively selected traits can be identified when we cannot observe the trait frequencies from the past, but only observe genomes and frequencies in the *current* population.

4.3.2   Identity by Descent Without Recombination

To understand the main idea to come, it is helpful to consider the situation where there has been no recombination, and to assume that there is only a single causal mutation for the beneficial trait. That mutation initially appears in a single individual, the *founder*, at a single site, c^*, on a particular chromosome. The founder's chromosomal sequence containing site c^* is called the "founding haplotype." Moreover, if at the time of the mutation, individuals in the population are not identical, then the founding haplotype will be distinguishable from the haplotypes on that chromosome possessed by other members of the population.

Without recombination, any descendant of the founder who inherits the trait, will also inherit a copy of the founding haplotype. Over time, additional mutations will occur, but overwhelmingly (again without recombination) the haplotype containing c^*, in an individual with the trait, will be highly similar to the founding haplotype, and distinguishable from the corresponding haplotypes (i.e., in the same chromosomal interval) of individuals without the trait. Therefore, in the current population, the haplotypes containing site c^*, in the set of individuals who have the trait, will be highly similar to *each other*, and will be distinguishable from the corresponding haplotypes of individuals who lack the trait. Individuals with the trait will have a haplotype (containing site c^*) that is said to be (nearly) *identical by descent (IBD)*.

Definition A *common haplotype* is a haplotype observed in multiple individuals in a population that is identical by descent, that is, a sequence on a chromosome that is transmitted to a set of individuals in the population, from some common ancestor of those individuals.

Combining the concept of identity by descent with the discussion of positive selection, it follows that without recombination, if c^* is a site that causes a positively selected trait, the frequency of the founding haplotype in a population will likely increase rapidly in successive generations. And, unless the trait becomes fixed in the population, there will be an identifiable subset of individuals with the trait, whose haplotypes containing site c^* are highly similar, and are distinguishable from the corresponding haplotypes of individuals who lack the trait. But these facts alone do not allow us to recognize that the trait was positively selected, because we don't know how rapidly the frequency of the trait increased. To allow that recognition, we need to introduce the effect of recombination.

4.3.3 Recombination and Haplotype Length: The Recombination Clock

We now consider how recombination changes the story, leading to the second component in the long-haplotype method. Recall that we have assumed that the positively selected trait is caused by a single mutation at site c^*. Without recombination, the frequency of the founding haplotype will likely increase rapidly, but since we cannot observe the past we don't know how rapidly the frequency increased. However, due to successive recombinations over time, the length of the haplotypes (containing c^*) that are highly similar in the individuals with the trait, and distinguishable from the corresponding haplotypes of individuals without the trait, will *decrease*. That is, successive recombinations *reduce* the length of any common haplotype in the population. The length of a common haplotype can then be used as a *measure* of the number of generations that have passed since the causal mutation occurred on the founding haplotype.

As a highly simplified example, see figure 4.10, where we assume that a mutation at SNP site 4 is the cause of a positively selected trait. That mutation occurs on the edge into node v. When the mutation occurs, the founding haplotype is 0001000. Further, to make the example simple, no other mutation occurs below node v, so the descendant sequences below v change only due to recombination. Note that the length of the segment that includes site 4 and is identical to the founding haplotype, decreases as recombinations occur in the ARG. In the end, the sequences that have state 1 at site 4 (at leaves a, b, c, d, and e), have only the haplotype 010 (at sites 3, 4, and 5) in common. So the individuals with the selected trait share the common haplotype 010 around site 4, while the individuals who do not have the trait share the haplotype 101 around site 4. In general, more than one haplotype will occur in the set of individuals without the trait, and the common haplotype shared by the individuals with the trait will not be *completely* identical.

If the time since the causal mutation is not too great, so that recombination has not reduced the length of the common haplotype containing the causal mutation to something too small to identify, there will still be some identifiable segment around the causal mutation where the haplotypes of the individuals with the trait will be highly similar. Further, the common haplotype will be distinguishable from the corresponding haplotypes in individuals who lack the trait. See figure 4.10.

How to recognize a common haplotype

The definition of a common haplotype is based on identity by descent, but does not make it easy to determine if a segment is a common haplotype, because we do not know the transmission history of the chromosomes. However, more constructive, operational definitions derive from this one. For example, a common haplotype can be defined as a maximal segment of one copy of a chromosome where high *LD* is observed, that is, a high correlation between the states observed at pairs of sites in the segment, calculated in a set of individuals in a population. Similarly, a common haplotype can be defined as a segment of one copy of a chromosome that is "highly similar" in a significantly large subset of individuals in a population, and dissimilar from the segments possessed by the individuals outside that subset.

Recombination and positive selection

The effect of recombination over time is to decrease the lengths of common haplotypes in a population. However, in general (with some exceptions) the *rate* of recombination is not influenced by the rate of selection of a trait. The segment around the causal mutation for a positively selected trait will experience recombinations at the same rate as a segment around a neutral trait (i.e., one not positively or negatively selected). Therefore, through recombination, the *length* of the haplotype around the causal mutation provides a natural

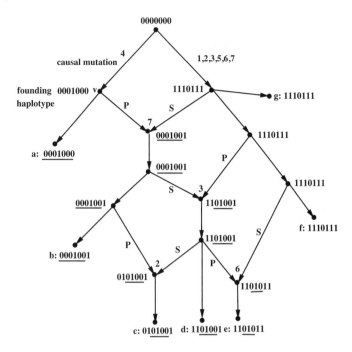

Figure 4.10 Breakdown of a common haplotype around site 4 as recombinations occur. The mutation at site 4, assumed to be the causal mutation for a positively selected trait, has founding haplotype 0001000. Due to recombination, descendants of node v, representing the founder, share only the central three characters, 010, of the founding haplotype. So, individuals a, b, c, d, and e, who are assumed to share a positively selected trait caused by the mutation at site 4, have the common haplotype 010. That haplotype is identical by descent (from the founder represented at node v), and is distinct from the haplotype 101, which contains site 4, in individuals f and g, that is, the individuals who do not have the selected trait. The lines under the sequences identify the haplotype that is identical to a segment of the founding haplotype.

clock that measures the time since the causal mutation occurred. This leads to the main idea behind the long-haplotype method for detecting recent positive selection:

> When we observe a haplotype that is much more frequent than is typical for a haplotype of that length, we can suspect that it contains a causal mutation for a positively selected trait.

Note that we haven't assumed that we know what the trait is, nor have we assumed that we initially know anything about the location of the causal mutation for the trait.

Recalling that an "allele" is a technical term for a genetic variant, or state of a character, the following quote summarizes this discussion (although with a small technical error):

> Positive selection is expected to more rapidly increase the frequency of an allele, and hence, the length of the haplotype (extent of DNA segment) associated with the selected allele, relative to those that are not under selection.[7] [406]

Implementation

The simplest implementation of the long-haplotype method is to empirically examine the SNP sequences of individuals in a population to identify distinct haplotypes, and to measure their lengths and frequencies in the regions examined. The data can then be used to identify any haplotypes that are unusually long for their frequencies, and well-established methods can determine the statistical significance of that deviation. Other techniques are based on theoretical derivations of the expected frequency of a haplotype as a function of time, in the absence of positive selection, and theoretical derivations of the expected length of a haplotype as a function of time. For example, theoretical deviations in [184] were central in the inference of positive selection in [185].

Adding cases and controls

The power of the long-haplotype method is increased when a specific trait of interest has already been identified, and sampled individuals in the population have been divided into those that have the trait of interest (the *cases*) and those that don't (the *controls*). Then, we look for a long haplotype (relative to its frequency) in the cases, that does not occur in the controls. The method is even more powerful when the location of the causal mutation for the trait is known. In that case, we look for a long haplotype (relative to its frequency) that encloses the causal mutation.

Different variants of the long-haplotype method have been used to identify numerous traits, and/or putative causal mutations, that are believed to have been positively selected in the recent past. In humans, some of the more notable recent traits (sweeping the population within the last several thousand years) identified in this way include the ability of adults in northern European populations, and in some African populations, to metabolize *lactose* [31, 409]; and the ability of Tibetans to more effectively utilize the limited oxygen at high elevations [33]. The following quotations summarize the situation of the *lactase-persistence allele*:

---

7 This statement is a bit imprecise. Positive selection does not increase the *length* of a haplotype; rather, it increases the *ratio* of the frequency of the positively selected allele, to haplotype length, relative to the ratio for an allele not under positive selection.

The long-range associations are seen as a long haplotype that has not been broken down by recombination. For example, the lactase-persistence allele at the LCT locus lies on a haplotype that is common (about 77%) in Europeans but that extends largely undisrupted for more than one million base pairs, much farther than is typical for an allele of that frequency. [361]

The striking pattern of genomic variability that is observed in this locus involves a long, high frequency haplotype that contains an allele associated with lactase persistence. The haplotypes that carry the allele are almost identical in regions close to the location of the causative SNP, whereas haplotypes that do not carry the allele show a normal level of variability. This is exactly the pattern we would expect to observe if the allele has recently increased in frequency as the result of positive selection. [312]

One other example of historical interest is the case of the rapid spread of the black form (appropriately called *carbonaria*) of the peppered moth in nineteenth-century Britain. The wild type of this moth is lightly colored, and the spread of the black form was correlated with the growth of dark air pollutants in industrial Britain. This change of color has been widely used as a textbook example of observed evolution, hypothesized to be a response to environmental change. However, the genetic and molecular basis of the change of color was only recently determined. In [177], the chromosomal sites responsible for the change of color were mapped to a 200-kilobase region. The black-colored moths were shown to contain a single haplotype in that region that differed from the haplotypes of the light-colored moths:

We have genetically mapped the *carbonaria* morph to a 200-kilobase region ... and shown that there is only one core sequence variant associated with the carbonaria morph, carrying a signature of recent strong selection. [177]

Technical problems

We have introduced the *logical basis* of the long-haplotype method, and observed the centrality of recombination. But, of course, the actual implementation of the long-haplotype method involves several technical issues. The most important ones are how to recognize haplotypes in practice and how to statistically define significant length and frequency associations. Another problematic issue [408] is that recombination rates can be highly variable across the genome (recombination *hotspots and coldspots* are known), although the recombination rate is not generally related to whether the haplotype contains a positively selected trait. Since recombination breaks down the length of the haplotype, recombination rates affect the rate at which haplotype lengths change. Well-done studies must take account of what is known about the varying rates of recombination. The more that is known about the parameters of recombination in different parts of a genome, the more these variations can be normalized, reducing their effect on the long-haplotype method. However, "the relative robustness of different methods to assumptions about recombination rates has not been systematically explored" [312].

The fact that variable recombination rates complicate the identification of positively selected traits is another motivation for building tools (of the type discussed in this book) to study recombination in populations. It also highlights the continued utility of classical genetic (i.e., linkage) maps [229], as well as recombination maps based on more modern data [300].

We also note that not all positive selection is likely to be detected through the long-haplotype method, and that other ideas are used to identify positive selection under a variety of assumptions and models [132, 312].

This highly simplified introduction to the long-haplotype method is intended to illustrate the central role that recombination plays in the important question of identifying traits that were positively selected, and also to motivate the general issue of understanding the patterns of recombination in a population. The long-haplotype method is one of several methods that are sometimes referred to as *shared-haplotype methods*. For related *haplotype sharing* ideas that can be used to estimate the *relative ages* of mutations, see [88]. For related ideas that have been used to deduce the history of migration and population admixture, see [169].

It is not yet clear how helpful explicit genealogical networks will be for shared-haplotype methods like the long-haplotype method. But ARGs explicitly display and expose the structure of common haplotypes, showing temporal locations of mutations and both spatial and temporal locations of recombination events. So it is plausible that explicit deduced histories (even if not completely correct), will be of value in many methods that are logically based on shared haplotypes.

4.4   Problem/Solution 3: An Idealized Introduction to Association Mapping

Finally, we discuss a third problem where the exploitation of recombination is central in the logic of the solution, and illustrate how knowledge of the historically correct ARG can be used.

Perhaps the most important practical exploitation of recombination is in the search for *causal loci and mutations* using population-based *association mapping*, which is sometimes (less frequently now) called "linkage *disequilibrium* mapping."[8]

Association mapping is a general method that has long been hoped to efficiently locate loci, genes, and mutations that contribute to genetic diseases or to important agricultural/commercial traits (see [353] for the introduction of the idea, and [72] for early comments on refining it). Small-interval association mapping has been widely used and has been very successful in *verifying* that specific *candidate genes* are associated with certain genetic traits (particularly Mendelian traits); and to finely map a site that has high association with a trait, when its general location is already known. For example, candidate-gene association mapping was used to verify the association of the *APOE* region on chromosome 19 with Alzheimer's disease [278], and was used to study flowering time in the plant *Arabidopsis thaliana* [101].

More ambitiously, in *genome-wide association studies* (GWAS), begun around 2007, the entire genome is searched for regions and loci associated with causal mutations, without any initial conjecture of where the causal mutations are. Most association mapping studies today begin with genome-wide scans [270, 271, 273, 295, 417], but often, once regions associated with a trait have been found in a GWAS, those regions will be more closely examined with candidate-gene association mapping methods. For example, a GWAS was used to identify a region of length one million bases, containing a causal locus for *white spotting* in boxers (dogs). Those million bases were next used as a candidate region to more finely map the causal locus to an interval of length 100,000 bases, which contained part of the pigmentation gene *MITF*. Finally, the *MITF* gene and its regulatory regions were sequenced in individual dogs, with and without the white spots (cases and controls), identifying three mutations that may underlie the white spotting trait [217].

---

8 This term should not be confused with "linkage mapping" discussed in section 4.2. In linkage mapping, one studies genetically *related* individuals, either in an existing family or (as in the example of Sturtevant's work) in an extended "family" created by the researcher. Further, in linkage mapping, one knows the details of the relationships (say in a pedigree). In linkage disequilibrium mapping, one studies *unrelated* individuals in a population. These are important operational differences. However, at a more abstract level, since all of humanity is related, it has been said that linkage disequilibrium mapping is just linkage mapping when the "family" relationships go far back in time, and one doesn't know what they are.

4.4.1 Association Mapping and ARGs

In this section we discuss a very simplified, idealized example of association mapping that illustrates the utility of knowing the true ARG that derived the extant SNP sequences. The logic exploits *common haplotypes* and *shared history*. The discussion here is informal and intuitive. In chapter 13, we will more formally and deeply discuss specific association mapping methods that use trees and ARGs, and also address the reality that we usually don't know the historically correct ARG that generated the current population. We will also discuss there some of the continuing controversies and questions about the validity and utility of GWAS association mapping.

4.4.2 Association Mapping of a Simple-Mendelian Trait

Definition A trait is *Mendelian* if it is caused by a particular allele at a *single*, fixed locus in the genome.

Definition A Mendelian trait is called *simple-Mendelian* if *everyone* who has that particular allele will have the trait, and *no one* else will have the trait.

Hence, if a trait is simple-Mendelian, then there is a single (causal) site c^* in the genome, and a single causal state (allele) i for c^*, such that any individual in the population will have the trait *if and only if* they have state i at site c^*.[9]

In the association mapping example to follow, we assume that the trait of interest is a simple-Mendelian trait. We also assume that the state of the causal site c^* mutated to i only once in the history of the population (reflecting the infinite-sites model). We further assume that we can correctly identify the individuals (the cases) who have the trait, and hence, we can also identify the individuals (the controls) who do not. These assumptions are idealized, but allow a simple example to introduce the *logical* basis of association mapping and to illustrate the central role of recombination.

We are given a set of SNP sequences M for the cases and the controls, and it is assumed that the causal mutation is at a site c^*, located somewhere in the genome spanned by the SNP sites in M; however, c^* need not be (and generally will *not* be) one of the SNP sites

---

9 There are diseases that are simple-Mendelian (or nearly so), for example, cystic fibrosis, sickle-cell anemia, hemophilia, and Huntington's disease. But most Mendelian diseases (i.e., caused by a mutation at a single site) are not simple-Mendelian in that some people with the mutation will avoid having the disease. There are several thousand such Mendelian diseases that are known, each typically occurring in a small number of individuals [313].

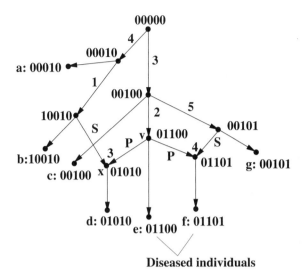

Figure 4.11 The assumed "true" ARG displaying the derivation of the SNP sequences for seven individuals. This ARG is similar, but not identical to, the ARG in figure 3.8: The choice of P and S edges into recombination node x has been reversed. We assume that individuals e and f have a simple-Mendelian disease, caused by a single causal mutation at an unknown site, c^*, in their genomes. In the following discussion, we will deduce the smallest interval containing c^* that is deducible from the given information.

in M. The association mapping goal is to use this population data to *bracket* the location of c^* in the genome as precisely as the given information allows.

To show the critical role of recombination in this goal, suppose we also know the true ARG $\mathcal{N}$ that derived the SNP sequences in M. ARG $\mathcal{N}$ can be used to deduce some of the evolutionary history of the unknown site c^*. Consider the ARG shown in figure 4.11, which is similar, but not identical to the ARG in figure 3.8. Individuals e and f have the trait (a disease, say), but none of the other individuals has it.

Recall that ARG $\mathcal{N}$ represents the evolution of complete DNA molecules that extant individuals receive, even though the only data we have is for particular SNP sites in those molecules. It is useful to think of a physical DNA molecule originating at the root of the ARG and then, by replications (represented by two out-edges leaving a node) and by mutations and recombinations, descending and evolving through the ARG, finally delivering the DNA molecules to the individuals represented at the leaves of the ARG.

Deducing the time of the mutation

The first thing we can deduce is the edge (and the interval of time that it represents) where site c^* must have mutated to the causal state. The mutation must have occurred on an edge that is ancestral to (leads to) leaves e and f, so it must either be on the edge where the state of site 2 mutated, or the edge where the state of site 3 mutated. However, if it were on the edge where site 3 mutated, then individual c (and g) would have the disease, since that edge is ancestral to leaf c via a path that does not contain a recombination node. Since c does not have the disease, we conclude that the mutation must have occurred on the edge labeled 2, and that the DNA molecule that arrives at node v in figure 4.11 must have the causal state at site c^*. Knowing the edge where site c^* mutates brackets the relative *time* of the mutation but not the *location* of c^* in the genome.

Deducing the location of c^*

To determine the location of c^*, we note that individual d does not have the disease, even though the edge labeled 2 is ancestral to d, and d receives part of its DNA via node v. Since the disease is simple-Mendelian, the fact that d does not have the disease *must be* due to the recombination event at node x, allowing d to receive DNA at site c^* from its suffix-contributing parent (who does not have the causal state at c^*), rather than from its prefix-contributing parent (at node v), who does have the causal state at c^*. Thus we can bracket the location of c^* in the genome by identifying the segment(s) of DNA that individuals d, e, and f obtain via node v. Site c^* *must be* located in a DNA segment that individual f receives via node v, and it *must not be* in the segment that individual d receives via node v. Individual e obtains all of its DNA via node v, and so information about e does not help bracket the location of c^*.

Individual d obtains DNA from v that starts at the left end of the chromosome and ends at some breakpoint between SNP sites 2 and 3 (recall that the physical location of a breakpoint can be strictly between two adjacent SNP sites). Therefore, c^* must be to the right of SNP site 2. Individual f obtains DNA from v that starts at the left end of the chromosome and ends at some breakpoint between SNP sites 3 and 4. So c^* must be to the left of SNP site 4 (see figure 4.12). Therefore, we can conclude that c^* must be in the interval of DNA between SNP sites 2 and 4. No finer localization of c^* is deducible from this data. We will return to this example in chapter 13, where we develop a more systematic association mapping method, which deduces the same interval.

What to do when the ARG is not known

The historically correct, true ARG that generated the extant SNP sequences, shows the locations (both in the genome and in time) of all mutations and recombinations, and so it explicitly reveals segments shared by cases but not controls, at least up to the resolution of

sites	1	2	3	4	5

```
individual d:              (—   —   —   —   —   —   —
individual f:    —   —   —   —   —   —   —)
common interval:           (—   —   —)
```

Figure 4.12 The intervals (deduced from individuals d and f) where c^* might be located, relative to the five SNP sites. Each dash indicates an entire interval between two adjacent SNP sites. The line for individual d (respectively f) indicates the interval of DNA that individual d (f) received via node v in the true ARG. Hence, the causal site c^* must be in the intersection of the intervals shown for d and f. So, c^* must be in the open interval $(2, 4)$. No finer localization of c^* is possible using the given information.

the SNP sites. Therefore, having the true ARG would greatly facilitate association mapping. However, we do not generally know the true ARG for the SNP sequences, and most current association mapping methods do not attempt to deduce a full genealogy, or even partially deduce features of the genealogy. Instead, they use features of the data that can be *explained* in terms of the true ARG, but can be determined without knowing any particular ARG (see [290] for a review of several such methods).

To explain the approach, we use another simple scenario, again assuming a simple-Mendelian disease. We also suppose that the causal mutation happened once in the history of the population, so that the cases must have inherited a segment of DNA that contains the causal mutation from a single common ancestor (the "founder"), while controls did not inherit their DNA segment from the founder. Thus, as explained in the discussion of signatures of positive selection (section 4.3), the cases should have a highly similar segment of DNA around the causal mutation. In contrast, the corresponding segment of DNA will be different and more diverse among the controls, since it did not originate from the founder, and may have originated from many different individuals.

In principle then, to bracket the location of the causal mutation, one looks for an interval in the SNP sequences where some maximal pattern occurs more frequently in cases than in controls. Because of mutations, the pattern might not be identical in all the cases, but it should be highly similar. Over time, recombination shortens the length of the shared pattern that contains the causal site, making it plausible that the casual site can be finely located. The shared pattern is actually a *common haplotype*, which is identified by the maximal shared pattern.

A related, but less direct reflection of a common haplotype, comes from the fact that the states of two SNPs that are physically close to the causal mutation should be more highly correlated among the cases than among the controls. Therefore, to help locate the causal mutation, one looks for pairs of SNPs whose states are highly correlated among the cases but not the controls. This kind of correlation is the basis for the notion of "linkage disequilibrium" (LD), and for measures of *LD* that are used in most association mapping methods [353, 422]. Measures of *LD* reflect, but do not require or expose, explicit genealogical networks, or necessarily correlate well with levels of recombination in different regions.[10]

An even more crude approach is to simply look for a single SNP site with a state that is highly correlated with the cases. In the most extreme situation, all the cases will have one state at that site, while none of the controls does. Such perfect separation of cases and controls is not generally expected, even for simple-Mendelian traits, unless the observed SNP site is the actual site of the causal mutation. Generally, to test the association of a single SNP site c with the cases, one uses the χ^2 Test for Independence in a two-by-two table. One dimension partitions the individuals into cases and controls, and the other dimension partitions the individuals according to the state they possess at site c. This single-marker approach has been adequate to locate some Mendelian traits, but generally has not been powerful enough for fine-scale mapping.

4.4.3 Association Mapping: In Summary

The idealized example discussed here shows the basis for association mapping, and the conceptual role of ARGs and recombination in the *logic* of association mapping. There are several key points: First, we assume that there is a founder (or a small set of founders) where the causal mutation for the trait of interest first occurs; all extant individuals who have the trait (i.e., cases) are descended from the founder(s); and in each, an interval of DNA around the causal mutation (i.e., a haplotype) has descended from the founder. Second, when we observe a haplotype whose states are highly correlated in the cases, but not in the controls, we can surmise that the causal mutation occurs in the interval spanned by that haplotype. Third, recombination reduces the size of contiguous intervals in the genome that descend intact to an individual, from any ancestor of that individual, in particular, the founder. It is this shortening of the haplotype containing the causal mutation that allows a more precise location of the causal mutation, and hence allows association mapping to be of value.

---

10 "It is customary in genomics for researchers to debate which measure of linkage disequilibrium to use to characterize the joint distribution of variation at linked sites. The correct answer is 'none of them' ... one needs a full coalescent calculation" [113], that is, a calculation that considers the space of all ARGs that could have generated the data. That requirement is another reflection of the need for combinatorial, as well as statistical, methods.

These three points explain the logic behind all association mapping methods that look for haplotypes that are common in the cases but not in the controls. This is summarized by a quotation from [440]:

> Association mapping relies on the assumption that the cases (or a significant fraction of them) share a genealogical history that is distinct from the history of the controls, and that over time, meiotic recombination has shortened the shared region(s) containing the causative mutation(s). It follows from these assumptions that SNP sites near a causative mutation will have states (alleles) that more highly correlate with the trait of interest than do sites that are far from a causative mutation.

Even though the genealogical history, and the history of recombination, are central to the logic of association mapping, most association mapping methods do not *explicitly* build an ARG or any part of one. They instead rely on statistical measures that derive from, and reflect, the unknown ARG. However, it is well understood that a full genealogical network contains more information than does any of the more indirect numerical and statistical reflections of it:

> The best information that we could possibly get about association is to know the full coalescent genealogy. [461]

Further,

> If the true ARG were known, it would provide the optimal amount of information for mapping—no extra information would be available from the genotypes. Not only would disease-associated regions be identified, but the ARG would give the ages of the causative mutations, would specify the haplotypic background of those mutations and so forth. It would also be possible to optimally impute missing data. [286]

So, in chapter 13, we more fully develop association mapping methods that do *explicitly* build ARGs or parts of ARGs.

5 First Bounds

Think left and think right and think low and think high. Oh, the thinks you can think up if only you try!
— Dr. Seuss

5.1 Introduction to Bounds

In chapter 3 we introduced $Rmin(M)$, the minimum number of recombination nodes used in any ARG to derive the set of sequences M, and we noted that the problem of computing $Rmin(M)$ is known to be NP-hard. Hence no provably correct, worst-case polynomial-time algorithm is known for computing $Rmin(M)$, and we do not expect there will be one. However, several worst-case polynomial-time algorithms have been developed that compute empirically good *lower bounds* on $Rmin(M)$, and there are several other lower-bound methods whose worst case time is not polynomially bounded, but are fast in practice. Some of the lower bounds apply only to ARGs, and some of them apply more generally to other kinds of genealogical and reticulate networks. The literature on computing lower bounds on $Rmin(M)$ is contained in many papers, including [17, 18, 19, 50, 150, 183, 302, 389, 390, 449].

Using methods that construct ARGs (which we will discuss in later chapters), we can compare the number of recombination nodes used in those ARGs to the computed lower bounds on $Rmin(M)$. We will see that some of the lower bound methods return values equal to $Rmin(M)$ or very close to $Rmin(M)$ on many datasets. Moreover, the ability to efficiently compute good lower bounds is central in *branch-and-bound* methods that construct good ARGs (discussed in chapter 9), and there are biological questions concerning recombination (for example, finding recombination *hotspots* [111]) that have been successfully addressed using lower bounds on $Rmin(M)$ rather than using $Rmin(M)$ itself [18, 19, 230, 422, 447]. Therefore, the development and study of lower bounds on $Rmin(M)$ has been a valuable contribution to the algorithmic understanding of recombination and reticulate networks.

In this chapter we first discuss several specific *lower* bounds on $Rmin(M)$, and a very effective general method for amplifying lower bounds. These bounds are either computable in worst-case polynomial time, or have been shown to be computable in practical time on datasets currently in use. In later chapters we discuss three additional lower bounds. These are deferred because additional background must be developed first, and because two of the lower bounds are less efficiently computed than the ones discussed in this chapter, although they give somewhat higher bounds. In the last section of the chapter, we will present a worst-case, tight *upper bound* on $Rmin(M)$.

5.2   Lower Bounds on *Rmin*

Before discussing any specific bounds, we clarify several different ways that the term "lower bound" is used.

Definition Given a specific binary matrix M a *lower bound on* $Rmin(M)$ is a *number* that is less than or equal to $Rmin(M)$. More generally, a *lower bound on* $Rmin(M)$ is a *function* of M, where M is considered to be a variable. A *lower-bound algorithm* for $Rmin(M)$ is an algorithm that computes a lower bound on $Rmin(M)$.

In practice, people often use the term "lower bound" for each of these three different meanings, letting the context indicate the intended use. We will follow that practice.

The first *trivial* lower bound, of one, is given by the four-gametes theorem: For any binary matrix M, $Rmin(M) \geq 1$ if and only if there is some incompatible pair of sites in M. Of course, we seek more effective, nontrivial lower bounds, when $Rmin(M)$ is larger than one.

5.2.1   The First Combinatorial Observations

We introduced the informal notion of a *recombination cycle* in section 1.2.2.3 (page 12). Now we give it a more formal definition.

Definition In an ARG $\mathcal{N}$, let v be a node such that two directed paths out of v meet at some recombination node x. Those two paths together define a *recombination cycle Q*. Node v is called the *coalescent node* of cycle Q.

For example, the nodes labeled 00000 and 01000 in figure 3.8 are the coalescent nodes of the two recombination cycles.

Lemma 5.2.1 *Let $\mathcal{N}$ be any ARG that derives a set of sequences M, and let (c, d) be a pair of incompatible sites in M. Then the edges, e_c and e_d, in $\mathcal{N}$ that are labeled with sites c and d, respectively, must be contained together in a common recombination cycle in $\mathcal{N}$.*

Proof: Without loss of generality, assume that the ancestral states of both c and d are zero, that is, at the root of $\mathcal{N}$. We examine first the case that e_d is not in the subnetwork of $\mathcal{N}$ below e_c, and e_c is not in the subnetwork below e_d. Then those two subnetworks cannot have a node x in common, since (with only one root node) edges e_c and e_d would be contained together in a recombination cycle with recombination node x. Therefore, in this case, the (c, d) state-pair (that is, the pair of ordered states for c and d) of $(1, 1)$ would not appear at any node in $\mathcal{N}$, and hence no sequence in M would have the (c, d) state-pair of $(1, 1)$.

Now consider the case that e_d is in the subnetwork of e_c. Then, the (c, d) state-pair at the head of edge e_c would be $(1, 0)$. If, through recombination, the state of c returns to zero at the tail of edge e_d, then the (c, d) state-pair of $(0, 1)$ will be created, but the state-pair of $(1, 1)$ will not be possible. Similarly, if the state of c is 1 at the tail of edge e_d, then the (c, d) state-pair of $(1, 1)$ will be created, but the state-pair of $(0, 1)$ will not be possible. The case where e_c is in the subnetwork of e_d is symmetric, and omitted. In all cases, only three (c, d) state-pairs would be created in $\mathcal{N}$, and so M would not have all four state combinations at sites c, d, so sites c and d would be compatible, a contradiction. Hence, any two sites that are incompatible must label edges in some common recombination cycle in $\mathcal{N}$. ∎

For example, in the data shown in figure 3.8 (page 75), site pairs (1,3), (1,4) and (2,5) are incompatible. Sites 1,3,4 are together in one recombination cycle, and sites 2,5 are together in another recombination cycle. Note that sites 3,4 are together in a recombination cycle, even though $(3, 4)$ is a compatible pair. Hence the converse of lemma 5.2.1 is not true in general.

5.2.2   HK: The First Nontrivial Lower Bound

The first published, and most basic, lower bound on $Rmin(M)$ is due to Hudson and Kaplan [183], and it is referred to as the *HK bound*. Recall that the definition of $Rmin(M)$ includes the assumption that only *single-crossover* recombination is allowed. The bound is obtained as follows:

Algorithm *HK-Bound*

Given an n by m binary matrix M, find all of the *incompatible* pairs of sites in M.

Consider the m sites of M to be *integer* points $1...m$ on the real line, and find a *minimum-sized* set of *non-integer* points, called $R^*(M)$, so that for every pair of incompatible sites (p, q) in M, there is at least one point in $R^*(M)$ that is (strictly) between p and q. The quantity $|R^*(M)|$ is called the *HK bound* for M, and is denoted *HK(M)*.

For example, in figure 3.8, where site pairs (1,3), (1,4), and (2,5) are incompatible, a single point strictly between sites 2 and 3 intersects the three intervals defined by those incompatible pairs, giving an *HK bound* of one. Of course, a lower bound of one is not of great interest, because by the four-gametes theorem, the existence of any incompatible pair means that there must be at least one recombination node in any ARG that derives the sequences. However, in general, the *HK bound* is more informative than in this specific example.

Theorem 5.2.1 *The HK bound is a lower bound on $Rmin(M)$. That is, $HK(M) \leq Rmin(M)$, for any M.*

Before proving theorem 5.2.1, we establish an important lemma.

Lemma 5.2.2 *For every pair of incompatible sites (p, q) in M, where $p < q$, every ARG that derives M must have a recombination node x whose crossover index b_x is greater than p and less than or equal to q. That is, b_x must be in the interval $(p, ..., q]$ where the left end is open and the right end is closed. So, in the underlying physical chromosome, the recombination breakpoint must occur somewhere in the interval between sites p and q.*

Proof: For contradiction, suppose there is an ARG $\mathcal{N}$ that derives M, where every recombination node x in $\mathcal{N}$ has a crossover index b_x such that $b_x \leq p$ or $b_x > q$.

Let $M(p, q)$ denote the sequences in M restricted to the sites p and q. We will modify $\mathcal{N}$ to obtain a perfect phylogeny T for the set of sequences $M(p, q)$. To do this, first remove all labels (mutations) on edges other than p and q. Next, at each recombination node x in $\mathcal{N}$, remove *exactly one* of the incoming edges as follows. If $b_x \leq p$, remove the P-labeled edge into x and retain the S-labeled edge. Conversely, if $b_x > q$, remove the S-labeled edge and retain the P-labeled edge into x. The resulting graph is now a directed tree T that derives the sequences in $M(p, q)$. See figures 5.1 and 5.2 for an example. There is one edge in T labeled with p and one edge in T labeled with q, so T is a perfect phylogeny for M with the same root node and ancestral sequence that $\mathcal{N}$ has. But, by theorem 2.3.2, $M(p, q)$ can have a perfect phylogeny only if p and q are a compatible pair, contradicting the assumption that p and q are incompatible. Hence, for any pair of incompatible sites in M, any ARG that derives M must have a recombination node with crossover index b_x, where $p < b_x \leq q$. ∎

Using lemma 5.2.2, we can now prove theorem 5.2.1.

Proof (Of theorem 5.2.1): Let $\mathcal{N}$ be an ARG that derives M, and let B be the set of crossover indices associated with recombination nodes in $\mathcal{N}$. For each recombination node x, let point $p_x = b_x - \epsilon$, where $0 < \epsilon < 1$. By lemma 5.2.2, the result is a set of non-integer points B' such that for every pair of incompatible sites (p, q) in M, there is at least one point in B' (strictly) between p and q. Therefore $|B| = |B'| \geq |R^*(M)|$, and in particular, when $\mathcal{N}$ is a MinARG using $Rmin(M)$ recombination nodes, we see that $Rmin(M) \geq |R^*(M)|$, so the *HK bound* is a lower bound on $Rmin(M)$. ∎

Note that the assumption of single-crossover recombination is critical in the proof of the *HK bound*. If multiple-crossover events are allowed, then a single recombination could satisfy more than one requirement for a crossover point between two specific sites. In that case, $|R^*(M)|$ could be larger than $Rmin(M)$.

Before going on, we note the relationship of theorem 5.2.1 to lemma 5.2.1. Lemma 5.2.1 shows that if sites p and q are incompatible in M, then they must be contained together in some recombination cycle in *every* ARG $\mathcal{N}$ that derives M. Theorem 5.2.1 shows that if p and q are incompatible, then any ARG that derives M must have a recombination node x whose crossover index is in the interval $(p, ..., q]$. In fact, both of these statements must hold *simultaneously*.

Lemma 5.2.3 *If c and d are incompatible in M, then they must be contained together in some recombination cycle in $\mathcal{N}$ whose recombination node has crossover index in the interval $(p, ..., q]$.*

We leave the proof to the reader.

5.2.2.1 *The Root-Known Version of the* HK *Bound*

The discussion of the *HK bound* so far did not assume that a fixed ancestral sequence was known. If a fixed ancestral sequence s_r is given, a lower bound on $Rmin_{s_r}(M)$ can be computed simply by adding s_r to M and computing the resulting bound on $Rmin(M)$.

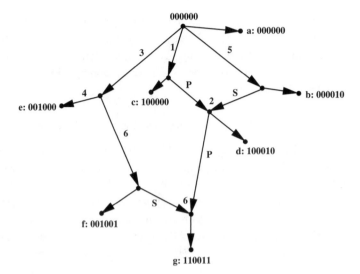

Figure 5.1 Network to illustrate the proof of lemma 5.2.2. For contradiction, $(3, 5)$ is assumed to be an incompatible pair, but as shown, there is no recombination node x with crossover index b_x in the interval $(3,...,5]$.

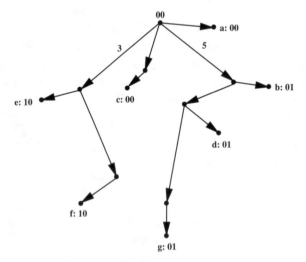

Figure 5.2 Continued illustration for the proof of lemma 5.2.2. The perfect phylogeny resulting from the network in figure 5.1 after removal of one edge into each recombination node, and removal of sites other than 3, 5. The tree correctly derives the sequences in M restricted to sites 3 and 5. Therefore, by theorem 2.3.2, sites 3 and 5 cannot be incompatible.

5.2.2.2 Computing the *HK* Bound

Let each incompatible pair (p, q) in M define the closed interval $[p, ..., q]$ on the real line, and let $\mathcal{L}$ denote the set of those intervals. Note that the endpoints of any interval in $\mathcal{L}$ are on integer points. The problem of finding $R^*(M)$ can be restated as:

> **The interval coverage problem**: Find a minimum-sized set of *non-integer* points $R^*(M)$ so that each interval in $\mathcal{L}$ contains at least one point in $R^*(M)$.

The *interval coverage problem* can be efficiently solved by the greedy, left-to-right scanning algorithm shown in figure 5.3. This type of scanning algorithm is used in many applications. For example, see the *interval-scheduling* problem and solution discussed in [228] (page 116).

Algorithm Interval-Scan (M)

Sort the intervals in $\mathcal{L}$ by their *right* endpoints, smallest (leftmost) first, where ties are broken arbitrarily.
Let L be the resulting sorted list of intervals in $\mathcal{L}$.

Set $R = \emptyset$ and $\mathcal{I} = \emptyset$, and let h be the *left* endpoint
of the interval I at the *head* of L (i.e., the first interval in L).

while (L is not empty) **do**

    Remove the interval I at the head of L.
    if (point h is not strictly in the interior of I) **then**
        place the point $q - \epsilon$ into R, where $0 < \epsilon < 1$,
        and q is the right endpoint of interval I; Set h to q, and place I into $\mathcal{I}$
        $\{\mathcal{I}$ is a set of intervals which will be used in the proof of correctness$\}$.
    endif

endwhile

Figure 5.3 Algorithm *Interval-Scan* solves the *interval coverage problem* and computes the *HK bound*, given the set of intervals $\mathcal{L}$ obtained from M.

See figure 5.4 for an example of an execution of algorithm *Interval-Scan*.

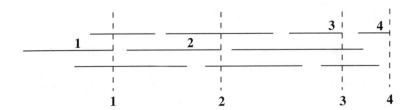

Figure 5.4   An illustration of the workings of the scan used in algorithm *Interval-Scan* to compute the *HK bound*. The intervals in $\mathcal{L}$ are represented by horizontal lines, and the points chosen for R are represented by dashed vertical lines. Each choice of ϵ is so small that the chosen points in R appear to be right-endpoints of intervals. The numbers labeling intervals identify the intervals in $\mathcal{I}$, and correspond to the points in R, in the order that points are added to R.

Theorem 5.2.2 *Algorithm Interval-Scan correctly computes the HK bound, HK(M). In particular, at the end of the scan, R is a minimum-sized set such that every interval I in $\mathcal{L}$ contains at least one point in R.*

Proof: By the way the scan works, the requirement that every interval I in $\mathcal{L}$ contains a non-integer point in R is clearly satisfied. To see that R is minimum-sized, suppose that I and I' are two intervals in $\mathcal{I}$, and that I was placed in $\mathcal{I}$ before I', so the right end of I is to the left of the right end of I', and the two intervals cannot share more than a single point (the right end of I and the left end of I'). If the two intervals did share more than a single point, then the point placed into R when I was placed into $\mathcal{I}$ would also be contained in I', which contradicts the fact that the algorithm placed both I and I' into $\mathcal{I}$. Therefore, no pair of intervals in $\mathcal{I}$ can have a nontrivial intersection, so no single point on the real line can both be strictly between the endpoints of one interval in $\mathcal{I}$, and also be strictly between the endpoints of a different interval in $\mathcal{I}$. It follows that $|\mathcal{I}|$ is a lower bound on $|R^*(M)|$. But $|R| = |\mathcal{I}|$, so $|R| = |R^*(M)| = HK(M)$, proving that this scanning procedure correctly computes the *HK bound*.   ∎

It should be clear that the time to compute the *HK bound*, given the sorted set L, is just $O(|L|)$, and so the *HK bound* is efficiently computable.

5.2.2.3   *Another Characterization of the* HK *Bound*

As an aside, we note another result that can be deduced from the reasoning in the proof of theorem 5.2.2. Let $\mathcal{I}^*(M)$ denote the *largest* set of intervals in $\mathcal{L}$ such that no pair of intervals in $\mathcal{I}^*(M)$ has a nontrivial intersection. Clearly, $|R^*(M)| \geq |\mathcal{I}^*(M)|$, since a distinct point must be chosen for each interval in $\mathcal{I}^*(M)$. But the interval scanning algorithm chooses a set of points R and a set of intervals $\mathcal{I}$, such that no pair has a nontrivial intersection, where $|R| = |\mathcal{I}|$. The next corollary then follows:

Corollary 5.2.1 *The set of intervals $\mathcal{I}$ found by the scanning algorithm has the maximum possible size. That is $|\mathcal{I}| = |\mathcal{I}^*(M)|$, and the output of algorithm Interval-Scan (and hence the HK bound) can be described as: The size of the largest set of intervals in $\mathcal{L}$ such that no pair of intervals have a nontrivial intersection.*

The characterization of the *HK bound* given in corollary 5.2.1 is closer to the original characterization given in [183] than is the definition of the *HK bound* in terms of $R^*(M)$.

The *HK bound* has been widely used in the biological literature, and the points in $R^*(M)$ have even sometimes been used as an estimate for where the historical recombination breakpoints occurred in the derivation of M. However, we will see that the *HK bound* is a relatively weak (low) lower bound[1] in comparison to other bounds on $Rmin(M)$ that were developed after it. Hence, using the points in $R^*(M)$ as estimates for the true recombination breakpoints is questionable.

5.2.3 The Haplotype Lower Bound

The *HK bound* was introduced in 1985, and very little progress was made in improving lower bounds on $Rmin(M)$ until the dissertation of Simon Myers in 2003 [299, 302]. There, the *haplotype lower bound* was introduced along with a "composite method" that dramatically improves the quality of that bound. We will explain the composite method after discussing the original *haplotype bound*.

Definition Consider the set of sequences M arrayed in a matrix. Let $D_r(M)$ and $D_c(M)$ be the number of distinct *rows* and distinct *columns* of M, respectively. The *haplotype lower bound* on M, denoted $H(M)$, is defined to be $D_r(M) - D_c(M) - 1$.

---

1 Ironically, the study of the *HK bound* in [183] was partly intended to show that the bound is often considerably lower than the true number of recombinations in simulated ARGs, and to therefore discourage its use. However, the *HK bound* continued to be widely used, since for some time it was the only nontrivial lower bound method known. There was also confusion in the literature, which sometimes implied (in different terminology, and incorrectly) that the *HK bound* is equal to $Rmin(M)$. In fact, the *HK bound* is typically far below $Rmin(M)$.

Actually, we assumed earlier that all the n rows of M are distinct, so we could have simplified the *haplotype bound* to $n - D_c(M) - 1$, but the statement of the bound as $D_r(M) - D_c(M) - 1$ emphasizes that it applies even when some rows are not distinct.

For example, consider the data M in figure 3.8 (page 75). In that example, there are 7 distinct rows and 5 distinct columns, so $H(M) = 1$, which is the same as the *HK bound*. However, if we modify M by adding the additional sequence $h = 00000$, creating input M', then the number of distinct rows becomes 8 and the number of distinct columns remains 5, so $H(M') = 2$. Note that the addition of sequence h does not create any more incompatible pairs in the data, and so the *HK bound* would remain equal to 1.

The intuitive basis for the *haplotype bound* is that any ARG for M must generate $D_r(M)$ distinct sequences, starting from a single ancestral sequence. Hence, the ARG must generate at least $D_r(M) - 1$ new sequences. Each mutation event generates at most one new sequence, so at most m of the $D_r(M) - 1$ new sequences can be accounted for by mutation events. Therefore, at least $D_r(M) - m - 1$ of the sequences must be generated at recombination nodes of the ARG. But each recombination node can generate at most one new sequence, so the number of recombination nodes must be at least $D_r(M) - m - 1$. This gives the intuition for the *haplotype bound*, and establishes it when $D_c(M) = m$. But when $D_c(M) < m$, then $D_r(M) - D_c(M) - 1 > D_r(M) - m - 1$. So, to fully establish the *haplotype bound*, we have to address the case that $D_c(M) < m$, that is, when there are duplicate columns. This is a small technical issue, handled in the formal proof of the following:

Theorem 5.2.3 *For any M, $H(M) \leq Rmin(M)$.*

Proof: First, if there are any nondistinct columns of M, arbitrarily remove duplicate columns from M so that all columns in the resulting matrix M' are distinct; so M' has exactly $D_c(M)$ columns. Note that $D_r(M') = D_r(M)$. Any ARG $\mathcal{N}$ that derives M also derives M': simply remove any site labeling an edge if that site is not in M'. It follows that $Rmin(M') \leq Rmin(M)$. That is, the number of recombination nodes needed in an ARG to derive M' cannot be larger than the number needed to derive M. So any lower bound on $Rmin(M')$ is a lower bound on $Rmin(M)$.

Let $\mathcal{N}'$ be a MinARG that derives M', that is, using $Rmin(M')$ recombination nodes. ARG $\mathcal{N}'$ must contain nodes labeled by $D_r(M')$ or more distinct sequences, since all the distinct sequences in M' must be generated in $\mathcal{N}'$. Note that the root of $\mathcal{N}'$ might be labeled with one of the sequences in M'. Since each character can mutate at most once, there can be at most $D_c(M')$ edges labeled by the sites in M', and so there are at most $D_c(M')$ distinct sequences that label tree-nodes or leaves whose entering edge is labeled with a site in M' (recall that edges into recombination nodes have no edge labels). So at least

$D_r(M') - (D_c(M') + 1)$ of the distinct sequences must label recombination nodes in $\mathcal{N}'$. Each recombination node has only one label, so there must be at least $D_r(M') - D_c(M') - 1 = D_r(M) - D_c(M) - 1 = H(M)$ recombination nodes in $\mathcal{N}'$. So, the actual number of recombination nodes in M' must be at least $H(M)$, and since this holds for any N', it follows that $Rmin(M') \geq H(M)$. But we first established that $Rmin(M) \geq Rmin(M')$, so $Rmin(M) \geq H(M)$. ∎

We can also establish the following rooted version of theorem 5.2.3, using a similar proof left to the reader.

Theorem 5.2.4 *If M is generated on an ARG $\mathcal{N}$ whose root sequence is not in M, then the number of recombination nodes in $\mathcal{N}$ must be at least the $D_r(M) - D_c(M) = H(M) + 1$.*

For simplicity, we will assume that the root sequence is specified and is in M. Close examination of the proof of theorem 5.2.3 leads to the following:

Theorem 5.2.5 *If $\mathcal{N}$ is an ARG for M that has only $H(M)$ recombination nodes (and hence is a MinARG), and every site in M is distinct, then every node in $\mathcal{N}$ is labeled by a sequence in M and no edge is labeled with more than one site.*

Proof: In addition to assuming that M does not contain any identical sites, we can also assume that no two non-leaf nodes in $\mathcal{N}$ can have the same node label, since we can always merge two such nodes. Similarly, we can assume that every tree edge in $\mathcal{N}$ is labeled with a site in M, since a tree edge that is not labeled can be contracted.

If M is generated on $\mathcal{N}$ with exactly $H(M)$ recombination nodes, then exactly $H(M)$ distinct sequences label recombination nodes. But since M contains no duplicate sites, $D_c(M)) = m$ and so $\mathcal{N}$ has at most $D_c(M)$ tree-nodes. It only has one root, so at most $H(M) + D_c(M) + 1 = [D_r(M) - D_c(M) - 1] + D_c(M) + 1 = D_r(M)$ distinct sequences can label the nodes of $\mathcal{N}$.

But the node labels must contain all of the $D_r(M)$ distinct sequences in M, so every node must be labeled with a sequence in M. Further, if some edge in $\mathcal{N}$ is labeled with two or more sites of M, then the number of tree nodes would be strictly less than $D_c(M)$ and so the number of distinct sequences would be strictly less than $D_r(M)$. Hence, no edge is labeled with more than one site. In fact, since we assumed that every tree edge is labeled, every tree edge is labeled with exactly one site, and every leaf edge that is labeled, is also labeled with exactly one site. ∎

Conversely, we have:

Theorem 5.2.6 *Let $\mathcal{N}$ be an ARG for M where all node labels are from M, and no two non-leaf nodes have the same label. If no edge is labeled by more than a single site, then $\mathcal{N}$ must have exactly $H(M)$ recombination nodes (and hence $\mathcal{N}$ must be a MinARG for M).*

Proof: Without loss of generality, assume that the ancestral sequence of $\mathcal{N}$ is the all-zero sequence. Since no edge is labeled by more than a single site, and every site in M is polymorphic, there are exactly m tree and leaf edges that are labeled by a site in M. Let H be the set of nodes at the heads of those m edges. The labels on the nodes in H must be distinct. This follows from the fact that a node label can only appear twice if it labels a leaf and its parent node, and the edge between them is unlabeled; in that case, the leaf is not in H. So all the sequences labeling nodes in H are distinct. The root is labeled with one sequence in M, so the number of distinct sequences labeling non-recombination nodes in $\mathcal{N}$ is exactly $m + 1$.

There are $D_r(M)$ distinct sequences in M, and so $D_r(M) - (m + 1) = D_r(M) - m - 1$ sequences must label recombination nodes. Since no two recombination nodes can have the same label, and all nodes are labeled by sequences in M, there must be exactly $D_r(M) - m - 1$ recombination nodes in $\mathcal{N}$. Now, $m \geq D_c(M)$, so $D_r(M) - m - 1 \leq D_r(M) - D_c(M) - 1 = H(M) \leq Rmin(M)$, and if $m > D_c(M)$, then $\mathcal{N}$ would have strictly fewer than $Rmin(M)$ recombination nodes, which is impossible. So $D_c(M) = m$, and $\mathcal{N}$ has exactly $D_r(M) - D_c(M) - 1 = H(M)$ recombination nodes.   ∎

Two additional observations

There are two observations about the *haplotype bound* that will be useful later. First, the *haplotype bound* is *unaffected* by the given *order* of the characters in M. We could permute the order of the columns of M and the numbers of distinct rows and columns would not change. Second, the *haplotype bound* is valid for biological processes other than single-crossover recombination. For example, it is valid for multiple-crossover recombination, and it is a valid lower bound on the number of reticulation nodes needed in a reticulation network (discussed in chapter 14). What is required for these and other applications of the *haplotype bound*, is that each mutation occurs at most once, and that each reticulation event (recombination, hybridization, lateral transfer, etc.) generates only one new sequence.

5.2.3.1 Removing Fully-Compatible Characters Usually Boosts the Haplotype Bound

The *haplotype bound* $H(M)$ can be efficiently computed, but the values of $H(M)$ for data generated by the program *ms* [182] show that $H(M)$ by itself is a very poor bound.

Often, it is a negative number! However, when used with the *composite bound method* (explained below), much improved lower bounds are obtained that are significantly higher than the *HK bound*. Before developing the composite bound method, we explain another significant way to boost $H(M)$ that also comes from [299, 302]. Recall that a character c is called "fully compatible" if c is not incompatible with any character in M.

Theorem 5.2.7 *Suppose c is a fully-compatible character, i.e., compatible with all other characters in M. Then the haplotype bound either increases or remains the same when character (column) c is removed from M. That is, letting M' denote the matrix M after removal of character c, $Rmin(M) \geq H(M') \geq H(M)$.*

Proof: If character c is a copy of some other character $\bar{c}$ in M, then certainly $D_c(M') = D_c(M)$, but it is also true that $D_r(M') = D_r(M)$. To see this claim, note first that any pair of rows that are identical in M remain identical in M'. Next, note that any pair of rows that are different in M are either different at a character $d \neq c = \bar{c}$, or are different only at characters c and $\bar{c}$. In either case, the pair of rows remain different in M': they differ at d in the first case, and at $\bar{c}$ in the second case. So $H(M') = D_r(M') - D_c(M') - 1 = D_r(M) - D_c(M) - 1 = H(M)$ if c is *not* a distinct column in M. So, we assume that character c is distinct in M.

Let $D_r(M')$ be the number of distinct rows in M'. We claim that $D_r(M) \geq D_r(M') \geq D_r(M) - 1$. To see the first inequality, note that the removal of character c from M can cause two nonidentical rows in M to become identical in M'. This happens if the rows only differ at character c in M. However, two rows that are identical in M remain identical in M'. Therefore, the number of *distinct* rows cannot increase with the removal of character c, so $D_r(M) \geq D_r(M')$.

To prove the second inequality, suppose for contradiction that $D_r(M') < D_r(M) - 1$, so $D_r(M) \geq D_r(M') + 2$. The rows in M' are partitioned into $D_r(M')$ maximal *classes* where all the rows in any class are identical to each other, and different from any row in a different class. Similarly, the rows of M are partitioned into $D_r(M) \geq D_r(M')$ maximal classes of identical rows. Consider adding column c back to M' to create M. The effect on each class in M' is either to leave it unchanged, or to split it into exactly two classes. If a class splits into two classes, then each row in one of the two new classes has a 0 for character c, and each row in the other new class has a one for character c, but all the rows in the two new classes are identical at each character other than c. See figure 5.5. No classes can merge through the addition of c to M'.

If $D_r(M) \geq D_r(M') + 2$, there must be at least two classes E_0 and E_1 in M' which each split when c is added back to M'. So, in M, at least one of the rows in E_0 has value zero at character c, and at least one of the rows has value one at c; the same holds for E_1.

$$c$$

0	1	0	0	1	0	0
0	1	0	0	1	0	0
0	1	0	1	1	0	0
0	1	0	1	1	0	0
0	1	0	1	1	0	0

Figure 5.5   A class in M' splits into two classes (in M) after the addition of character c.

Moreover, since in M', no row in E_0 is identical to a row in E_1, and all rows in E_0 (or E_1) are identical, there must be a character $c' \neq c$ where all values in E_0 differ from all values in E_1. Without loss of generality, say all rows in E_0 have value zero at character c' and all rows in E_1 have value one at c'. See figure 5.6. But then there exist four rows in M, two from E_0 and two from E_1, which contain all of the four binary combinations 0,0; 1,0; 0,1; and 1,1 for characters c, c'. By the four-gametes theorem (theorem 2.3.2, page 51), characters c and c' in M are incompatible, contradicting the assumption that character c is fully compatible in M. Hence $D_r(M') \geq D_r(M) - 1$.

Since column c is assumed to be distinct from all other columns in M, $D_c(M') = D_c(M) - 1$, so $H(M') = D_r(M') - D_c(M') - 1 = D_r(M') - [D_c(M) - 1] - 1 = D_r(M') - D_c(M) \geq D_r(M) - 1 - D_c(M) = H(M)$.

We have shown that $H(M') \geq H(M)$, but we still need to prove that $H(M')$ is a lower bound on $Rmin(M)$, that is, that $Rmin(M) \geq H(M')$. Let $\mathcal{N}$ be an ARG that generates M using $Rmin(M)$ recombination nodes. If we simply remove c from the edge that it labels in $\mathcal{N}$, and remove character c from all of the node labels, then we have an ARG $\mathcal{N}'$ that generates M' and has the same number of recombination nodes as $\mathcal{N}$. Therefore $Rmin(M) \geq Rmin(M')$. Applying the *haplotype bound* to M', $Rmin(M') \geq H(M')$, and we conclude that $Rmin(M) \geq H(M') \geq H(M)$. Hence, the removal of c can boost the lower bound on $Rmin(M)$ but can never reduce it.   ∎

Myers and Griffiths [299, 302] suggest that the removal of all fully compatible characters can speed up computations by reducing the size of the matrix. That is one benefit. But the more important benefit that we have observed is the substantial increase in the resulting lower bound on $Rmin(M)$, when all fully compatible characters are removed. As we will see, this is particularly true when used locally in the composite bound method, introduced next.

				c			c'
	0	1	0	0	1	0	0
	0	1	0	0	1	0	0
E_0	-	-	-	-	-	-	-
	0	1	0	1	1	0	0
	0	1	0	1	1	0	0
	0	1	0	1	1	0	0
	0	1	0	1	1	0	1
	0	1	0	1	1	0	1
	0	1	0	1	1	0	1
E_1	-	-	-	-	-	-	-
	0	1	0	0	1	0	1

Figure 5.6   Assuming that $D_r(M') < D_r(M) - 1$, two classes in M', E_0 and E_1, split into two classes each after the addition of character c. Character c' must also exist as shown, creating four gametes, contradicting the assumption that character c is fully compatible in M.

5.2.4   The Composite Bound Method

Here we introduce a general method developed by Myers and Griffiths [299, 302], called the *composite bound method*, that can (and usually does) substantially boost several lower bounds on $Rmin(M)$. We will explain the method in general, and also discuss in particular how it boosts the haplotype lower bound.

As noted earlier, the *haplotype bound*, when applied to a *whole* dataset M, is often very low and can even be a negative number. Intuitively, the *haplotype bound* will tend to be low when the number of columns of M is large relative to the number of rows of M. This is not guaranteed, but tends to be the case. Conversely, the *haplotype bound* tends to be higher when the number of columns is small relative to the number of rows. We would like to take advantage of this intuition to obtain a better lower bound on $Rmin(M)$. This is what the composite method does.

Recall that the sites in M have a *fixed linear order*, so we can unambiguously specify an interval of sites. The fact that the sites have a fixed linear order is critical in the composite method. The composite method *combines* several (*local*) lower bounds computed (by any method) over a family $\mathcal{L}$ of intervals of sites. For simplicity of exposition, we assume that the m sites of M are arrayed on the real line, on the *integer* points $1, ..., m$. Hence, when we speak of an interval of sites I, the left and right ends of I are at integer points.

Definition Given an interval of sites I, let $M(I)$ denote the matrix M restricted to the sites in I, and let $b(I)$ denote a lower bound computed (somehow) for $Rmin(M(I))$. The bound $b(I)$ is called a "local bound."

Definition Given a family $\mathcal{L}$ of intervals of sites, let $\mathcal{B} = \{b(I) : I \in \mathcal{L}\}$, i.e., the set of local bounds, one for each interval in $\mathcal{L}$.

Definition Given $\mathcal{L}$ and $\mathcal{B}$, $R^*(\mathcal{L})$ is defined as the *minimum-sized* set of points such that for each interval $I \in \mathcal{L}$, there are at least $b(I)$ points in $R^*(\mathcal{L})$ that fall strictly in the interior of interval I. The number $|R^*(\mathcal{L})|$ is called a *composite bound*.

The reader should see that the composite bound is a generalization of the *HK bound*. In particular, if each pair of incompatible sites (p, q) in M defines a closed interval $I = [p, ..., q]$, and for each such interval I, $b(I)$ is set to one, then the composite bound, $|R^*(\mathcal{L})|$, is exactly the *HK bound* for M.

Theorem 5.2.8 *For any M, if $b(I)$ is a lower bound on $Rmin(M(I))$, for each interval $I \in \mathcal{L}$, then $|R^*(\mathcal{L})| \leq Rmin(M)$. That is, the composite bound is a valid lower bound on $Rmin(M)$.*

Proof: The proof is a simple extension of the proof used to show that the *HK bound* is a valid lower bound on $Rmin(M)$. We first show that for every interval $I = [p, ..., q]$ in $\mathcal{L}$, every ARG that derives M must have at least $b(I)$ recombination nodes whose crossover indexes are in the interval $(p, ..., q]$, which is open on the left end and closed on the right end. For contradiction, suppose there is an ARG $\mathcal{N}$ which derives M, where this is not the case. Then modify $\mathcal{N}$ to obtain an ARG $\mathcal{N}'$ just for the set of sequences $M(I)$. We do this by first removing all labels on edges that fall outside of I. Second, at each recombination node x in $\mathcal{N}$, where the crossover index b_x is *not* in the interval $(p, ..., q]$, we remove exactly one of the incoming edges as follows. If $b_x \leq p$, remove the P-labeled edge into x and retain the S-labeled edge; conversely, if $b_x > q$ remove the S-labeled edge, and retain the P-labeled edge into x. Note that at any recombination node x where $b_x \in (p, ..., q]$, nothing is changed. The resulting network is an ARG $\mathcal{N}'$ that derives the sequences $M(I)$ and where every recombination node x has a crossover index $b_x \in (p, ..., q]$. But $\mathcal{N}'$ has strictly fewer than $b(I)$ recombination nodes, contradicting the assumption that $b(I)$ is a lower bound on $Rmin(M(I))$.

Now, for any ARG $\mathcal{N}$ that derives M, let $R(\mathcal{N})$ be the number of recombination nodes in $\mathcal{N}$, and let $\mathcal{B}$ be the set of crossover indexes at the recombination nodes of $\mathcal{N}$. Clearly, $|\mathcal{B}| \leq R(\mathcal{N})$. Next, for each recombination node x, create the point $p_x = b_x - \epsilon_x$, where $0 < \epsilon_x < 1$ and ϵ_x is different from ϵ_y for any other recombination node y. The result is a set of $|\mathcal{B}|$ distinct points such that for every interval $I \in \mathcal{L}$, there are at least $b(I)$ points in the set that are strictly in the interior of I. Therefore $|R^*(\mathcal{L})| \leq |\mathcal{B}| \leq R(\mathcal{N})$. In particular, when $\mathcal{N}$ is a MinARG for M, $R(\mathcal{N}) = Rmin(M)$, so $|R^*(\mathcal{L})| \leq Rmin(M)$, so the composite bound is a valid lower bound on $Rmin(M)$.   ∎

5.2.4.1   Computing and Using the Composite Bound

As probably anticipated by the reader, given the sets $\mathcal{L}$ and $\mathcal{B}$, the composite bound can be efficiently computed by an extension of algorithm *Interval-Scan* used to compute the *HK bound*. As in algorithm *Interval-Scan*, let L be the sorted list of right endpoints of the intervals in $\mathcal{L}$, smallest (leftmost) first. Then $R^*(\mathcal{L})$ can be found efficiently by a single scan of L, corresponding to left-to-right sweep of the points in L. In particular, when the right endpoint q of an interval $I = [p, ..., q]$ is examined, if z is the number of points strictly in I that have already been chosen to be in $R^*(\mathcal{L})$, and $z < b(I)$, place an additional $b(I) - z$ copies of point $q - \epsilon$, for $0 < \epsilon < 1$, into $R^*(\mathcal{L})$.

The proof of correctness of this method is a simple extension of the proof of correctness of algorithm *Interval-Scan*, and we leave the formal proof to the reader. Also, as in the computation of the *HK bound*, after the sorted list L is known, the time required for the method is linear in the size of L.

The simplest use of the composite bound method

The simplest way to apply the composite bound method is to first compute a local bound for each of the $\binom{m}{2}$ intervals in a dataset M with m sites, and then compute the composite bound using those local bounds. This can be done with any local bound, but this approach has been shown to be particularly effective when the local bound for each interval is the *haplotype bound*.

5.2.4.2   The Composite Bound Greatly Boosts the Haplotype Bound

As mentioned earlier, the *haplotype bound* is generally not very large when computed on data with a large number of sites compared to the number of taxa. But by computing local *haplotype bounds* for all the intervals, or for the smaller intervals only, and then computing the composite bound using those local *haplotype bounds*, the resulting lower bound on $Rmin(M)$ can be greatly increased. This has been consistently shown empirically. Additional increases are obtained through the use of theorem 5.2.7 in each interval where the *haplotype bound* is computed. That is, for each interval I where a local *haplotype bound* is to be computed, first remove every site that is compatible with every other site *in interval* I. Note that a site might be removed in an interval I, even though it is incompatible with some site outside of I. We call the resulting lower bound the *Interval RecMin bound*. We will explain the name for this bound shortly.

An example of the effectiveness of the composite bound method using local *haplotype bounds* is shown in figure 5.7. The *haplotype bound* computed from the entire data is -3 (a useless bound), but the *interval RecMin* bound is 6. The data used in that example is the widely studied SNP data obtained by Kreitman [230] in 1983, which was one of the first significant SNP datasets published. The *HK bound* is five, and $Rmin(M)$ for this data is 7, as we will establish in sections 9.1.4 and 9.1.5.3.

```
00000000011000000001101110111100000000000000
00100000000000000001101110111100000000000000
00000000000000000000000000000000000010000101
00000000000000001100000000000000000010011000
00011000101100111100000000000000000001000000
00100000000000001000000000000000101011000010
00100000000000001000000000000011111101000000
11111000101110010000000000000011111101100000
11111000101110010000000000000011111101100000
11111000101110010000000000000011111101100000
11111111100001010001000100001111101000000
```

Figure 5.7   In this data (Kreitman's data), the *HK bound* is 5; the *haplotype bound* computed on the entire data is -3; the composite bound computed using the *haplotype bound* as the local bound in each of the intervals is 5; the composite bound computed using the *haplotype bound* as the local bound, when in each interval, columns that are compatible with all other columns in that interval are removed (i.e., the *interval RecMin bound*) is 6. $Rmin(M)$ is actually 7.

Why, intuitively, does the composite bound method help raise the haplotype bound?

It is worth considering why the composite bound is effective when used together with the *haplotype bound*. In addition to the fact that the *haplotype bound* tends to be low when intervals are large, and the composite method uses intervals of all sizes, the key point is that the *haplotype bound* is *unaffected* by the *linear order* of the sites. That is, we could *permute* the order of the sites in M and the resulting *haplotype bound* would be unchanged. So, the *haplotype bound* does not incorporate any constraints imposed by the physical reality of the fixed linear order of the sites, and the fact that a crossover is an event that takes place at a particular location among the sites. The effect of a recombination event is constrained and influenced by a given linear order of the sites, but that constraint is not incorporated into the *haplotype bound*. Intuitively, the composite bound approach is effective when combined with the *haplotype bound* because composite approach imposes *ordered* constraints (that at least $b(I)$ recombinations must occur *inside* each local interval I) and hence reflects the fixed linear order of the sites, and the spatial aspect of a recombination crossover.

The same intuition holds for other local lower bounds that are not affected by the linear order of the sites. For example, we will discuss the *connected-component bound* in section 7.1 and see that it is also unaffected by the order of the sites. Again, the composite bound method improves the resulting lower bound on $Rmin(M)$ when each local lower bound is a *connected-component lower bound*. In contrast, the composite method might not be as effective when the local bounds are already affected by the linear order of the sites. For example, when each local bound is an *HK bound* (which is already affected by the linear order of the sites) the composite bound is just the *HK bound* applied to the entire data, and so the composite bound method offers no improvement at all in the case of the *HK bound*.

5.2.4.3 *Applying the Haplotype Bound to Subsets of Sites: A Major Improvement*

Myers and Griffiths [299, 302] introduced another way to significantly boost the basic *haplotype bound*, at the price of an increase in computation time. The basic idea is to compute the *haplotype bound* on *subsets* of columns in an interval, rather than on all of the columns in the interval. We formalize that idea here.

Definition For a *subset* of sites S (*not* necessarily contiguous) in M, let $M(S)$ be the sequences in M restricted to the sites in S, and let $H(M(S))$ be the *haplotype bound* computed on $M(S)$.

Lemma 5.2.4 *Let S be any subset of sites in M, whose leftmost point is p and whose rightmost point is q. $H(M(S))$ denotes the haplotype bound applied to the submatrix $M(S)$. Then $H(M(S)) \leq Rmin(M(I))$, where $I = [p, ..., q]$. So, any ARG that derives M must have at least $H(M(S))$ recombination nodes whose crossover indexes are in the interval $(p, ..., q]$.*

Proof: Similar to the first part of the proof of theorem 5.2.8, consider a MinARG $\mathcal{N}$ for the sequences $M(I)$, and then remove all of the sites from $\mathcal{N}$ that are not in S. The result is an ARG for $M(S)$ with $Rmin(M(I))$ recombination nodes. But, a recombinant sequence might now be equal to one of its parental sequences, in which case that recombination event can be eliminated. So, $Rmin(M(S)) \leq Rmin(M(I))$. Since $H(M(S)) \leq Rmin(M(S))$, the lemma follows. ∎

	c_1	c_2	c_3	c_4	c_5	c_6
r_1	0	0	1	0	0	0
r_2	0	0	1	0	1	0
r_3	0	1	1	0	0	0
r_4	0	1	1	1	1	0
r_5	0	0	1	1	0	1
r_6	1	0	1	1	1	1
r_7	1	1	0	0	0	1
r_8	1	1	0	0	1	1

Figure 5.8 Dataset M with 8 distinct rows and 6 distinct sites, so $H(M) = 1$. However, if sites 1, 3 and 4 are removed, the resulting dataset still has 8 distinct rows, but now only has three distinct sites, so the resulting *haplotype bound* would be 4.

Does this help?

Lemma 5.2.4 says that a valid lower bound on $Rmin(M(I))$ can be computed by using any subset of the columns in I, but can this actually help, that is, obtain *higher* lower bounds? It may not be intuitive, but the subset approach *does* often yield a local lower bound that is larger than $H(I)$, the *haplotype bound* computed for the entire set of sites in interval I. In fact, we have already seen one indication of this, in theorem 5.2.7, when fully-compatible sites are removed from M. For a more general illustration, consider the example in figure 5.8 where there are 8 distinct rows and 6 distinct columns, giving a *haplotype bound* of one. Removal of sites 1, 3, and 4 decreases the number of distinct columns by three, but does not reduce the number of distinct rows, and so the *haplotype bound* computed on the remaining columns is increased to four. This type of situation will be more formally treated in theorem 5.2.9.

Given lemma 5.2.4, we could compute a valid local lower bound $b(I)$ for an interval I by computing $H(M(S))$ for *each* subset S of sites in I (with length denoted $|I|$), and then taking $b(I)$ to be the *maximum* of those $2^{|I|}$ *haplotype bounds*. Clearly, increasing local lower bounds cannot reduce the resulting composite lower bound, and may lead to a larger composite bound for $Rmin(M)$. We will examine this in more detail later.

5.2.5   The Optimal RecMin Bound

Definition Let $S^*(I)$ be a subset of sites in interval I that maximizes $H(M(S))$ over all subsets S of sites in I. Subset $S^*(I)$ is called the *optimal subset* for I, and $H(M(S^*(I)))$ is called the *optimal haplotype bound* for I.

Definition When, for every interval I in M, the local lower bound $b(I)$ is set to the *optimal haplotype bound* for I, the resulting composite lower bound on $Rmin(M)$ is called the *optimal RecMin bound* (ORB) on $Rmin(M)$.

Clearly, the *optimal RecMin bound* is at least as high as the *interval RecMin bound*. It has been empirically demonstrated that the *optimal RecMin bound* is typically much larger than the *interval RecMin bound*, reflecting the utility of considering subsets of sites inside of each interval. We will see an example of this later.

Efficient organization
Since a subset of sites S in an interval I is also a subset in any interval that contains I, it is more efficient to compute the *optimal RecMin bound* by first enumerating every subset of sites S, and then computing $H(M(S))$ for each subset. Then, those bounds can be used to determine the local bound $b(I)$ for each interval I. Still, because all $2^m - 1$ subsets of sites must be explicitly enumerated in this approach, the time required will generally be prohibitive. Consistent with the fact that full enumeration is computationally infeasible, Bafna and Bansal [18, 19] proved that the problem of computing the *optimal RecMin bound* is NP-hard. Given these realities, we will discuss two alternative approaches, the first called Program *RecMin* that enumerates only *some* subsets of sites, and a second that computes the *optimal RecMin bound* without explicitly enumerating any subsets of sites, although it again involves a computation that requires worst-case exponential time.

5.2.6   Program RecMin

Myers and Griffiths [299, 302] encapsulated the use of local *haplotype bounds* computed over *subsets* of sites, together with the composite bound method, in a program they call *RecMin*. However, because enumerating all subsets of sites is generally impractical (taking days on even moderate sized data), *RecMin* generally computes local bounds for a *restricted* set of subsets of sites.

In *RecMin*, the user specifies two parameters s and w, and *RecMin* computes the *haplotype bound* $H(M(S))$ for every subset S of M, with s or fewer sites, provided that no pair of sites in S is more than w positions apart. The parameter s is called the *subset size* parameter, and the parameter w is called the *subset width* parameter. Let S be the family of subsets that obey the restrictions imposed by s and w. *RecMin* computes $H(M(S))$ for every subset S in $\mathcal{S}$, and then for every interval I, the local lower bound $b(I)$ is set to the largest of value $H(M(S))$ where S is contained in I. *RecMin* then computes the composite bound using those local bounds. *RecMin* also uses heuristics to avoid the explicit examination of some of the specified subsets.

The default settings for *RecMin* were initially $s = 8$ and $w = 12$, but we have found that *RecMin* gives better bounds in reasonable time when we set $s = w = 20$. Overall, *RecMin* is a very impressive, efficient program for computing lower bounds on $Rmin(M)$, and far superior to any of the practical alternatives that came before it. However, as noted earlier, for problem instances of the size already encountered in genomic data, *RecMin* cannot compute the *optimal RecMin bound*, and be sure it has been computed. The only way that *RecMin* can *guarantee* to compute the *optimal RecMin bound* is to set $s = w = m$, which is usually prohibitive.

We next discuss an alternative to *RecMin* that is able to compute the *optimal RecMin bound* efficiently on data that is much larger than the data for which *RecMin* can compute the *optimal RecMin bound*.

5.2.7   Program HapBound

Practical computation of the optimal RecMin bound, and beyond
We first discuss the integer linear programming (ILP) approach developed in [390] to efficiently compute, in practice, the *optimal RecMin bound* on tested biological datasets.

Following the definition, in order to compute the *optimal RecMin bound*, we want to find the optimal subset $S^*(I)$ for each interval I in M. But how can we do that efficiently? Theorem 5.2.9 begins to explain the answer.

Definition For an interval I, let $\widetilde{M}(I)$ denote any matrix obtained from $M(I)$ by successively removing a duplicate row, until no duplicate rows remain. When there are choices of duplicate rows to remove, the choice can be made arbitrarily.

Note that $\widetilde{M}(I)$ is *not* the matrix created by first removing duplicate rows from M, and then restricting to the sites in I.

Definition For any subset of columns S of I, let $\widetilde{M}(I, S)$ denote matrix $\widetilde{M}(I)$ restricted to the columns in S.

All the rows in $\widetilde{M}(I)$ are distinct by definition, but that might not be true for a submatrix $\widetilde{M}(I, S)$. For example, if S is a single column, the rows cannot be distinct if there are three or more rows. Hence, there is a well-defined smallest subset of columns where row-distinctness remains. The following theorem was suggested by S. Myers [298].

Theorem 5.2.9 *Any* smallest *subset of sites* S *in* I, *such that every row in* $\widetilde{M}(I,S)$ *is* distinct, *is an optimal subset,* $S^*(I)$.

Proof: First, observe that for any subset of sites S in I, two columns in $M(S)$ are distinct if and only if they are distinct in $\widetilde{M}(I,S)$. Similarly, if two rows are identical in $M(I)$, then they are identical in $M(S)$, so the number of distinct rows in $M(S)$ is equal to the number of distinct rows in $\widetilde{M}(I,S)$. Hence, $H(M(S)) = H(\widetilde{M}(I,S))$. It follows that the removal of duplicate rows in $M(I)$ (creating $\widetilde{M}(I)$) does not affect which subsets of columns in I maximize the *haplotype bound*, and so $S^*(I)$ is an optimal subset for I in both $M(I)$ and $\widetilde{M}(I)$. Let $\tilde{n}(I)$ denote the number of rows in $\widetilde{M}(I)$.

We next show that all rows in $\widetilde{M}(I,S^*(I))$ are distinct, or an equally good subset of sites can be found where this is true. Suppose for contradiction that two rows r_1, r_2 in $\widetilde{M}(I,S^*(I))$ are identical. Since r_1, r_2 are not identical in $\widetilde{M}(I)$, there must be a site c in $I - S^*(I)$ such that the state of r_1 at c differs from the state of r_2 at c. So, if we add c to $S^*(I)$, we increase the number of distinct rows by *at least* one, while increasing the number of distinct sites by *at most* one. Therefore $H(\widetilde{M}(I,S^*(I)\cup\{c\})) \geq H(\widetilde{M}(I,S^*(I)))$. But, by the choice of $S^*(I)$, $H(\widetilde{M}(I,S^*(I) \cup \{c\})) > H(\widetilde{M}(I,S^*(I)))$ is not possible, so $H(\widetilde{M}(I,S^*(I)\cup\{c\})) = H(\widetilde{M}(I,S^*(I)))$. If $\widetilde{M}(I,S^* \cup \{c\})$ still contains two identical rows, we can again find a new column to add, and repeat this step until we finally have an optimal subset of sites $S^*(I)$ where all the rows of $\widetilde{M}(I,S^*(I))$ are distinct.

Similarly, we can assume that all the *columns* in $\widetilde{M}(I,S^*(I))$ are distinct, for if not, the removal of any duplicate columns cannot change the number of distinct columns or the number of distinct rows in the resulting matrix. Therefore, we assume that all the $\tilde{n}$ rows and all the $|S^*(I)|$ sites of $\widetilde{M}(I,S^*(I))$ are distinct, so $H(\widetilde{M}(I,S^*(I))) = \tilde{n}(I) - |S^*(I)| - 1$.

Now let S be *any* smallest subset of sites in I such that the rows in $\widetilde{M}(I,S)$ are distinct. Then, $|S| \leq |S^*(I)|$, so $H(M(S)) = H(\widetilde{M}(I,S)) = \tilde{n} - |S| - 1 \geq \tilde{n} - |S^*(I)| - 1 = H(\widetilde{M}(I,S^*(I))) = H(M(S^*)))$. So, $H(M(S)) = H(M(S^*(I)))$, proving the theorem. ∎

Definition A subset of sites is called a *minimum* $S^*(I)$ if it is a smallest subset of sites S in I such that every row in $\widetilde{M}(S)$ is distinct.

As an example of theorem 5.2.9, consider the dataset M in figure 5.8, and let I in this case be the entire interval $[1, ..., 6]$. The eight rows are distinct, so $\widetilde{M}(I) = M$. The set of sites $\{2, 5, 6\}$ is a smallest set S such that the eight rows are distinct in $M(S)$. So those three sites yield the highest *haplotype bound* possible in the interval $[1, ..., 6]$, and hence form a minimum $S^*(I)$ for M.

5.2.7.1   *Finding $S^*(I)$ by Integer Linear Programming*

Theorem 5.2.9 tells us precisely what to look for in order to find an optimal subset, $S^*(I)$, for I, but it does not tell us *how* to look for it. In this section we discuss the integer linear programming solution to finding $S^*(I)$, developed in [390]. For an introduction to *integer linear programming*, see appendix A.

Definition We say that a site c in $\widetilde{M}(I)$ *distinguishes* two rows r_1 and r_2 if $\widetilde{M}(I)[r_1, c] \neq \widetilde{M}(I)[r_2, c]$. We say that a subset of sites S distinguishes the rows in $\widetilde{M}(I)$, if for each pair of rows r_1, r_2 in $\widetilde{M}(I)$, there is a site c in S that distinguishes r_1 and r_2.

   The following lemma is almost self-evident, and a formal proof is left to the reader.

Lemma 5.2.5 *Let S be a subset of sites in interval I. Then, the rows in $\widetilde{M}(I, S)$ are distinct if and only if S distinguishes the rows of $\widetilde{M}(I)$.*

   Given theorem 5.2.9 and lemma 5.2.5, in order to find an optimal subset for an interval I, we want to find a smallest set of sites, S, that distinguishes the rows of $\widetilde{M}(I)$. That task can be formulated and solved as an Integer Linear Program, as follows.

   For any pair of rows r_1, r_2, let $D(r_1, r_2)$ denote the set of sites in I that distinguish rows r_1 and r_2 of $\widetilde{M}(I)$. Since the rows of $\widetilde{M}(I)$ are distinct, $D(r_1, r_2)$ cannot be empty. For each site c in I, let $X(c)$ be a binary integer linear programming variable, that is, one that is only allowed to take on values zero or one. In a solution, variable $X(c)$ will be set to 1 to indicate that site c should be taken into S, and set to 0 to indicate it should not be taken into S. The integer program will have the inequality

$$\sum_{c \in D(r_1, r_2)} X(c) \geq 1$$

for each pair of rows r_1, r_2 in $\widetilde{M}(I)$. The effect of these inequalities is to force the selection of a subset of sites that distinguish the rows of $\widetilde{M}(I)$. Finally, the objective function is

$$\text{Minimize} \sum_{c \in I} X(c).$$

A solution to the integer linear program sets each variable $X(c)$ to either 0 or 1, and hence specifies a smallest set S that distinguishes every pair of rows in $\widetilde{M}(I)$. Thus, by theorem

5.2.9 and lemma 5.2.5, a solution to the integer program identifies $S^*(I)$, an optimal subset for I.

As an aside, for those familiar with the *set cover* problem or the *minimum test-set* problem [123], one can view this integer program as solving an instance of the set cover or test-set problems, where each pair of rows (r_1, r_2) defines the set of sites $D(r_1, r_2)$, and the problem is to choose the smallest set of sites to cover all of those sets.

Basic HapBound

Summarizing the discussion, the *optimal RecMin bound* for M can be found by using integer programming to find an optimal subset $S^*(I)$ for each interval I in M. Then for each interval I, the local lower bound $b(I)$ is set to $H(M(S^*(I)))$, and a global bound is computed by the composite bound method. This approach has been implemented in a program called *HapBound* [390]. In *HapBound* each of the integer programs is solved by using the GNU ILP solver, GLPK. GNU allows GLPK to be incorporated into other programs, and using GLPK allows a complete version of *HapBound* to be released to users. With GLPK, *HapBound* is able to find $S^*(I)$ efficiently (in fractions of a second, to several seconds) for many problem sizes of current interest. This is robust enough to illustrate the practicality of the integer programming approach. However, for large problem instances, the commercial ILP solver, CPLEX, is significantly faster than GLPK. *HapBound* is not built around CPLEX because users would then need to have a CPLEX license to run *HapBound*. We will discuss *HapBound* in more detail in the next section, and some empirical results in section 5.2.8.

The Bafna-Bansal method

A different alternative approach to improving on *RecMin*, developed by Bafna and Bansal in [18, 19], formulates the problem of finding $S^*(I)$ (in different notation) in each interval I, and also uses the composite method to find a global lower bound on $Rmin(M)$. However, instead of using a method that is guaranteed to find $S^*(I)$, they use a fast heuristic algorithm to find a set S that distinguishes every pair of rows.

In more detail, their method first cleans up the input M to reduce the size of the problem in a way that does not change the number of needed recombination events (see algorithm *Clean* in section 9.1.1), and that results in a matrix $\widetilde{M}$ with no duplicate rows. Then the algorithm computes a lower bound on $Rmin(\widetilde{M})$ as follows:

> Until a set of columns has been selected that distinguishes all pairs of rows in $\widetilde{M}$, successively select the column that distinguishes the *most* pairs of rows that are not already distinguished by any previously selected column.

Let S denote the set of selected columns, and let q denote the number of distinct rows in $\widetilde{M}(S)$. The lower bound used is $q - |S| - 1$.

Although the lower bound computed in this way is not guaranteed to be as large as the *optimal RecMin bound*, and cannot ever be larger than it, empirical tests reported in [18, 19] show that the computed bound consistently produced lower bounds that are higher than the *HK bound* and the bound returned by *RecMin* with its default settings (see section 5.2.8).

Missing data

The Bafna-Bansal method in [19] extends naturally to the common problem that biological data has missing, unknown entries. In molecular sequence data the rate of missing entries can be 1 to 5 percent, while in phylogenetic data the percentage can often be as high as 35 percent. It is a common practice to remove columns or rows of M so that the remaining submatrix, M', has no cell with a missing value. A lower bound on $Rmin(M')$ can then be computed. However, this approach can substantially reduce the resulting lower bound. Instead, we can define the following problem:

Problem MDOR Given a matrix M with missing entries, fill in the missing entries with binary values in the way that *minimizes* the *optimal RecMin bound* on the resulting matrix.

It may seem incorrect to *minimize* the resulting lower bound rather than *maximizing* it. However, it is only through minimizing the lower bound that we obtain a value that is guaranteed to be a valid lower bound on $Rmin(M_t)$, where M_t is the correct original matrix, with no missing values, from which M was derived.

Of course, we don't know how to solve problem MDOR efficiently, but if we could solve it, the *optimal RecMin bound* on the resulting matrix would be larger or equal to the *optimal RecMin bound* computed on M', the matrix resulting from removal of all columns containing missing values.

The effect of including missing entries in the computation of lower bounds can be considerable, even if the lower bound method used is not the best lower bound method available for use on *complete* data. Table 5.1 shows several lower bound values for nine datasets, computed by Program *HapBound* (discussed in detail in the next section), which is guaranteed to produce a value that is as high or higher than the *optimal RecMin bound*. Each of the nine datasets corresponds to a human subpopulation and a region in the LPL gene *after* sites were deleted to remove any cells with missing entries. In contrast, the weaker lower bound method from [19] (discussed above) was applied to the whole dataset, including cells with missing entries. In many cases, the lower bounds obtained were higher than the ones shown in table 5.1, and even higher than the number of recombination nodes in an ARG computed for the *cleaned-up* data. For example, table 5.1 shows a lower bound of 13, and reports that there is an ARG with 16 recombination nodes, for cleaned-up data in the Jackson population in LPL region 3. But the Bafna-Bansal method computed a lower bound of 17 for the data when sites with missing entries were not removed. In the N. Karlia population in region 3, the table shows a lower bound of 8, while the lower bound from [19] when missing data was not removed, is 13. More impressive, the combined data over the three populations in region 3 gives a lower bound of 36 when sites with missing data are included in the analysis, while the lower bound after removing those sites is only 25. However, there are also cases where *HapBound*, which runs only on complete data, gives a higher bound on cleaned-up data than did the Bafna-Bansal method on the data with missing values included.

5.2.7.2   Program HapBound: Speedups and Extensions

The basic ideas behind program *HapBound* were introduced in the prior section. However, *HapBound* incorporates additional ideas that make it run faster, and often allows it to compute lower bounds that are higher than the *optimal RecMin bound*. We first discuss the major way to speedup *HapBound*.

Speeding up the computation

In the discussion so far, the computation of the *optimal RecMin bound* requires that we explicitly find $b(I) = H(\widetilde{M}(I, S^*(I)))$ for *every* interval I of M. If M has m sites, this approach executes $\binom{m}{2}$ integer programs. That number grows quadratically in m, rather than exponentially in m, which characterizes the number of subsets that Program *RecMin* needs to examine in order to find, and be sure it has found, the *optimal RecMin bound*. However, in the worst case, the time to solve integer programming problems grows exponentially with increasing problem size. Thus, compared to *RecMin*, *HapBound* does a quadratic number of (worst case) exponential-time computations, instead of a guaranteed exponential number of simple, polynomial-time computations. It was an empirical question whether this substitution would work in practice to compute the *optimal RecMin bound* efficiently. The empirical results are that it does work, but additional speedups are also desired.

The main speedup is the reduction in the number of intervals that *HapBound* needs to explicitly examine, and thus a reduction in the number of required ILP computations. Suppose we find an optimal subset of sites $S^*(I)$ for interval $I = [1, m]$, and the leftmost and rightmost points of $S^*(I)$ are p and q respectively. Then $S^*(I)$ will also be an optimal subset for any interval I' that is strictly contained in $[1, ..., m]$, but strictly contains $[p, ..., q]$. It follows that $b(I') = b(I) = b([p, ..., q])$ and so there is no need to solve an ILP problem for any such I'. Further, since $[p, ..., q]$ is strictly contained in I', the local lower bound $b(I')$ can be ignored in obtaining the overall composite bound. Of course, the interval $[p, ..., q]$ and its local bound $b(I)$ are used in the composite bound computation. With this approach, we can exclude many intervals from consideration, speeding up the total time needed to compute the composite bound.

Excluding the intervals that contain $[p, ..., q]$ is helpful, but that does not (yet) exclude the need to examine all of the subintervals contained in $[1, q-1]$, $[p+1, m]$, $[p+1, q]$, or $[p, ..., q-1]$. But the number of those subintervals that have to be examined can be reduced by recursively applying the same exclusion idea: For each of the four intervals $I = [1, q-1]$, $[p+1, m]$, $[p+1, q]$, and $[p, ..., q-1]$, find an optimal subset $S^*(I)$ for interval I, and then recurse on four new subintervals defined from interval I and the span of $S^*(I)$. In this way, over the entire computation, fewer intervals are explicitly examined, and fewer integer programming computations are needed.

In our simulations, this simple idea greatly reduces the number of ILP problems that need to be solved. When m and n are about the same size, we typically need to solve about 25 percent of the $\binom{m}{2}$ possible problems, and when m is several times larger than n, the percentage typically falls to under 5 percent. As a consequence, on simulated problems of size of current biological interest, *HapBound* runs in seconds to minutes.

5.2.7.3 HapBound Can Often Compute Larger Bounds Than the Optimal RecMin Bound

As described, Program *HapBound* is guaranteed to compute the *optimal RecMin bound*. However, *HapBound* has an option (-S) that typically produces an even higher lower bound. The option increases the running time, but not beyond the range of practicality.

In the method described so far, if $S^*(I)$ is an optimal subset found for interval I, then $b(I)$ is set to $H(S^*) = n - |S^*(I)| - 1$, where M is assumed to have n distinct rows. But an increase in the local bound for I may increase the resulting composite bound. *Hap-Bound* implements a test to determine whether the sequences in $M(S^*(I))$ can actually be generated on an ARG with *only* $H(S^*)$ recombinations. If not, then $b(I)$ can be set to $H(S^*(I)) + 1$. *HapBound* can also test if that bound is tight, or if $b(I)$ should again be increased to $H(S^*(I)) + 2$. These are small increases, but small increases in several local bounds can result in a large increase in the overall composite bound. To explain how *HapBound* determines these local increases, we introduce the concept of *self-derivability*, which will be central to several other topics discussed later in the book.

5.2.7.4   *Self-Derivability and Its Use*

Definition Given a set of sequences M, we say that M is *self-derivable* (SD) if M can be generated on an ARG $\mathcal{N}$ where *all* the node labels in $\mathcal{N}$, including the ancestral sequence, are in M. We then say that $\mathcal{N}$ *self-derives* M.

That is, M is self-derivable if the sequences in M can be generated without generating any Steiner sequences. Clearly, if M is self-derivable, then it can be self-derived on an ARG where all of the node labels are distinct. Note that *not* every set of sequences is self-derivable.

Definition We say that M is *unary-self-derivable* if it is self-derivable on an ARG $\mathcal{N}$ where no edge is labeled with more than one site. We then say that $\mathcal{N}$ *unary-self-derives* M.

Using this terminology, theorem 5.2.5 (page 137) can be restated as:

Theorem 5.2.10 *If a set of sequences M, with no duplicate sites, can be generated on an ARG using exactly $H(M)$ recombination nodes, then M is unary-self-derivable.*

We now state the following result, which addresses the converse direction. However, it is more general than the converse of theorem 5.2.10, because it does not need to assume that M has no duplicate sites. Although we state the theorem here, we defer the proof until section 13.5.1, since the theorem will not be used until then.

Theorem 5.2.11 *If M is unary-self-derived on an ARG $\mathcal{N}$, then $\mathcal{N}$ contains exactly $H(M)$ recombination nodes. Hence $\mathcal{N}$ is a MinARG for M. Moreover, if M is unary-self-derivable, then every MinARG for M must unary-self-derive M.*

However it is not always true that when M is self-derivable (without the *unary* condition) on an ARG $\mathcal{N}$, then $\mathcal{N}$ has exactly $H(M)$ recombination nodes.

Applying theorem 5.2.10 to the set of sequences $M(S^*(I))$ used in *HapBound*, we have the following:

Lemma 5.2.6 *As above, let $S^*(I)$ be an optimal subset of sites found for interval I. If the sequences $M(S^*(I))$ can be generated on an ARG $\mathcal{N}$ using exactly $H(M(S^*(I)))$ recombination nodes, then $M(S^*(I))$ must be unary-self-derivable.*

Proof: Clearly, $S^*(I)$ cannot contain any pair of sites that are identical, for the removal of a duplicate site does not change the number of distinct rows or distinct sites in $M(S^*(I))$, so if $S^*(I)$ had a duplicate site, it would not be an optimal subset for I. Therefore, theorem 5.2.10 applies.   ∎

Using lemma 5.2.6

Lemma 5.2.6 is useful because it says that if we determine that in an interval I, the set of sequences $M(S^*(I))$ is *not* self-derivable, or not unary-self-derivable, then $b(I)$ should be set to $H(M(S^*(I))) + 1$, rather than to $H(M(S^*(I)))$. We next discuss how we can determine whether or not a set of sequences is unary-self-derivable. In section 9.3 we will need to determine whether a set of sequences M is self-derivable, allowing multiple sites per edge, and will discuss the solution to that problem there.

Algorithm DP-USD $(M(S))$

    Let K denote a subset of the n rows in $M(S)$,
    and let $SD(K)$ be a Boolean variable that will be set to **true** if the sequences
    in K are unary-self-derivable, and will be set to **false** otherwise.

    To start, set $SD(K)$ to **true** for every singleton set K, i.e., where $\|K\| = 1$.

    In order of size (breaking ties arbitrarily), consider each subset K:

    if (there is a sequence $s \in K$ such that $SD(K - \{s\}$
    has been set **true**) **then**

        if (s can be created by a recombination of two sequences in $K - \{s\}$) **then**
          set $SD(K)$ to **true**.
        endif

        if (s differs from some sequence in $K - \{s\}$ by a single mutation at a site c)
        and (all of the sequences in $K - \{s\}$ have the same state at site c) **then**
          set $SD(K)$ to **true**.
        endif
    endif

Figure 5.9   Algorithm DP-USD determines whether $M(S)$ is unary-self-derivable.

Algorithm DP-USD: A first algorithm for unary-self-derivability

Algorithm *DP-USD*, shown in figure 5.9, is a dynamic programming algorithm that determines if $M(S)$ is unary-self-derivable. It was developed by Yufeng Wu [438]. It works by examining every subset of rows, K in $M(S)$ to determine if K is unary-self-derivable.

The correctness of algorithm *DP-USD* is based on the observation that if K is unary-self-derivable on an ARG $\mathcal{N}$, then there is at least one sequence $s \in K$ that is not further mutated or used in a recombination. Therefore, the removal of the leaf with label s (and its parent if it is also labeled s) from $\mathcal{N}$, leaves an ARG that self-derives $K - \{s\}$. Further, s must be a recombinant sequence created by the recombination of two sequences in $K - \{s\}$; or s must differ from some sequence in $K - \{s\}$ by a single mutation at a site c, where all of the sequences in $K - \{s\}$ have the same state at site c. The latter condition ensures that no mutation at site c has occurred in the creation of $K - \{s\}$. A full proof of the correctness and time analysis is left to the reader.

A crude analysis establishes that the worst-case time for algorithm *DP-USD* is $O(n^2 m^2 2^n)$, although the polynomial terms can be reduced with the use of appropriate preprocessing and data structures [218]. Therefore, this test for unary-self-derivability runs in time that is polynomial in m, but exponential in n.

Algorithm USD: A second algorithm for unary-self-derivability

Reversing the roles of n and m in the time bound, we can also test whether $M(S)$ is unary-self-derivable with an algorithm that runs in time that is polynomial in n, but exponential in m. It is observed to be practical for large datasets. To describe the algorithm, we first consider a simplified situation.

When only recombinations are permitted, the test for self-derivability of a set of sequences $M(S)$ from a fixed *pair* of ancestral sequences has an efficient solution [218]: To start, a "reached set" of sequences consists of the two ancestral sequences alone. Then at each step, we try to expand the reached set by finding a pair of sequences (f, g) in the reached set that can recombine to create a sequence h that is in $M(S)$ but not yet in the reached set. This is repeated until either the reached set contains all the sequences in $M(S)$, or until no further expansion of the reached set is possible. In the first case, $M(S)$ is generated by recombinations, starting from an ancestral pair, and only using sequences in $M(S)$. In the second case, it is not hard to prove that $M(S)$ cannot be generated from the chosen ancestral pair, only using recombinations and only generating sequences in $M(S)$. This algorithm can be sped up by preprocessing [218] using a suffix-tree [137], but clearly only takes polynomial time even without advanced data structures.

The preceding algorithm, for the simplified situation where there are two ancestral sequences and no mutations are allowed, is not a test for whether $M(S)$ is self-derivable, or unary-self-derivable. To become such a test, it must be modified so that there is only *one* ancestral sequence, rather than two, and so that mutations are allowed. The key insight for this modification is the following:

If $M(S)$ is unary-self-derivable, then for every site $i \in S$, during any self-derivation of $M(S)$, a mutation at site i must occur in some sequence that is definitely in $M(S)$ and the mutation must generate another sequence that is definitely in $M(S)$.

Therefore those two sequences differ by *exactly* one site, namely site i. (It also follows that if there is a site i such that no pair of sequences in $M(S)$ differ only at i, then $M(S)$ is not unary-self-derivable.)

To exploit this key insight, let MUT_i be the set of sequence pairs that differ at exactly site i. We modify the algorithm for the simplified situation by now starting the reached set with just a single ancestral sequence. We also allow the reached set to expand in two different ways. As before, the reached set can expand by a recombination of two sequences in the reached set if the recombination creates a sequence in $M(S)$ that is not yet in the reached set. But the reached set can also be expanded to include a sequence h in $M(S)$, if some sequence f is already in the reached set, and (f, h) is in MUT_i for some i, where no prior expansion of the reached set used a pair in MUT_i. Clearly, if $M(S)$ is unary-self-derivable, then starting from the chosen ancestral sequence, there is a generation of $M(S)$ by this algorithm, where for every site i in S, exactly one pair in MUT_i is used to expand the reached set.

Algorithm *USD*, the unary-self-derivability test, tries all sequences in $M(S)$ as the ancestral sequence, and all ways to choose exactly one pair from each set MUT_i. The number of choices is $n \times \prod_{i \in S} |MUT_i|$ which is bounded by $n \times [\frac{n}{2}]^m$, but is generally much smaller. Further, some combinations of choices can be immediately ruled out and additional effective heuristics reduce the number of choices (details left to the reader). In summary, we have:

Theorem 5.2.12 *Let $M(S)$ be a set of sequences with n taxa and m sites. We can test whether $M(S)$ is unary-self-derivable in time that is polynomial in n and exponential in m, or that is polynomial in m and exponential in n. Further, if $M(S)$ is unary-self-derivable, then the two algorithms produce an ARG that unary-self-derives $M(S)$.*

Further increases

The unary-self-derivability test is used to determine if the set of reduced sequences $M(S^*(I))$ can be generated on an ARG that has only $H(M(S^*(I))$ recombination nodes. We claim that if not, then there cannot be any ARG $\mathcal{N}$ that generates the unreduced sequences $M(I)$, using only $H(M(S^*(I))$ recombination nodes. To see this, suppose ARG $\mathcal{N}$ generates $M(I)$ using only $H(M(S^*(I))$ recombinations. We could then remove sites from $M(I)$, and modify $\mathcal{N}$ to generate $M(S^*(I))$, without increasing the number of recombinations. That would create an ARG for $M(S^*(I))$ using $H(M(S^*(I))$ recombination nodes, and by theorem 5.2.10, the sequences $M(S^*(I))$ would then be unary-self-derivable, a contradiction. Therefore, if $M(S^*(I))$ is not unary-self-derivable, then the local bound $b(I)$ should be set to $H(M(S^*(I)) + 1$ rather than to $H(M(S^*(I)))$.

To test if a local bound should be increased by *two*, we note that it should be if $M(S)$ is not self-derivable, and for no sequence $s \notin M(S)$ is $M(S) \cup \{s\}$ self-derivable. So, we could generate each candidate sequence, and test if $M(S) \cup \{s\}$ is self-derivable. But how many candidate sequences are there, and how can they be efficiently generated?

If one additional sequence s did allow the expanded set to be self-derivable, then sequence s must either be generated by the recombination of two sequences in $M(S)$, or must differ from a sequence in $M(S)$ at *exactly* one site in S. There are only a polynomial number of such candidate sequences, and we can efficiently generate each new candidate sequence in turn and test the resulting set for self-derivability. If the sequences are not self-derivable in any of these tests, then the local bound should be increased by two. We can continue in this way to determine if the local bound should be increased by three, or even more, but the time needed increases too rapidly for practical implementation.

Program *HapBound*, with option -S, tests each minimum $S^*(I)$ subset it finds, to see if $M(S^*(I))$ is self-derivable, and if not, to see if the local bound should be increased by one or by two. For the datasets that we have examined, the extra computation time for the -S option does not reduce the practicality of the algorithm, and frequently results in a lower bound on $Rmin(M)$ that is higher than the *optimal RecMin bound* (some comprehensive test results are shown in the next section). For example, for M shown as the leaf sequences in figure 7.6 (page 207) the *optimal RecMin bound* is two, but *HapBound* -S returns the value of three.

Program *HapBound* has another option (-M), that is also based on the self-derivability test but is more time consuming, and it typically increases the lower bound by only a small amount. Still, for small datasets, Program *HapBound* with option -M has produced lower bounds that were higher than any other lower bound program we have tested. An example will be shown in the next section. See [390] for ideas behind the -M option.

5.2.8 Lower Bounds for the LPL and ADH Datasets

As an illustration of the efficiency and efficacy of program *HapBound*, we discuss here the SNP data from [74]. This seminal data is from the LPL locus in humans and contains 88 rows and (coincidentally) 88 sites before removing sites with missing or non-SNP data. That dataset was also examined in [302]. (The first paper uses 42 sites from the full data in their recombination analysis, while the second paper uses 48 sites. For clearer comparison, we discuss results that included the same 48 sites.)

Using CPLEX to solve the ILP problems, *HapBound* computed the *optimal RecMin bound* of 75 in 31 seconds, and *HapBound* -S computed a higher bound of 78 in 1,643 seconds, on a 2 GHz machine. Using the GNU ILP solver on the same machine, the times were 871 and 3,326 seconds respectively. Program *RecMin* with the default settings of $s = 8$ and $w = 12$ produced the lower bound of 59 in 3 seconds. It found the *optimal RecMin bound* of 75 with parameters $s = w = 25$, in 7,944 seconds. As mentioned earlier, a user would not know that this was the *optimal RecMin bound*. To simulate what the user would need to do in order to be sure of getting the *optimal RecMin bound*, we set $s = w = 48$ but *RecMin* did not finish within five days of execution. The analysis in [302], based on *RecMin*, reports a lower bound on $Rmin(M)$ for this data of only 70, rather than 75. This is due to running *RecMin* with parameters that are too low [298]. This illustrates a central point of this section, that with *RecMin* one does not know which parameter settings are high enough, and illustrates the utility of program HapBound. For comparison, the *HK bound* is only 22, illustrating the major advance that *RecMin* made, and the importance of using it or *HapBound* in place of the *HK bound*.[2]

As mentioned above, *HapBound* has another option, the -M option, that runs efficiently on moderate size datasets and produces the highest lower bounds on those data. For example, the benchmark dataset [230] for the ADH locus in humans (shown in figure 5.7) has 11 sequences and 43 sites. Song and Hein [389] established that $Rmin(M)$ is exactly 7 for this data. The lower bound method in [389] was used to establish a matching lower bound of 7. HapBound -M ran in about three seconds on this data and also computed the lower bound of 7. The *History* lower bound developed in [18, 19], which we will discuss in section 10.1.5, also produced a lower bound of 7. All other implemented lower bound methods that we know of, nine in total, produce lower bounds of only 5 or 6.

---

2 Additional comparisons of the *HK bound* to the lower bound produced by *RecMin* (with the default parameter settings) [19] further demonstrate the weakness of the *HK bound*, particularly as the recombination rate increases. For example, averaged over 100,000 datasets with a high level of recombination, the mean *RecMin* lower bound is 36.80, while the mean *HK bound* is 12.07. The Bafna-Bansal lower bound [19] (discussed on page 151) which can never be larger than the *optimal RecMin bound*, gave mean value of 49.69 on the same 100,000 datasets. We don't know what the average *optimal RecMin bound* would be on that data, but it would be at least as high as 49.69. So, on these data, we see extensive examples where $Rmin(M)$ is three to four times as large as the *HK bound* for M.

Population	HapBound -S -M			The optimal RecMin bound		
	reg 1	reg 2	reg 3	reg 1	reg 2	reg 3
Jackson	11 (13)	10 (10)	13 (16)	10	9	12
N. Karelia	2 (2)	15 (17)	8 (10)	2	13	7
Rochester	1 (1)	14 (14)	8 (8)	1	12	7
All	13 (14)	21 (23)	25 (31)	12	21	22

Table 5.1 Lower and upper bounds for the LPL data, where the sites are partitioned into three regions. The numbers in parentheses are "upper bounds" on $Rmin$, i.e., the minimum number of recombination nodes used in ARGs (not necessarily MinARGs) constructed for the data. The lower bounds on the left-hand side were computed using *HapBound* -S -M.

5.2.8.1   The Human LPL Data in More Detail

Using human LPL data from [311], we computed lower bounds on $Rmin(M)$, using *Hap-Bound*, and compared those numbers to the numbers of recombination nodes used in ARGs for that data. The ARGs were constructed by methods that we will discuss in chapter 9. Unless the lower bound exactly matches the number of recombination nodes in the corresponding ARG, we do not know $Rmin(M)$ exactly, and so we refer to the number of recombination nodes used in an ARG as an "upper bound" on $Rmin(M)$.

The sequences examined were sampled from three populations—namely, Jackson, North Karelia, and Rochester. In the analysis, sites with missing or non-SNP data was removed. This is the treatment of the data that was used in [311]. Following Myers and Griffiths, sites of the LPL data were partitioned into three regions (see table 5 of [302]).

The left-hand side of table 5.1 gives a summary of the lower bounds produced by *HapBound* -S -M, and of the upper bounds on $Rmin(M)$, i.e., the minimum number of recombination nodes used in ARGs (not necessarily MinARGs) constructed for the data. The three populations are considered separately as well as together. HapBound -S and HapBound -S -M produced similar lower bounds. The only difference was in region 2 of the Jackson population; HapBound -S produced 9, whereas HapBound -S -M produced 10. The lower and upper bounds are generally quite close. In particular, they exactly match in each of the three regions for the Rochester population, and hence we know $Rmin(M)$ exactly for those regions. Moreover, the lower bounds computed by *HapBound* were generally higher than the *optimal RecMin bound*. *Optimal RecMin bounds* are shown on the right-hand side of table 5.1 for comparison.[3]

---

3 This table differs from table 5 of [302], because insertion/deletion sites were not deleted for the analysis in that paper.

5.2.9   Advanced Material: The Rank Bound

In this section we discuss an efficiently computed lower bound on $Rmin_0(M)$, called the *rank bound*, that is provably as good as the *haplotype bound* and is sometimes better. The *rank bound* can be used in place of the *haplotype bound* in any application where the *haplotype bound* is used. It follows that if the *rank bound* is used in place of the *haplotype bound* in the *interval RecMin bound*, and the resulting composite bound may increase. However, it has been claimed [236] that if we use the *rank bound* on every *subset* of sites, in place of the *haplotype bound*, as in the *optimal RecMin bound*, we will obtain exactly the same composite lower bound. Therefore, the *rank bound* would not be helpful in that context. The *rank bound* was derived by Dan Brown and Ian Harrower [49].

Recall that $Rmin_0(M)$ is the minimum number of recombination nodes used in any ARG that generates M, using the all-zero ancestral sequence. Recall from linear algebra that a subset of rows $\vec{R}$ in an n by m matrix M is *linearly dependent* if the m-length zero vector can be expressed as a linear combination of the rows of $\vec{R}$, where at least one of the coefficients in the expression is not zero. A subset that is not linearly dependent is called *linearly independent*. Recall that the *rank* of a matrix M, denoted rank(M), is the size of the largest linearly independent subset of rows of M. The rank(M) can be computed in time that is polynomial in the size of M, by well-known methods in matrix algebra. Linear dependence and independence can also be defined in terms of the *columns* of M, and **rank**(M) is also equal to the size of the largest linearly independent subset of columns of M. Recall that $D_r(M)$ and $D_c(M)$ denote the number of distinct rows and columns in M, respectively.

Lemma 5.2.7 *If there is a perfect phylogeny T (with all-zero ancestral sequence) for a binary matrix M, then* rank$(M) \geq D_r(M) - 1$.

Proof: We prove this by induction on $D_r(M)$. If $D_r(M)$ is one, then rank(M) is either 0 (if the single distinct row of M is the all-zero row) or 1, so the basis of the induction holds. Suppose the lemma holds for all matrices with $D_r(M) \leq t$, and consider a matrix M with $D_r(M) = t + 1$. Let T be a perfect phylogeny for M, and let e be an edge labeled with site c, such that no edges below e are labeled with any site. It follows that character c is possessed by only one taxon, say f, and copies of f.

If we remove all copies of f from M, we obtain a matrix M' with t distinct rows, and if we remove edge e and the subtree below e from T, we obtain a perfect phylogeny for M'. So rank$(M') \geq t - 1$, by the induction hypothesis. Let $\vec{R}$ be a set of $t - 1$ linearly independent rows of M'. We claim that the set $R \cup \{f\}$ is linearly independent. To prove this, consider a linear combination of the rows in $R \cup \{f\}$ that expresses the

m-length zero vector. None of the rows in $\vec{R}$ has a one in column c, but f does, so the linear combination of $R \cup \{f\}$ that expresses the zero vector must have coefficient 0 for f. It follows that the remaining linear combination of $\vec{R}$ must also express the zero vector. But, by the assumption that $\vec{R}$ is linearly independent, all the coefficients used for rows in $\vec{R}$ must be zero, proving that all the coefficients in the linear combination of $R + f$ must be zero. Hence, the set of rows $R \cup \{f\}$ is linearly independent. Therefore, $\text{rank}(M) \geq |R| + 1 = t - 1 + 1 = t = D_r(M) - 1$, and the induction is proved. ∎

Notice that in the case that there is a perfect phylogeny (with all-zero ancestral sequence) for M, $Rmin_0(M) = 0$ and the inequality in lemma 5.2.7 can be stated as $\text{rank}(M) \geq D_r(M) - Rmin_0(M) - 1$. Equivalently, $Rmin_0(M) \geq D_r(M) - \text{rank}(M) - 1 \geq H(M)$. This latter inequality makes explicit the application of $\text{rank}(M)$ to a lower bound on $Rmin_0(M)$. We next prove that this inequality holds for all M, even when there is no perfect phylogeny for M, that is, when $Rmin_0(M) > 0$.

Theorem 5.2.13 *For all M, $Rmin_0(M) \geq D_r(M) - \text{rank}(M) - 1 \geq H(M)$.*

Proof: First, $\text{rank}(M) \leq D_c(M)$, since any subset of $D_c(M)+1$ columns must have two identical columns, in which case that subset must be linearly dependent (use a coefficient $C \neq 0$ for one of the identical columns, and $-C$ for another copy, and 0 for all other columns). Therefore $D_r(M) - \text{rank}(M) - 1 \geq D_r(M) - D_c(M) - 1 = H(M)$, proving the second inequality of the theorem.

We next establish the first inequality of the theorem by induction on $Rmin_0(M)$. The base case of $Rmin_0(M) = 0$ (when M has a perfect phylogeny), follows from lemma 5.2.7. So, assume the theorem holds for all matrices where $Rmin_0(M) \leq t$, and consider a matrix M and an ARG $\mathcal{N}$ for M, with ancestral sequence zero, containing $Rmin_0(M) = t+1$ recombination nodes. We want to prove that $Rmin_0(M) \geq D_r(M) - \text{rank}(M) - 1$.

Let x be a recombination node in $\mathcal{N}$ such that no recombination node is reachable from x. Then, the subnetwork of $\mathcal{N}$ that is reachable from x is a tree, T, rooted at x. Note that tree T includes node x, and that if x is labeled by a taxon f in M, then there is an edge from x that goes directly to a leaf labeled f. Let $L(T)$ be the submatrix of M defined by the subset of rows of M that label the leaves of T, and let $C(T)$ be the columns of M that label edges in T. $C(T)$ is the set of sites of M that change state in T. Clearly, the state at node x of each site in $C(T)$ is zero. Finally, let $M(T)$ be the submatrix of $L(T)$ restricted to the columns of $C(T)$. It follows that T defines a perfect phylogeny for $M(T)$, and the ancestral sequence of that perfect phylogeny is the zero vector of length $|C(T)|$. Hence, by lemma 5.2.7, $\text{rank}(M(T)) \geq D_r(M(T)) - 1$. Note that the number of distinct rows in $M(T)$ is exactly the number of distinct rows in $L(T)$, so $\text{rank}(M(T)) \geq D_r(L(T)) - 1$. Let $R(T)$ be a set of $|D_r(M(T))| - 1$ linearly independent rows in $M(T)$. Clearly, the rows of M corresponding to $R(T)$ must also be linearly independent. We will show how to add rows to $R(T)$ to create a larger set of linearly independent rows of M.

Let $\vec{R}$ denote a largest set of linearly independent rows of $M - L(T)$. We claim that the set of rows $R(T) \cup R$ is linearly independent. To see this, note that the rows of $\vec{R}$ are generated in the network formed by removing T from $\mathcal{N}$, and so none of the rows in $\vec{R}$ have a one in any column in $C(T)$. Therefore, any linear combination of the rows in $R(T) \cup R$ that expresses the zero vector (and in particular, produces a zero in positions in $C(T)$), must use coefficient 0 for every row in $R(T)$. But then that linear combination, restricted to the rows of $\vec{R}$ must also produce zero, and since the rows of $\vec{R}$ are linearly independent, *all* the coefficients of the linear combination must be zero for those rows. It follows that the rows of $R(T) \cup R$ are linearly independent. Therefore,

$$
\begin{aligned}
\texttt{rank}(M) \;\geq\; & |R| + |R(T)| = \texttt{rank}(M - L(T)) + \texttt{rank}(M(T)) \\
=\; & D_r(M - L(T)) - Rmin_0(M - L(T)) - 1 + \texttt{rank}(M(T)) \\
\geq\; & D_r(M - L(T)) - Rmin_0(M - L(T)) - 1 + D_r(L(T)) - 1,
\end{aligned}
$$

by the induction hypothesis, lemma 5.2.7, and the fact that $D_r(L(T)) = D_r(M(T))$, respectively. But $Rmin_0(M - L(T)) \leq t$, since $M - L(T)$ can be generated by the sub-network of $\mathcal{N}$ that remains after the removal of tree T, and T contains the recombination node x. So,

$$
\begin{aligned}
\texttt{rank}(M) \;\geq\; & D_r(M - L(T)) - t - 1 + D_r(L(T)) - 1 \\
=\; & D_r(M - L(T)) + D_r(L(T)) - (t + 1) - 1 \\
=\; & D_r(M) - (t + 1) - 1 = D_r(M) - Rmin_0(M) - 1,
\end{aligned}
$$

which is equivalent to $Rmin_0(M) \geq D_r(M) - \texttt{rank}(M) - 1$, and the induction step is proved.   ∎

It is easy to find data M where $D_r(M) - \texttt{rank}(M) - 1 > H(M)$, meaning that the replacement of $D_c(M)$ by $\texttt{rank}(M)$ *strictly* raises the lower bound on $Rmin_0(M)$. As one reflection of this, note that $D_r(M) - \texttt{rank}(M) - 1 \geq -1$ no matter what M is, while $H(M)$ can be made arbitrarily negative.

5.2.10   Advanced Material: Lower Bounds with Gene Conversion

In this section we move from a focus on $Rmin(M)$, which is defined for single-crossover recombination, to incorporate *gene-conversion* events, which can be viewed as a form of two-crossover recombination. The material in this section is adapted from [387].

Definition We define $Tmin(M)$ as the minimum total number of recombination events needed to derive M from an ancestral sequence (either known or unknown in different versions of the problem) where a recombination event is *either* a single-crossover recombination *or* a gene-conversion event.

Recall that in gene-conversion, both a prefix and a suffix of the new recombinant sequence come from one of the parental sequences, s_1, while a midpart of the recombinant sequence, called the *conversion tract* or "tract" for short, comes from the other parental sequence, s_2. Because gene-conversion tract length is typically small (between 50 and 500 nucleotides), we will often bound its permitted length.

Definition We define $Tmin(M, t)$ as the minimum number of recombination events needed to derive M, where each gene-conversion has tract length at most t (in some unit of distance, for example nucleotides). Two adjacent sites in M might be *physically* separated by many positions that are not included as sites in M.

We are interested in methods to compute *lower bounds* on $Tmin(M, t)$. Since $Tmin(M) = Tmin(M, t)$, when t is sufficiently large (for example, the physical distance between the first and the last sites in M), these methods can be used to compute bounds on $Tmin(M)$. So, the problem of computing a lower bound on $Tmin(M)$ can be considered as a special case of the problem of computing bounds on $Tmin(M, t)$, for arbitrary t.

An obvious lower bound

Since the effect of a single gene-conversion can be modeled as two single-crossover events, $LBCO(M)/2$ is a valid lower bound on $Tmin(M, t)$, where $LBCO(M)$ is any lower bound on $Rmin(M)$, for example the lower bounds already discussed in this chapter. We will prove that when t is unconstrained, the most natural extensions of these methods yield only weak lower bounds on $Tmin(M)$. However, we will introduce additional ideas that increase these lower bounds, and show how to obtain improved lower bounds when t is bounded. The methods to compute lower bounds on $Tmin(M, t)$ are again based on the general composite bound method discussed in section 5.2.4: Compute local lower bounds on $Tmin(M(I), t)$ for each interval I, and then use those local bounds to obtain a *global composite* lower bound on $Tmin(M, t)$.

Composite bound: Problems and solutions

The *haplotype* and *rank* bounds (applied to all of M) extend immediately to the case that gene-conversions are allowed, but the *HK bound* does not. Further, the extension of the composite method to the case of gene-conversion is not immediate. Even when *haplotype* and *rank bounds* are used for local lower bounds, the composite method for combining them into a global bound on $Tmin(M, t)$ may be more involved. We next address this issue. We will concentrate on the use of the haplotype lower bound, but other specific lower bound methods could be treated in a similar way.

Recall that in the argument showing that $H(M(I))$ is a lower bound on $Rmin(M(I))$, the key points were that each recombination event, and each mutation, can create at most one new sequence. These two properties also hold when gene-conversion is allowed, and therefore $H(M(I))$ is also a lower bound on $Tmin(M(I), t)$ for any t. Similarly, $H(M)$ is a lower bound on $Tmin(M)$. This is an improvement over the bound of $H(M)/2$, obtained from the obvious bound when $LBCO(M)$ is $H(M)$. However, we saw earlier that $H(M)$ was not a good lower bound on $Rmin(M)$ (when the *haplotype bound* is applied to all of M). Similarly, although $H(M)$ is a valid lower bound on $Tmin(M)$, it is not generally a good bound. But we also saw that the *haplotype bound* can be used as a local lower bound in each interval I, and that these local bounds can be combined by the composite bound method to create a much larger lower bound on $Rmin(M)$.

Following the composite bound approach, we want to combine local lower bounds on $Tmin(M(I), t)$ for each interval I, into a global lower bound on $Tmin(M, t)$. But the composite bound method must be modified because gene-conversion (although involving two breakpoints) is biologically a *single* event.

The composite global lower bound for $Tmin(M, t)$

It is useful to somewhat recast and restate the problem of computing a composite lower bound on $Rmin(M)$ from local lower bounds.

Definition We say a point p "covers" an interval I, containing several sites of M, if p is contained in I with at least one site of M on each side of p. This leads to

The crossover coverage problem:

Given M, and a set of intervals $\mathcal{L}$ in M, find the smallest set of points $R^*(\mathcal{L})$, so that each interval $I \in \mathcal{L}$ is covered by at least $H(M(I))$ points of $R^*(\mathcal{L})$.

Clearly, we obtain a composite global bound on $Rmin(M)$ by solving the crossover coverage problem.

$|R^*(\mathcal{L})|$ is a valid lower bound on $Rmin(M)$, but it is *not* necessarily a valid lower bound on $Tmin(M, t)$. The problem is that it may be *too high*. The reason is that each gene-conversion can be viewed as two crossovers. If the two crossovers occur in different intervals, I and I' in $\mathcal{L}$, then the single gene conversion can "contribute to" or "count as" a single crossover in *both* I and I' in a solution to the crossover coverage problem. So, fewer than $|R^*(\mathcal{L})|$ recombination events suffice to cover the intervals. Hence, at this point we are only allowed to conclude that $|R^*(\mathcal{L})|/2$ is a valid lower bound on $Tmin(M,t)$.

To use the local *haplotype bounds* to obtain a better composite bound on $Tmin(M,t)$, we must formulate a generalization of the crossover coverage problem.

Definition We say a line segment *covers* an interval I if at least one end p of the line segment covers I.

Note that a line segment that strictly contains I does not cover I, and that a line segment covers I only *once*, even if both of its endpoints are in I. We obtain a composite lower bound on $Tmin(M,t)$ by solving:

> **The gene-conversion coverage problem**: Given a set of intervals $\mathcal{L}$, find the smallest set consisting of points $R^*(\mathcal{L})$ and line segments $S^*(\mathcal{L},t)$ with physical length at most t, so that each interval $I \in \mathcal{L}$ is covered by at least $H(M(I))$ elements of $R^*(\mathcal{L}) \cup S^*(\mathcal{L},t)$.

Each point in $R^*(\mathcal{L})$ represents a single-crossover and each line segment in $S^*(\mathcal{L},t)$ represents a gene-conversion. The intuitive meaning is that a line segment covers an interval I only if the action of the gene-conversion it represents *could* create a new sequence in $M(I)$. The reason that a gene-conversion covers I only once even if both endpoints are in I is that a gene-conversion can create only *one* new sequence in $M(I)$, even if both crossover indexes are in I. It then follows that:

Theorem 5.2.14 *If $R^*(\mathcal{L})$ and $S^*(\mathcal{L},t)$ solve the gene-conversion coverage problem, then $|R^*(\mathcal{L}) \cup S^*(\mathcal{L},t)|$ is a valid lower bound on $Tmin(M,t)$.*

The gene-conversion coverage problem would also lead to a valid lower bound if we used the *HK*, or the *connected-component* local bounds for the local lower bounds. However, the correctness of the approach needs somewhat different arguments when *HK* or *connected-component* local lower bounds are used. There is no general argument that shows that any local bounds for single-crossover recombination can be used together with the gene-conversion coverage problem to obtain a global lower bound on $Tmin(M,t)$ or $Tmin(M)$.

5.2.10.1   Solving the Gene-Conversion Coverage Problem: The Special Case of Unbounded Tract Lengths

In the previous section we established the framework for applying the composite bound method to gene-conversion, via the gene-conversion coverage problem. However, we have not shown how to solve the gene-conversion coverage problem; nor have we considered how large the solution can be. Here, we first show that when the tract length t is *unbounded*, the gene-conversion coverage problem has a simple, yet disappointing solution. After that, we will see how to boost the lower bound for $Tmin(M, t)$, when t is of bounded length.

For the following discussion, let $R^*(M)$ be a minimum-sized set of crossovers that solves the crossover coverage problem for M, that is, the problem when only single-crossover recombination events are allowed; and let z denote $|R^*(M)|$. Number the crossover indexes in $R^*(\mathcal{L})$ left to right, choosing an arbitrary ordering of crossover indexes that lie on the same point.

Definition For any $k \leq \lfloor \frac{z}{2} \rfloor$, let $P(R^*(\mathcal{L}), k)$ be a pairing of the leftmost k crossover indexes to the rightmost k crossover indexes of $R^*(\mathcal{L})$ under the mapping $i \to z - k + i$, and create a line segment between the two endpoints of each pair in $P(R^*(\mathcal{L}), k)$. Let S be the set of these k line segments, and let P be the set of $z - 2k$ unpaired points in $R^*(\mathcal{L})$. We will show that there is a solution of the gene-conversion coverage problem that has this form for some k, and show that the best k can be easily obtained.

Note that in the pairing $P(R^*(\mathcal{L}), k)$, if $i < j \leq k$, then i maps to a crossover index to the left of the crossover index that j maps to, a property that we call *monotonicity*.

Definition Define $L(I)$ as the number of crossover indexes in $R^*(\mathcal{L})$ to the left of interval I, and define $R(I)$ as the number of crossover indexes in $R^*(\mathcal{L})$ to the right of I. We say a line segment is "contained in" I if both of its ends are contained in I. Define the "coverage" of interval I as the *number* of elements of $P \cup S$ that cover I.

Lemma 5.2.8 *Let I be any interval where some line segment in S is contained in I. Then exactly $k - (L(I) + R(I))$ line segments in S are contained in I.*

Proof: First, if a line segment (a, b) in S is contained in I, then $k \geq L(I) + 1$, so crossover index $L(I) + 1$ (the leftmost crossover index in I) must be the left end of some line segment in S. Moreover, by monotonicity, the right endpoint of that segment (which is at crossover index $z - k + L(I) + 1$) must be at or to the left of b, and hence in I. Since the pairing $P(R^*(\mathcal{L}), k)$ involves the rightmost k crossover indexes in $R^*(\mathcal{L})$, all the crossover indexes in I to the right of $z - k + L(I) + 1$ must be right endpoints of some line segment in S; again by monotonicity, their paired left endpoints must be to the right of $L(I) + 1$, and hence must be in I. The rightmost crossover index in I is $z - R(I)$, so there are exactly $z - R(I) - [z - k + L(I) + 1] + 1 = k - (L(I) + R(I))$ line segments in S that are contained in I. ∎

Lemma 5.2.9 *For $k \leq z - \max_I(H(M(I)))$, the coverage of any interval I is at least $H(M(I)))$.*

Proof: Let $R^*(\mathcal{L})(I)$ be the number of crossover indexes in $R^*(\mathcal{L})$ that are contained in I. The coverage of I is exactly $R^*(\mathcal{L})(I)$ minus the number of line segments in S contained in I. Since, $R^*(\mathcal{L})(I) \geq H(M(I)))$ for all I, we only need to examine intervals where some line segment in S is contained in the interval. Let I be such an interval. By assumption, $k \leq z - \max_I H(M(I))) \leq z - H(M((I)))$, so

$$k - (L(I) + R(I)) \leq z - H(M((I))) - (L(I) + R(I)) = R^*(\mathcal{L})(I) - H(M((I))).$$

Therefore, $H(M((I))) \leq R^*(\mathcal{L})(I) - [k - (L(I) + R(I))]$, and by lemma 5.2.8, the coverage of I is at least $H(M(I)))$. ∎

Corollary 5.2.2 *If $k = \min(\lfloor \frac{z}{2} \rfloor, z - \max_I H(M(I)))$, then the coverage of I is at least $H(M(I))$, for each interval I.*

Theorem 5.2.15 *If $R^*(\mathcal{L})$ is a minimum sized set of crossover indexes (of size z) solving the crossover coverage problem, then the optimal solution to the gene-conversion coverage problem has size exactly $\max(\lceil \frac{z}{2} \rceil, \max_I H(M(I)))$.*

Proof: By corollary 5.2.2, if we set k to $\min(\lfloor \frac{z}{2} \rfloor, z - \max_I H(M(I)))$, then every interval I has coverage at least $H(M(I))$, and $|S \cup P|$ is exactly $z - k = \max(\lceil \frac{z}{2} \rceil, \max_I H(M(I)))$. But both of those terms are trivial lower bounds on the number of needed line segments and single crossover indexes in any solution to the gene-conversion coverage problem, and hence that choice of k gives the optimal solution. ∎

So when t is unbounded, we have a simple, efficient algorithm for the gene-conversion coverage problem:

> Solve the crossover coverage problem, yielding set $R^*(\mathcal{L})$, and then apply theorem 5.2.15.

Note that theorem 5.2.15 holds regardless of which (optimal) solution $R^*(\mathcal{L})$ is used, and provides a lower bound on $Tmin(M, t)$ for any t, as well as for $Tmin(M)$. It can also be shown that theorem 5.2.15 holds even if we use *HK* or *connected-component* local lower bounds, instead of *haplotype* local lower bounds.

5.2.10.2   Improving the Bounds

Theorem 5.2.15 proves that the natural extension of the way that good lower bounds on $Rmin(M)$ were obtained, will yield only *trivial* lower bounds when gene-conversion tract length is *unbounded*. To get higher bounds we have to use additional constraints. The first such constraint is to bound the permitted tract length to t in any solution to the gene-conversion coverage problem. Bounding the tract length is biologically valid, and was first ignored with the hope that good lower bounds would still be obtained. We do not have a polynomial-time algorithm for the gene-conversion coverage problem when tract length is bounded, but we next show how to effectively solve it using integer linear programming (ILP).

5.2.10.3   An ILP Formulation for Bounded t

We define $\phi(i)$ as the physical position in the chromosome of site i. Given an input matrix M with m sites, and a bound t, we define an integer-valued, linear programming variable $K_{i,j}$ for each pair of integers i, j where $0 < i \leq m - 1$, $0 < j \leq m - 1$, $i \leq j$, and either $i = j$ or $\phi(j) - \phi(i+1) < t$. The value that variable $K_{i,j}$ takes on in the ILP solution specifies the number of line segments $[i, j]$ (whose two endpoints are between sites $i, i+1$ and sites $j, j + 1$) that will be used in the solution. For an interval $I = [a, b]$, we define the set $A(I) = \{K_{i,j} : a \leq i < b \ or \ a \leq j < b\}$. Set $A(I)$ is the set of the variables that can specify a line segment that covers I. We allow $i = j$ to indicate a single point. Then the following integer programming formulation solves the gene-conversion coverage problem when t is bounded:

Minimize $\sum_{(i,j)} K_{i,j}$

Subject to $\sum_{K_{i,j} \in A(I)} K_{i,j} \geq H(M(I))$, for each interval I.

Thus, the composite bound on $Tmin(M, t)$ is found by solving an ILP problem that models the gene-conversion coverage problem with bounded t. See [387] for details on the empirical practicality of this approach.

Further improvements

We can often increase the composite global lower bound on $Tmin(M, t)$ by considering *three* sites at a time, similar to the way that three sites helped in linkage mapping, discussed in section 4.2. For example, consider the following four sequences: $000, 011, 110, 101$. All three sites are pairwise incompatible. Let c_1, c_2, c_3 denote the first, second, and third sites, respectively. Intervals $(c_1, c_2]$, $(c_2, c_3]$, and $(c_1, c_3]$ all have a local lower bound of 1. We can cover those intervals using a *single* line segment ℓ with one endpoint between c_1 and

c_2, and the other endpoint between c_2 and c_3. That single segment ℓ covers all three of the intervals, and hence only contributes one to the solution of the gene-conversion coverage problem.

However, the gene-conversion coverage problem is intended to reflect gene-conversion and single-crossover recombination events that could actually create the observed incompatibilities, and the gene-conversion corresponding to line segment ℓ would not do that. To see this, suppose there is a single gene-conversion involving sequences s and s', where s' contributes the conversion-tract, which lies inside the open interval $(c_1, ..., c_3)$. Then the recombinant sequence created will be identical to s at site c_1 and at site c_3. Therefore, the gene-conversion event will not create an additional (c_1, c_2) state-pair. Hence, a history with exactly one gene-conversion inside $(c_1, ..., c_2)$ could not be sufficient to generate the observed data.[4] Thus, there must be at least *two* gene-conversion or single-crossover events for this example. Moreover, the gene-conversion tract must have one end in $(c_1, ..., c_2)$, and the other end outside of that interval. These additional constraints can increase the resulting global lower bound, and can be incorporated into the ILP formulation by adding the following inequalities:

$$\sum_{c \leq a < d,\ b \geq d} K_{a,b} + \sum_{a < c,\ c \leq b < d} K_{a,b} + \sum_{c \leq a < d} K_{a,a} \geq 1$$

for each pair of incompatible sites c, d. Note that $K_{a,a}$ defines a single-crossover event rather than a gene-conversion.

HapBound-GC

The preceding ILP formulation can be solved reasonably efficiently for tested data. See [387] for specific empirical results. This approach has been implemented in the program called *HapBound-GC*. A demonstration version of *HapBound-GC* uses the free GNU GLPK package to solve the ILP [387].

A further boost Another way to raise the composite lower bound involves the interaction of local bounds and global bounds that use those local bounds.

The ILP formulation imposes constraints based on a set of intervals, with each interval having a local lower bound. The formulation ensures all the intervals are properly covered by points and line segments. Recall that the local bound $b(I)$ for an interval is computed based on a subset of sites within the interval. For single-crossover recombination, the composite bound does not depend on exactly where the subset of sites are located

---

4 This is essentially the same issue discussed in section 4.2, where we observed that in linkage mapping, only an *odd* number of crossovers between two sites, during meiosis, will create a gamete that shows a recombination event between those sites.

in the intervals. But with gene-conversion and bounded tract length, the composite bound may be affected by which subsets of sites give the local bound for the intervals. For example, consider a long interval whose local bound is computed from sites c_1, c_2, and c_3, which are located near the left end of the interval. Also suppose there are three sites c'_1, c'_2, and c'_3, within the interval, that are located near the right end of the interval. Those two subsets impose different constraints on where line segments (which are of bounded length) must be placed, and to obtain higher composite bounds, it is desirable to create constraints by using both subsets. That is, we create multiple ILP constraints for a single interval, based on different subsets of sites. These subsets of sites need not all give the same bound for the interval since the physical locations of the sites may affect the composite bound even if its local bound is lower than possible. Therefore, a way to further raise the composite global bound is to enumerate each subset of sites, S, in I up to a certain size; and compute the haplotype lower bound on each sequence $M(S)$. From those bounds, we generate constraints for the ILP requiring that the number of points and line segments covering S must be at least the computed *haplotype bound* for $M(S)$, and requiring that the selection of covering points and line segments be constrained as described earlier. These additional ideas result in larger lower bounds at the cost of increasing the size of the ILP and the time needed to solve it. Our experience shows that when CPLEX is used, the ILP formulation can still be solved reasonably fast when enumerating all size three (or even four) subsets of sites.

5.2.11   More Lower Bounds Later

In sections 7.1, 10.1, and 10.2, we will discuss three additional lower bounds, called the *connected-component bound*, the *history bound*, and the *forest bound*. The first bound is very efficiently computed, but was not included in this chapter because the proof that it is a valid lower bound uses a fundamental combinatorial theorem that we will develop in chapter 6. The latter two bounds are not as efficiently computed, but have been shown empirically to often give better (higher) lower bounds than the bounds discussed in this chapter. There are also some theoretical results to support the observation that those bounds will typically be higher. We defer the discussion of those two bounds to a later chapter for two reasons. First, the *history bound* is conceptually related to general methods to construct an ARG from binary sequence data, and so it is more natural to discuss the *history bound* after discussing construction methods in chapter 9. Second, since the *history bound* cannot be computed efficiently, in the worst case, and even in practice, the *history bound* is effective only for very small data, although it is also of conceptual and theoretical importance. That is in contrast to all the bounds discussed in this chapter, and to the *connected-component bound*. Those bounds can all be computed efficiently either in provable worst-case time, or in practice. The *forest bound* is another theoretically-interesting bound, related to the *history bound*, and so the discussion of it is deferred until the *history bound* has been fully discussed.

We will also note in section 8.7 that the non-existence of a *galled tree* can efficiently establish a lower bound of *two* recombinations. Another lower bound method, which we will briefly discuss, in section 9.2.4, is based on *tree-scanning* developed in [394]. Those methods can all be used as local lower bounds, for use in computing a composite, global lower bound.

5.3 Advanced Material: A Sharp Analytical Upper Bound on $Rmin(M)$

> Reality is something you rise above.
> — Liza Minnelli

Theorem 3.2.1 (in section 3.2.4) established that for any n by m input matrix M, $Rmin(M) \leq nm$. Here we improve that result as follows:

Theorem 5.3.1 *For any n by m input matrix M, $Rmin(M) \leq \frac{2nm}{\log_2 n}$.*

Theorem 5.3.1 was established in [441]. In order to prove theorem 5.3.1, we first state and prove the following:

Lemma 5.3.1 *For any k, all possible 2^k binary sequences of length k can be generated on an ARG with at most $2^k - k - 1$ recombination nodes.*

Proof: We set the ARG's ancestral sequence, s_r, to the all-zero sequence of length k. Then, using the one allowed mutation per site, the ARG generates all of the k sequences that have exactly *one* site set to state one. At that point, the ARG will have all of the $k + 1$ binary sequences of length k, with either zero or one site(s) set to state one. Let s_c denote the sequence that has state one at site c, and state zero at every other site. Then, for i from two to k, the ARG generates every binary sequence of length k, with exactly i sites set to state one. Moreover, each of those sequences will be generated with one additional recombination event.

In detail, let s be a binary sequence that has exactly i sites set to state one, and let c be the leftmost site in s with state one. By construction, the ARG has already generated all the sequences with $i - 1$ sites set to state one, hence it has generated the sequence s' that is identical to s, except at site c (i.e., s has state one, and s' has state zero at site c). Therefore, sequence s can be created by a single-crossover recombination between s_c (contributing the prefix through site c) and s', hence with crossover index equal to $c+1$. Hence, all of the $2^k - k - 1$ binary sequences that have two or more sites set to state one, can be generated on an ARG using exactly $2^k - k - 1$ single-crossover recombinations. ∎

We can now present the

Proof (Of theorem 5.3.1): We will construct an ARG $\mathcal{N}$ for M that achieves the stated bound. For simplicity, but without loss of generality, assume that m is a multiple of $\log_2 n$, and that the ancestral sequence of $\mathcal{N}$ is specified to be the all-zero sequence.

We start with an ARG $\mathcal{N}'$ that constructs a set $\mathcal{S}$ of $\frac{nm}{\log_2 n}$ binary sequences of length m each. These sequences will be used to create, by recombination alone, all of the sequences in M. To describe $\mathcal{S}$, we define intervals of length $\log_2 n$ in $\mathcal{S}$: For i from 1 to $\frac{m}{\log_2 n}$ define I_i as the interval of $\log_2 n$ sites in $\mathcal{S}$ from position $1 + [(i - 1) \times \frac{m}{\log_2 n}]$ to $i \times \frac{m}{\log_2 n}$. Note that the number of distinct binary sequences of length $\log_2 n$ is $2^{\log_2 n} = n$. We next construct $\frac{m}{\log_2 n}$ blocks in $\mathcal{S}$, where each block contains n sequences, each of length m. In block I_i, interval I_i will contain *all* of the n binary numbers of length $\log_2 n$, and will contain zeros in all other locations of block I_i. See figure 5.10.

The total number of sequences in $\mathcal{S}$ is $\frac{nm}{\log_2 n}$. By an extension of lemma 5.3.1, in block i, the m-length sequences for $\mathcal{S}$ can be generated by an ARG with $O(n - \log_{2n} -1) < n$ recombinations nodes, with all-zero ancestral sequence. Hence, the total number of recombination nodes required for an ARG $\mathcal{N}'$ to generate all of the sequences in $\mathcal{S}$ is less than $\frac{nm}{\log_2 n}$.

The ARG $\mathcal{N}$ that will generate M starts with ARG $\mathcal{N}'$, followed by additional recombinations below the leaves of $\mathcal{N}'$. Any sequence s in M can be created using the leaf sequences in $\mathcal{N}'$, by a succession of at most $\frac{m}{\log_2 n}$ recombinations. In more detail, the first recombination is between the leaf sequence in $\mathcal{N}'$ that agrees with s in interval I_1, and the leaf sequence in $\mathcal{N}'$ that agrees with s in interval I_2. That creates a sequence that agrees with s in intervals I_1 and I_2. Inductively, we will have a sequence that agrees with s in intervals I_1 through I_i, and we can then recombine that sequence with the leaf sequence in $\mathcal{N}'$ that agrees with s in interval I_{i+1}. Hence, sequence s will be generated after $\frac{m}{\log_2 n} - 1$ such recombinations. These recombinations occur at additional recombination nodes placed into the growing $\mathcal{N}$. Repeating this for each of the n sequences in M, the resulting ARG $\mathcal{N}$ generates M and has at most $\frac{nm}{\log_2 n}$ more recombination nodes than does $\mathcal{N}'$, so at most $\frac{2nm}{\log_2 n}$ recombination nodes in total. ∎

Note that the ARG $\mathcal{N}'$ described in the proof of theorem 5.3.1 can generally be simplified. For example, the all-zero sequence is generated several times in $\mathcal{N}'$, and not every leaf sequence in $\mathcal{N}'$ is needed in the construction of M. Simplifying $\mathcal{N}'$ simplifies $\mathcal{N}$.

	I_1		I_2		I_3	
Block 1	0	0	0	0	0	0
	0	1	0	0	0	0
	1	0	0	0	0	0
	1	1	0	0	0	0
Block 2	0	0	0	0	0	0
	0	0	0	1	0	0
	0	0	1	0	0	0
	0	0	1	1	0	0
Block 3	0	0	0	0	0	0
	0	0	0	0	0	1
	0	0	0	0	1	0
	0	0	0	0	1	1

Figure 5.10 The leaf sequences of $\mathcal{N}'$ derived from an input matrix M with $n = 4$ taxa and $m = 6$ sites. Hence, each interval is $\log_2 4 = 2$ sites long, and there are $\frac{6}{2} = 3$ intervals and blocks. Each block has $2^2 = 4$ sequences. Interval I_i of block i contains all four binary sequences of length two.

The upper bound in theorem 5.3.1 is tight, or almost tight; a construction for M is given in [441], where a lower bound on $Rmin(M)$ is $\frac{nm}{\log_2 n}$.

6   Fundamental Combinatorial Structure and Tools

> Do you wish to rise? Begin by descending. You plan a tower that will pierce the clouds? Lay first
> the foundation...
> — Saint Augustine

This chapter is primarily technical, defining fundamental combinatorial objects (*incompatibility graphs* and *conflict graphs*) and developing powerful structural theorems about their nontrivial connected components, which are used as essential tools to derive deep results about recombination and ARGs. In a similar vein, we present in this chapter a surprising algorithmic result that will allow rapid computation of the most important feature of incompatibility and conflict graphs, that is, the number of nontrivial connected components, and how the nodes of the graphs *partition* into connected components.

6.1   Incompatibility and Conflict Graphs

Recall the definitions in chapter 2 (page 58) of what it means for two sites to be *incompatible*, and to *conflict relative to a sequence s*.

Definition We define the *incompatibility graph $G(M)$* for M as the graph containing one node for each site in M, and an edge connecting two nodes c and d if and only if sites c and d are incompatible in M.

Similarly,

Definition Given a sequence s, we define the "conflict graph" $G_s(M)$ for M (relative to s) as a graph containing one node for each column in M, and an edge connecting two nodes c and d if and only if columns c and d conflict relative to s.

Figure 6.2 shows the incompatibility graph $G(M)$ for M from figure 6.1.

	1	2	3	4	5
r_1	0	0	0	1	0
r_2	1	0	0	1	0
r_3	0	0	1	0	0
r_4	1	0	1	0	0
r_5	0	1	1	0	0
r_6	0	1	1	0	1
r_7	0	0	1	0	1

Figure 6.1 The dataset M derived by the ARG in figure 3.8 (page 75). Site-pairs (1,3); (1,4); (2,5) are incompatible. The incompatibility graph for this data is shown in figure 6.2.

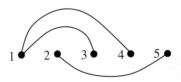

Figure 6.2 The incompatibility graph $G(M)$ for the sequences M from figure 6.1. Note that the *HK bound* is one, and the *haplotype bound* for this data is also $1 = 7 - 5 - 1$.

It is easy to see that $G_s(M) = G(M)$ if s is in M, and $G_s(M) = G(M \cup \{s\})$, if s is not in M. Therefore, through this reduction, results and algorithms developed for incompatibility graphs can often be used to obtain results and algorithms for conflict graphs. Equivalently, root-known ARG problems that specify that a fixed ancestral sequence s_r must be used in the ARG generating M, can often be solved by adding s_r to M, reducing to the root-unknown problem on input $M \cup \{s_r\}$. So in this chapter, we only address the *root-unknown* case.

Definition A "connected component" (or "component" for short), C, of a graph is a *maximal* subgraph such that for any pair of nodes (u, v) in C there is at least one path between u and v in the subgraph C. A "trivial" component has only one node, and no edges. The single node in a trivial connected component corresponds to a *fully-compatible* site.

For example, the incompatibility graph for the data given in figure 6.1 is shown in figure 6.2. It has two nontrivial connected components.

The nontrivial components are very informative

The nontrivial connected components of $G(M)$ encode a great deal of information about possible recombinations, and constraints on recombination, that can occur in any ARG that generates the sequences in M. We will see this throughout the book, particularly in the next two chapters, where we will prove that the number of nontrivial connected components of $G(M)$ is a lower bound on $Rmin(M)$, and study the limits of how ARG construction problems can be decomposed into smaller problems, and establish efficient algorithms for building restricted ARGs called *Galled Trees*. But first, we develop two fundamental structural facts about M, and the connected components of $G(M)$.

6.2 Connected Components and the Structure of M

We develop here a fundamental structural relationship between the sites in two different connected components of the incompatibility graph. The structural relationship is encapsulated in theorem 6.2.4 and corollary 6.2.2.

6.2.1 A (Hopefully) Motivating Preview

The main result in this section, theorem 6.2.4, is technical but essential to many later topics in the book. Its development and proof require a fair amount of new terminology and preliminary results. So, in order to both motivate the theorem, and to give a plausibility argument for it, we take some time here to motivate it and suggest why it was conjectured. In short, theorem 6.2.4 is a *stronger, more informative* version of a fact (stated as theorem 6.2.1) that is easy to observe and prove. In this section we develop that easy fact.

We start by looking at an example. Consider the set of sequences M given in figure 6.1. The incompatibility graph $G(M)$ for M is shown in figure 6.2. $G(M)$ has two connected components, C and C', containing the nodes $\{1, 3, 4\}$ and $\{2, 5\}$, respectively. Now look at the ARG $\mathcal{N}$ for M shown in figure 7.5 (page 206), but ignore the caption there. Notice that there is a cut edge, (u, v), that separates the edges labeled by the sites $\{1, 3, 4\}$, from the edges labeled by the sites $\{2, 5\}$. That is, (u, v) separates all the edges labeled with sites in C, from all the edges labeled with sites in C'.

Next, let (R, R') denote the bipartition of the taxa in M, induced by removing the cut edge (u, v) from $\mathcal{N}$. So, $R = \{r_1, r_2, r_3, r_4\}$ and $R' = \{r_5, r_6, r_7, r_8\}$ in the example. Then, observe that when restricted to the sites in C', every taxon in R is labeled with the *same* sequence, namely 00; and when restricted to the sites in C, every taxon in R'

is labeled with the *same* sequence, namely 010. See figure 6.3 (page 181). This is not an accident, and it is easy to see why this happens.

Recall that s_r denotes the ancestral sequence in $\mathcal{N}$, and for any node z and set of sites S in M, $s_z(S)$ denotes the sequence s_z, restricted to the sites in S. Restricted to the sites in C', all taxa in R must be labeled by $s_r(C') = 00$, because all the taxa in R are on the u-side of the directed cut edge (u, v) and all the edges labeled with sites in C' are on the v-side of the cut edge (u, v). This is the consequence of the directed cut edge (u, v) that separates R from R', and also separates all edges labeled with sites in C from all edges labeled with sites in C'. Similarly, restricted to the sites in C, all taxa in R' must be labeled by $s_u(C) = 010$, because all the taxa in R' are on the v-side of (u, v) and all the edges labeled with sites in C are on the u-side.

What we have demonstrated in this example is the following fact, which is easy to prove:

Theorem 6.2.1 *Let C and C' be any two connected components in the incompatibility graph, $G(M)$, for M, and let $\mathcal{N}$ be an ARG for M. Suppose that there is a directed cut edge, (u, v) in $\mathcal{N}$, that separates all the edges labeled with sites in C, from all the edges labeled with sites in C'. Removing the edge (u, v) from $\mathcal{N}$ defines a bipartition, (R, R'), of the leaves of $\mathcal{N}$, and hence of the taxa. Then, restricted to sites in C', all taxa in R have the same sequence; and restricted to the sites in C, all taxa in R' have the same sequence.*

What we really want

But, what we really observe in practice is a *stronger* fact that does not need to suppose the cut edge (u, v) in theorem 6.2.1. We observe that for *every* M, there is such a bipartition (R, R'). In particular:

Theorem 6.2.2 *For any M, and any two connected components C, C' in $G(M)$, there exists a bipartition of the taxa in M, (R, R'), so that, restricted to sites in C', all taxa in R have the same sequence; and restricted to the sites in C, all taxa in R' have the same sequence.*

The punchline

In this chapter, we will prove theorem 6.2.2 in an even stronger form—theorem 6.2.4. Then in the next chapter, we will use theorem 6.2.2 to prove that the premise of theorem 6.2.1 is true for any M. That is, for any M and for any two connected components C, C' in $G(M)$, there is an ARG $\mathcal{N}$ for M with a directed cut edge (u, v) that separates all the edges labeled with sites in C, from all the edges labeled with sites in C'.

	C			C'	
	1	3	4	2	5
r_1	0	0	1	0	0
r_2	1	0	1	0	0
r_3	0	1	0	0	0
r_4	1	1	0	0	0
r_5	0	1	0	1	0
r_6	0	1	0	1	1
r_7	0	1	0	0	1

Figure 6.3 Component C contains sites $\{1, 3, 4\}$, and connected component C' contains sites $\{2, 5\}$. The columns are permuted to group together the sites in C, and to group together the sites in C'. This makes it easier to see the sequences in M, restricted to the sites in C, and to the sites in C', respectively.

6.2.2 Back to the Formal Treatment

We will assume in this chapter that we have removed any duplicate *fully-compatible* sites. That is, if two identical sites c and d form two separate trivial connected components of $G(M)$, then we will remove one of those duplicate sites. In general, because of the different effect of recombinations on different sites, we cannot remove duplicate sites in M. But we will see that for the uses we make of the main results in this chapter, it is acceptable to remove duplicate fully-compatible sites, and this simplifies the exposition.

Recall the definition, in chapter 2 (page 56), of a *split* (bipartition) of a set of elements. Recall also that for a site c, $(O_c, \overline{O_c})$ is the split of the rows of M defined by the 0 and 1 states of sites c, respectively. Let c and c' be two sites in two distinct connected components C and C', respectively, of the incompatibility graph $G(M)$. Hence, (c, c') is a compatible pair of sites. The two splits, $(O_c, \overline{O_c})$ and $(O_{c'}, \overline{O_{c'}})$, cannot be identical. This is true by assumption if C and C' are both trivial connected components; if at least one is nontrivial and the splits were identical, then sites c and c' would have exactly the same incompatibilities and so would be in the same connected component.

Definition The split $(O_c, \overline{O_c})$ is called the c-split, and the split $(O_{c'}, \overline{O_{c'}})$ is called the c'-split.

Definition Each of the four subsets $O_c, \overline{O_c}, O_{c'}, \overline{O_{c'}}$ is called a "class" of the split it is part of.

Lemma 6.2.1 *One class of the c-split must strictly contain one class of the c'-split. Symmetrically, one class of the c'-split must strictly contain one class of the c-split.*

Proof: Since (c, c') is a compatible pair of sites, one of the four binary pairs $((1, 0)$, without loss of generality) does not appear in any row of M in column-pair (c, c'). But then every row that contains a 0 in column c' must contain a 0 in column c, and so $O_c \supseteq O_{c'}$. But if $O_c = O_{c'}$, then c and c' are identical, so $O_c \supset O_{c'}$. ∎

Clearly, when one class, say O_c, of the c-split strictly contains a class, say $O_{c'}$, of the c'-split, then the second class, $\overline{O_{c'}}$, of the c'-split must strictly contain the second class, $\overline{O_c}$, of the c-split. That is, $O_c \supset O_{c'}$ implies $\overline{O_{c'}} \supset \overline{O_c}$. Further $\overline{O_c} \cap O_{c'} = \emptyset$. See figure 6.4.

Definition If $O_c \supset O_{c'}$ or $O_c \supset \overline{O_{c'}}$, we say that 0 is the *containing* state of c *with respect to c'*, and we say that 1 is the *contained* state of c *with respect to c'*. Otherwise, we say that 1 is the containing state, and 0 is the contained state, of c, with respect to c'.

Since exactly one class of the c-split must strictly contain exactly one class of the c'-split, the containing state of c, with respect to c', is uniquely defined. Similarly, the contained state of c, with respect to c', is uniquely defined.

For example, in figure 6.2, $C = \{1, 3, 4\}$ and $C' = \{2, 5\}$. The matrix M for that graph is in figure 6.1, and we can see that for site $c = 1$ in C, $O_c = \{r_1, r_3, r_5, r_6, r_7\}$ and $\overline{O_c} = \{r_2, r_4\}$. For site $c' = 2$ in C', $O_{c'} = \{r_1, r_2, r_3, r_4, r_7\}$ and $\overline{O_{c'}} = \{r_5, r_6\}$. We see that $O_c \supset \overline{O_{c'}}$, so 0 is the containing state of c, and 1 is the contained state of c, with respect to c'. The fact that $O_c \supset \overline{O_{c'}}$ implies $O_{c'} \supset \overline{O_c}$, is also seen in this example. So 0 is the containing state of c', and 1 is the contained state of c', with respect to c.

Since the containment relation depends on two sites in two different connected components of $G(M)$, we can only identify the containing (or contained) state of a site c, *with respect to* some other site(s) in connected components that don't contain c.

For the remainder of the exposition, and without loss of generality, we will assume that $O_c \supset O_{c'}$. Hence, 0 is the containing state of c, and 1 is the contained state of c, with respect to c'; it is then also true that 1 is the containing state of c', and 0 is the contained state of c', with respect to c.

Lemma 6.2.2 *No row f in M can have the contained state (with respect to a site $c' \in C'$) at a site $c \in C$, and also have the contained state (with respect to c) at a site $c' \in C'$.*

Proof: Since $O_c \supset O_{c'}$, $\overline{O_c} \cap O_{c'} = \emptyset$. But $\overline{O_c}$ is the set of all the rows which have the contained state of c, with respect to c', and $O_{c'}$ is the set of all the rows that have the contained state of c', with respect to c. ∎

The next lemma is the key technical observation of this section.

Lemma 6.2.3 *Let $C, C', c, c', O_c, \overline{O_c}, O_{c'},$ and $\overline{O_{c'}}$ be defined as before. Let d' be any site in C'. Then the containing state of c, with respect to site d', is the same as the containing state of c, with respect to site c'.*

Proof: This is trivially true if C' is a trivial connected component, so that c' is the only site in C'. So assume that C' is nontrivial.

Recall that, without loss of generality, $O_c \supset O_{c'}$, and so $\overline{O_{c'}} \supset \overline{O_c}$. Let $(O_{d'}, \overline{O_{d'}})$ be the split of the rows defined by the 0 and 1 states of site d'. We need to prove that $O_c \supset O_{d'}$ or $O_c \supset \overline{O_{d'}}$.

Assume first that d' is a site in C' that is incompatible with c'. Such a site must exist since C' is nontrivial and connected, and each edge represents the incompatibility of a pair of sites. If 0 is not the containing state of c, with respect to d', then either $\overline{O_c} \supset O_{d'}$ or $\overline{O_c} \supset \overline{O_{d'}}$. Suppose first that $\overline{O_c} \supset \overline{O_{d'}}$. Since $\overline{O_{c'}} \supset \overline{O_c}$, as above, it follows that $\overline{O_{c'}} \supset \overline{O_{d'}}$, so $O_{c'} \cap \overline{O_{d'}} = \emptyset$. But then c' and d' can't be incompatible, which is a contradiction. Similarly, if $\overline{O_c} \supset O_{d'}$, then $\overline{O_{c'}} \supset O_{d'}$, so $O_{c'} \cap O_{d'} = \emptyset$, again contradicting the assumption that c' and d' are incompatible. So 0, the containing state of c with respect to c', must also be the containing state of c with respect to d', where d' is any site that is incompatible with c'. The lemma now follows by transitivity, because C' is a connected component. That is, it is possible to reach any node in C' by a series of incompatibility relations, starting from c'. ■

For example, in figure 6.1, $O_c = \{r_1, r_3, r_5, r_6, r_7\}$ for site $c = 1$ in C, and 0 is the containing state of site c in C with respect to site $d' = 5$ in C', as well as the containing state of site c with respect to site $c' = 2$ in C'.

Lemma 6.2.3 establishes that for any $c \in C$, *one* state of c is the containing state of c with respect to *all* sites in C', and the other state is the contained state with respect to *all* sites in C'. Symmetrically, for any $c' \in C'$ one state of c' is the containing state with respect to all sites in C, and the other state is the contained state with respect to *all* sites in C. We summarize as follows:

Theorem 6.2.3 *Each site in C (respectively C') has a well defined containing state, and a well defined contained state, with respect to the* entire *connected component C' (respectively C).*

For example, in the data given in figure 6.1, 0 is the containing state of site $c = 1$, with respect to C', and 1 is the contained state of c, with respect to C'. Then, lemma 6.2.2 can be recast as:

Corollary 6.2.1 *No row has the contained state, with respect to C', at a site in C, and also has the contained state, with respect to C, at a site in C'.*

6.2.3   The Key Structural Result

Definition For any connected component C of $G(M)$, $M(C)$ denotes the sequences in M, restricted to the sites in C.

Definition The sequence consisting of the containing state of each site c in C, with respect to C', is called *the constant* sequence of C, with respect to C', or more simply as "the constant sequence of $< C, C' >$."

Note that the order of C and C' in $< C, C' >$ matters; "the constant sequence of $< C', C >$" refers to the constant sequence of C', with respect to C.

Since each site in C has a well defined containing state, with respect to C', the constant sequence of $< C, C' >$ is well defined and unique. Similarly, the constant sequence of $< C', C >$ is well defined and unique.

For example, for the data in figure 6.5, the constant sequence of $< C, C' >$ is 010, and the constant sequence of $< C', C >$ is 00.

Although well defined, we have not established yet that the constant sequence of $< C, C' >$ (or of $< C', C >$) is an actual sequence in $M(C)$ (or $M(C')$). We next show that it must be.

Lemma 6.2.4 *The constant sequence of $< C, C' >$ is one of the sequences in $M(C)$. That is, there is a sequence in $M(C)$ where every site $c \in C$ has the containing state of c, with respect to C'.*

Proof: Consider any row f where some site $c' \in C'$ has the contained state of c', with respect to C. By lemma 6.2.2, in row f, no site $c \in C$ can have the contained state of c, with respect to C', so in row f every site $c \in C$ must have the containing state of c with respect to C'. ∎

Definition Any sequence in $M(C)$ (or C') that is not the constant sequence of $< C, C' >$ (or of $< C', C >$) is called *a variable* sequence of $< C, C' >$ (or of $< C', C >$).

Note that a variable sequence of $< C, C' >$ is simply a sequence in $M(C)$ such that at some site $c \in C$, the state of c is the contained state of c. From the proof of lemma 6.2.4, we see the **main result** of this section:

Theorem 6.2.4 *Every row in* M *that contains a* variable *sequence of* $< C', C >$ *in* $M(C')$, *must contain the* constant *sequence of* $< C, C' >$ *in* $M(C)$. *Symmetrically, every row that contains a variable sequence of* $< C, C' >$ *in* $M(C)$ *must contain the constant sequence of* $< C', C >$ *in* $M(C')$.

It immediately follows that:

Corollary 6.2.2 *Let* C *and* C' *be two nontrivial connected components of* $G(M)$. *No row of* M *contains a variable sequence of* $< C, C' >$ *in* $M(C)$, *and a variable sequence of* $< C', C >$ *in* $M(C')$.

In the running example, the variable sequences of $< C, C' >$ appear in rows $\{r_1, r_2, r_4\}$ and the variable sequences of $< C', C >$ appear in rows $\{r_5, r_6, r_7\}$. These are shown in figure 6.5, which illustrates lemma 6.2.3, theorem 6.2.4, and corollary 6.2.2. Another example appears in figure 7.14 (page 232).

Note that a row can have both the constant sequence of $< C, C' >$ and the constant sequence of $< C', C >$. Row r_3 in figure 6.5 is an example of this. Theorem 6.2.4 is a more refined version, and with a simpler proof, of a result first proved in [17]. The approach in this book builds on the approach in [144].

Theorem 6.2.4 and corollary 6.2.2 capture the key structural relationship between the sites in two different connected components of the incompatibility graph $G(M)$. Note that these results do not use the structure of the edges inside any of the connected components. The results only use the way that the nodes of the $G(M)$ are *partitioned* into connected components. That point will be important in section 6.3.

Note also that the results do not depend in any way on the biological phenomena that caused the site incompatibilities or conflicts. Hence, they apply to any phenomena that cause incompatibilities or conflicts.

6.2.4   Adding New Sequences

Here we develop an observation that will be needed in section 11.2.2.

Lemma 6.2.5 *Suppose* C *and* C' *are nontrivial components of* $G(M)$. *If we add a new sequence to* M, *and no incompatibility is created between a site in* C *and a site in* C', *then the constant sequences of* $< C, C' >$ *and* $< C', C >$ *remain unchanged.*

Proof: Consider two sites $c \in C$ and $c' \in C'$, where C and C' are two distinct nontrivial connected components of $G(M)$. By assumption, every site is polymorphic, that is, has both states 0 and 1. Also, no two sites in different nontrivial components can be identical, or can be exact complements of each other. If they were, then they would both be incompatible with the same sites in C and C', and would then be in the same component. Now if site-pair (c, c') had only one pair of states, then c and c' would be identical; and if they have only two distinct pairs of states, but both sites are polymorphic, then c and c' would be exact complements. So, site-pair (c, c') must have at least *three* distinct pairs of binary characters. But since c and c' are in different connected components, they are a compatible pair of sites, and so they cannot have all four distinct binary pairs. So, they must have exactly three distinct binary pairs.

If we add new sequences to M, creating the set of sequences $\overline{M}$, and c and c' are *not* incompatible in $\overline{M}$, then any binary pair in $\overline{M}$ at sites c, c', must already have been a binary pair at those sites in M. Therefore, the containing state of c with respect to c', is the same in both M and $\overline{M}$. Similarly, the containing state of c' with respect to c, is the same in M and $\overline{M}$. Extending this observation to all sites in C and C' proves the lemma.   ∎

6.2.5   The Results Apply in Many Contexts

We will use theorem 6.2.4 and corollary 6.2.2 in our study of ARGs and recombination, but it is worth noting here that the theorems themselves and their proofs do not rely in any way on properties of networks or recombination. They rely *only* on the conflicts and incompatibilities in M. Similarly, the conflict graph and the incompatibility graph only rely on the incompatibilities in M. Incompatibilities in M can occur through many biological mechanisms, and the theorems can be applied to any of those biological contexts. In particular, we note that the theorems hold if incompatibilities are created through multiple-crossover recombination as well as single-crossover recombination, or by recurrent or back mutation, or by lateral transfer, or hybrid speciation and so on. Moreover, since conflicts and incompatibilities do not depend on the order of the sites in M, the results in this chapter apply to problems of representing clusters, discussed in chapter 14.

6.3   Surprisingly Fast Identification of the Components of the Conflict and Incompatibility Graphs

In numerous places in this book, we will see the central role played by the connected components of the incompatibility graph $G(M)$, and of the conflict graph $G_0(M)$. However, in all of those applications we only use $G(M)$ and $G_0(M)$ in order to find how the nodes are *partitioned* into connected components; no listing of the edges inside a connected component is ever needed. Similarly, we noted earlier that theorem 6.2.4 and corollary 6.2.2 only use the partition of the nodes of $G(M)$ into connected components. Hence, if the partition

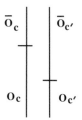

Figure 6.4 $O_c \supset O_{c'}$ implies that $\overline{O_{c'}} \supset \overline{O_c}$ and $\overline{O_c} \cap O_{c'} = \emptyset$.

		C			C'	
		1	3	4	2	5
	r_1	0	0	1	0	0
variable sequences of $< C, C' >$	r_2	1	0	1	0	0
	r_4	1	1	0	0	0
	r_3	0	1	0	0	0
	r_5	0	1	0	1	0
variable sequences of $< C', C >$	r_6	0	1	0	1	1
	r_7	0	1	0	0	1

Figure 6.5 The incompatibility graph in figure 6.2 (page 178), derived from the matrix M in figure 6.1 (page 178), has connected component C containing sites $\{1, 3, 4\}$, and connected component C' containing sites $\{2, 5\}$. The constant sequence of $< C, C' >$ in $M(C)$ is 010, and the constant sequence of $< C', C >$ in $M(C')$ is 00. The rows containing variable sequences of $< C, C' >$ are $\{r_1, r_2, r_4\}$, and the rows containing variable sequences of $< C', C >$ are $\{r_5, r_6, r_7\}$. The rows and columns have been permuted from their original order to more clearly illustrate the structure of M. Note that row r_3 has both the constant sequence of $< C, C' >$ in $M(C)$ and the constant sequence of $< C', C >$ in $M(C')$.

of nodes into connected components (in $G(M)$ or $G_0(M)$) could be found more efficiently than by explicitly constructing all the edges in the incompatibility or conflict graph, then the overall running time of the applications would be reduced.

In this section we show the surprising result, that given M, we can find the partition of the nodes of the conflict graph $G_0(M)$, induced by the connected components of $G_0(M)$, in $O(nm)$ time. This is a reduction of the straightforward bound of $O(nm^2)$ time to explicitly build $G_0(M)$, although fast Boolean matrix multiplication can be used[1] to identify all the conflicting pairs of sites in $O(nm^{1.52})$ time [20]. So, without explicitly knowing the conflict graph, the essential feature of interest (the partition of the nodes of $G_0(M)$ into connected components) can be found in *linear* time. This result[2] was first shown in [83] and later simplified in [63]. However, we will show the stronger result that a *subgraph* of $G_0(M)$ with the same connected components as $G_0(M)$ can be found in $O(nm)$ time. The ability to identify a component-preserving subgraph was discussed only in [63]. After deriving that result, we show how to obtain the same result for the incompatibility graph. Our exposition of the $O(nm)$-time result for the conflict graph mixes elements of both expositions from [63] and [83], and the algorithm presented here differs (a bit) from both of the prior algorithms.

Recall that sites c and d in M are said to *conflict relative to ancestral sequence* s_r if they are incompatible in $M \cup \{s_r\}$. Recall also that the *conflict graph* relative to s_r is the incompatibility graph for $M \cup \{s_r\}$. Finally, recall that we can assume that s_r is the all-zero sequence, since problems involving a different ancestral sequence can be reduced to the case of the all-zero ancestral sequence. These definitions and facts were established in chapters 2 and 6. Therefore, we will assume that s_r is the all-zero sequence, so sites c and d conflict if and only if they contain all three ordered binary pairs $1, 1; 1, 0;$ and $0, 1$. We will use the term "conflict graph" here to mean the conflict graph $G_0(M)$ relative to the all-zero sequence. Also, since each node in $G_0(M)$ is associated with a site in M, we will refer to any node in $G_0(M)$ by its associated site.

Note that two identical sites in M will be incompatible with, or in conflict with, exactly the same sites. Hence for the purposes of finding $G(M)$ or $G_0(M)$, we can assume (and will) that M has no duplicate columns.

---

1 If we compute the matrix $U = M^t M$, where M^t is the transpose of M, then $U(c, c')$ will be greater than 0 if and only if there is some taxon f where $M(f, c) = M(f, c') = 1$. If we use Boolean matrix multiplication, then replace "greater than 0" with "equal to 1." Similarly, if we exchange all the 0s and 1s in M^t, calling the result matrix W, and then compute $U = WM$, then $U(c, c') > 0$ (or $U(c, c') = 1$, if Boolean multiplication is used) if and only if there is a taxon f where $M(f, c) = 1$ and $M(f, c') = 0$. In this way, we can use four (Boolean) matrix multiplications to find all of the incompatible pairs of sites in M.

2 Actually, both papers show that the time bound can be reduced further to be proportional to the number of rows plus the number of ones in M.

Recall that for any site c, O_c denotes the set of taxa where the state of site c is one (see page 46). Recall also that if two sites c and d do *not* conflict then either $O_c \cap O_d = \emptyset$ or $O_c \subseteq O_d$ or $O_d \subseteq O_c$.

To start, we sort the columns of M by the *number* of ones they contain, largest number first, breaking ties arbitrarily. This can be done in $O(nm)$ time by *bucket-sort* or *counting-sort*. So, in the rest of the discussion, we assume that $|O_c| \geq |O_{c+1}|$, for c from 1 to $m - 1$.

Definition For any site c in M, $L(c)$ is the smallest-index (Leftmost) site $c' < c$ such that sites c and c' conflict. Note that $L(c)$ might not exist.

The following is the main technical lemma used to develop the algorithm and establish its correctness.

Lemma 6.3.1 *Let $\tilde{c}$ be a site such that $L(\tilde{c})$ exists. For any site c in M where $L(\tilde{c}) < c < \tilde{c}$ and $O_c \cap O_{\tilde{c}} \neq \emptyset$, either site c conflicts with site $\tilde{c}$, or site c conflicts with site $L(\tilde{c})$.*

Proof: Suppose that c does not conflict with $\tilde{c}$. Then, since $O_c \cap O_{\tilde{c}} \neq \emptyset$, either $O_c \subseteq O_{\tilde{c}}$ or $O_{\tilde{c}} \subseteq O_c$. But, $|O_{\tilde{c}}| \leq |O_c|$, so it must be $O_{\tilde{c}} \subseteq O_c$. In that case, $O_c \cap O_{L(\tilde{c})} \neq \emptyset$, since $O_{\tilde{c}} \cap O_{L(\tilde{c})} \neq \emptyset$. Now if c also does not conflict with $L(\tilde{c})$, then by the same reasoning $O_c \subseteq O_{L(\tilde{c})}$, and it follows that $O_{\tilde{c}} \subseteq O_{L(\tilde{c})}$. But that contradicts the assumption that $\tilde{c}$ conflicts with $L(\tilde{c})$. So either c conflicts with $\tilde{c}$, or c conflicts with $L(\tilde{c})$. ∎

Note that lemma 6.3.1 shows that c either conflicts with $L(\tilde{c})$ or with $\tilde{c}$, or both, but it does not specifically indicate which one(s). Later we will prove an extension of lemma 6.3.1 that will allow the algorithm to easily identify one conflicting pair of sites $(L(\tilde{c}), c)$ or $(c, \tilde{c})$.

Graph $G'(M)$

Our next goal is to show that the node partition of the connected components of $G_0(M)$ can be found in a *sparse subgraph*, $G'(M)$, of $G_0(M)$. We show that $G'(M)$ can be constructed in $O(nm)$ time. We assume for now that for any site c, site $L(c)$ is known, if it exists. We will show later how to find $L(c)$ for each c. We also assume here that given $L(\tilde{c}), c$, and $\tilde{c}$, as in lemma 6.3.1, we can identify one conflicting pair, $(L(\tilde{c}), c)$ or $(c, \tilde{c})$, in constant time. We will later show how that can be done.

The construction

Graph $G'(M)$ contains one node for each site of M. The edges of $G'(M)$ are created by algorithm *Build-G'(M)*, shown in figure 6.6. See figure 6.7 for an example. Algorithm

Build-G' begins by creating an edge for every pair $(c, L(c))$, where $L(c)$ exists. Then, the second *FOR* loop of the algorithm loops through the rows of M. In the third *FOR* loop, the algorithm initiates a scan of the entries in a row f, from right to left. Each scan takes $O(m)$ time, assuming that the algorithm can identify in constant time a conflicting pair $(L(\tilde{c}), c)$ or $(c, \tilde{c})$, for any encountered site c between $\tilde{c}$ and $L(\tilde{c})$, when $f \in O_c$. Therefore, the graph $G'(M)$ is built in $O(nm)$ time, and the number of edges in $G'(M)$ will be at most $2n(m-1)$, since the algorithm initially creates at most $m-1$ edges and then adds at most $m-1$ edges during the scan of any row. In fact, the total number of edges in $G'(M)$ is less than $m+t$, where t is the number of ones in M.

Lemma 6.3.2 *$G'(M)$ is a subgraph of $G_0(M)$.*

Proof: Note that during the scan of row f, if $\tilde{c}$ is defined, then $f \in O_{\tilde{c}}$ and $L(\tilde{c})$ exists. This is true by the two ways that the value of $\tilde{c}$ is changed in algorithm *Build-G'(M)*. So, during the processing of site c, if $\tilde{c}$ is defined and $L(\tilde{c}) < c < \tilde{c}$, the conditions of lemma 6.3.1 are satisfied and hence c conflicts with at least one of the sites $L(\tilde{c})$ or $\tilde{c}$. Therefore, every edge added to $G'(M)$ connects two nodes whose associated sites are in conflict, establishing that $G'(M)$ is a subgraph of $G_0(M)$.   ∎

After $G'(M)$ has been built, its connected components can be found in $O(m+t) = O(nm)$ time by any linear-time search, for example by breadth-first search. We next show that the partition of nodes into connected components in $G'(M)$ is the same as in $G_0(M)$.

Theorem 6.3.1 *The partition of the nodes into connected components of $G_0(M)$ is identical to the partition of the nodes into connected components of $G'(M)$.*

Proof: Trivially, the nodes in any connected component of $G'(M)$ are contained in a single connected component of $G_0(M)$, because $G'(M)$ is a subgraph of $G_0(M)$.

Conversely, we will prove that if (b, d) is an edge in $G_0(M)$, then nodes b and d are in the same connected component of $G'(M)$. That implies, by transitivity, that all the nodes in any connected component of $G_0(M)$ are contained in a *single* connected component of $G'(M)$, which proves the theorem.

Assume that $b < d$. By the definition of $L(d)$, $L(d) \leq b$ since b and d conflict. Also, since b and d conflict, there is a taxon f in $O_b \cap O_d$. Consider the scan of row f during algorithm *Build-G'(M)*. Note that during the scan of row f, if $\tilde{c}$ is defined, then $f \in O_{\tilde{c}}$. This is true because $\tilde{c}$ is only set to a site c when f is in O_c.

Define $\tilde{c}_0$ to be the value of $\tilde{c}$ immediately *after* algorithm *Build-G'(M)* finishes processing site d. We will see that $\tilde{c}$ is defined at that point, so $\tilde{c}_0$ exists. We make the following

---

Algorithm Build-G$'$ (M)

   for (each site c such that $L(c)$ exists) **do**
      Create the edge $(c, L(c))$ in $G'(M)$.
   endfor

   for (each row $f \in M$) **do**
      Set $\tilde{c}$ to be undefined
      for (c from m to 1) **do**
         if ($f \in O_c$) **then**
            Process site c with the following block of three "$\mathtt{if}$" statements:

            if (($\tilde{c}$ is defined) AND ($L(\tilde{c}) < c < \tilde{c}$)) **then**
               Identify one conflicting pair $(L(\tilde{c}), c)$ or $(c, \tilde{c})$ and
               add an edge to $G'(M)$ between the nodes for
               the identified conflicting pair of sites.

            endif

            if (($\tilde{c}$ is defined) AND ($L(c)$ exists) AND ($L(c) < L(\tilde{c})$)) **then**
               Set $\tilde{c}$ to c.
            endif

            if (($\tilde{c}$ is undefined) AND ($L(c)$ exists)) **then**
               Set $\tilde{c}$ to c.
            endif

         endif
      endfor
   endfor
   return $G'(M)$

Figure 6.6   Algorithm *Build-G$'$(M)*

Connectedness claim: Immediately after the algorithm has processed site d, sites d and $\tilde{c}_0$ are in the same connected component of the partially constructed $G'(M)$.

To prove the connectedness claim, we first consider the case where $\tilde{c}$ is already defined at the point when the scan of row f *reaches* site d. If $L(\tilde{c}) < d < \tilde{c}$, then the conditions for lemma 6.3.1 hold, so d conflicts with at least one of the sites $\tilde{c}$ or $L(\tilde{c})$. Hence, either edge $(d, \tilde{c})$ or $(L(\tilde{c}), d)$ will be added to $G'(M)$ at that point. But $L(\tilde{c})$ and $\tilde{c}$ are already

	1	2	3	4	5	6
r_1	1	1	1	0	1	0
r_2	1	1	1	0	0	0
r_3	1	1	0	0	1	1
r_4	1	0	1	1	0	1
r_5	0	0	0	1	0	0

Figure 6.7 In this data there are eight conflicting pairs, so $G_0(M)$ has eight edges. $L(1)$ and $L(2)$ do not exist, $L(3)$ and $L(6)$ are 2; $L(4)$ is 1; and $L(5)$ is 3. The creation of edges in $G'(M)$ for the pairs $(c, L(c))$ puts four edges into $G'(M)$. Additional edges are added during scans of rows in M. The scans of rows r_1 and r_2 do not create any additional edges. The scan of row r_3 immediately sets $\tilde{c}$ to 6 ($L(6) = 2$). Then site 5 is reached, and 5 conflicts both with site $L(\tilde{c}) = 2$ and with site $\tilde{c} = 6$. The constant time method described in section 6.3.2 chooses to add the edge $(5, 6)$. The scan of row r_4 again sets $\tilde{c}$ to 6. Then site 4 is reached, and 4 conflicts with $\tilde{c} = 6$, but not with $L(\tilde{c}) = 2$, so the edge $(4, 6)$ is added. Then site 3 is reached, and 3 conflicts with both $\tilde{c} = 6$ and $L(\tilde{c}) = 2$, but the algorithm chooses to add the edge $(3, 6)$. In total, $G'(M)$ has seven edges.

connected by an edge, so nodes $d, \tilde{c}, L(\tilde{c})$ will all be together in one connected component of $G'(M)$. Next, during the processing of site d, the algorithm either changes $\tilde{c}$ to d or does not change it. If no change is made, then $\tilde{c}_0$ is equal to the value of $\tilde{c}$ when the scan reached site d, and the claim holds by the fact just established. If $\tilde{c}$ is changed to d, then $\tilde{c}_0 = d$, and the claim holds trivially. So if $\tilde{c}$ is defined when the scan of row f reaches site d, and $L(\tilde{c}) < d < \tilde{c}$, then nodes d and $\tilde{c}_0$ will be in the same connected component of $G'(M)$ immediately after the algorithm finished processing site d.

Now consider the case that $\tilde{c}$ is defined when the scan of row f reaches site d, but $d \le L(\tilde{c}) < \tilde{c}$. Suppose $d = L(\tilde{c})$; then edge $(d, \tilde{c})$ exists before the scan of row f reaches d. If $\tilde{c}$ is not changed during the processing of d, then $\tilde{c}_0 = \tilde{c}$ when the processing of d is finished, so the claim holds. If $\tilde{c}$ is changed during the processing of d, then $\tilde{c}_0 = d$ when the processing of d ends, so the claim holds trivially. Finally, suppose that $d < L(\tilde{c}) < \tilde{c}$. In that case, the algorithm will set $\tilde{c}$ to d, since $b < d < L(\tilde{c}) < \tilde{c}$, and d conflicts with b. So $L(d) \le b$. Therefore, $\tilde{c}_0 = d$ when the processing of d is finished, so the claim again holds.

Next, consider the case that $\tilde{c}$ is not defined when the scan of f reaches d. Since $b < d$ and b and d conflict, $L(d)$ exists, and $\tilde{c}$ will be set to d, and so $\tilde{c}_0$ equals d when the processing of d is finished. In this case, the claim again holds trivially, finishing the proof of the connectedness claim.

Returning to the proof

Returning to the proof of the theorem, note that when the algorithm finishes processing site d, either $\tilde{c}_0 = d$ or $d < \tilde{c}_0$, and $L(\tilde{c}_0) \leq b$, since b and d conflict. Let $[\tilde{c}_0 > \tilde{c}_1 > \ldots > \tilde{c}_k]$ be the ordered list of values of $\tilde{c}$, from the time the scan of row f *finishes* processing site d, until the time the scan *reaches* site b. So, $\tilde{c}_k$ is $\tilde{c}$ when the scan reaches site $b < \tilde{c}_k$. It is possible that k is still 0 at that point. But, if $k > 0$, then by the working of algorithm *Build* $G'(M)$, and the fact that b and d conflict, $L(\tilde{c}_i) < L(\tilde{c}_0) \leq b$, for each $0 < i \leq k$.

If $k = 0$ when the scan reaches b, then d and $\tilde{c}_k$ are in the same connected component of $G'(M)$, since we have shown that d and $\tilde{c}_0$ are. If $k > 0$, then assume inductively that when $\tilde{c}_h$ is set, for some $h < k$, edges have been added to $G'(M)$ so that $\tilde{c}_0, \ldots, \tilde{c}_h$ are all in the same connected component of $G'(M)$. Next, consider the point in the scan of row f when site $\tilde{c}_{h+1} > b$ is reached. $L(\tilde{c}_h) < b$ as noted above, hence $L(\tilde{c}_h) < b \leq \tilde{c}_{h+1} < \tilde{c}_h$, so the conditions for lemma 6.3.1, are satisfied for $L(\tilde{c}_h), \tilde{c}_{h+1}$, and $\tilde{c}_h$. Therefore, the algorithm will either add the edge $(L(\tilde{c}_h), \tilde{c}_{h+1})$ or the edge $(\tilde{c}_{h+1}, \tilde{c}_h)$ to $G'(M)$. In either case, since there is also an edge between $L(\tilde{c}_h)$ and $\tilde{c}_h$, nodes $\tilde{c}_{h+1}$ and d will be in the same connected component of $G'(M)$. So by induction, $\tilde{c}_k$ is in the same connected component as d. Finally, consider the processing of site b in the scan of row f. For the same reasons as above, $L(\tilde{c}_k) \leq b < \tilde{c}_k$. If $L(\tilde{c}_k) = b$ then there is an edge in $G'(M)$ between b and $\tilde{c}_k$; if $L(\tilde{c}_k) < b$ then by lemma 6.3.1, the algorithm will either add the edge $(L(\tilde{c}_k), b)$ or add the edge $(b, \tilde{c}_k)$ to $G'(M)$. In all cases, nodes b and d will be in the same connected component of $G'(M)$, proving the theorem. ∎

Summarizing, we have established that the partition of nodes into connected components is the same in $G'(M)$ as it is in $G_0(M)$, so $G'(M)$ can be used in place of $G_0(M)$ in order to find that node partition. Further, we have established that $G'(M)$ is a subgraph of $G_0(M)$, with at most $2nm$ edges. The algorithm to build $G'(M)$ runs in linear time as a function of the size of M, which is nm.

In order to establish that algorithm *Build-$G'(M)$* runs in linear time, we have to show how all of the $L(c)$ values can be found in $O(nm)$ time, and when required, how the algorithm can identify a conflicting pair of sites, in constant time.

6.3.1 Computing All the $L(c)$ Values in $O(nm)$ Total Time

Recall that the columns of M are sorted by the number of ones they contain, largest first. In addition, we will now consider each row of M to define a *binary* number with the most significant bit on the left. We then sort those binary numbers, largest first, so that if taxon f appears above taxon g in the sorted M, then the binary number given by row f is greater than or equal to the binary number given by row g. For example, the rows in the data in figure 6.7 have already been sorted in this way. The sorting can be done in $O(nm)$ time by using *radix-sort* on the rows of M. From this point, we assume that M has been sorted in this way.

Definition For a site c, $first(c)$ is the topmost row in M where site c has entry 1; similarly, $last(c)$ is the bottommost row where site c has entry 1.

Definition For a site c, let $I(c)$ be the closed interval of rows $[first(c) \ldots last(c)]$, and let $\tilde{I}(c)$ be the closed interval of rows $[first(c) \ldots last(c) - 1]$.

Note that by definition, $M(first(c), c) = M(last(c), c) = 1$. Also, the rows in $I(c)$ contain all the 1 entries at site c, but they might also contain some 0 entries.

Definition Let $b(f)$ be defined as the leftmost site c' such that $M(f, c') \neq M(f + 1, c')$.

Note that by definition, $M(f, c'') = M(f + 1, c'')$ for all $c'' < b(f)$ and $f < n$.

Theorem 6.3.2 $L(c)$ *exists if and only if there is a row* $f \in \tilde{I}(c)$ *such that* $b(f) < c$. *Further, if $L(c)$ exists, it is equal to the* minimum $b(f)$ *such that* $f \in \tilde{I}(c)$.

Proof: If $L(c)$ exists, then because $L(c)$ conflicts with c, $L(c)$ must contain both a zero and a one in the rows in $I(c)$. Therefore, there is a row $f \in \tilde{I}(c)$ such that $M(f, L(c)) \neq M(f + 1, L(c))$, and hence $b(f) \leq L(c) < c$.

Conversely, suppose there is a row $f \in \tilde{I}(c)$ such that $b(f) < c$. Let $f^* \in \tilde{I}(c)$ be the row that minimizes $b(f)$ over all $f \in \tilde{I}(c)$. Therefore, for every site $c' < b(f^*)$ all entries at site c' in rows in $I(c)$ must be identical, and hence $b(f^*)$ is the leftmost site where entries in the rows of $I(c)$ differ. Since the rows of M, considered as binary numbers, are sorted largest first, all rows in $I(c)$ that contain a 1 at site $b(f^*)$ must appear above all rows in $I(c)$ that contain a 0 at site $b(f^*)$. In particular, $M(first(c), b(f^*)) = 1$ and $M(last(c), b(f^*)) = 0$. Now, $M(first(c), c) = M(last(c), c) = 1$, so site-pair $(b(f^*), c)$ contains the ordered binary pairs $(1, 1)$ and $(0, 1)$. But $|O_{b(f^*)}| \geq |O_c|$, and site-pair $(b(f^*), c)$ contains the ordered pair $(0, 1)$, so they must also contain the ordered pair $(1, 0)$ (not necessarily in $I(c)$). It follows that $b(f^*)$ and c conflict. Therefore, since $b(f^*) < c$, $L(c)$ exists, and $L(c) \leq b(f^*)$.

Now we prove that if $L(c)$ exists, then $L(c) = b(f^*)$. From the argument in the first paragraph, if $L(c)$ exists, then $L(c) \leq b(f^*)$. We also observed that when there is a row $f \in \tilde{I}(c)$ such that $b(f) < c$, then at any site $c' < b(f^*)$, all entries at site c' in rows in $I(c)$ must be identical. Therefore, no site $c' < b(f^*)$ can conflict with c, since all the 1 entries at site c are in the rows of $I(c)$, and in those rows no site c' can have both entries 0 and 1. Hence, $L(c) \geq b(f^*)$. We conclude that when $L(c)$ exists, $L(c) = b(f^*)$, the minimum $b(f)$ such that $f \in \tilde{I}(c)$. ∎

From the proof of theorem 6.3.2, we can also establish:

Lemma 6.3.3 *If $L(c)$ exists, then $M(first(c), L(c)) = 1$ and $M(last(c), L(c)) = 0$.*

Proof: In the proof of theorem 6.3.2, we saw that if $L(c)$ exists, then $L(c) = b(f^*) < c$, and that $M(first(c), b(f^*)) = 1$ and $M(last(c), b(f^*)) = 0$. ∎

The algorithmic importance of theorem 6.3.2 is clear. For any row $f \in M$, $b(f)$ can be found by a simple scan of rows f and $f + 1$ in $O(m)$ time. So all of the $b(f)$ values can be found in order and put into a table, indexed by row number, in $O(nm)$ time. Then for any site c, the minimum $b(f)$ for $f \in \tilde{I}(c)$ can be found in $O(n)$ time by scanning through the $b(f)$ values for $f \in \tilde{I}(c)$. By theorem 6.3.2, if the minimum of those $b(f)$ values is less than c, then $L(c)$ equals that minimum. Clearly, each interval $I(c)$ can be found in $O(n)$ time, hence we have:

Theorem 6.3.3 *All the $L(c)$ values can be found in $O(nm)$ time.*

6.3.2 Identifying a Conflicting Pair $(L(\tilde{c}), c)$ or $(c, \tilde{c})$ in Constant Time

We now address the one remaining issue in algorithm *Build-G'(M)*.

Lemma 6.3.4 *During the scan of row f in algorithm Build-G'(M), suppose site c is reached, where $f \in O_c$, $L(\tilde{c}) < c < \tilde{c}$, and $O_c \cap O_{\tilde{c}} \neq \emptyset$. Then one conflicting pair, $(L(\tilde{c}), c)$ or $(c, \tilde{c})$, can be identified in* constant *time.*

Proof: Suppose $M(first(\tilde{c}), c) = M(last(\tilde{c}), c) = 1$. Then by lemma 6.3.3, site-pair $(L(\tilde{c}), c)$ has the ordered binary pairs $(1, 1)$ and $(0, 1)$. But since $|O_{L(\tilde{c})}| \geq |O_c|$, and sites $\tilde{c}$ and c are not identical, site-pair $(L(\tilde{c}), c)$ must also have the ordered binary pair $(1, 0)$, and hence c must conflict with $L(\tilde{c})$.

Conversely, if it is not the case that $M(first(\tilde{c}), c) = M(last(\tilde{c}), c) = 1$, then either $M(first(\tilde{c}), c) = 0$, or $M(last(\tilde{c}), c) = 0$, in which case, the site-pair $(c, \tilde{c})$ has the ordered binary pair $(0, 1)$. But, by assumption, $O_c \cap O_{\tilde{c}} \neq \emptyset$, so the site pair also has the binary pair $(1, 1)$. Finally, as above, since $|O_c| \geq |O_{\tilde{c}}|$, and sites $\tilde{c}$ and c are not identical, the site-pair $(c, \tilde{c})$ must also have the ordered binary pair $(1, 0)$, and hence sites c and $\tilde{c}$ conflict.

The algorithmic consequence is that when site c is encountered in the scan of row f, and $f \in O_c$, we test whether $M(first(\tilde{c}), c) = M(last(\tilde{c}), c) = 1$. If so, then c conflicts with $L(\tilde{c})$, and if not, then c conflicts with $\tilde{c}$. ∎

In summary we conclude

Theorem 6.3.4 *A partition of the nodes into connected components of the conflict graph* $G_0(M)$ *can be found in* $O(nm)$ *time. In fact, a subgraph of* $G_0(M)$ *that preserves the connected components can be found in* $O(nm)$ *time.*

The second statement in theorem 6.3.4 was established in [63]. It is also noted there that a *spanning tree* of each connected component of $G_0(M)$ can be found in $O(nm)$ time, and having such a spanning tree is of use in several applications. Also, since any fully-compatible site forms a connected component of its own, we have:

Corollary 6.3.1 *All of the sites that are fully compatible in* M *can be found in* $O(nm)$ *time.*

A different algorithm was developed in [47] that finds all of the fully-compatible sites in time proportional to the number of rows plus the number of ones in M.

6.3.3 The Case of the Incompatibility Graph

A related version of theorem 6.3.4 also holds for the incompatibility graph.

Theorem 6.3.5 *The partition of the nodes into connected components of the incompatibility graph* $G(M)$ *can be found in* $O(nm)$ *time.*

Proof: One way to see this is to reduce the case of the incompatibility graph to the case of the conflict graph by transforming the input matrix M to a new matrix M'. Let f be any taxon in M. For every site c where taxon f has state 1, change every 0 to a 1, and every 1 to a 0, and let M' be the new matrix. Row f in M' is now the all-zero row, and it is easy to verify that any pair of sites c, d are incompatible in M, if and only if they conflict in M'. So to find the partition of the nodes into connected components of the incompatibility graph $G(M)$, we transform M to M' (in $O(nm)$ time) and then apply theorem 6.3.4 to M'. ∎

7   First Uses of Fundamental Structure

No theory is good except on condition that one use it to go on beyond.
— André Gide

In this chapter we discuss two uses of the fundamental combinatorial results developed in chapter 6. The first use establishes a graph-theoretic lower bound on $Rmin(M)$, called the *connected-component lower bound*. As with other lower bounds, the *connected-component lower bound* is useful when used to obtain local lower bounds in the composite bound method, but it is also very useful as a mathematical tool, particularly in the discussion of *galled trees* in chapter 8. The second use of fundamental structure is to establish a key structural result about ARGs, called the *full-decomposition theorem*. That theorem will be central in discussing constructive methods for ARGs, again, particularly in the construction of galled trees.

7.1   The Connected-Component Lower Bound on $Rmin$

Definition Let $cc(M)$ be the number of *nontrivial* connected components in the incompatibility graph $G(M)$ for M.

Recall that s_r denotes the ancestral sequence of an ARG. The main result of this section is:

The connected-component (cc) lower bound

Theorem 7.1.1 *For a set of sequences M, $Rmin(M) \geq cc(M)$, and $R^m min(M) \geq cc(M)$. That is, $cc(M)$ is a lower bound on both $Rmin(M)$ and $R^m min(M)$. Similarly,* *if a fixed ancestral sequence s_r is known, then $Rmin_{s_r} \geq cc(M \cup \{s_r\})$ and $R^m min_{s_r} \geq cc(M \cup \{s_r\})$.*

For example, for the data M in figure 6.1, the *connected-component lower bound* is two, as established in figure 6.2. Figure 3.8 (page 75) shows an ARG that generates this data using only two recombination nodes. So we can now conclude that the ARG in Figure 3.8 is a MinARG for M. The *connected-component lower bound* was derived independently in [17, 149]. The result from [149] was published in [150].

In order to prove the correctness of the *connected-component lower bound*, we establish a more basic combinatorial result.

Definition A pair of incompatible sites (c, d) in M may become compatible after the removal of a single sequence from M. In that case, we say that a sequence removal *breaks* the (c, d) incompatibility.

For example, in figure 6.1 (page 178), the removal of sequence r_1 breaks the $(1, 3)$ incompatibility and the $(1, 4)$ incompatibility, but not the $(4, 5)$ incompatibility. Note that site-pairs $(1, 3)$ and $(1, 4)$ are in the same connected component of $G(M)$, and that pair $(4, 5)$ is in a different connected component. The next lemma shows that this must be the case.

Lemma 7.1.1 *Suppose $G(M)$ has more than one nontrivial connected component. The removal of a single sequence from M can break incompatibilities in at most one nontrivial connected component of $G(M)$. That is, all of the incompatible pairs of sites in M that become compatible after a single sequence removal, must be represented by edges in the same connected component of $G(M)$.*

Proof: Let C and C' be two nontrivial connected components of $G(M)$. We will show that if the removal of a sequence f from M breaks an incompatibility between sites c, d in C, then that removal does not break any incompatibility in C'.

We claim that $M(C)$ must have at least three rows containing variable sequences of $< C, C' >$. This is because all four binary combinations appear in $M(C)$ at sites c, d, but at c, d only a single binary combination can appear in the rows containing the constant sequence of $< C, C' >$. Similarly, $M(C')$ must have at least three rows containing variable sequences of $< C', C >$. By corollary 6.2.2, every row of M that has a variable sequence of $< C, C' >$, in $M(C)$, must have the constant sequence in $M(C')$, so there are at least three rows with the constant sequence of $< C', C >$ in $M(C')$. Similarly, there must be at least three rows with the constant sequence of $< C, C' >$ in $M(C)$.

Since the removal of sequence f from M (and hence $M(C)$) breaks the c, d incompatibility, the binary pair in row f that appears at sites c, d *cannot* appear in any other row in $M(C)$ at sites c, d. If it did, then the same binary pair would remain in sites c, d after the removal of sequence f, and the incompatibility between sites c and d would remain. Therefore, f must contain a variable sequence of $< C, C' >$ in $M(C)$, and hence f must contain the constant sequence of $< C', C >$ in $M(C')$. But because the (same) constant sequence of $< C', C >$ in $M(C')$ must appear in at least three rows, the removal of f from $M(C')$ cannot break the incompatibility of any pair of sites in C'. ∎

Now we can establish the *connected-component lower bound* theorem.

Proof (Of theorem 7.1.1): We will prove the theorem for $Rmin(M)$. The proof for $R^m min(M)$ holds by replacing $Rmin$ with $R^m min$ everywhere.

Suppose the claim in the theorem is untrue. Among all counterexamples, select one, M, for which $Rmin(M)$ is smallest, and among those, choose M to be minimal. That is, the theorem holds for any data M' where $Rmin(M') < Rmin(M)$, or where $Rmin(M') = Rmin(M)$ but M' is a proper submatrix of M. Since M is a counterexample to the theorem, it must be that $cc(M) \geq Rmin(M) + 1$, i.e., $G(M)$ must have at least $Rmin(M) + 1$ nontrivial connected components.

Let $\mathcal{N}$ be a MinARG for M. Because $\mathcal{N}$ is a directed acyclic graph, if it has any recombination nodes, then it has one, x, which does not lead to another recombination node. So the subgraph of $\mathcal{N}$ rooted at x must be a directed tree, denoted T_x. From lemma 5.2.1 (page 129), any two incompatible sites c and d in M must occur together on some common recombination cycle in $\mathcal{N}$. Hence, no site labeling an edge in T_x can be incompatible with any other site, and hence no site labeling an edge in T_x can be in a nontrivial connected component of $G(M)$. Therefore, any site c that labels an edge in T_x can be removed from M and from $\mathcal{N}$ without changing $G(M)$, or the number of recombination nodes in the resulting ARG. Therefore, the resulting matrix M' would also be a counterexample to the theorem. But then M' would violate the assumed minimality of M, hence there are no mutations on T_x, and T_x can be assumed to be a single directed edge from x to a leaf.

Recall that s_x denotes the sequence that labels recombination node x, and hence labels the leaf that x points to in $\mathcal{N}$. If s_x is not a sequence in M, then we can remove x and have fewer recombination nodes and so $\mathcal{N}$ would not be a MinARG for M. Therefore, s_x must be a sequence in M. If we delete s_x from M; and delete the leaf labeled s_x, and the node x, from $\mathcal{N}$, we have a set of sequences $M - \{s_x\}$, that can be derived on an ARG using at most $Rmin(M) - 1$ recombination nodes. By the minimality assumption, the theorem holds for $M - \{s_x\}$, so the incompatibility graph $G(M - \{s_x\})$ must have at most $Rmin(M) - 1$ nontrivial connected components. But graph $G(M)$ had at least $Rmin(M) + 1$ nontrivial connected components, so the removal of sequence s_x from M must have resulted in the removal of edges in at least two distinct nontrivial connected components of $G(M)$. That is, the removal of s_x must have broken incompatibilities represented in at least two distinct connected components of $G(M)$. But lemma 7.1.1 makes this impossible, and so theorem 7.1.1 is proved.   ∎

By using theorem 6.3.5, we have:

Theorem 7.1.2 $cc(M)$ *can be computed in* $O(nm)$ *time.*

7.1.1   A Root-Known Version of the Connected-Component Lower Bound

Recall that $Rmin_{s_r}(M)$ is defined as the minimum number of recombination nodes in any ARG that generates M and has ancestral sequence s_r. When the problem specifies that the ancestral sequence must be a specific sequence s_r, $Rmin_{s_r}(M)$ might be larger than $cc(M)$, and it may be possible to establish a higher lower bound on $Rmin_{s_r}$, than $cc(M)$. To address this, we simply add s_r to M and apply theorem 7.1.1 to the set of sequences $M \cup \{s_r\}$. The number of nontrivial connected components in $G(M \cup \{s_r\})$ is a lower bound on $Rmin_{s_r}$. It is interesting to note, however, that $G(M \cup \{s_r\})$ might have more nontrivial connected components than $G(M)$, but it is also possible that it has *fewer* nontrivial connected components. Examples of this are easy to create. Hence, we have:

Theorem 7.1.3 *When a specific sequence* s_r *must be used as the ancestral sequence in any ARG that derives* M, $Rmin_{s_r}(M) \geq \max[cc(M), cc(M \cup \{s_r\})]$.

7.1.2   Use of the Connected-Component Lower Bound in the Composite-Bound Method

The number of nontrivial connected components of $G(M)$ is not affected by permuting the order of the sites in M. Therefore, as discussed in section 5.2.4.2, if we use the connected-component lower bound as a local lower bound in each interval of M, and then use the composite method to obtain a global lower bound, the resulting bound will generally be much higher than $cc(M)$. We can also use the *connected-component lower bound* on sub-sets of sites in an interval, as discussed in section 5.2.4.3, to obtain higher local lower bounds. Further, in each interval I, we can use the maximum of $cc(I(M))$ and $H(I(M))$ as the local lower bound, and this approach extends to subsets of sites as well.

7.1.3   Extensions to Other Biological Phenomena

We noted earlier that $G(M)$ and theorem 6.2.4 and corollary 6.2.2 rely only on the incompatibilities in M and not on any particular biological phenomena that created those incompatibilities. This property holds for lemma 7.1.1 as well. The proof of theorem 7.1.1 also does not rely on any particular biological phenomena, but only on the definitions of an ARG and the fact that at most one new sequence is generated by a single recombination event. It is therefore easy to see that theorem 7.1.1 would also hold if general reticulation events are allowed. More generally, since multiple-crossover recombination can be used to model other biological phenomena (see section 14.3), the connected-component lower bound applies to those phenomena as well.

7.2 The ARG Full-Decomposition Theory

We now examine the second use of the fundamental structure developed in chapter 6. We will state and prove a central result about ARGs, called the *full-decomposition theorem*, developed in [144, 145]. We believe the full-decomposition theorem and insights obtained from its proof are fundamental, and extend the theory of perfect phylogeny from trees to general reticulation networks and ARGs. The full-decomposition theorem will also be used in chapter 8 to show how a *galled tree* for M (when it exists) can be efficiently constructed. We now begin the needed definitions and facts that lead to the statement and proof of the full-decomposition theorem.

7.2.1 Cycles That Share Edges Form Blobs in ARGs

Consider a recombination cycle Q_1 in an ARG $\mathcal{N}$ that shares at least one *edge* with some other recombination cycle Q_2 (see figure 7.1). If one or both of Q_1 or Q_2 also shares an edge with another recombination cycle Q_3, we can add Q_3 to Q_1 and Q_2 to obtain a bigger complex of cycles (see figure 7.2). Continuing in this way, iteratively adding one more recombination cycle to the growing complex, if and only if it shares an edge with at least one cycle already in the complex, we ultimately get a well defined, *maximal* complex of recombination cycles in $\mathcal{N}$. We call such a maximal set of recombination cycles a "blob." A more formal and graph-theoretic definition of a blob is obtained through the concept of a *biconnected component* in an undirected graph, defined on page 24.

Definition A "blob" in an ARG $\mathcal{N}$ is a biconnected component of the undirected graph formed by removing directions from all the edges in $\mathcal{N}$.

Because of maximality, the blobs in a phylogenetic network $\mathcal{N}$ are well defined. That is, each blob can be found by the preceding iterative procedure, starting from any recombination cycle in the blob. Figure 7.3 shows a tree with two blobs, one blob containing two interconnected cycles, and one blob containing only one cycle.

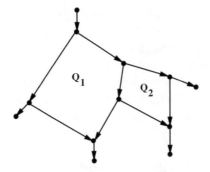

Figure 7.1   Recombination cycles Q_1 and Q_2 share an edge and are therefore contained in the same blob of $\mathcal{N}$.

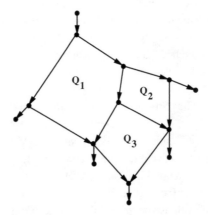

Figure 7.2   Recombination cycle Q_3 can be added to the growing blob containing recombination cycles Q_1 and Q_2.

Node disjoint assumption

For ease of exposition, we will also assume that two blobs are not only edge-disjoint but also *node-disjoint*. If two blobs in $\mathcal{N}$ share a node (but of course do not share an edge) then we can separate them by adding an edge without a label between the two blobs, labeling both ends of the new edge with the sequence labeling the shared node, and directing the new edge away from the root of $\mathcal{N}$. The result is an ARG that derives the same sequences as $\mathcal{N}$ and contains the same blobs as $\mathcal{N}$, but the blobs are now node disjoint.

Figure 7.3 A blobbed tree with several recombination cycles. How many cycles and blobs do you count? I call it four cycles and two blobs, but four and one is also defensible.

7.2.2 The Backbone Tree of an ARG

Since each blob is node-disjoint from all other blobs, if we contract each blob in an ARG $\mathcal{N}$ to a single node, the resulting graph must be a *directed tree*. To see this, note that if the contracted graph continued to have a cycle (in the underlying undirected graph) that cycle would correspond to a recombination cycle in a blob of $\mathcal{N}$ and therefore should have been contracted.

Definition Given any ARG $\mathcal{N}$ and its well defined, node-disjoint blobs, the directed tree obtained by contracting each blob to a single node is denoted $T_{\mathcal{N}}$ and is called the *backbone tree* of $\mathcal{N}$. When the edges of $T_{\mathcal{N}}$ are made undirected, the resulting tree is called the "undirected backbone tree" of $\mathcal{N}$.

Note that the edges of $T_{\mathcal{N}}$ are exactly the edges of $\mathcal{N}$ that are not contained in any recombination cycle in $\mathcal{N}$. These edges are the cut edges of $\mathcal{N}$.

Blobbed-tree Given these definitions, and since the blobs of an ARG $\mathcal{N}$ are well defined, any ARG $\mathcal{N}$ can be viewed as a well defined set of blobs connected together by a well defined (directed) backbone tree $T_{\mathcal{N}}$. So conceptually, we can view any ARG as a *directed tree of blobs* or a *blobbed-tree*. See figure 7.4.

Definition For each blob B in $\mathcal{N}$, there is a single node which must be on every path from the root of $\mathcal{N}$ to any node in B. Such a node in B is called the *root node* of B, or the *coalescent node* of B.

7.2.3 The Full-Decomposition Theorem and Its Reverse

Recall that a site that is not contained in any incompatible pair of sites is called a *fully-compatible* site. We can now state and prove the

Full-decomposition theorem:

Theorem 7.2.1 *Let $G(M)$ be the incompatibility graph for the set of sequences M. Then there is an ARG $\mathcal{N}$ that derives M, where each blob in $\mathcal{N}$ contains* all and only *the sites of a* single *nontrivial connected component of $G(M)$, and every fully-compatible site is on a cut edge of $\mathcal{N}$. The result holds no matter what constraints, if any, are placed on the number of crossovers that are allowed at a recombination node.*

Stated another way, for any input M, there is a blobbed-tree that derives M, where the blobs are in *one-to-one correspondence* with the nontrivial connected components of $G(M)$, and if B_C is the blob corresponding to nontrivial connected component C in $G(M)$, then B_C contains all and only the sites in C.

Definition An ARG $\mathcal{N}$ is called *fully decomposed* if it has the structure specified in theorem 7.2.1.

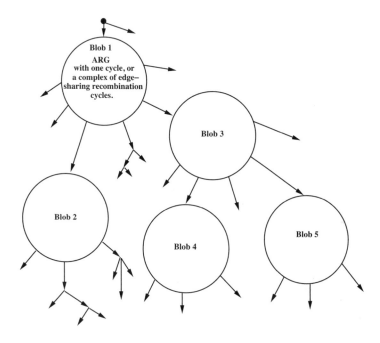

Figure 7.4   Cartoon of a Blobbed tree with five blobs.

Definition For any fully decomposed ARG for M and any nontrivial connected component, C, of $G(M)$, we use B_C to denote the blob in $\mathcal{N}$ containing all, and only, the sites of C.

Note that the number of blobs in a fully decomposed ARG for M is exactly $cc(M)$, the number of nontrivial connected components of $G(M)$. This is one of the key *invariant* features of a fully decomposed ARG.

Figure 7.5 shows a fully decomposed ARG for the sequences M from figure 6.1 (page 178). An ARG for this M that is not fully decomposed is shown in figure 3.8 (page 75).

Figure 7.6 shows another ARG with a single blob, and figure 7.7 shows a fully decomposed ARG with two blobs that generate the same set of sequences. The ARG in figure 7.6 is a MinARG for the sequence it generates. These two ARGs will be used later in this chapter. These ARGs generate a set M of six sequences, shown at the leaves. In M, sites $\{2,3,5\}$ form a connected component of the incompatibility graph $G(M)$, and sites $\{1,4,6\}$ form another connected component.

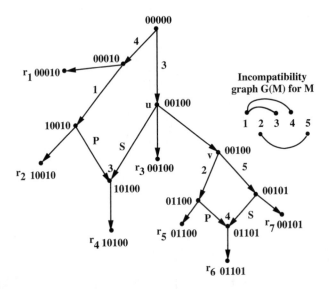

Figure 7.5 A fully decomposed ARG generating the same set of sequences as were generated by the ARG in figure 3.8 (page 75). The ARG in figure 3.8 has two recombination cycles that share the edge labeled 2. In this fully decomposed ARG, there are two blobs corresponding to the two nontrivial connected components of the incompatibility graph. One of those components, C, contains sites $\{1, 3, 4\}$ and the other component, C', contains sites $\{2, 5\}$. In this example, each blob consists of a single recombination cycle, but this is not true in general.

The reverse

The strict "converse" of theorem 7.2.1 is not true. A strict converse would say that all ARGs *must be* fully decomposed, which is false. However, some properties of the converse are true, and we call those properties the "reverse" of theorem 7.2.1, stated as follows:

Theorem 7.2.2 *In any ARG $\mathcal{N}$ that derives M, all sites from the same nontrivial connected component C of $G(M)$ must appear on the same blob in $\mathcal{N}$, and this does not depend on the number of crossovers used at recombination nodes.*

The proof of theorem 7.2.2 is easy, and will be presented after the proof of the full-decomposition theorem. An immediate consequence, which also follows from lemma 5.2.1, is that if a blob contains one incompatible site, it must contain at least one pair of incompatible sites.

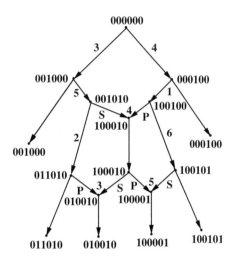

Figure 7.6 An ARG that is not fully decomposed, deriving six sequences with three recombination nodes. Note that in the first recombination, the *S*-edge into the recombination node is shown on the left and the *P*-edge is shown on the right. Program HapBound -S gives a lower bound of three for the sequences generated by this ARG, showing that the ARG is a MinARG.

Why the term?

Given theorem 7.2.2, it is not possible to split up the sites from one connected component of $G(M)$ into two or more blobs in $\mathcal{N}$. Therefore a fully decomposed ARG is as highly decomposed into distinct blobs as possible, justifying the term "fully decomposed."

A rooted version

There is an analogous theorem to theorem 7.2.1 in the case that the ancestral sequence s_r is known in advance. In that case, there is an ARG $\mathcal{N}$ that derives M, with ancestral sequence s_r, where the blobs in $\mathcal{N}$ are in one-to-one correspondence with the nontrivial connected components of $G_{s_r}(M)$, and any fully-compatible site is on a cut edge of $\mathcal{N}$. Again, no finer decomposition is possible. This follows from theorem 7.2.2, and the fact that two sites conflict relative to s_r if and only if they are incompatible in $M \cup \{s_r\}$.

We will prove the full-decomposition theorem and its reverse in the next section and (with some technical qualification) show that the undirected backbone tree is *invariant* over all the fully decomposed ARGs for M.

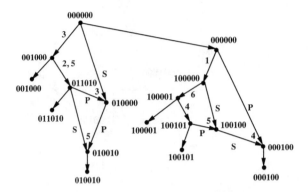

Figure 7.7   A fully decomposed ARG generating the same sequences as the ARG in figure 7.6. Note that this ARG has four recombination nodes in two blobs.

7.3   Proof of the Full-Decomposition Theorem and Its Reverse

We first prove theorem 7.2.1, the full-decomposition theorem. We will do this constructively, through several lemmas and corollaries.

7.3.1   The Superstates of M and the Matrix W

We begin by defining a new set of binary sequences W created from M and $G(M)$, and we represent the set W as a matrix. Recall that for any connected component C of $G(M)$, $M(C)$ is the set of sequences in M restricted to the sites in C. **Definition** A sequence s is called a *superstate* of M if s is a sequence in $M(C)$ for some connected component C of $G(M)$. We say that superstate s *originates* from C.

For example, figure 6.5 (page 187) shows that the superstates of connected component C are $\{001, 101, 110, 010\}$ and the superstates of connected component C' are $\{00, 01, 10, 11\}$. Those superstates come from the data M in figure 6.1 (page 178) and its incompatibility graph $G(M)$ in figure 6.2 (page 178).

Definition W is a binary matrix with one row for each taxon in M and one column for each superstate of M. If superstate s originates from connected component C, then the column of W associated with s has state 1 for each row of M that contains sequence s in $M(C)$, and otherwise has state 0. The new column defines a binary *character* derived from M and $G(M)$.

	1	2	3	4	5	6	7	8
r_1	1	0	0	0	1	0	0	0
r_2	0	1	0	0	1	0	0	0
r_3	0	0	1	0	1	0	0	0
r_4	0	0	0	1	1	0	0	0
r_5	0	0	1	0	0	1	0	0
r_6	0	0	1	0	0	0	1	0
r_7	0	0	1	0	0	0	0	1

Figure 7.8 Matrix W derived from matrix M, shown in figure 6.1 (page 178), which is derived from the ARG shown in figure 3.8 (page 75). The incompatibility graph $G(M)$ is shown in figure 7.5 (page 206). The superstates of M originating from C are 001, 101, 010, 110, and that the superstates of M originating from C' are 00, 10, 11, 01. These are also show in figure 6.5 (page 187). The columns (characters) of W are ordered to correspond to those ordered lists of superstates of M.

Figure 7.8 shows the matrix W representing the superstates shown in figure 6.5.

Each column in W encodes a bipartition of the taxa of M where one side of the bipartition contains all the taxa in M that contain a particular sequence in $M(C)$, and the other side of the bipartition contains the remaining taxa of M. If C is a trivial connected-component, so that it only contains one site, then W will have two columns derived from that one site, one for the "sequence" 0 and one for the "sequence" 1, but those two columns define the same bipartition of the taxa. That will cause no problem. We will use the characters of W to build a tree in order to prove theorem 7.2.1. To begin that process, we first establish:

Lemma 7.3.1 *There are no incompatible pairs of sites in W.*

Proof: Consider two sites c, c' in W, representing superstates s and s', respectively. If s and s' originate from the same connected component C in $G(M)$, then by construction, no taxon can have state 1 for both sites, and therefore the sites are compatible.

Now suppose s and s' in W originate from two different connected components C and C' in $G(M)$. Sequence s is either a variable sequence of $< C, C' >$, or it is the constant sequence of $< C, C' >$. A similar statement holds for s' and $< C', C >$. We therefore have four cases to consider.

If s and s' are both variable sequences of $< C, C' >$ and $< C', C >$ respectively, then by corollary 6.2.2, there is no taxon in M that contains both sequences, and hence there is no row in W that contains the state-pair (1,1) for the site-pair (c, c'). It follows that c

and c' cannot be incompatible in this case. Symmetrically, if s and s' are each the constant sequence of $< C, C' >$ and $< C', C >$ respectively, then there is no taxon f in M that lacks both sequences, for then f would contain variable sequences of both $< C, C' >$ and $< C', C >$, which is not possible. Hence, in this case also there is no row in W that contains the state-pair $(0, 0)$ for the site-pair (c, c').

Now suppose s is the constant sequence of $< C, C' >$ and s' is a variable sequence of $< C', C >$. Then there can be no row f of W that contains the state-pair $(0, 1)$ for site-pair (c, c'), for otherwise taxon f would contain variable sequences of both $< C, C' >$ and $< C', C >$. Symmetrically, if s is a variable sequence of $< C, C' >$ and s' is the constant sequence of $< C', C >$, then there is no row f in W that has state-pair $(1, 0)$ for site-pair (c, c').

In all cases, one of the four binary combinations 0,0; 1,1; 0,1; 1,0 is missing from the site-pair (c, c'), and so s and s' are compatible. It follows that no pair of sites in W is incompatible.   ∎

Since W is a binary matrix, we can apply the splits-equivalence theorem (theorem 2.4.2, page 57), establishing:

Corollary 7.3.1 *There is an undirected tree $\overline{T}(W)$, whose leaves are in one-to-one correspondence with the rows of M; where each site c of W labels exactly one edge e_c in $\overline{T}(W)$; where the split in $\overline{T}(W)$ defined by edge e_c exactly specifies the bipartition of rows of M encoded by column c; and where each internal edge is labeled by one or more sites of W. An edge touching a leaf node might not be labeled.*

We can also use the perfect-phylogeny theorem to obtain corollary 7.3.1.

Converting to $\overline{T}(M)$

Since each site of W represents a superstate of M, and the set of taxa of M and W are the same, we can equivalently think of $\overline{T}(W)$, whose labeled edges are labeled by sites of W, as a tree whose labeled edges are labeled by the corresponding superstates of M. This leads to:

Definition $\overline{T}(M)$ is the undirected tree obtained from $\overline{T}(W)$ by replacing each edge label c by the superstate of M that c represents.

For example, figure 7.9 shows $\overline{T}(W)$ and $\overline{T}(M)$ for matrix W in figure 7.8.

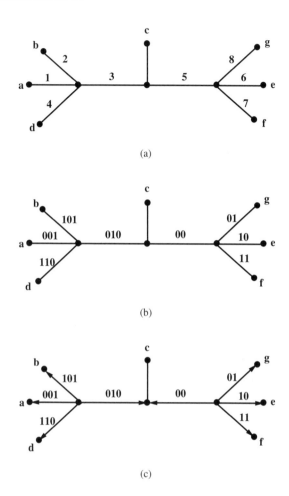

(a)

(b)

(c)

Figure 7.9 (*a*) Tree $\overline{T}(W)$ derived from matrix W in figure 7.8.
(*b*) The tree $\overline{T}(M)$ converted from $\overline{T}(W)$, where edges are labeled by the superstates of M corresponding to the characters of W. The superstates of length three originate from C, and the superstates of length two originate from C'.
(*c*) Each edge is directed toward the 1-side of the superstate labeling the edge.

Clearly, we can constructively find $\overline{T}(M)$ by building W from M and then finding the unique undirected perfect phylogeny for W. It may seem that $\overline{T}(M)$ could be obtained more simply by building an undirected perfect phylogeny T, using one site from each connected component of $G(M)$. However, this is not correct, because the edge structure of T may be very different from that of $\overline{T}(M)$.

7.3.2   Using $\overline{T}(M)$ to Create a Fully Decomposed ARG for M

The next step in establishing the full-decomposition theorem is to show a mapping from the nontrivial connected components of $G(M)$ to internal nodes of $\overline{T}(M)$. After that, we will *inflate* each mapped internal node of $\overline{T}(M)$ into a blob containing all and only the sites from the associated nontrivial connected component. We may assume that each internal node of $\overline{T}(M)$ has degree of at least three.

Definition Let edge e in $\overline{T}(M)$ be labeled by a superstate s originating from connected component C in $G(M)$. We define the *1-side* of s as the subtree of $\overline{T}(M) - e$ that contains the taxa of M (at leaves) which have sequence s in $M(C)$ (equivalently, have superstate s in W). Symmetrically, the "0-side" of s is the subtree of $\overline{T}(M) - e$ containing the taxa that do *not* have sequence s in $M(C)$.

For example, in figure 7.9c, the edges are directed toward the 1-side of the superstate labeling the edge.

Lemma 7.3.2 *If s and s' are superstates that originate from the same connected component, then the 1-sides of s and s' are disjoint.*

Proof: Let edges e and e' be labeled with superstates s and s', that originate from the same connected component, C, of $G(M)$. By definition, any taxon on the 1-side of s (or of s') in $\overline{T}(M)$, has sequence s (or s' in $M(C)$. But a taxon is associated with only one row in M, and hence only one distinct sequence in $M(C)$. By definition, two superstates are distinct, and hence the 1-sides of s and s' must be disjoint.   ∎

Given lemma 7.3.2, if we direct edges e and e' in the direction of the 1-sides of s and s' respectively, the directions of the edges must "point away" from each other. See figure 7.9c. Also, in figure 7.10, superstates s_1 and s_2 are assumed to originate from the same connected component, and label edges e_1 and e_2, respectively.

Lemma 7.3.3 *Let C be a nontrivial connected component of $G(M)$. All the edges labeled by superstates of M that originate from the same connected component C are incident with a single node, denoted v_C, of $\overline{T}(M)$. That is, those edges form a star around v_C. Further,*

Figure 7.10 When e_1 and e_2 are not incident with each other, the situation must be as pictured here. Superstates s_1 and s_2 are assumed to originate from the same connected component, and hence their 1-sides must point away from each other.

every edge incident with v_C is labeled by exactly one superstate s that originates from C, and v_C is on the 0-side of s.

Note that lemma 7.3.3 does *not* say that v_C is incident only with edges labeled by superstates that originate from C. In fact, an edge can be labeled with two superstates, originating from two different connected components of $G(M)$. We will more completely characterize this situation in the following proof and discussion.

Proof (Of lemma 7.3.3): For any nontrivial connected component C, the submatrix $M(C)$ contains at least four distinct sequences, since there must be at least one incompatible pair of sites in C, and hence all four binary pairs must appear in $M(C)$. Therefore, there are at least four superstates s_1, s_2, s_3, s_4 of M that originate from C. Each of those four superstates is possessed by a distinct set of taxa and therefore these four superstates label distinct edges e_1, e_2, e_3, e_4, respectively in $\overline{T}(M)$. Consider any two of those superstates, say s_1 and s_2. We will first show that edges e_1 and e_2 must be incident with each other.

Suppose e_1 and e_2 are not incident with each other. Then there is a unique shortest path P from an endpoint of e_1 to an endpoint of e_2, and it must have at least one edge, since e_1 is not incident with e_2. By the "shortest" requirement, path P does not contain edge e_1 or e_2. Path P also cannot go through edge e_3 or e_4. If it did, then some leaf on the 1-side of s_3 or s_4 would also be on the 1-side of s_1 or s_2, violating lemma 7.3.2. So, there is an edge e that is incident with e_1 on path P, where e is not labeled by a superstate of M that originates from C. We will show that such an edge e cannot exist, and so e_1 and e_2 must be incident with each other.

Every internal edge in $\overline{T}(M)$ is labeled by some superstate of M, so e is labeled by a superstate that originates from a connected component $C' \neq C$. Let v be the common endpoint of e_1 and e. Since v has degree three or more, there is a leaf l_v that is reachable from v without going through e_1 or e. See figure 7.10.

Recall that each superstate of M that originates from C or C' is a sequence in $M(C)$ or $M(C')$ respectively. Recall also that there is a unique constant sequence s_C of $< C, C' >$ in $M(C)$, and a unique constant sequence $s_{C'}$ of $< C', C >$ in $M(C')$. Now either $s_1 = s_C$ or $s_1 \neq s_C$, and either e is labeled by $s_{C'}$ or not, so we have four cases to consider. The proof in each case uses similar ideas, but the details are explicitly given for completeness.

Case 1: Suppose $s_1 = s_C$ and e is labeled by $s_{C'}$. Since the 1-side of s_1 and the 1-side of s_2 point away from each other, $l(v)$ must be on the 0-side of s_1, so it represents a taxon that has a variable sequence of $< C, C' >$ in $M(C)$. Therefore, $l(v)$ must be on the 1-side of $s_{C'}$ by corollary 6.2.2, and so edge e_2 must be on the 0-side of $s_{C'}$. But then, all leaves on the 0-side of $s_{C'}$ will be on the 0-side of both s_C and $s_{C'}$, and so all the taxa at those leaves will have a variable sequence of $< C, C' >$ in $M(C)$ and also a variable sequence of $< C', C >$ in $M(C')$. That contradicts corollary 6.2.2 (page 185), so e cannot exist in this case.

Case 2: Suppose $s_1 = s_C$, but e is not labeled by $s_{C'}$, so e is labeled with a superstate s' from $M(C')$ that is a variable sequence of $< C', C >$. Again, $l(v)$ is on the 0-side of s_C, so it represents a taxon that has a variable sequence of $< C, C' >$. Therefore, it cannot also be on the 1-side of s', by corollary 6.2.2. But then, all the leaves on the 1-side of s_2 will be on the 1-side of s'. Since e_1 is labeled with the constant sequence of $< C, C' >$, e_2 must be labeled with a variable sequence of $< C, C' >$, so all the taxa on the 1-side of s_2 will have a variable sequence of $< C, C' >$ and also a variable sequence of $< C', C >$. That again contradicts corollary 6.2.2, so edge e cannot exist in this case.

Case 3: Suppose $s_1 \neq s_C$, and e is not labeled by $s_{C'}$, so s_1 is a variable sequence of $< C, C' >$, and e is labeled by a variable sequence s' of $< C', C >$. As before, edge e is on the 0-side of s_1, so by corollary 6.2.2, edge e_2 must be on the 1-side of s'. Then, by corollary 6.2.2, e_2 must be labeled by s_C, the constant sequence of $< C, C' >$. Now let u be the endpoint of e_2 on the e side of e_2, and let l_u be a leaf reachable from u without going through e_2 or e. Such a leaf l_u must exist, and it is on the 1-side of s'. The taxon at l_u must have a variable sequence of $< C, C' >$, since l_u is on the 0-side of s_2. But then the taxon at l_u has a variable sequence $< C, C' >$ and also a variable sequence of $< C', C >$, which is not possible. So e cannot exist in this case.

Case 4: Suppose $s_1 \neq s_C$, but e is labeled by $s_{C'}$. Then, by corollary 6.2.2, edge e_1 must be on the 1-side of $s_{C'}$, so if e_2 were labeled by a variable sequence of $< C, C' >$, any leaf on the 1-side of s_2 would be on the 1-side of both $s_{C'}$ and s_2, and the taxon at that leaf would have a variable sequence for $< C, C' >$ and $< C', C >$, which is impossible. So, e_2 must be labeled by the constant sequence of $< C, C' >$, and leaf l_u must be labeled with a variable sequence of $< C, C' >$. But then the taxa at l_u would have a variable sequence for both $< C, C' >$ and a variable sequence for $< C', C >$. Again, that is not possible, so e cannot exist in this case.

We have shown that e cannot exist in any case, and hence e_1 and e_2 must be incident with each other. Since e_1 and e_2 were arbitrary edges labeled by superstates that originate from C, every pair of edges labeled by superstates that originate from C must be incident with each other. But in a tree, that is only possible if all those edges share exactly one endpoint, and so form a star around a single node. That node is the claimed node v_C. We have also established that if there are two distinct edges labeled with superstates that originate from C, then the 1-sides of those superstates point away from each other. This holds for any pair of edges labeled with superstates that originate from C, so v_C is on the 0-side of every such superstate.

Now we show that every edge incident with v_C is labeled by exactly one superstate from C. If an incident edge e is not labeled by a superstate originating from C, or is labeled by two or more such superstates, then any leaf reachable from v_C through edge e would either have state 0 for each superstate from C, or would have state 1 for two or more such superstates. But each leaf represents a taxon in M, and for each taxon f in M there is *exactly* one superstate c that originates from C such that f has state 1 for c. ∎

Definition A node v_C in $\overline{T}(M)$, obeying the properties described in lemma 7.3.3 is called a *star node*.

So for each nontrivial connected component C in $G(M)$ there is exactly one star node v_C in $\overline{T}(M)$ such that all of the edges incident with v_C are labeled by a superstate from $M(C)$, and the degree of v_C is exactly the number of superstates that originate from C.

We next show that there is a distinct star node v_C for each nontrivial connected component C.

Lemma 7.3.4 *If C and C' are distinct, nontrivial connected components in $G(M)$, then $v_C \neq v_{C'}$.*

Proof: For contradiction, suppose $v_C = v_{C'}$, so each edge incident with v_C is labeled by one superstate that originates from C, and by one superstate that originates from C', and the 1-sides of both of these characters point away from v_C.

There is at least one pair of sites p, q in C' that are incompatible, so there must be at least three variable sequences of $< C', C >$, since the constant sequence of $< C', C >$ has only one binary pair in site-pair (p, q). By theorem 6.2.4, every taxon in M that has a variable sequence of $< C', C >$ in $M(C')$ has the constant sequence of $< C, C' >$ in $M(C)$. So there must be at least three taxa in M with distinct sequences in $M(C')$, which all have the constant sequence of $< C, C' >$ in $M(C)$.

Now let e be the edge incident with v_C that is labeled by the constant sequence, s_C, of $< C, C' >$, and let s' denote the sequence of $M(C')$ that labels edge e. Since the 1-sides of s_C and s' point away from v_C, the set of taxa (at leaves of $\overline{T}(M)$) that have sequence s_C in $M(C)$ must also have sequence s' in $M(C')$. But that contradicts the fact that there must be at least three taxa that have s_C in $M(C)$ and have distinct sequences in $M(C')$. ∎

Lemmas 7.3.3 and 7.3.4 establish the claimed mapping from the nontrivial connected components of $G(M)$ into the internal nodes of $\overline{T}(M)$. The mapping is *injective*, that is, some internal node(s) of $\overline{T}(M)$ are not star nodes. For example, see figure 7.9c. Note that the lemmas also establish that an edge in $\overline{T}(M)$ can be labeled with superstates that originate from *trivial* connected components (each of which corresponds to a site that is fully compatible), and with two superstates originating from different *nontrivial* connected components. In the latter case, the 1-sides of those two superstates must point in opposite directions.

7.3.3   Completion of the Proof

Recall that for any taxon f in M, s_f denotes the sequence in M for f, and for any node v in an ARG $\mathcal{N}$, s_v denotes the sequence labeling v.

Definition For any *taxon* f and any connected component C in $G(M)$, $s_f(C)$ denotes the sequence s_f restricted to the sites in C. Similarly, for any *node* v, $s_v(C)$ denotes the sequence s_v restricted to the sites in C.

Definition If the sites in C are a proper subset of the sites in M, when we say that a node v is *C-labeled by* $s_v(C)$, we mean that its full label, s_v, restricted to the sites in C, is $s_v(C)$.

To finish the proof of the full-decomposition theorem, we need to *inflate* each star node v_C of $\overline{T}(M)$ to become a subARG that contains all and only the sites of its corresponding connected component C. That subARG must create all of the sequences in $M(C)$, that is, the superstates of M that originate from C. Formally, this is done in the following:

Proof (Of theorem 7.2.1): We arbitrarily select a leaf f of $\overline{T}(M)$ to be the root of $\overline{T}(M)$, where f identifies a taxon in M, and we direct all the edges in $\overline{T}(M)$ away from the root. So s_f becomes the ancestral sequence of the directed tree $\overline{T}(M)$.

Next, we need to inflate each star node v_C in $\overline{T}(M)$, mapped to by a nontrivial connected component C in $G(M)$, into a subARG. We define the *ancestral sequence of v_C* to be $s_f(C)$. Hence, the ancestral sequence of v_C is a sequence in $M(C)$ and is the superstate that labels the unique edge directed into v_C in the directed tree $\overline{T}(M)$.

By theorem 3.2.1, any superstate in M that originates from C can be derived from the ancestral sequence of v_C using at most one mutation in each site of C, and no mutations in sites outside of C, if an unlimited number of recombination nodes are allowed. This is true if only single-crossovers are allowed, or if multiple-crossovers are permitted.[1] So, each star node v_C can be inflated into a blob B_C with ancestral sequence $s_f(C)$, where B_C generates all the sequences in $M(C)$. The directed tree $\overline{T}(M)$ specifies how these blobs should be connected to create the fully decomposed ARG for M. In particular, for each superstate s that originates from C, we connect the node in B_C labeled s to the edge labeled s in the directed $\overline{T}(M)$.

After inflating each star node in $\overline{T}(M)$, the end result is a fully decomposed ARG $\mathcal{N}$ for M. Note that no fully-compatible site in M is contained in a blob of $\mathcal{N}$, and so each fully-compatible site labels a cut edge of $\mathcal{N}$. This completes the proof of theorem 7.2.1. ∎

7.3.4 Summary of the Procedure for Finding a Fully Decomposed ARG for M

The proof of the existence of a fully decomposed ARG for any set of data M contains in it a constructive procedure to *build* a fully decomposed ARG for M. For clarity, in figure 7.11 we summarize the constructive procedure implied in those discussions.

The causes of incompatibility are not relevant
Note that the existence and the topology of $\overline{T}(M)$ depends only on the *partition* of the nodes of $G(M)$ into connected components. This means that the particular incompatibilities inside of a connected component are not needed. It also means that the topology of $\overline{T}(M)$ does not depend on the biological causes of the incompatibilities in M. In particular, it does not depend on whether or not multiple-crossovers are allowed at recombination nodes. However, the networks inside each blob of $\mathcal{N}$ *do* depend on which biological events (such as single-crossover versus multiple-crossover recombination, or recurrent mutation, etc.) occur there.

---

1 It is also true that any sequence can be derived from the ancestral sequence without using recombination, if an unlimited number of *recurrent* or *back mutation* events are allowed. Recurrent mutation occurs when the state of a site mutates from its ancestral state more than once in an evolutionary history. Back mutation occurs when the state mutates from the derived state back to the ancestral state. Therefore, the full-decomposition theorem holds for the problem of computing a phylogenetic *tree* that generates M allowing back and recurrent mutations.

Algorithm FD-ARG(M)

(1) Find the incompatible pairs of sites in M and build the incompatibility
graph $G(M)$. Alternatively, find the partition of the nodes of $G(M)$ into
connected components, as discussed in section 6.3.
{For each connected component C of $G(M)$, the distinct sequences in $M(C)$
are the *superstates* originating from C. The superstates of M are
the union of the superstates originating from each connected-component of $G(M)$.}

(2) Create the matrix W containing one row for each taxon in M, and one column for
each superstate of M.
{If column c represents superstate s, and s originates from
connected component C, then entry (f, c) of W has state 1 if taxon f
has sequence s in $M(C)$, and has state 0 otherwise.}

(3) Create the undirected perfect phylogeny $\overline{T}(M)$ for matrix W.
Contract any node of degree two, so that all nodes in $\overline{T}(M)$ have degree at least three.
{Each superstate of M labels one edge in $\overline{T}(M)$.}

(4) Arbitrarily select one leaf, associated with taxon f say, of $\overline{T}(M)$
to be the root node r, and direct all the edges in $\overline{T}(M)$ away from r.
Add a new edge directed from r to a new leaf labeled by sequence s_f.

(5) For each nontrivial connected component C of $G(M)$, identify the star node v_C in
$\overline{T}(M)$ with that property that each of the superstates that originates from C labels
an edge incident with v_C. Let $s_f(C)$ be the superstate that originates from C
and labels the unique edge directed into v_C.
{$s_f(C)$ is called the *ancestral-sequence for v_C*.}

(6) For each nontrivial connected component C of $G(M)$, create an ARG B_C with
root sequence $s_f(C)$, that generates all the sequences in $M(C)$.
{Note that all and only the sites in C label edges in B_C.}

(7) Replace star node v_C in $\overline{T}(M)$ with the ARG B_C,
connecting the node in B_C labeled with sequence s of $M(C)$, to the
edge in $\overline{T}(M)$ labeled with s.
This creates the DAG and the edge labels of an ARG $\mathcal{N}$ for M.

Determine the node labels of $\mathcal{N}$ from the ancestral sequence, s_f
at the root of $\mathcal{N}$, and the edge labels, as described in the definition
of a genealogical network (page 70).

Figure 7.11   Algorithm *FD-ARG* for building a fully decomposed ARG for M.

Time analysis

We will use algorithm *FD-ARG* when discussing certain ARG construction methods, such as for *galled trees* in chapter 8. Those methods use steps (1) through (5), and step (7), without modification, but they specialize step (6) to build the subARGs for the blobs more efficiently than through the use of theorem 3.2.1. So it is useful to establish here the running times of steps (1) through (5). The time for step (7) is clearly bounded by the size of ARG constructed, which is bounded by the time for steps (1) through (5). We will also incorporate an additional optimization: since no fully-compatible site labels an edge inside any blob of the ARG constructed by algorithm *FD-ARG*, two identical fully-compatible sites in M must label the same edge (which is outside of any blob) in N, and so duplicate fully-compatible sites can be removed from M. We will therefore assume that the fully-compatible sites of M are distinct. The time needed to group together identical sites in M, and remove any duplicates, is at most $O(nm)$ by executing a radix sort of the columns of M.

The time analysis of algorithm *FD-ARG* can then be done as follows. First, FD-ARG finds all incompatible pairs of sites. This can be done explicitly in a straightforward way in $O(nm^2)$ time. However, as noted earlier, the explicit construction of the connected components of $G(M)$ is not needed; all that is required is the partition of the nodes into connected components. By theorem 6.3.5, that partition can be found in $O(nm)$ time.

Finding all of the superstates can be done in $O(nm)$ time by several different methods. For example, all the distinct sequences in $M(C)$ can be found by entering them into a *Lex-Tree*, or by *Radix-Sorting* the sequences, to group identical sequences together. The number of superstates is bounded by $O(n)$. To see this, recall that every superstate labels some edge in $\overline{T}(M)$, and no edge is labeled by more than two superstates that originate from nontrivial connected components. No edge is labeled by more than one superstate that originates from a trivial connected component, so the number of superstates is bounded by three times the number of edges of $\overline{T}(M)$. $\overline{T}(M)$ has n leaves and each internal node has degree three, so $\overline{T}(M)$ has at most $2n$ edges. Hence, there are at most $6n$ superstates of M, and at most $6n$ columns in W. Therefore, W and $\overline{T}(M)$ can be built in $O(n^2)$ time. In summary,

Theorem 7.3.1 *Steps (1) through (5) of algorithm FD-ARG can be implemented to run in* $O(nm + n^2)$ *time.*

If we build each blob B_C as specified in the proof of theorem 3.2.1, then the time for step (6) is bounded by $O(n^2m)$, so we have the following corollary to the full-decomposition theorem:

Corollary 7.3.2 *For any input M, a fully decomposed ARG for M can be constructed in $O(nm + n^2m)$ time.*

7.3.5   Invariant Properties of Fully Decomposed ARGs

In the discussion of the perfect-phylogeny problem, we noted the biological importance of the fact that when there is a perfect phylogeny for a set of sequences M, there is only *one, unique* perfect phylogeny for M. With a similar motivation, we are interested in properties of a fully decomposed ARG M that are *invariant* over all fully decomposed ARGs for M. For example, we have already seen that all fully decomposed ARGs for M have exactly $cc(M)$ blobs, and by the definition of a fully decomposed ARG, the partition of the sites of M into blobs and cut edges is invariant over all fully decomposed ARGs for M. There are additional invariants as well. Unfortunately, there are some degenerate ways to violate invariance, and excluding those requires adding some more conditions on a fully decomposed ARG.

Definition Let C be a connected component of the incompatibility graph $G(M)$ for M. If $s_r(C) \in M(C)$, then $s_r(C)$ is called a *visible segment*. Otherwise, it is an *invisible segment*.

Definition A node v in a blob B is an *external node* of B if v is adjacent (without regard to edge direction) to some node v' not in B. Note that edge (v, v') must be a cut edge.

   If v is an external node of blob B, then (v, v') is directed from v' to v, if and only if v is the root node of B. Otherwise, it is directed from v to v'.

Definition An ARG $\mathcal{N}$ is *canonical* if:
(a) No blob contains two nodes labeled with the same sequence.
(b) Each external node v (in a blob B) is incident with *exactly one* cut edge (which cannot be in B).
(c) If the root node v of a blob B is an external node, then the cut edge incident with v is directed into v.
(d) Every blob contains at least one incompatible pair of sites.

   For example, the ARG shown in figure 7.7 (page 208) is a canonical, fully decomposed ARG.

Theorem 7.3.2 *For any M, there is a canonical ARG for M.*

Proof: Let $\mathcal{N}$ be any ARG for M. If any blob in $\mathcal{N}$ has two nodes labeled with the same sequence, the two nodes can be merged to satisfy condition (a). If any external node v, which is not the root of a blob, is incident with more than one cut edge, we can disconnect those edges from v; create a new edge (v, w), directed from v; and reconnect all of the disconnected cut edges to w, satisfying condition (b). If node v is the root of a blob B, and there is a directed edge (v, v') from v to a node v' not in B, we add a new node w, and the edge (w, v) directed into v; then remove (v, v'); and create the directed edge (w, v') labeled with whatever label was on edge (v, v'). This satisfies condition (c). If $\mathcal{N}$ has a blob B that has no pair of incompatible sites, then there is a rooted perfect phylogeny T with the same ancestral sequence as the root of B, such that T generates the sequences in B. Blob B can then be replaced by T. Repeating this for every blob that contains no pair of incompatible sites, we satisfy condition (d). The result is an ARG that generates the same sequences as $\mathcal{N}$, satisfying the conditions for a canonical ARG. ∎

For example, the ARG shown in figure 7.5 (page 206) is not canonical; however, it can be made canonical by applying the above operations at the first node (of two nodes) labeled 00100. Note that the two nodes labeled 00100 are not in the same blob, so the duplicate node labeling does not violate the definition of a canonical ARG.

Clearly, if $\mathcal{N}$ is fully decomposed, but not canonical, the modifications described in the proof of theorem 7.3.2 maintain the fully decomposed property. So we have:

Corollary 7.3.3 *For any M, there is a canonical, fully decomposed ARG for M.*

Properties
Before directly addressing invariant properties, we need a few more tools. The following lemma establishes a central technical fact.

Lemma 7.3.5 *Let (v, v') be a cut edge directed from v to v' in an ARG $\mathcal{N}$ for M. Let $\mathcal{N}_{v'}$ be the subnetwork of $\mathcal{N}$ consisting of every node and edge reachable by some directed path from v', and suppose that site c of M does not label any edge in $\mathcal{N}_{v'}$ (i.e., does not mutate in $\mathcal{N}'_v$). Then, the state of site c at every node in $\mathcal{N}_{v'}$ is the same as at node v'. Further, if c does not label edge (v, v'), then the state of c at every node in $\mathcal{N}_{v'}$ is the same as it is at node v.*

Proof: Suppose that at some node in $\mathcal{N}_{v'}$, the state of c is different from that at v'. Let u be such a node with the property that at every ancestor of u in $\mathcal{N}_{v'}$, the state of c is the same as at v'. By construction, u cannot be v'. Since c only mutates once in $\mathcal{N}$, and not on an edge in $\mathcal{N}_{v'}$, the state of c can only change in $\mathcal{N}_{v'}$ due to recombination. So by the choice of u, node u must be a recombination node. Hence, at least one of the parents of u must be *in $\mathcal{N}_{v'}$.*

If both parents of u are in $\mathcal{N}_{v'}$, then by the choice of u, the state of c at both parents would be the same as at v', and so the state of c would be unchanged by recombination at u, regardless of what crossover index, b_u, is specified. So, one parent of u must be inside $\mathcal{N}_{v'}$, and the other parent of u, call it w, must be outside of $\mathcal{N}_{v'}$. Note that the edge (w, u) is not in B_C, and since u is not v', edge (w, u) cannot be the edge (v, v'). See figure 7.12 (page 223).

Let P be any path from the root of $\mathcal{N}$ to w, and let y be the last ancestor of v (possibly v) on path P. It can happen that all three nodes $\{y, w, v\}$ are actually the same node, or that any one pair of them is the same node, distinct from the remaining node. In all cases, the path from y to u that goes through edge (w, u), together with the path from y to u through (v, v'), form a recombination cycle in $\mathcal{N}$ that contains edge (v, v'). That contradicts the assumption that (v, v') is a cut edge. Hence, there is no node u that contradicts the statement of theorem, and so the state of site c at every node in $\mathcal{N}_{v'}$ must be the same as it is at node v'. Since an ARG has only one root node, the head of a cut node cannot be a recombination node, so if c does not change on edge (v, v'), then the state of c at v is the same as at v', and the theorem is proved.   ∎

Lemma 7.3.5 will be applied in several different contexts. One is when c is in a nontrivial connected component C of $G(M)$, and v is an external node on blob B_C in an ARG $\mathcal{N}$ for M, incident with cut edge (v, v') directed from v to v'. A second is when c labels a cut edge (v, v') (so c must be a fully-compatible site). A third is when v is the root r of an ARG, and the directed edge (r, v') is not part of any blob.

Lemma 7.3.6 *The root node of a canonical, fully decomposed ARG cannot be an external node.*

Proof: Either r is not in any blob, in which case it cannot be external by definition; or it is in a blob, in which case it is the root of the blob, and so can only be incident with a cut node if the cut node is directed into it, contradicting the definition of the root of $\mathcal{N}$.   ∎

Lemma 7.3.7 *Let $\mathcal{N}$ be a canonical, fully decomposed ARG for M. Sequence $s_r(C)$ is a visible segment (i.e., is in $M(C)$), if and only if r is* not *in the blob B_C.*

Proof: Suppose r is in B_C. By lemma 7.3.6, r is not external, so if $s_r(C)$ is visible, there would have to be some external node $v \neq r$ in B_C, where $s_v(C) = s_r(C)$. But only sites in B_C can change state in B_C, so s_v and s_r would be identical at any site not in C. So, if $s_r(C)$ is a visible segment, then $s_v = s_r$, violating condition (c) of the definition of a canonical ARG.

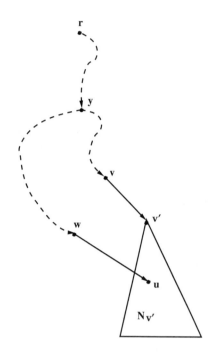

Figure 7.12 Figure for the proof of lemma 7.3.5. The dashed curves represent paths. Any pair of $\{y, w, v\}$, or all three, could be the same node. In the latter case, they would all be node v.

Conversely, if r is not in B_C, then since r has degree at least two, and there are at least three external nodes in any blob, there must be a path from r to a leaf f that does not contain any node in B_C. Therefore, by lemma 7.3.5 $s_r(C) = s_f(C) \in M(C)$, so $s_r(C)$ is a visible segment. ∎

Corollary 7.3.4 *Let $\mathcal{N}$ be a canonical, fully decomposed ARG for M, and let C be a nontrivial connected component in $G(M)$. Then the number of external nodes of B_C is exactly the number of distinct sequences in $M(C)$, and each external node of B_C is C-labeled by a distinct sequence in $M(C)$.*

Proof: If r is in the blob B_C, then by lemma 7.3.7 $s_r(C)$ is not in $M(C)$, and by lemma 7.3.6 r is not an external node. If r is not in the blob B_C, then the root, v, of B_C is not r, and hence it is external. Clearly, v must be C-labeled by $s_r(C)$, since no edge on the path from r to v is labeled with a site in C. Further, by lemma 7.3.7 $s_r(C)$ is in $M(C)$. Hence,

$s_r(C)$ labels the root v of the blob B_C, and v is external, and $s_r(C) \in M(C)$, if and only if v is not r. This agrees with the corollary for $s_r(C)$.

Now let v be any external node in B_C, other than the root of B_C, and let s_v denote its label. Since v is an external node, say incident with the cut edge (v, v'), there must be a path from v' to a leaf f of $\mathcal{N}$ representing some taxon f. By lemma 7.3.5, the sequence $s_f(C)$ must equal $s_v(C)$, and since f is a leaf, $s_f(C)$ is in $M(C)$. So, every external node in B_C, other than the root, must be labeled by a sequence in $M(C) - s_r(C)$.

Conversely, since c only mutates in B_C, for any leaf f of $\mathcal{N}$, sequence $s_f(C)$ must be $s_r(C)$ unless f is reachable from some external node v in B_C, not the root of B_C. In that case, $s_f(C) = s_v(C)$ by lemma 7.3.5. Therefore, every sequence in $M(C)$ must be $s_v(C)$ for some external node v in B_C. ∎

More invariants

Note that when the root, r, of $\mathcal{N}$ is in a blob B_C, then it is C-labeled by $s_r(C)$, even though $s_r(C)$ is not in $M(C)$. So corollary 7.3.4 says that s_r, $G(M)$, and $M(C)$ completely determine which sequences, restricted to the sites in a nontrivial connected component C of $G(M)$, must label the external nodes of B_C in *any* canonical, fully decomposed ARG for M: sequence $s_r(C)$ and the remaining sequences of $M(C)$, *must* label the external nodes of B_C; and each sequence in $M(C) - s_r(C)$, must be created *inside* B_C, and *exported* out of B_C, down to some leaf of $\mathcal{N}$.

Each sequence in $M(C)$ is a *superstate* of M. So, each external node in $\mathcal{N}$ is C-labeled by a superstate of M. In fact, the full label, s_v, for any external node v in $\mathcal{N}$, is a sequence that is the concatenation of superstates of M, one from each of the set of superstates originating from the different connected components of $G(M)$. This is an *invariant property* of *every* canonical, fully decomposed ARG for M.

Corollary 7.3.5 *Let $\mathcal{N}$ be a canonical, fully decomposed ARG for M, with ancestral sequence s_r; let v be any external node in a blob B_C in $\mathcal{N}$; and let $e = (v, v')$ be the unique cut edge incident with v (where the direction is unspecified). Removing edge e from $\mathcal{N}$ creates a split of the leaves of $\mathcal{N}$, corresponding to the two connected components of $\mathcal{N} - e$. Then, all leaves C-labeled by $s_v(C)$ are on one side of the split, and all leaves not labeled by $s_v(C)$ are on the other side. Similarly, if $e = (v, v')$ is a directed edge outside of any blob, labeled by the fully-compatible site c, then all the leaves with state 0 at c are on one side of the $\mathcal{N} - e$, and all leaves with state 1 at c are on the other side.*

Proof: As observed earlier, in a canonical ARG $\mathcal{N}$, the root of $\mathcal{N}$ cannot be an external node. If v is the root of blob B_C, but not the root of $\mathcal{N}$, then v is incident with one cut edge $e = (v', v)$, directed into v. Clearly, every leaf on the v'-side of $\mathcal{N} - e$ is C-labeled by $s_r(C)$. Conversely, since $\mathcal{N}$ is canonical, v is only incident with one cut edge, and no other external node in B_C is labeled with $s_v(C)$. So, by corollary 7.3.4 and lemma 7.3.5, no leaf reachable from a node in B_C is C-labeled by $s_r(C)$. This proves the theorem when v is the root of a blob, but not the root of $\mathcal{N}$.

Now consider an external node v in B_C, where v is not the root node of B_C, so v is incident with one cut edge (v, v') directed from v to v'. Again, by corollary 7.3.4 and lemma 7.3.5, all the leaves in $\mathcal{N}$ reachable from v' will be C-labeled by $s_v(C)$. Conversely, since sites in C mutate only inside B_C, a leaf f can be C-labeled by $s_v(C)$ only if there is a path from the root of $\mathcal{N}$ to f that enters B_C. But because $\mathcal{N}$ is canonical, no external node in B_C, other than v, is C-labeled by $s_v(C)$, and hence the only leaves C-labeled by $s_v(C)$ are the ones reachable from v'. This proves the theorem. ∎

Algorithmically, the important point is that we can efficiently determine those invariant features of any canonical, fully decomposed ARG for M. Note that theorem 7.2.1 and corollary 7.3.4 are true regardless of the biological, or other, causes of the incompatibilities in M. In fact, these results are results about M alone, and extend into other models of phylogenetic networks. We will make this point more explicitly in section 14.3, particularly in theorem 14.6.2 (page 514).

7.3.5.1 A Major Invariant

The most important invariant property of a fully decomposed ARG $\mathcal{N}$ for M is implied by its definition—that each blob in $\mathcal{N}$ contains all and only the sites of a single nontrivial connected component of $G(M)$, and every fully-compatible site is on a cut edge of $\mathcal{N}$. But there is another major invariant that is stated in the following *full-decomposition invariance theorem*.

Theorem 7.3.3 *All canonical, fully decomposed ARGs for M have the same undirected backbone tree, which is necessarily the tree $\overline{T}(M)$ defined in the proof of theorem 7.2.1.*

Proof: Let $\mathcal{N}$ be a canonical, fully decomposed ARG for M. For each nontrivial connected component C of $G(M)$, and each external node v in blob B_C in $\mathcal{N}$, label the unique cut edge incident with v by the superstate of M, that originates from C, and labels node v. Note that any edge not in a blob, labeled by a fully-compatible site c, is already labeled by a superstate of M, namely c.

Recall the standard assumption from section 3.2.3.4 (page 80), that no internal node in an ARG has degree two. Now, consider contracting each blob in $\mathcal{N}$ to a single node, creating the rooted backbone tree $T_\mathcal{N}$ whose leaves are labeled by the taxa of M. It is easy to see that every blob, B_C, in $\mathcal{N}$ has at least three external nodes, so no node of $T_{\mathcal{N}}$ corresponding to the contracted blob B_C, will have degree two. Hence, no internal nodes in $T_{\mathcal{N}}$ will have degree two. Clearly, $T_{\mathcal{N}}$ is a perfect phylogeny for the matrix W, defined by the superstates of M (page 208). Similarly, if we start from $\mathcal{N}'$, the rooted backbone tree $T_{\mathcal{N}'}$ for $\mathcal{N}'$, is a rooted perfect phylogeny for W, where no internal node has degree two. By theorem 2.1.2, $T_{\mathcal{N}}$, $T_{\mathcal{N}'}$ and $\overline{T}(M)$ have the same leaf-labeled topology, so all canonical, fully decomposed ARGs for M have the same undirected backbone tree, $\overline{T}(M)$.   ∎

7.3.5.2   *The Number of Recombination Nodes in a Fully Decomposed ARG*

Recall that $G_{s_r}(M)$ is the conflict graph for M, relative to sequence s_r, and that C is a connected component in G_{s_r}. The preceding discussion leads to the following conclusion:

Theorem 7.3.4 *For any sequence s_r, there is a canonical fully decomposed ARG with ancestral sequence s_r that generates M, using at most*

$$\sum_{C \in G_{s_r}(M)} R^1 min_{s_r(C)}(M(C))$$

recombination nodes, if only single-crossover recombination is allowed; and at most

$$\sum_{C \in G_{s_r}(M)} R^m min_{s_r(C)}(M(C))$$

recombination nodes, if multiple-crossover recombination is allowed.

A natural but incorrect conjecture
In theorem 7.3.4, the ancestral sequence s_r is specified. Of course, there may be a different ancestral sequence that results in fewer needed recombination nodes in some canonical fully decomposed ARG for M. Therefore, it is tempting to conjecture that in the *root-unknown* case, there is a canonical fully decomposed ARG using only

$$\sum_{C \in G(M)} R^1 min(M(C))$$

recombination nodes in the single-crossover case, and

$$\sum_{C \in G(M)} R^m min(M(C))$$

```
11110000
11000000
11100000
00110000
01110000
00001111
00001100
00001110
00000011
00000111
```

Figure 7.13 Example where $R^1 min(M) > \sum_{C \in G(M)} R^1 min(M(C))$.

recombination nodes in the multiple-crossover case. If that were true, we could solve the *root-unknown* MinARG problem for $M(C)$, separately for each nontrivial connected component C, finding the MinARGs that individually minimize the number of recombinations in each blob, and then sum those numbers to get the total number of recombinations in a canonical fully decomposed ARG for M. However, the conjecture is wrong.

The reason is that the MinARGs for the separate problems (each defined by a connected component of $G(M)$) may have different ancestral sequences, so that the separate MinARGs cannot be combined into a *single*, fully decomposed ARG. For example, consider the set of sequences M shown in figure 7.13. The incompatible pairs are $(1, 3), (1, 4), (2, 3), (2, 4), (5, 7), (5, 8), (6, 7), (6, 8)$, so there are two connected components in $G(M)$: component C_1 containing the first four sites of M, and component C_2 containing the last four sites of M. We can establish, using the program *Beagle* (discussed in section 9.1.5), that $R^1 min(M(C_1)) = R^1 min(M(C_2)) = 1$, but $R^1 min(M) = 3$. If the two MinARGs could have been combined, then $R^1 min(M)$ would be 2, instead of 3.

7.3.6 The Reverse of the Full-Decomposition Theorem

Next we prove theorem 7.2.2 (stated on page 206), the reverse of theorem 7.2.1.

Proof (Of theorem 7.2.2): Let $\mathcal{N}$ be an ARG for M. By lemma 5.2.1, we know that any two sites c, c' that are incompatible must appear together in some recombination cycle Q in $\mathcal{N}$, proving the theorem in the case that C only contains two sites. Otherwise, let c'' be a site in C that is incompatible with either c or c', say c'. Such a site c'' must exist since C is connected and the edges in C represent pairwise incompatibilities. Then by lemma 5.2.1, sites c' and c'' must appear together in some recombination cycle Q', which might be Q but need not be. Since a site only appears on one edge in $\mathcal{N}$, cycles Q and Q' must share the edge labeled by c', and so c, c' and c'' are all in the same blob in $\mathcal{N}$. C is connected, so we can repeat this argument, successively adding in a site that is incompatible with some site already included in the growing blob, until all sites in C have been included, proving the theorem.   ∎

7.3.7   The Root-Known Full-Decomposition Theorem

The full-decomposition theorem was stated for the case that no specific ancestral sequence was specified. The full-decomposition theorem continues to hold in the case that a specific ancestral sequence s_r is required, but the decomposition into blobs may be different. To handle this case we simply add s_r to M and form the incompatibility graph for $M \cup \{s_r\}$ (or equivalently, we form the conflict graph for M relative to s_r). Matrix W and tree $\overline{T}(M \cup \{s_r\})$ are constructed as before, *but* now we root $\overline{T}(M \cup \{s_r\})$ at the leaf corresponding to s_r. Applying the full-decomposition theorem to $G(M \cup \{s_r\})$ leads to the following:

Theorem 7.3.5 *For any M and any sequence s_r, there is an ARG $\mathcal{N}$ with ancestral sequence s_r that derives M, where every blob in $\mathcal{N}$ contains all and only the sites of a single nontrivial connected component of $G(M \cup \{s_r\})$, and every fully-compatible site is on a cut edge of $\mathcal{N}$. Moreover, all canonical fully decomposed ARGs with ancestral sequence s_r have the same directed backbone tree.*

7.4   The Utility of Full Decomposition

The full-decomposition theorem exposes both permitted and nonpermitted combinatorial structure in the set of ARGs for an input set M. It relates structure in the set of incompatibilities in M to structure in a set of ARGs for M. This is of interest in itself, as a combinatorial result. However, we will also see in our discussion of galled trees that the full-decomposition theorem is a useful technical tool. But there are two other uses we can point out here, and one caveat.

7.4.1   Decomposing the MinARG Problem

The problem of finding a MinARG for a set of sequences M is an NP-hard problem, and hence we do not have an algorithm to build a MinARG in worst-case polynomial time. In fact, we do not even have an algorithm that runs in purely *exponential* time (see section

9.3). However, there are *super-exponential-time* algorithms to find a MinARG, which we will discuss in chapter 9. Since the running time of those algorithms grows extremely rapidly with the size of the input, we would like to execute those algorithms on as small an input as possible. The full-decomposition theorem and theorem 7.3.4 show that for an input M, where $G(M)$ contains several nontrivial connected components, we can separately solve a root-known MinARG problem on the sequences in each $M(C)$, and then compose these solutions into an ARG from M.

Definition $F^1(M)$ denotes the minimum number of recombination nodes used in any *fully decomposed* ARG that generates M, using single-crossover recombination. $F^m(M)$ denotes the minimum number of recombination nodes used in any fully decomposed ARG that generates M, using multiple-crossover recombination. Similarly, $F_s^1(M)$ and $F_s^m(M)$ denote the minimum number of recombination nodes used in any fully decomposed ARG for M with ancestral sequence s, when only single-crossover or multiple-crossover recombination, respectively, is allowed.

Computing $F_s^1(M)$ and $F_s^m(M)$

We can follow a general approach to compute $F_s^1(M)$ or $F_s^m(M)$, and to find a canonical fully decomposed ARG with that many recombination nodes. First, the backbone tree $\overline{T}(M \cup \{s\})$ is constructed from $M \cup \{s\}$; s is chosen as the root sequence of $\overline{T}(M \cup \{s\})$; and for each $C \in G_s(M)$, $s(C)$ is made the ancestral sequence for the blob associated with component C. Then, find a MinARG for $M(C)$ with ancestral sequence $s(C)$, for each connected component C in $G_s(M)$, and connect the ARGs as specified by $\overline{T}(M \cup \{s\})$. This will find a fully decomposed ARG with ancestral sequence s that uses $F_s^1(M)$ or $F_s^m(M)$ recombination nodes, where only single-crossover or multiple-crossover recombination, respectively, is allowed.

The approach is a bit more involved in the root-unknown case. Again, we start by finding the undirected backbone tree $\overline{T}(M)$, but we have to be more careful about the choice of the root node. Suppose we choose a star-node v_C to be the root of $\overline{T}(M)$. However, we still need to select the root of the ARG created by inflating the blob, B_C, for v_C. That selection determines the ancestral sequence for each blob in the ARG constructed from $\overline{T}(M)$, and induces a rooted problem for each connected component. Unlike the situation in the proof of theorem 7.2.1, where the label on v_C can be used for the ancestral sequence of the inflated blob B_C, we now have to be more careful in selecting the ancestral sequence for B_C. To determine the ancestral sequence for B_C, and hence for the entire ARG (since v_C is chosen as the root of $\overline{T}(M)$), we determine the minimum number of recombination nodes needed to derive $M(C)$ (obeying the chosen crossover model) allowing any possible ancestral sequence s. That optimal s will be used for the ancestral sequence of B_C. The correctness of the procedure follows from theorem 7.3.3.

In summary, the best ancestral sequence for the ARG is found by repeating the above computation for each interior node v_C in $\overline{T}(M)$, to find a root node and ancestral sequence for the ARG for M with the minimum number of recombinations, over all fully decomposed ARGs for M. The result is a fully decomposed ARG that uses $F^1(M)$ or $F^m(M)$ recombination nodes, depending on the type of recombination that is allowed. The procedure can obviously be sped up by taking advantage of repeated computations.

Simulations show that the fully decomposed ARG for M constructed in this way is usually a MinARG for M, with the corresponding type of allowed recombination, but there are examples where this is not the case. We will discuss that issue in section 7.4.3.

7.4.2   Building the "Most Treelike" ARGs

When a set of sequences M fails the four-gametes test and hence cannot be generated on a perfect phylogeny, one would still like to derive the sequences on an ARG that is the "most treelike." There is no accepted definition of "treeness," and under many natural definitions, the problem of finding the most treelike network would likely be computationally difficult. But theorem 7.2.1 does suggest one approach.

Recall that we can view any ARG $\mathcal{N}$ for M as a tree of blobs, and that when each blob is contracted to a single node and the edges are made undirected, the resulting graph is the undirected backbone tree T of $\mathcal{N}$. After contracting any nodes of degree two in T, the number of edges in T is one measure of the "treeness" of $\mathcal{N}$. In other words, the "treeness" of $\mathcal{N}$ is measured by the size of its undirected backbone tree (assuming we have contracted nodes of degree two). For example, if all the sites in M are in a single blob in $\mathcal{N}$, then $\mathcal{N}$ is less treelike than a network where the sites are distributed between several blobs, connected by several edges in a tree structure. With this definition of "treeness," theorems 7.2.1 and 7.2.2 imply:

Theorem 7.4.1 *For any input M, the canonical fully decomposed ARG for M is "the most treelike" ARG for M.*

Clearly, the definition of "most treelike" used here is somewhat crude because it does not consider any details of the ARG inside of a blob; but it has the advantage of being easy to compute and allowing a clear identification of the most treelike networks. Further, it seems reasonable that any other natural definition of "most treelike" would identify a *subset* of the networks identified by the definition given here.

7.4.3 Is There Always a Fully Decomposed MinARG?

We showed above that for any M, there is always a canonical fully decomposed ARG for M. But, since our major interest is in constructing ARGs with the *fewest* possible recombination nodes, it is natural to ask:

> Is it true that for every M there is a MinARG for M that is fully decomposed?

A "yes" answer would make the decomposition approach discussed in section 7.4.1 more useful. However, the answer to the question is "no," as detailed in figure 7.14 (page 232). Further, it was established in [145] that:

Theorem 7.4.2 *For any positive integer k, there exists a set of sequences M such that $F^1(M) - Rmin(M) \geq k$. So, the absolute difference between the number of recombinations in the best fully decomposed ARG, and in a MinARG, for M, can be made arbitrarily large.*

However, the construction that establishes theorem 7.4.2 requires that $Rmin(M)$ grows as k grows. It is an open question whether, for any k, there is a set of sequences, M, such that $\frac{F^1(M)}{Rmin(M)} \geq k$.

Despite this negative result, in chapter 8 we will discuss a class of ARGs where the MinARG *is* always fully decomposed. Moreover, empirical investigations have shown that very frequently there is a MinARG that is fully decomposed. Further, there are several natural sufficient conditions that *guarantee* the existence of a MinARG that is fully decomposed. Those results will be discussed in chapter 11.

7.5 Broader Implications and Applications

7.5.1 Superstates Succeed and Succeed

We want to emphasize that superstates are not just a successful technical device for proving results about decomposition. Rather, what has been established in this chapter is that superstates are the successful natural successors of perfect characters – they succeed and succeed.

Superstates of M define the binary characters of W. So we can think of them either as states or characters. As characters in W, they are *perfect characters* (proved in lemma 7.3.1). Hence superstates are the natural generalization of perfect characters, in the situation where the original binary characters of M are *not* perfect. Superstates expose the remaining underlying *tree structure* that is hidden in a matrix M that does not allow a

	C_1			C_2		
	2	3	5	1	4	6
r_1	0	1	0	0	0	0
r_2	1	1	1	0	0	0
r_3	1	0	1	0	0	0
r_4	0	0	0	0	1	0
r_5	0	0	0	1	1	1
r_6	0	0	0	1	0	1

Figure 7.14 The six sequences derived on the (not fully decomposed) ARG in figure 7.6 with three recombination nodes. The columns have been permuted to bring together the sites in connected components C_1 and C_2 of the incompatibility graph. The incompatible pairs of sites in C_1 are $\{(2,3),(3,5)\}$ and in C_2 are $\{(1,4),(4,6)\}$. The *HK bound* applied to the sites in C_1 is two, and applied to the sites in C_2 is also two. Therefore, the minimum number of recombination nodes possible in any fully decomposed ARG for these six sequences is at least four. But, from figure 7.6 (page 207), three recombinations are sufficient in an ARG that is not fully decomposed. Figure 7.7 shows a fully decomposed ARG for these six sequences, using four recombination nodes. As an aside, we note that the sequences obey the structure described in theorem 6.2.4. The horizontal line under row r_3 separates the variable sequences of $< C_1, C_2 >$ from the constant sequence of $< C_1, C_2 >$.

perfect phylogeny. And, by theorem 7.2.2, they expose that structure in a "maximal" way. Thus, the superstates of M play a role in the theory of reticulate networks that binary characters play in the theory of phylogenetic trees.

7.5.2 Superstates Play a General Role

In the proof of the full-decomposition theorem, there was no mention of recombination until close to the end of the proof, when discussing the inflation of the star nodes. Therefore, all the results proved to that point in the proof hold for any incompatible characters of M, independent of any specific cause of the incompatibilities. Also, the proof of uniqueness of the undirected backbone tree did not depend on recombination. Hence, the existence, structure, and uniqueness of $\overline{T}(M)$ holds for any M and any cause of incompatible characters. This again reflects the general role of superstates.

Another view

Another way to see the generality of the results proved here, is to note that *multiple* crossover recombination can be considered as a mathematical operation on binary sequences, rather than a biological event, and can be used to model biological events that don't explicitly involve recombination. For example, we established in section 3.2.3.3 that any occurrence of *back mutation* or *recurrent mutation* can be modeled as a two-crossover recombination. Modeling back and recurrent mutations in this way creates recombination cycles and blobs, and shows explicitly how theorem 7.2.1 applies when back mutation and/or recurrent mutation cause incompatibilities.

Generally, when back or recurrent mutation is the cause of incompatibility, we seek an evolutionary tree that derives a given set of sequences using as few back or recurrent mutations as possible. Such a tree is called a "maximum parsimony tree" and it is a solution to the maximum parsimony problem [112, 373]. Hence, the full-decomposition theorem holds in the context of the *maximum parsimony problem*. See also section 11.4 for a related discussion of the maximum parsimony problem. See section 14.6.1 for another example of how multiple-crossover recombination can be used to *model* other biological phenomena, illustrating again the generality of mathematical structure in very different biological contexts.

8   Galled Trees

Almost every shrub and tree has its gall, oftentimes esteemed its chief ornament...
— Henry David Thoreau

In the previous chapters, we examined algorithms to compute good *lower bounds* on $Rmin(M)$, and we showed polynomial-time constructive methods (via theorem 3.2.1 and corollary 7.3.2) to build an ARG for any input M, but those methods were not guaranteed to build a MinARG and generally use many more recombination nodes than may be needed. We now begin the discussion of algorithms that construct ARGs with the goal of limiting the number of recombination nodes used. The number of recombination nodes used in a constructed ARG gives an *upper bound* on $Rmin(M)$.

Ideally, we would like to construct MinARGs or ARGs that use a number of recombination nodes close to $Rmin(M)$. But recall that the problems of computing $Rmin(M)$, and of finding a MinARG for M, are NP-hard, so we do not have, and do not expect to have, polynomial-time algorithms for those problems. Instead, we must concentrate on *special cases* of MinARG construction problems that do allow polynomial-time methods, or on *heuristics* for ARG construction problems in general, or on *computationally expensive*, but exact, methods. Here we will discuss special cases, and in the next chapter we will discuss the construction problem for general ARGs.

8.1   Introduction to Galled Trees

In this chapter we discuss *Galled Trees*, the most deeply developed special case of a MinARG. A galled tree is an ARG that deviates from a perfect phylogeny in a limited way, allowing only *constrained* recombination cycles (made precise below). Galled trees are appropriate for situations of small or modest rates of recombination. We will develop a polynomial-time algorithm to determine whether a given set of sequence M can be generated on a galled tree, and prove that if there is a galled tree for M, then the one produced by the algorithm is actually a MinARG for M. Moreover, when there is a galled tree T for M, $R^m min(M) = Rmin(M)$, so that even if multiple-crossover recombinations are allowed, there is *no* ARG (whether a galled tree or not) that derives M using fewer recombination nodes than T.

We will also prove that when there is a galled tree T for M, T is "essentially unique," allowing only easily characterized variation: most of the features of T are shared by every (reduced) galled tree that generates M. Thus, if the true history of the sequences can be described by a galled tree, the algorithm will produce one that correctly captures much of the essential features of that history.

We will also discuss three additional results: First, any set of sequences that can be derived on a galled tree can be derived on a *true* tree (without recombination cycles), where at most one *back mutation* per site is allowed. Second, that the *site compatibility* problem (which is NP-hard in general) can be solved in polynomial—even linear—time for any set of sequences that can be derived on a galled tree. Finally, we will prove a *concise necessary and sufficient* condition for a set of sequences to be derivable on a galled tree. Having such a condition can be more useful for further deductions than having only an efficient algorithm, and also provides additional insight into the structure of galled trees.

As introduced earlier, and detailed in this chapter, galled trees have strong combinatorial and algorithmic properties that do not hold for ARGs in general. For additional special properties of galled trees, see [13, 49, 59, 122]. In [122], galled trees are called *level-1* trees. For a discussion of some related networks, and how they differ from the galled trees discussed here, see [59, 358]. The latter paper additionally establishes which network comparison methods are metrics when applied to different variations of galled trees.

Historical note

In the published galled tree literature, the first results concerned the simplest case, the *root-known* case [146, 147, 426], where the ancestral sequence is specified as part of the input; next the *root-unknown* case was addressed [142], developing ideas that hinted at the full-decomposition theorem (theorem 7.2.1); then those ideas were extended to produce the full-decomposition theorem [144, 145]. This historical progression reflected a growing understanding of the combinatorial structure of ARGs, leading to deeper and more powerful insights. But in this book, we have used our current, deeper understanding, to explain galled trees in a way that does not follow the chronological development of the field, but reverses it. Moreover, the first papers on galled trees came before the discovery of theorem 7.1.1, and hence had more involved proofs that a (reduced) galled tree is a MinARG. In this chapter, we use the full-decomposition theorem, and the constructive method developed for full decomposition, and theorem 7.1.1 to analyze and solve galled-tree problems.

8.1.1 Galled Trees: A Biologically and Algorithmically Motivated Structural Restriction

Given the NP-hardness of the problems of computing $Rmin(M)$ and of constructing a MinARG, in a seminal paper, Wang et al. [426] studied the problem of constructing an ARG with all-zero ancestral sequence, with the added constraint that all of the recombination cycles must be *node-disjoint*. We call such an ARG a galled tree and give a formal definition next.

Definition A recombination cycle Q in an ARG that shares no node with any other recombination cycle is called a *gall*.

Definition We say a site c is "on" or "appears on" a gall Q, if c labels one of the edges of Q. When c appears on Q, we also say that "Q contains c."

Note that a gall Q is a blob that contains exactly *one* recombination cycle, and hence exactly one recombination node.

Definition An ARG $\mathcal{N}$ is called a *galled tree* if every recombination cycle in $\mathcal{N}$ is a gall. Equivalently, an ARG is a galled tree if every blob is a gall.

For example, the ARG shown in figure 7.5 (page 206) is a galled tree that generates the same sequences generated in the ARG (not a galled tree) shown in figure 3.8 (on page 75). A biological example of a galled tree is shown in [195] (pages 247–248). The galled tree shown there has a single gall, and derives 64 SNP sites in 25 strains of the *TRI101* gene in the fungus *F. lunulosporum*, studied in [318]. See [10] for discussion of a program that takes in aligned sequences and determines which of four explicit phylogenetic networks (including a true tree) can generate the sequences. Galled trees are one of the types of networks that are recognized by the program. In the study, the overwhelming majority of networks that were not true trees were galled-trees, although the mean number of galls was close to one.

Note that the definition of a galled tree given here allows any sequence to be the ancestral sequence of the ARG. We will later discuss the case where a specified ancestral sequence is required. We also note that when the galls are *edge disjoint*, but not necessarily node-disjoint, the underlying, undirected graph of a galled tree is known as a *cactus graph* in the graph-theory literature. Since a galled tree is an ARG, all terminology that was developed for ARGs, or blobbed-trees in general, can be applied to galled trees. A needed specific case is:

Definition A *canonical* galled tree is a galled tree that obeys the conditions specified in the definition of a canonical ARG.

It is easy to see that in any canonical galled tree T, every node on a gall must have degree *exactly* three, except for the root of T, if it is on a gall, in which case it has degree at least two. We also continue to assume that any internal node of T that is not on any gall has degree at least three. Also, every tree-edge in a gall must be labeled by a site, since otherwise the two endpoint would be labeled by the same sequence, violating the assumption that T is canonical.

Tree terminology

Although a galled tree has recombination nodes and recombination cycles, and hence is a *network* rather than a tree, we will denote a galled tree by using the symbol T instead of the more general $\mathcal{N}$, to emphasize the near-tree aspects of a galled tree. We point out that the paper [426] (which introduced this topic) does not use the term galled tree. The biological inspiration for the term came from the tree pictured in figure 8.1.

Also, there are contexts where it would be helpful to have defined a gall as a recombination cycle that shares no *edge* with any other recombination cycle. And since two cycles that are node disjoint are also edge disjoint, a definition of a gall based on edge-disjointness is more general and includes more networks. However, there are some technical advantages to keeping the original node-disjoint definition of a gall. We have chosen to use the original definition, while noting that it is possible to convert all of the results about galled trees to the more general case that requires only that recombination cycles be edge-disjoint.

8.1.2 Motivation for Galled Trees

There are several reasons for interest in galled trees. A galled tree generates a set of sequences using a *modest* number of recombination nodes,[1] at most the minimum of n and $m/2$, in a topology that is more "treelike" than any ARG for the same sequences that is not a galled tree. Also, galled trees form a class of MinARGs that are necessarily fully decomposed.

---

1 Recall from section 5.3 that there are some MinARGs that use $\theta\left(\frac{nm}{\log n}\right)$ recombination nodes.

Figure 8.1 Part of a true galled tree with three galls.

A set of sequences is likely to be derivable on a galled tree if the recombination rate was low in the true history that derived the sequences, or if most of the *observable* recombinations are very recent. In human populations, both conditions are frequently, but not always, thought to hold. Other examples of galled trees arise in the data reported in [253]. The simplest situation where a galled tree arises is the case of an interval in the genome where only a *single* recombination event has occurred. In that case, the true history of the sequences in that interval takes the form of a galled tree, and the algorithm (to be presented) will correctly reconstruct the essential features of that history. More generally, it is important (in disease association studies, for example) to find regions of the genome where the sequences in a population exhibit moderate levels of recombination, and the galled tree algorithm can be used to search for such regions.

The strongest sales pitch

Perhaps the most compelling motivation for interest in galled trees comes from the fact that there is an efficient, polynomial-time algorithm to determine whether a set of sequences M can be derived on a galled tree, and if they can, to construct a galled tree that derives M. Further, the galled tree constructed by the algorithm will use the *minimum* number of recombination nodes possible, over any ARG for M, even ARGs that are not galled trees and allow multiple-crossover recombinations. That is, the galled tree found will be a MinARG for M, and it will be true that $R^m min(M) = Rmin(M)$. In fact, any *canonical* galled tree for M will be a MinARG for M. This result holds for either the root-known or the root-unknown versions of the MinARG problem. This provides the only known nontrivial case where optimal (with respect to the number of recombinations) genealogical networks can be efficiently constructed from a set of sequences.

Further motivation for galled trees comes from the fact that if M can be derived on a galled tree, then most of the features of a canonical galled tree for M are *invariant* over the set of all canonical galled trees for M. That is, those features appear in every canonical galled tree that generates M. We will prove all these claims in this chapter. Thus, as in the case of perfect phylogeny, if the mathematical assumptions underlying the ARG model are biologically valid, and there is a canonical galled tree T for M, then most of the features of T are guaranteed to correctly reflect the historically correct derivation of M.

8.1.3   The Galled-Tree Problems

As in other ARG problems discussed so far, questions about galled trees either assume that some ancestral sequence is known, or allow it to vary. The questions can also incorporate different assumptions about the number of crossovers allowed at a recombination event. We will see that for galled trees, the number of allowed crossovers is not important, but assumptions about the ancestral sequence can be important. Therefore, we have the following two central galled-tree problems:

> **The root-known galled-tree problem** Given a set M of n binary sequences and a sequence s_r, all of length m, determine whether there exists a galled tree T, with ancestral sequence s_r, that derives M, and if there is one, construct one.

Wang et al. [426] give an $O(nm + n^4)$-time algorithm that was intended to solve the root-known galled-tree problem in the special case that s_r is the all-zero sequence. That work was seminal as it was the first paper to introduce a biologically motivated, structural restriction on reticulate networks that allows a polynomial time construction algorithm. Unfortunately, the algorithm in [426] is incomplete, and only provides a sufficient (but not a necessary) test for the existence of a galled tree for M, with an all-zero ancestral sequence.

The root-unknown galled-tree problem Given a set M of n binary sequences, each of length m, find a sequence s_r such that there is a galled tree for M with ancestral sequence s_r, or determine that there is none.

It is easy to construct examples where there is a galled tree for M with one specified ancestral sequence, but no galled tree for M with a different specified ancestral sequence. Moreover, since there are choices for the ancestral sequence, and more than one ancestral sequence may be possible, if there is a galled tree for M it may seem natural to seek one that *minimizes* the number of recombination nodes, over all galled trees for M, and all choices of the ancestral sequence. However, we will see that this is not necessary: theorem 8.2.2 (to be proved later) establishes that if T is a canonical galled tree for M, then T is a MinARG for M, no matter what its ancestral sequence is. And, as stated in theorem 8.2.3 (to be proved later), if there is a galled tree for M with ancestral sequence s_r, then there is a canonical galled tree for M, with ancestral sequence s_r. So the choice of the ancestral sequence (as long as that choice allows a galled tree for M) does not affect the minimum number of recombination nodes that must be used. Further, any canonical galled tree for M with ancestral sequence s_r is a MinARG for M even if multiple-crossover recombinations are allowed in competing ARGs.

8.2 First Major Results

In this section we establish that every canonical galled tree for M is a MinARG for M. We will also show that the main features of any canonical galled tree for input M are invariant over all canonical galled trees for M.

Recall that $cc(M)$ is the number of nontrivial connected components in the incompatibility graph $G(M)$ for M.

Theorem 8.2.1 *If T is a galled tree (not necessarily canonical), then it has* exactly $cc(M)$ *galls that have at least one pair of incompatible sites. Further, every such gall contains all of the sites of exactly one nontrivial connected component of $G(M)$ and does not contain any sites from any other nontrivial connected component of $G(M)$. However, a gall might contain fully-compatible sites.*

Proof: From theorem 7.1.1, $cc(M) \leq R^m min(M) \leq Rmin(M)$, so the number of recombination nodes in any ARG for M must be at least $cc(M)$. In a galled tree, each gall contains exactly one recombination node, so any galled tree for M must contain at least $cc(M)$ galls.

If galled tree T has a gall Q that does not contain a pair of incompatible sites, we can replace Q with a *tree*, as detailed in the proof of theorem 7.3.2, producing the same sequences, but reducing the number of recombinations by one. This process can be continued until each remaining gall has at least one incompatible pair of sites. Since any galled tree for M has at least $cc(M)$ galls, the process will stop before the number of galls falls below $cc(M)$, so at least $cc(M)$ galls of T must each have an incompatible pair of sites.

Conversely, from theorem 7.2.2, if a gall Q in an ARG contains any site from a nontrivial connected component, C, of $G(M)$, then Q must contain *all* sites in C. Since, in any ARG, a site labels only one edge, no site is contained in more than one gall, so there can be at most $cc(M)$ galls that contain any pair of incompatible sites. This proves the theorem. ∎

Corollary 8.2.1 *The partition of incompatible sites of M into galls (that is, which set of sites appear together on a gall) is* invariant *over all galled trees for M. And in any galled tree with $q > cc(M)$ galls, exactly $q - cc(M)$ galls will contain only fully-compatible sites.*

Corollary 8.2.2 *If a set of sequences M can be generated on a canonical galled tree T with q recombination nodes, then $Rmin(M) = R^m min(M) = q$. That is, T is a MinARG for M, and even if a competing ARG (which need not be a galled tree) is allowed to use multiple-crossover recombination, it cannot have fewer recombination nodes than T has.*

Proof: The corollary follows from the established facts that $cc(M) \leq R^m min(M) \leq Rmin(M) \leq q = cc(M)$. The first two inequalities are from theorem 7.1.1 and the definitions of $R^m min(M)$ and $Rmin(M)$. The third inequality is from the definition of $Rmin(M)$ and q, and the equality is from theorem 8.2.3. ∎

For emphasis, corollary 8.2.2 can be restated as:

Theorem 8.2.2 *Every canonical galled tree for M is a MinARG for M.*

A central theorem

We will establish a central theorem after one additional definition.

Definition Suppose T is a galled tree for M. For any connected component C in $G(M)$, let Q_C be the unique gall in T containing sites in C.

Theorem 8.2.3 *If there is a galled tree for M with ancestral sequence s_r, then there is a canonical, fully decomposed galled tree for M with ancestral sequence s_r.*

Proof: First, the operations described in the proof of theorem 7.3.2, which convert an ARG to a canonical ARG, do not change a gall into a non-gall; they do not remove any gall that contains an incompatible pair of sites; and they do not change the ancestral sequence. Hence, any galled tree can be converted into a canonical galled tree with the same ancestral sequence.

By definition, every blob in any canonical ARG for M must contain at least one pair of incompatible sites of M. So, from theorem 8.2.1, *any* canonical galled tree T for M contains *exactly* $cc(M)$ galls, and any gall in T must contain all the sites of one nontrivial connected component of $G(M)$, and no sites from any other nontrivial connected component of $G(M)$. This almost proves that T is a fully decomposed ARG for M, but the definition of a fully decomposed ARG (page 204) also requires that every *fully-compatible* site can put on a cut edge of T. We address that issue next.

Let M' be M after the removal of all fully-compatible sites. Clearly $G(M')$ consists precisely of all the nontrivial connected components of $G(M)$. Also, removing any edge label of a fully-compatible site in M, from a canonical galled tree T for M, produces a canonical galled tree T' for M'. Hence, by what is established in the first paragraph, T' is a canonical, *fully decomposed* ARG for M'.

Let s'_r be the ancestral sequence in T'. For any connected component C in $G(M')$, let Q'_C be the gall in T' containing sites in C. By corollary 7.3.4, the number of nodes in Q'_C is equal to the number of distinct sequences in $M(C)$, and each one is C-labeled with a distinct sequence in $M(C)$. Further, the root node of Q'_C is C-labeled with $s'_r(C)$. This establishes that the sequences in $M(C)$, for any C in $G(M')$, can be derived on a *single* recombination cycle, whose ancestral sequence is $s'_r(C)$. Therefore, in step (6) of algorithm *FD-ARG(M)* (page 218), which constructively builds a fully decomposed ARG for M, we can use the single recombination cycle Q'_C for the required blob B_C. The result is a canonical, fully decomposed galled tree for M. ■

So, everything proved about canonical, fully decomposed ARGs in chapter 7 applies and specializes to canonical, fully decomposed galled trees. Given the importance of being both canonical and fully decomposed, we have:

Definition A galled tree that is both canonical and fully decomposed is called a *reduced* galled tree.

From theorem 7.3.4 and 7.2.1, we have:

Corollary 8.2.3 *If there is a galled tree for M, then in any reduced galled tree for M, and for any nontrivial connected component, C, in $G(M)$, every node in the gall Q_C is labeled with a distinct sequence in $M(C)$. Also, in any reduced galled tree T for M, every gall in T contains* all and only *the sites of a single nontrivial connected component of $G(M)$, and every fully-compatible site is on a cut edge of T.*

The invariant backbone tree of a reduced galled tree

It follows from the *full-decomposition invariance theorem* (theorem 7.3.3 on page 225), that:

Theorem 8.2.4 *If M can be derived on a galled tree, then the undirected backbone tree for any reduced galled tree for M is $\overline{T}(M)$, and this is true whether the ancestral sequence is specified in advance or not.*

Due to possible degeneracies in galled tree (avoided in *reduced* galled trees), theorem 8.2.4 does not hold generally for galled trees that are not reduced.

8.3   An Efficient Algorithm for the Root-Known Galled-Tree Problem

Given theorem 8.2.3, the *root-known galled-tree problem* can be simply restated as:

> Given M and sequence s_r, is there a reduced, fully decomposed ARG $\mathcal{N}$ for M, with ancestral sequence s_r, such that each blob in $\mathcal{N}$ is a gall?

8.3.1   The Algorithm

Given theorem 7.3.3, we know that the backbone tree of a galled tree for M (if there is one) must be $\overline{T}(M)$, and we know how to find it from M efficiently. Note that $\overline{T}(M)$ does not depend on any specified ancestral sequence for the galled tree. Further, from corollary 7.3.4, for each nontrivial connected component, C, in $G(C)$, we know that the external nodes of Q_C, the gall we want to create for C, must be C-labeled with the distinct sequences of $M(C)$. Hence the *root-known galled-tree problem* reduces to:

> Given M and s_r, can the distinct sequences in $M(C)$ be generated on a gall with ancestral sequence $s_r(C)$, for each nontrivial connected component C in $G(M)$?

We will next develop an efficient algorithm for this problem, for the case of *single-crossover* recombination. However, the solution can be easily extended to handle multiple-crossover recombination.

Definition Let M be a set of sequences, and s_r the specified ancestral sequence, which need not be in M. The set of sequences $M \cup \{s_r\}$ will be denoted M_r.

Note that when s_r is in M, $M_r = M$. Also, if $s_r \notin M$, then there may be a galled tree for M, but no galled tree for M_r, and hence no galled tree for M rooted at s_r.

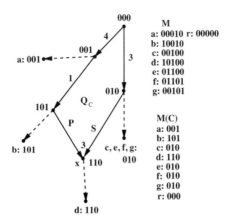

Figure 8.2 Q_C is the gall for the connected component C in figure 7.5 containing sites 1, 3, and 4. In this example, s_r is the all-zero sequence, and $s_x(C) = s_d(C) = 110$, corresponding to taxon d in M. The node labels shown on Q_C are the node labels from figure 7.5, restricted to the sites in C. Note that the node labels on Q_C are exactly the sequences in $M_r(C)$, including $s_r(C)$. Note also that the edge labels $\{1, 3, 4\}$ and the crossover index 3 refer to the original indices for sites in M.

8.3.1.1 Testing Whether a Single Recombination Cycle Suffices

To prepare for the method to come, we do another thought experiment. Suppose there is a reduced galled tree T for M_r, with ancestral sequence s_r. Let C be a nontrivial connected component of $G(M_r)$, and focus on the arrangement of sites of C on gall Q_C, in isolation of the rest of T. Let x be the recombination node of Q_C. Now consider Q_C as a directed graph, separate from the rest of the galled tree, and imagine removing the recombination node x, and the two edges entering x. The resulting graph is a *tree* consisting of one or two directed paths starting at the root node of Q_C. All of the edges in this tree will be labeled with a site in C, every site in C will label some edge in this tree, and no sites outside of C will label any edge. Moreover, by corollary 8.2.3, all, and only, the sequences in $M(C) - s_x(C)$ will be derived on this tree.

   If the tree contains only one path, we denote the root node of the tree as y, and denote the other end node of the path as u; otherwise let u and y be the two end nodes (furthest from the root) of the two paths. For every internal node v, add an edge directed from v and label its leaf end with $s_v(C)$. The result is a *perfect phylogeny* T_C, with ancestral sequence $s_r(C)$, that derives exactly the sequences $M(C) - s_x(C)$. See figures 8.2 and 8.3.

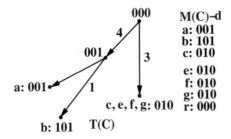

Figure 8.3   After removal of the recombinant node x from gall Q_C of figure 8.2, and the addition of the edge to a leaf for taxon a, the resulting graph is a perfect phylogeny for the sequences $M(C) - s_d(C)$. Moreover, the mutations on the perfect phylogeny are organized into two paths, and the sequence $s_d(C) = 110$ can be created by recombining the two sequences at the ends of those paths; Q_C then consists of those paths, together with a recombination node with crossover index b_x.

Further, $s_x(C)$ can be formed by a recombination of the sequences $s_u(C)$, and $s_y(C)$. And, since the root-known perfect phylogeny for a set of sequences and a fixed ancestral sequence is *unique*, tree T_C is uniquely determined by $M(C), s_r(C)$, and $s_x(C)$. That means that sequence s_u and s_y are also *uniquely determined* from $M(C), s_r(C)$, and $s_x(C)$. This fact is the key to how we can efficiently find T_C knowing only $M(C)$. Hence, we have:

Theorem 8.3.1 *Suppose the sequences in $M(C)$ can be generated on a gall Q_C with root sequence $s_r(C)$, and crossover index b_x. Then, there is a taxon f such that after removal of all copies of $s_f(C)$ from $M(C)$, there is a unique perfect phylogeny for the resulting sequences, with root sequence $s_r(C)$; the labeled edges of that perfect phylogeny contain all sites in C organized into one or two paths; a recombination of the two "end" sequences (from either the root sequence of the perfect phylogeny and the single leaf sequence, or from the two leaf sequences) creates sequence $s_f(C)$. The desired gall, Q_C, consists of those paths, together with the described recombination, with crossover index b_x.*

The converse

Clearly, the converse of theorem 8.3.1 also holds. That is, if there is a perfect phylogeny with the properties enumerated in the statement of theorem 8.3.1, then the sequences in $M(C)$ can be generated on a gall Q_C with root sequence $s_r(C)$. Theorem 8.3.1 and its converse therefore suggest a polynomial-time method to test whether the sequences in $M(C)$ can be generated on a gall Q_C with root sequence $s_r(C)$. This method is detailed in algorithm *Arrange-Sites* shown in figure 8.4 (page 247).

Algorithm Arrange-Sites $(M, C, s_r(C))$

(1) From M and C, create the sequences $M(C)$.

(2) For each distinct sequence $s_f(C)$ in $M(C)$
(other than $s_r(C)$, if s_r is in M):

Test if there is a perfect phylogeny $T(C)$ with ancestral sequence $s_r(C)$
that generates the sequences $M(C) - s_f(C)$.
{This test can be efficiently implemented using methods from chapter 2.}

(3) If the test in step (2) succeeds,
check if all sites on C are contained in one or two directed paths on $T(C)$
whose end sequences, $s_u(C)$ and $s_y(C)$ (at nodes u and y), can be recombined,
with crossover index denoted b_x, to create $s_f(C)$.

(4) If the check in step (3) succeeds,
create the desired gall Q_C by removing any edges into
leaves of $T(C)$, and adding a recombination node x
with crossover index b_x, whose parent nodes are u and y.
Label node x with the recombinant sequence $s_f(C)$.

{For the analysis to follow, call the sequence $s_f(C)$ a
successful recombinant sequence for Q_C. }

Figure 8.4 Algorithm *Arrange-Sites* takes in M, C and $s_r(C)$, and determines whether the sites in C can be arranged on a gall Q_C to generate the sequences in $M(C)$, assuming that the ancestral sequence of the galled tree is s_r.

More details

Step (3) of the algorithm *Arrange-Sites* can be implemented efficiently by checking if all nodes in $T(C)$ have degree one or two. If so, then there will be exactly two nodes, u and y, of degree one. That check can also be incorporated into the building of $T(C)$. To check if $s_u(C)$ and $s_y(C)$ can recombine to form $s_f(C)$, find the longest prefix that is common to $s_u(C)$ and $s_f(C)$; set b_x to the length of that prefix minus one; and check if $s_y(C)$ and $s_f(C)$ match from position b_x to their ends. There are also alternative algorithmic ideas that can replace step (3). For example, the algorithm could search for two sequences s and s' in $M(C)$ that can recombine to create $s_f(C)$. If they exist, then all sites on C must be contained in two directed paths whose end sequences are s and s'. The two approaches can be implemented within the same time bounds.

Note that if there is a perfect phylogeny for M, algorithm *Arrange-Sites* must find a successful recombinant sequence, and must also find an associated arrangement of sites on $Q(C)$ for some choice of $s_f(C)$ in step (2). However, it is possible that the algorithm will succeed for more than one choice in step (2).

Theorem 8.3.2 *Assuming that there is a galled tree for M with ancestral sequence s_r, every arrangement of the sites of C on Q_C that is possible in some reduced galled tree for M, can be found by algorithm Arrange-Sites. That is, each such arrangement corresponds to a successful arrangement pair.*

Proof: Since $M(C)$ and $s_r(C)$ are fixed, independent of the choice of $s_f(C)$ in step (2), and $M(C), s_r(C)$, and $s_f(C)$ uniquely determine the arrangement of sites on Q_C, each successful choice of $s_f(C)$ maps to a unique arrangement of sites on Q_C, and determines the direction of the edges on Q_C. Further, $s_f(C)$ cannot be $s_r(C)$, so there is nothing lost by never choosing $s_r(C)$ in step (2). The converse direction of the theorem follows directly from theorem 8.3.1. ∎

In summary, we have established:

Theorem 8.3.3 *There is a reduced galled tree for M, rooted at a given ancestral sequence s_r, if and only if for each nontrivial connected component C of $G(M)$, there is at least one taxon f in M such that the sequences $M(C) - s_f(C)$ can be generated on a perfect phylogeny rooted at $s_r(C)$, where all the sites are contained in one or two paths, and where $s_f(C)$ can be generated by a recombination of the two end sequences on those paths.*

The approach given here to algorithm *Arrange-Sites* follows the exposition in [147]. Later, Gupta et al. [135] observed that the approach could be simplified in the case of *single-crossover* recombination. They showed the following:

Theorem 8.3.4 *Suppose that for nontrivial connected component C of $G(M)$, there is an integer k such that for every incompatible pair of sites $c < d$ in C, $c < k \leq d$. Suppose also that there is a perfect phylogeny, with ancestral sequence $s_r(C)$, that generates $M(C) - s_f(C)$, for some $s_f(C) \neq s_r(C)$ in $M(C)$. Then the sequences in $M(C)$ can be generated on a single gall, with one single-crossover recombination, and ancestral sequence $s_r(C)$.*

The importance of theorem 8.3.4 is that if there is an integer k as described in theorem 8.3.4, the test in step (3) of algorithm *Arrange-Sites* will *necessarily* succeed. Note that if the sites in C can be generated on a gall with ancestral sequence $s_r(C)$, and crossover index b_x, then for every incompatible pair of sites $c < d$ in C, it must be that $c < b_x \leq d$. So, the existence of an integer k with the property stated in theorem 8.3.4 is a necessary condition for a gall to generate the sequences in $M(C)$. Therefore, algorithm *Arrange-Sites* remains correct if step (3) is removed, and a step is added that tests for an integer k with the property stated above. If no such k is found, then the sequences in $M(C)$ cannot be generated on a gall.

The proof of theorem 8.3.4 given in [135] is fairly complex. Also, algorithm *Arrange-Sites* can be easily generalized to the case of *multiple-crossover* recombination (see section 8.6), while no version of theorem 8.3.4 that handles multiple-crossovers is known. Therefore, we have chosen in this book to first follow the original exposition of algorithm *Arrange-Sites* from [147], rather than incorporating the simplification implied by theorem 8.3.4. However, we will prove theorem 8.3.4 in section 8.10.1, with a proof that is different from the one in [135].

Putting it all together: Algorithm Root-Known-Galled-Tree

Putting all the pieces together, we can now summarize the complete algorithm to test for the existence of a *root-known* galled tree for M, and to create one if there is one. Algorithm *Root-Known-Galled-Tree* is shown in figure 8.5.

Note that in algorithm *Root-Known-Galled-Tree*, algorithm *Arrange-Sites* is run *independently* for each nontrivial connected component. Given the ancestral sequence s_r, any combination of permitted arrangements of galls, one for each nontrivial connected-component of $G(M)$, can be used to create a reduced galled tree for M. The directed backbone tree, obtained from s_r and $\overline{T}(M_r)$, ties the galls together and determines the root of each gall. This point will be important in the time analysis, and in the speedup presented in section 8.8.1.

Initial time analysis

We now establish a first time bound for algorithm *Root-Known-Galled-Tree*. We established in section 7.3.4 that the first five steps of algorithm *FD-ARG* can be implemented to run in $O(nm + n^2)$ time, so the main task here is to analyze the time for step (3) of algorithm *Root-Known-Galled-Tree*. For each connected component C with $m(C)$ sites, the time for algorithm *Arrange-Sites* is $O(n \times nm(C))$, which is the number of taxa, n, times the time bound, $O(nm(C))$, for solving the perfect-phylogeny problem on data with n rows and $m(C)$ columns. Since no site in M is contained in more than one nontrivial connected component, the total time for all executions of algorithm *Arrange-Sites* (over all nontrivial connected components) is $O(n^2m)$. Therefore:

Theorem 8.3.5 *Without further improvement or analysis, the total time for algorithm Root-Known-Galled-Tree is $O(nm + n^2m)$.*

Algorithm Root-Known-Galled-Tree (M, s_r)

Set M_r to $M \cup \{s_r\}$.

Create the undirected tree $\overline{T}(M_r)$ from M_r and $G(M_r)$,
using the first five steps of algorithm *FD-ARG* (page 218).
However, replace M with M_r, and in step (4), do not choose a leaf
for the root arbitrarily; rather choose the leaf in $\overline{T}(M_r)$
associated with the sequence s_r in M_r.

For each nontrivial connected component C in $G(M)$,
test (using algorithm *Arrange-Sites*) if the sequences in $M(C)$
can be generated on a gall Q_C whose ancestral sequence is $s_r(C)$.

If the answer is "no" for some nontrivial connected component,
then M cannot be generated on a galled tree with ancestral sequence s_r.

Otherwise, create the ARG as in steps (6) and (7) of algorithm *FD-ARG*,
using gall Q_C for the required blob B_C associated with C.
The resulting ARG is a reduced galled tree for M with ancestral sequence s_r.

Figure 8.5   Algorithm *Root-Known-Galled-Tree* takes in M and s_r and determines whether there is a galled tree for M with ancestral sequence s_r. If there is one, the algorithm constructs a reduced galled tree for M with ancestral sequence s_r.

In section 8.8, we will show that algorithm *Root-Known-Galled-Tree* can be implemented to run in $O(nm + n^3)$ time, an improvement in the common case that $m > n$.

8.4   The Essential Uniqueness of Reduced Galled Trees

We have previously demonstrated that the major features of the reduced galled trees for a set of sequences M are invariant over the set of reduced galled trees for M. In particular, all reduced galled trees for M have exactly $cc(M)$ galls and have the same undirected backbone tree $\overline{T}(M)$; the mapping of the nontrivial connected components to nodes in $\overline{T}(M)$ is invariant over all reduced galled trees for M; for every nontrivial connected component C, the gall Q_C associated with C contains all and only the sites in C, and precisely the sequences in $M(C)$ are generated on Q_C. Further, if the ancestral sequence

s_r is fixed, then the root sequence for gall Q_C must be $s_r(C)$, and all reduced galled trees for M with ancestral sequence s_r must have the same directed backbone tree.

So in the *root-known* case, the only variability in the set of reduced galled trees for M is in the detailed ways that the sites on each gall are *arranged*. In the root-unknown case, there is additional variability in the choice of ancestral sequence and in the placement of the root node on the galled tree. We will show that even these variations are quite limited. In particular, in the root-known case, there are at most *three* ways that sites can be arranged on a gall, no matter how many sites are on the gall. In the root-unknown case there are at most *four* arrangements of sites on any gall, and the choices of ancestral sequence and root placement are limited. We now begin detailing these claims.

Theorem 8.4.1 *Let C be a nontrivial connected component of $G(M)$ whose sites can be arranged on a gall Q_C in a galled tree for M, with ancestral sequence s_r. Then, the sites in C can be arranged on Q_C in at most three distinct ways, as long as no ordering is given to multiple sites on the same edge.*

Proof: Assume, without loss of generality, that all of the sequences in $M(C)$ are distinct. We have already established, in theorem 8.3.2, that each distinct arrangement of sites on Q_C is associated with one distinct sequence $s_f(C)$ in $M(C)$, with the property that when $s_f(C)$ is removed from $M(C)$, all incompatibilities between pairs of sites in C are broken. Further, $s_f(C)$ is not the root sequence of Q_C (which is $s_r(C)$). Therefore, we can bound the number of distinct arrangements of sites on Q_C by bounding the number of distinct sequences in $M(C)$ whose removal breaks all incompatibilities in C.

For convenience, assume that the ancestral sequence s_r of Q_C is the all-zero sequence. So, the root sequence of Q_C is also the all-zero sequence. Let c, d be an incompatible pair of sites in C, and suppose without loss of generality, that sequence s_f has state-pair $(0, 1)$ at sites (c, d). In order for the removal of s_f to break the c, d incompatibility, no other sequence in M can have the state-pair $(0, 1)$ at sites (c, d). More generally, any sequence whose removal breaks the c, d incompatibility must have a state-pair at sites c, d that differs from the (c, d) state-pair of any other sequence in M. It follows that there can be at most three sequences in $M(C)$ whose removal can break the c, d incompatibility, and hence there can be *at most three* distinct arrangements of the sites in C on Q_C. ∎

Figure 8.6 shows the converse of theorem 8.4.1, namely, that there *can be* three distinct arrangements of sites on a gall. Note that in this example the gall only contains two sites. We will show in section 8.9 that this is a structural necessity, and that when C has more than two sites, there can only be *two* possible arrangements of the sites on a gall.

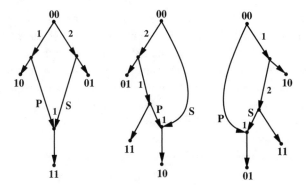

Figure 8.6   The three distinct arrangements of two sites on a gall.

8.5   An Efficient Algorithm for the Root-Unknown Galled-Tree Problem

In the previous section we assumed that an ancestral sequence s_r was given as part of the input, and the solution to the galled-tree problem given there relied heavily on that assumption. In this section we consider the galled-tree problem where *no* ancestral sequence is specified in the problem input. In this case, there are choices for the ancestral sequence. Recall:

The root-unknown galled-tree problem: Given M, find a sequence s_r such that there is a galled tree for M with ancestral sequence s_r, or determine that there is none.

Note that if there is a galled tree for M with ancestral sequence s_r, then there is a *reduced* galled tree for M with ancestral sequence s_r. Further, by theorem 8.2.2, any reduced galled tree for M is a MinARG for M. So, there is no secondary question of finding an ancestral sequence that minimizes the number of recombination nodes over all choices of ancestral sequences that allow a galled tree for M.

We will efficiently solve the root-unknown galled-tree problem with an algorithm that runs in $O(nm + n^3)$ time. The algorithm will be developed for single-crossover recombination; as in the case of the root-known galled-tree problem, the solution can be extended to handle more complex biological phenomena.

8.5.1   The Root-Unknown Problem Reduces to the Root-Known Problem

Given the development of an efficient algorithm to solve the root-*known* galled-tree problem, the most natural approach to solving the root-*unknown* galled-tree problem is to find a way to efficiently *reduce* the root-unknown problem to the root-known problem. This can be done through the following:

Theorem 8.5.1 *If there is a reduced galled tree for M, then there is a reduced galled tree for M where the ancestral sequence is one of the sequences in M.*

Proof: Let T be any reduced galled tree for M where the ancestral sequence is not in M. Since T is reduced, each gall Q in T must contain at least two incompatible sites, and therefore must have at least two external nodes that are *neither* the root nor the recombination node of Q. So each of these nodes is adjacent to a cut edge directed off of Q. We will call these two nodes *exit nodes*. Now create a directed path P from the root of T to some leaf v of T, by walking along any directed path from the root, and if ever the walk enters a gall Q, exit Q via one of its exit nodes. It follows that P is a path from the root to some leaf v of T, that does not contain any recombination nodes. Since T is a galled tree for M, the sequence s_v, labeling v, is in M.

Now consider rerooting T at leaf v, making s_v the new ancestral sequence and reversing the directions of all edges on path P. Each such reversal of an edge also changes the direction of the mutation on the edge. For example, if the original mutation had been from 0 to 1, it is now from 1 to 0. These reversals do not change any of the labels of nodes in T, nor do they change which node is the recombination node on any gall (see figures 8.7 and 8.8). Hence, the modified galled tree, T', also derives M. In order to make T' a proper galled tree (where the root has degree at least two, and each internal node has degree exactly three), we add an edge from the v to a new leaf labeled s_v, and if the old root r has degree two, we contract r, merging the two edges incident with r. This causes no problem, since s_r is not in M, by assumption. The result is a reduced galled tree for M with ancestral sequence $s_v \in M$.   ∎

Although theorem 8.5.1 establishes that some sequence in M can always be used as the ancestral sequence (if there is a galled tree for M), it does *not* establish that *every* sequence in M can be used as an ancestral sequence in some galled tree for M. Since a recombination node cannot be the root node of any gall, T cannot be rerooted at a leaf v, if every path from the root r of T to leaf v goes through a recombination node. For example, the galled tree in figure 8.7 cannot be rerooted at leaves f or d.

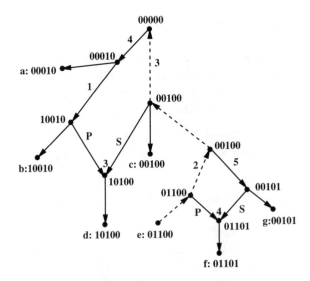

Figure 8.7   The ancestral sequence 00000 is assumed not to be in M. Path P is shown in reverse from leaf e to the root, using dashed lines.

Still, theorem 8.5.1 does imply that the root-unknown galled-tree problem can be solved in $O(n^2m + n^4)$ time by trying each sequence in M as the ancestral sequence. Algorithm *Root-Known-Galled-Tree* determines which sequences in M work as the root, allowing a galled tree for M. In this approach, the n executions of algorithm *Root-Known-Galled-Tree* are completely independent. However, some work involved in these n executions can be shared. By eliminating redundancy, we can speed up this approach to obtain an $O(nm + n^3)$ time bound. In preparation for that improvement, we prove the following:

Theorem 8.5.2 *For any sequence s_r, if there is a reduced galled tree T for M with ancestral sequence s_r, then the incompatibility graphs $G(M)$ and $G(M_r)$ partition the sites of M into connected components in exactly the same way.*

Proof: The theorem is vacuously true if $s_r \in M$, so suppose s_r is not in M, and reroot the galled tree at a leaf, as in the proof of theorem 8.5.1, creating the new galled tree T'. The addition of the leaf labeled s_v does not change the labeling of any edge in a gall in T. If r is not in a gall in T, then no site labeling an edge (u, r) in T is in a gall and this remains true if r is contracted. Similarly, if r is in a gall Q in T, then since r is not external in T (by lemma 7.3.6 (page 222)), any site labeling an edge (u, r) is in Q, and continues to label

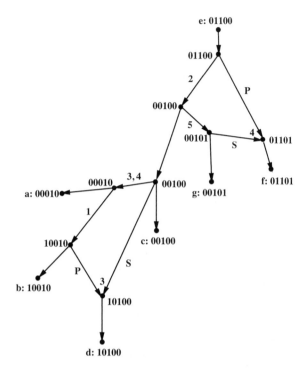

Figure 8.8 The galled tree for M rerooted at e. The galled tree could also have been rerooted at a, b, or g.

an edge in Q after r is contracted. Hence, the partition of the sites into galls, is the same in T and T'. By corollary 8.2.3, the partition of sites that are in galls in T (respectively, T') is the same as their partition into nontrivial connected components of $G(M_r)$ (respectively, $G(M)$), and the theorem follows. ∎

The algorithmic importance of theorem 8.5.2 is that we do not need to know a specific ancestral sequence s_r that allows a galled tree T for M, in order to know which sites of M appear together in galls in T, and which sites are on cut edges. Also, since the undirected backbone tree of an ARG for M is determined only by the partitioning of sites into connected components in $G(M)$, this leads again to the conclusion that the undirected backbone tree is *invariant* over all the reduced galled trees for M.

8.5.2 A Faster Solution to the Root-Unknown Problem

We obtain a faster solution to the root-unknown galled-tree problem by using theorems
8.2.4 and 8.5.2, closely examining the approach discussed in the previous section: solve n
instances of the *root-known* galled-tree problem, using each of the n sequences in M as
the specified ancestral sequence s_r.

When solving an instance of the root-known problem with ancestral sequence s_r, where
s_r is the sequence for taxon f say, algorithm *Root-Known-Galled-Tree* (page 250) first
finds the connected components of the incompatibility graph $G(M_r)$; it then builds the
undirected backbone tree $\overline{T}(M_r)$, and roots it at the leaf associated with taxon f. But
theorem 8.2.4 says that for every ancestral sequence s_r that allows a galled tree for M, the
undirected backbone tree $\overline{T}(M_r)$ is identical to the undirected backbone tree $\overline{T}(M)$. So
it is unnecessary to build $\overline{T}(M_r)$ in each of the n executions of algorithm *Root-Known-
Galled-Tree*; tree $\overline{T}(M)$ can be built once, and then rooted differently in each of the n
executions. That eliminates some redundant work, but not the major work of finding a
successful recombinant sequence for each gall, in each instance of the root-known problem.
We next show how to speed up those computations.

The major task

Given a specified ancestral sequence, s_r, the major task for algorithm *Root-Known-Galled-
Tree* (other than determining $\overline{T}(M_r)$), is to search for a successful recombinant sequence
for each nontrivial connected component C of $G(M_r)$. This suggests that the algorithm
must find the nontrivial connected components of $G(M_r)$ for each of the n different
choices of s_r. But theorem 8.5.2 says that this is unnecessary: the partition of sites into
connected components of the incompatibility graph is invariant over all ancestral sequences
that allow a galled tree for M. Moreover, that invariant partition is the partition in $G(M)$,
which is a function of M alone. So, we can compute the partition without having to specify any particular ancestral sequence. That eliminates more redundancy, but not all: the
search for a successful recombinant sequence for C is done by algorithm *Arrange-Sites*,
where a given ancestral sequence $s_r(C)$ *does* have an affect on which successful recombinant sequences are found. To address that issue, consider the following modification of
algorithm *Arrange-Sites* shown in figure 8.9.

Algorithm *Unrooted-Arrange-Sites* differs from algorithm *Arrange-Sites* in that no
ancestral sequence s_r is input to it, and in step (2) an *unrooted, undirected* perfect phylogeny is sought, rather than a rooted perfect phylogeny. Still, we next show that the results
returned from algorithm *Unrooted-Arrange-Sites* can be used to determine the results that

---

Algorithm Unrooted-Arrange-Sites (M, C)

(1) From M and C, create the sequences $M(C)$.

(2) For each distinct sequence $s_f(C)$ in $M(C)$:

Test if there is an *unrooted* perfect phylogeny $T(C)$
that generates the sequences $M(C) - s_f(C)$.

(3) If the test in step (2) succeeds,
check if all sites on C are contained in one undirected path on $T(C)$
whose end sequences $s_u(C), s_y(C)$ (at nodes u and y) can be recombined,
with crossover index denoted b_x, to create $s_f(C)$.

(4) If the check in step (3) succeeds, then add a recombination node x,
and directed edges into x from parent nodes u and y.
Label x with the recombinant sequence $s_f(C)$.
The result is a cycle Q_C where the edges into x are directed, but the other
edges are undirected, because no node on Q_C has been specified as the root node.

{If the algorithm reaches step (4), the sequence $s_f(C)$ is called
a *successful unrooted recombinant sequence* for Q_C.}

Figure 8.9 Algorithm *Unrooted-Arrange-Sites* takes in M and C, and determines whether the sites in C can be arranged on a cycle Q_C to generate the sequences in $M(C)$. If there is such a cycle, the algorithm constructs it. Note that the cycle Q_C is not a gall because the no root node has been determined.

would be returned from algorithm *Arrange-Sites*, for any specified ancestral sequence $s_r \in M$.

Lemma 8.5.1 *Let s_r be a sequence in M. Then algorithm Arrange-Sites (with input M, C, and s_r) finds that sequence $s_f(C) \in M(C)$ is a successful recombinant sequence for gall Q_C with ancestral sequence $s_r(C)$, if and only if algorithm Unrooted-Arrange-Sites (with input M and C) finds that $s_f(C)$ is a successful unrooted recombinant sequence for Q_C, and $s_r(C) \neq s_f(C)$.*

Proof: Suppose algorithm *Arrange-Sites* finds that $s_f(C)$ is a successful recombinant sequence for Q_C, and let $T(C)$ be the perfect phylogeny, with root node labeled $s_r(C)$, found in step (2). By construction, $s_r(C) \neq s_f(C)$. Removing the direction of any edge on $T(C)$ yields the unique unrooted perfect phylogeny for $M(C) - s_f(C)$ that algorithm

Unrooted-Arrange-Sites would find in step (2). Clearly, since the check in step (3) of algorithm *Arrange-Sites* succeeds, the check in step (3) of algorithm *Unrooted-Arrange-Sites* will also succeed, and so $s_f(C)$ will be found to be a successful unrooted recombinant sequence for Q_C. Further, $s_f(C) \neq s_r(C)$.

Conversely, suppose algorithm *Unrooted-Arrange-Sites* finds that $s_f(C)$ is a successful unrooted recombinant sequence for Q_C, and $s_f(C) \neq s_r(C)$. Let $T(C)$ be the unrooted perfect phylogeny found in step (2). Since $s_r \in M$, there will be an interior node on $T(C)$ labeled $s_r(C)$. Directing all edges on $T(C)$ away from $s_r(C)$ yields a perfect phylogeny for $M(C) - s_f(C)$, with ancestral sequence $s_r(C)$. Since $s_r(C) \neq s_f(C)$, algorithm *Arrange-Sites* will test $s_f(C)$ to see whether it is a successful recombinant sequence. Clearly, since the check in step (3) of algorithm *Unrooted-Arrange-Sites* succeeds, the check in step (3) of algorithm *Arrange-Sites* will also succeed, and so $s_f(C)$ will be found to be a successful recombinant sequence for Q_C.   ∎

Lemma 8.5.1 shows how to remove additional redundancy from the solution given in section 8.5.1. Note that algorithm *Unrooted-Arrange-Sites* will find *all* successful unrooted recombinant sequences for Q_C. The conclusion is that we can solve the *Unrooted Perfect-Phylogeny Problem* by using the logic of the approach given in section 8.5.1, but speedup the computation with the algorithm shown in figure 8.10.

8.5.2.1   Correctness and Time Analysis

Theorem 8.5.3 *Algorithm Root-Unknown-Galled-Tree correctly determines if there is a galled tree for M, and if there is one, constructs one with an ancestral sequence in M.*

Proof: Theorem 8.5.1 shows that if there is a galled tree for M then there is a reduced galled tree for M with an ancestral sequence, s_r, in M. So, when s_r is examined in step (3) of algorithm *Root-Unknown-Galled-Tree*, the result will be the same as when algorithm *Root-Known-Galled-Tree* is executed with input M and s_r. That is what theorem 8.5.2 and lemma 8.5.1 establish. Therefore, by theorem 8.3.3, when there is a galled tree T for M with ancestral sequence $s_r \in M$, algorithm *Root-Unknown-Galled-Tree* will correctly find it.

Conversely, it is not hard to see that if the test in step (3a) is successful for each C, then $\overline{T}(M)$ can be used along with the arranged galls created in step (3a) to create a correct galled tree for M with ancestral sequence s_r.   ∎

Algorithm Root-Unknown-Galled-Tree (M)

(1) From M, construct $G(M)$ and find $\overline{T}(M)$.

(2) For each nontrivial connected component C of $G(M)$,
use algorithm *Unrooted-Arrange-Sites* to find and store *all*
of the successful unrooted recombinant sequences for Q_C.

If for some C there are no successful unrooted recombinant sequences for Q_C,
then there is no galled tree for M, and the algorithm can be stopped.

(3) For each sequence $s_r \in M$, consider each nontrivial connected component
C of $G(M)$:

(3a) Determine if there is a successful unrooted recombinant sequence for Q_C
that is not equal to $s_r(C)$. This can be done very efficiently, because
the answer is yes if there is more than one successful unrooted recombinant
sequence for Q_C; if there is only one, $s_f(C)$ say, then compare $s_r(C)$
to $s_f(C)$ to verify that they are unequal.

(3b) If the test in step (3a) is successful for each C,
then there is a galled tree for M with ancestral sequence s_r.
Otherwise, there is no galled tree for M with ancestral sequence s_r.

(4) If a successful ancestral sequence s_r was identified in step (3b),
use $\overline{T}(M)$ and s_r and the gall arrangements found in step (3a)
to build a galled tree for M with ancestral sequence s_r,
as detailed in algorithm *Root-Known-Galled-Tree*.

If no successful ancestral sequence was identified in step (3b),
then there is no galled tree for M.

Figure 8.10 Algorithm *Root-Unknown-Galled-Tree* takes in M, but no specified ancestral sequence, and determines if there is a galled tree for M. If so, it constructs a reduced galled tree for M.

Time analysis

Similar to the initial time analysis for algorithm *Root-Known-Galled-Tree* (in section 8.3.1.1), the time to build $\overline{T}(M)$ and to find the nontrivial connected components of $G(M)$ is $O(nm)$. Similarly, the total time to run algorithm *Unknown-Arrange-Sites*, to find and store all of the successful unrooted recombinant sequences for all of the nontrivial connected components of $G(M)$, is $O(n^2m)$. That covers steps (1) and (2). Step (3) runs in

$O(nm)$ time because for each sequence s_r in M, step (3a) requires only $O(m(C))$ time for connected-component C, and hence $O(m)$ total time over all of the connected components. The construction required in step (4) clearly can also be done in $O(nm + n^2 m)$ time (the time bound for algorithm *Root-Known-Galled-Tree*). So we have:

Theorem 8.5.4 *Algorithm Root-Unknown-Galled-Tree runs in $O(nm + n^2 m)$ time, which is the same as the initial time bound established for algorithm Root-Known-Galled-Tree.*

Algorithm *Root-Unknown-Galled-Tree* has been implemented in a program *galledtree.pl*. We stated earlier that it can be improved to run in $O(nm + n^3)$ time; this will be established in section 8.8. We will see that this improvement can also be made for algorithm *Root-Known-Galled-Tree*.

8.6   Extensions to Other Biological Phenomena and Structured Recombination

The solutions to both the root-known and the root-unknown galled-tree problems were developed only for single-crossover recombination. However, those algorithms are easily extended to allow multiple-crossover recombination at any recombination node, and multiple-crossover recombination can be used to model many complex biological phenomena. When multiple-crossovers are allowed at recombination nodes, but all recombination cycles are node disjoint, we call the resulting ARG a "multiple-crossover galled tree."

To modify algorithms *Root-Known-Galled-Tree* and *Root-Unknown-Galled-Tree*, we simply change step (3) of the algorithms *Arrange-Sites* and *Unrooted-Arrange-Sites* so that instead of testing whether $s_f(C)$ can be created by a single-crossover recombination of sequences $s_u(C)$ and $s_y(C)$, we allow $s_f(C)$ to be created by a *multiple-crossover recombination*.

We can test whether $s_f(C)$ can be created by a multiple-crossover recombination of $s_u(C)$ and $s_y(C)$, starting with a *prefix* of $s_u(C)$, using the algorithm shown in figure 8.11. If $s_f(C)$ can be created, the algorithm minimizes the number of crossovers used, when starting with a prefix of $s_u(C)$. We leave the proof of that claim to the reader. The algorithm should also be executed, starting with a prefix of $s_y(C)$, and the best solution chosen.

The time for the modified step (3) is clearly $O(n)$. For different biological applications, we can put a bound (for example, two) on the number of crossovers allowed.

Clearly, when the modified algorithm produces a multiple-crossover galled tree for M, the number of recombination nodes used is still just $cc(M)$, and so the ARG produced is a MinARG for M. The time bound for the algorithm is the same as for single-crossover recombination. So we have:

Algorithm Multiple-Crossover-Test $(s_f(C), s_u(C), s_y(C))$

Set i to 1, and set χ to $s_u(C)$.

while i is less than or equal to the length of $s_f(C)$) **do**

    Find the longest substring of χ starting at position i
    that matches a substring of $s_f(C)$ starting at position i.
    if (there is no matching substring) **then**
        return No
    else
        set i to one position past the right end of those matching substrings.
    endif

    if χ is $s_u(C)$ **then**
        set χ to $s_y(C)$
    else
        set χ to $s_u(C)$.
    endif

endwhile
return Yes

Figure 8.11 Algorithm *Multiple-Crossover-Test* determines if $s_f(C)$ can be created by crossovers involving $s_u(C)$ and $s_y(C)$, starting with a prefix of $s_u(C)$. If $s_f(C)$ can be created, the algorithm minimizes the number of crossovers used.

Theorem 8.6.1 *If there is a multiple-crossover galled tree for M, then the modified algorithm will find one, and it will use the minimum number of recombination nodes over all ARGs for M, and all choices of ancestral sequence.*

The algorithm to find a multiple-crossover galled tree for M, or to determine that there is none, has been implemented as the program *multicross.pl* which can be found through the link to software discussed in the preface.

8.6.1 Multiple Crossovers Model Complex Biological Phenomena

Hybrid speciation and *lateral gene-transfer* cause the movement of genetic material between two sequences (often between two species). The biological mechanisms for that movement of material are very different from the mechanisms for meiotic recombination. However, mathematically, these phenomena look like what we have defined as multiple-crossover recombination. Hence, the algorithm to find multiple-crossover galled trees can be used to derive a set of sequences believed to have been created by mutation and hybrid speciation or lateral gene-transfer. The galled tree model also has application in areas outside of biology, such as in determining the history of languages that share words and structure [306].

Multiple-crossover recombination can also be used to model *back mutation* or *recurrent mutation*, as shown in section 3.2.3.3. If the number of back mutations is small, then the recombination cycles created by modeling back mutations as recombinations may be node disjoint, and if so, the theorems and methods of galled trees will apply in this biological context. We can modify step (3) of algorithm *Arrange-Sites* to only allow a recombinant sequence to be derived from the end sequences of $T(C)$ by a single back mutation (or perhaps several back mutations at different sites, if that is meaningful). One can again prove that *if* there is a genealogical network with back mutations where all the recombination cycles are node disjoint, then the modified algorithm will find one that minimizes the number of back mutations over all evolutionary histories that allow back mutations, and all choices of ancestral sequence. Recurrent mutations can also be handled in a similar way.

8.7   A Comment on Galled Trees and Lower Bounds

We have discussed several lower bounds on $Rmin(M)$ that are generally effective (i.e., that are efficient in practice, and produce nontrivial bounds), but it is interesting to note that *none* of the methods discussed are *guaranteed* to give a lower bound larger than *one*. When $Rmin(M)$ is 1, the *HK bound*, *connected-component bound*, and the *optimal RecMin bound* are all guaranteed to have value one; but if $Rmin(M)$ is strictly greater than one, these lower bounds *might* still only be one. Now, with the ability to efficiently test if M can be generated on a galled tree, we have a *slight* improvement in the situation. If $Rmin(M) \geq 2$, then certainly M cannot be generated on a tree, or on a galled tree with only a single gall. Conversely, note that when $Rmin(M) = 1$, any MinARG for M is necessarily a galled tree, so if M cannot be generated on a tree or on a galled tree with only one gall, then $Rmin(M) \geq 2$. This suggests a polynomial-time method that is guaranteed to produce a lower bound on $Rmin(M)$ of *at least two*, when $Rmin(M) \geq 2$: test if there is galled tree that can generate M; if there is, then $Rmin(M)$ equals the number of recombinations in the galled tree; if there is no galled tree for M, then $Rmin(M)$ must be at least two.

Use with the composite method

The increase in guaranteed lower bound from one to two (when $Rmin(M) \geq 2$) is not much of an improvement when applied to all of M, but the improvement can be very helpful when used to compute *local* lower bounds in the composite method. Having several intervals where the local bound is increased from one to two can result in a meaningful increase in the resulting composite lower bound.

A further application of galled trees in the context of the composite method is the following: for any interval I, if there is a reduced galled tree with q recombination nodes that generates sequences $M(I)$, then the local lower bound for I should be set to q, since we are guaranteed, by corollary 8.2.2, that $Rmin(M(I)) = cc(M(I)) = q$. Of course, since q would also be equal to $cc(M(I))$ there would be no increase in the local lower bound, compared to using the *connected-component lower bound*. However, by knowing for *sure* that $Rmin(M(I)) = q$, we would know that any additional (perhaps costly) efforts to increase the local lower bound would be unsuccessful.

8.8 Advanced Topic: Further Speeding Up the Galled-Tree Algorithms

We will show here how to reduce the time needed for algorithm *Root-Known-Galled-Tree* and algorithm *Root-Unknown-Galled-Tree* from $O(nm + n^2m)$ (established in theorems 8.3.5 and 8.5.4) to $O(nm + n^3)$, which is an improvement for the common case that m is larger than n. We establish these bounds through several lemmas.

8.8.1 The Key Observations

Lemma 8.8.1 *Assume that all columns in M (with n rows and m columns) are distinct. If there is galled tree that derives M, then $m \leq 4n$.*

Proof: Let T be a reduced galled tree for M. At each recombination node x in T, remove one of the two edges entering x and contract the other edge. Since the root of T has at least two children, and each internal node has degree three, the result is a directed tree T' with n leaves, where all non-leaf nodes have at least two children. Therefore the number of non-leaf nodes is at most n, so the number of edges in T' is at most $2n$. In any ARG, no edge into a recombination node is labeled. It follows that all sites in M appear on T', so that the number of edges in T that are labeled with some site(s) is bounded by $2n$.

Next we show that every edge in T can be labeled by at most two sites. By lemma 7.3.5, two sites that label a cut edge must be identical, and since we have assumed all sites are distinct, every cut edge in T can be labeled by at most one site. If two sites $c < d$ label the same edge on a gall with recombination node x, then the sites must be identical in M, unless the crossover index b_x is between c and d, i.e., $c < b_x \leq d$. This also follows from lemma 7.3.5. Therefore, if more than two sites label the same edge on a gall, at least two of those sites must be identical, since at least two of the sites must be on the same side of b_x. But, that would contradict the assumption that all columns in M are distinct. So no edge in T can be labeled with more than two sites. Then since there are at most $2n$ labeled edges in T, it follows that $m \leq 4n$. ∎

Corollary 8.8.1 *If M can be derived on a galled tree, then M can have at most $4n$ distinct sites.*

Proof: If M can be derived on a galled tree, then so can any subset of the sites of M. Consider a subset M' consisting of one copy (chosen arbitrarily) of each duplicated site in M. Then all sites in M' are distinct, so by lemma 8.8.1, there can be at most $4n$ sites in M', and at most $4n$ distinct sites in M. ∎

Corollary 8.8.1 shows that the number of distinct sites in M must be $O(n)$, if there is a galled tree for M. This was first claimed in [426], with a different constant, and a different argument that unfortunately does not seem complete.

We want to remove duplicates to speed up the algorithm

The running time of algorithm *Root-Known-Galled-Tree* is analyzed in theorem 8.3.5 to be bounded by $O(nm + n^2m)$, suggesting the utility of *removing* duplicate sites from M when $n < m$. Duplicate sites can be efficiently found, for example by using *radix sort*, treating the entries in each column of M as a binary number. If we just replace m by $4n$ in the second term of the time bound, we get $O(nm + n^3)$.

A glitch

But there is a problem in removing duplicate sites. In general, if M' is M after removal of duplicate sites, the set of ARGs that generate M and M', respectively, might be different. That is, the removal of duplicate sites alters the set of ARGs that can generate the data. For example, see figure 8.12, constructed by Y. Wu. So for many ARG problems, duplicate sites *must not* be removed. However, we will show that in the context of galled-tree problems, duplicate sites can be removed *if* additional checks are included in the algorithms.

Lemma 8.8.2 *If sites c and d in M are identical, then they must appear together on the same edge in any reduced galled tree T for M (assuming there is a galled tree for M).*

M	c_1	c_2	c_3
$r_1\ (r_s)$	0	0	0
r_2	0	1	0
r_3	0	0	1
r_4	1	1	0
r_5	1	0	1

(a)

M'	c_1	c_2	c_3	c_4
$r_1\ (r_s)$	0	0	0	0
r_2	0	1	0	0
r_3	0	0	1	0
r_4	1	1	0	1
r_5	1	0	1	1

(b)

Figure 8.12   Matrix M in panel (a) has a galled tree with all-zero ancestral sequence, and a single gall with recombinant sequence either 101 or 110. Matrix M' in panel (b) is M with an additional copy of site 1, at site 4. The incompatibility graph for M' contains one component, but the *HK bound* is two, so M' cannot be derived on a galled tree. Hence the set of ARGs differs, and $Rmin$ changes, with or without duplicate sites.

Proof: Without loss of generality, suppose that the ancestral sequence s_r for T is the all-zero sequence. Suppose site c is incompatible with some other site(s). Since sites c and d are identical, they will be incompatible with exactly the same set of sites, and hence must be in the same nontrivial connected component of $G(M)$, and must be together on some gall Q. Suppose that site c is on edge $e = (u, v)$ in Q, and that d is *not* on edge e. We can assume that d is also not on the (unique) path from the root of Q to node u; otherwise, switch the roles of c and d in the proof. Since node v is not the root of Q, it must be an external node with degree three, so there must be a directed edge (v, v') where v' is not on Q. But then by lemma 7.3.5, all of the taxa at leaves reachable from v' in T will have state 1 for c and will have state 0 for d, contradicting that assumption that sites c and d are identical. So, when c is incompatible with some site(s), sites c and d must be on the same edge e on T. The case that c is a fully-compatible site is similar, and is left to the reader. ∎

Note that lemma 8.8.2 is proved only for galled trees and does not apply to general ARGs. Recombination is an operation that acts on an *ordered* set of sites, so even if sites c and d are identical, a recombination between c and d could affect the sites differently. The consequence is that on an ARG for M that is not a galled tree, we have no assurance that two identical sites will be on the same edge.

Lemma 8.8.2 is the key to seeing how duplicate columns in M can be removed, and hence the key to speeding up algorithm *Root-Known-Galled-Tree* and algorithm *Root-Unknown-Galled-Tree*.

Fixing the glitch

Suppose sites c and d are identical and one of them, say d, is removed from M, creating the smaller matrix M'. We do not specify whether site c is a fully-compatible site or not. If there is no galled tree for M', then certainly there cannot be a galled tree for M. However, if there is a galled tree for M', it is not certain that there is a galled tree for M. Figure 8.12 illustrates this point. So when there is a galled tree for M', we cannot yet be sure that there will be one for M.

Lemma 8.8.3 *Every reduced galled tree T for M, if there is one, is obtained from some reduced galled tree T' for M', by simply adding site d to the edge in T' that contains site c.*

Proof: Clearly, if there is a reduced galled tree T for M, then the removal of site d from M, and from T, yields a reduced galled T' for M'. By lemma 8.8.2, T can be reconstructed from T' by adding d to the edge in T' labeled by site c.   ∎

More generally, suppose there is a reduced galled tree T for M, and there are $D_c(M)$ distinct sites in M. Let M' be a submatrix of M consisting of $D_c(M)$ distinct columns of M. Then a reduced galled tree T for M induces a reduced galled tree T' for M', by removing any sites in $M - M'$ from T. So, if there is a reduced galled tree for M, then it must be obtainable from *some* reduced galled tree for M'. This suggests an approach for determining if there is a reduced galled tree for Mr: construct a reduced galled tree for M', and see whether it can be used to obtain a reduced galled tree for M.

But which reduced galled tree for M should we use? The backbone tree is the same for all reduced galled trees for M', but there might be $\Theta(3^{cc(M)})$ ways that sites can be arranged on the $(cc(M))$ galls. Fortunately, the individual problems on the galls can be *decoupled*.

We assume that there is a reduced galled tree T' for M'. Since two identical sites will be incompatible with exactly the same set of sites in M, $G(M')$ is obtained from $G(M)$ by removing the nodes for sites in $M - M'$. This specifies a one-to-one correspondence between the connected components of $G(M)$ and $G(M')$. Let C' be a nontrivial connected component in $G(M')$ corresponding to nontrivial connected component C in $G(M)$. All of the sites in C' will be on a single gall $Q_{C'}$ in T', so if T can be obtained from T', all the sites in C will be on a single gall Q_C obtained from $Q_{C'}$, by adding the sites in $M - M'$ to $Q_{C'}$. Now recall (page 249) that in algorithm *Root-Known-Galled-Tree*, algorithm *Arrange-Sites* is run *independently* for each nontrivial connected component. Given s_r, there is a galled tree for M if and only if, for each nontrivial connected component C in $G(M)$, there an arrangement of sites of C on gall Q_C which generates all of the sequences $M(C) - s_r(C)$. So:

Theorem 8.8.1 *A reduced galled tree T for M can be obtained from a reduced galled tree T' from M' if and only if, for each gall $Q_{C'}$ in T', the addition of sites in $(M - M') \cap C$ to $Q_{C'}$ (as above), creates a gall, with root sequence $s_r(C)$, that generates $M(C) - s_r(C)$.*

Since there are at most three arrangements of sites on any gall in T', theorem 8.8.1 says that for any reduced galled tree T' for M', at most $3cc(M)$ arrangements of sites must be examined to determine if there is an arrangement that will work for M. Finally, since all reduced galled trees for M' have the same backbone tree $\overline{T}(M')$, any T' suffices; they only differ in the way the sites on the galls are independently arranged. For example, in figure 8.12, remove site 4 from the data in panel (b). The result is matrix M shown in panel (a), which has two galled trees (with recombinant sequence 101 or 110). However, when site 4 is added to the edge with site 1, neither of the resulting galled trees creates the sequences from panel (b).

8.8.2 Summarizing the Improved Time Bound

The time for algorithm *Root-Known-Galled-Tree* was previously established to be $O(nm + n^2m)$. Using radix sort on the columns of M (considering each column of M as a binary number), we can collect together and remove duplicate columns in M in $O(nm)$ time. If there are more than $4n$ sites in the resulting matrix M', then by corollary 8.8.1, there cannot be a galled tree for M. Otherwise, running the galled tree algorithm on M' takes $O(n^3)$ time, since M' has only $O(n)$ sites. If there isn't a galled tree for M', then there is none for M. Otherwise, given any reduced galled tree T' for M', we can determine independently for each nontrivial connected-component C in $G(M)$, and its corresponding C' in $G(M')$, whether gall $Q_{C'}$ can be expanded to create the sequences in $M(C)$. This takes $O(nm(C))$ time, for any C, where $m(C)$ is the number of sites in C. Since no site is in more than one connected-component, the total time over all the galls is $O(nm) = O(n^2)$. In summary:

Theorem 8.8.2 *Algorithm Root-Known-Galled-Tree can be modified, as discussed above, so that it runs in $O(nm + n^3)$ time.*

The same speedup can be obtained for algorithm *Root-Unknown-Galled-Tree*, with a small modification of the discussion. We leave that to the reader.

8.9   Advanced Topic: Further Limitations on the Number of Site Arrangements on a Gall

In section 8.4, we saw that in the root-known galled-tree problem the sites on a gall Q_C can be arranged in at most three ways, and that three arrangements are actually possible when C contains exactly two sites. We next show that three arrangements are possible *only* when C contains exactly two sites. Otherwise, Q_C can be arranged in only *two* ways. Moreover, with an additional condition that typically holds, Q_C can be arranged in only *one* way. First, we need the following two lemmas.

Lemma 8.9.1 *If there is a galled tree T for M, then every nontrivial connected component C of $G(M)$ must be a bipartite graph, and the bipartition of the sites will be unique: the sites on one side of the bipartite graph must be strictly smaller (i.e., smaller in the numbering of the sites) than the sites on the other side.*

Proof: Let c, d be an incompatible pair of sites in C, with $c < d$. Let Q_C denote the gall containing the sites of C, with recombination node x and crossover index b_x. Then $c < b_x \leq d$, by lemma 5.2.3.[2] Therefore, each edge in C connects one site that is less than b_x to one site that is greater than or equal to b_x.   ∎

As an aside, it is algorithmically easy to find the bipartition described in lemma 8.9.1: let c be the largest site in C that is connected only to larger sites than c, and let d be the smallest site in C that is connected only to smaller sites. Then all the sites in C that are less than or equal to c are on one side of the bipartite graph, and all the other sites are on the other side. Moreover, the crossover index b_x for Q_C, can be chosen to be any integer in the interval $[c + 1, d]$.

Recall that when the ancestral sequence is assumed to be the all-zero sequence, two sites c, d are said to *conflict* if they contain all three binary state-pairs 0,1; 1,0; and 1,1.

Lemma 8.9.2 *Let M be a set of sequences that can be generated on a galled tree, with all-zero ancestral sequence s_r, and let C be a connected component in $G_0(M)$, the conflict graph for M. Suppose C contains at least three sites c, d, d', where c is incompatible with*

---

2 Lemma 5.2.3 holds for any ARG, but can be established by a simpler argument in the case of a gall: In a gall, if $b_x \leq c < d$ or $c < d < b_x$, then the recombination event at node x leaves the states of c and d equal to their states at one of the parents of node x. In that case, only three distinct state-pairs for (c, d) can appear at the nodes of Q_C, and hence, by theorem 7.3.5 (page 221), only three distinct state-pairs for (c, d) appear at the leaves of T. Equivalently, only three distinct state-pairs for (c, d) occur in M, contradicting the assumption that they are incompatible.

both d and d'. Let $s_1 \neq s_r$ *be a sequence whose removal from* M *breaks both the* (c, d) *and* (c, d') *incompatibilities. If the* (c, d) *state-pair in* s_1 *is* $(0, 1)$*, then there is only one way to arrange the sites of* C *on gall* Q_C.

Proof: From the discussion of algorithm *Arrange-Sites*, since there is a galled tree for M, there is a sequence $s_1 \neq s_r$ whose removal breaks all conflicts in $M(C)$. We can assume, given lemma 8.8.2, that all sites in input M are distinct. Since the (c, d) state-pair in s_1 is $(0, 1)$, the (c, d') state-pair in s_1 must also be $(0, 1)$ for otherwise it would be $(0, 0)$, and the removal of a $(0, 0)$ pair would not break the (c, d') conflict.

Now suppose s_2 is another sequence whose removal breaks all conflicts in $M(C)$. The state of s_2 at site c must be 1, since if it were 0, then the (c, d) state-pair for s_2 would either be $(0, 1)$ or $(0, 0)$. The first possibility would contradict the assumption that removing s_1 breaks the c, d conflict, and the second possibility would contradict the assumption that removing s_2 breaks the c, d conflict. So let $1, x, y$ denote the c, d, d' state-triple for s_2. Since c and d are in conflict, there must also be another sequence s_3 in $M(C)$ whose (c, d) state-pair is $(1, \bar{x})$, where $\bar{x}$ is $x + 1 \bmod 2$. Also, the (c, d') state-pair in s_3 must be $(1, \bar{y})$, since otherwise it would be $(1, y)$, and then the removal of s_2 would not break the c, d' conflict.

If $x = y$, then sites d and d' are identical in sequences s_1, s_2, and s_3. Any other row s must have the (c, d') state-pair of $(0, 0)$ or $(\bar{x}, \bar{y})$. To see this, consider the state of c in row s. When c is 0, the (d, d') state-pair must be $(0, 0)$ for otherwise either (c, d) or (c, d') will be $(0, 1)$ and the removal of s_1 would not break all the conflicts. When c is 1 in s, the (c, d') state-pair must be $(\bar{x}, \bar{y})$, or else one of the (c, d) and (c, d') state-pairs in s would be identical to the corresponding pair in s_2, and the removal of s_2 would not break all conflicts in $M(C)$. So when $x = y$, sites d and d' are identical, which contradicts the assumption that all sites in the input are distinct.

But if $x \neq y$ (and hence $\bar{x} \neq \bar{y}$), then site-pair (d, d') would have all three state-pairs 1,1; 0,1; and 1,0 and so would be in conflict. That is a contradiction, because d and d' are both in conflict with c, and hence must be on the same side of the conflict graph $G_0(M)$, since it is bipartite.

The conclusion is that s_1 is the only sequence in M whose removal breaks all conflicts in $M(C)$, and since there is a unique arrangement of sites in C on gall Q_C, for each sequence whose removal breaks all conflicts in $M(C)$, the arrangement of Q_C is unique. ∎

Theorem 8.9.1 *Let* C *be a connected component of the conflict graph whose sites can be arranged on a gall* Q_C *in a galled tree for* M. *If* C *has at least three sites, then the sites can be arranged on* Q_C *in at most two distinct ways, and if* C *has at least two sites on each side of* $G_0(M)$ *(which is bipartite), then the sites can be arranged on* Q_C *in only one way.*

Proof: When there are at least three sites in C, at least two sites are together on one side of $G_0(M)$, and since $G_0(M)$ is connected, there must be at least one site, c, that is connected to two distinct sites d and d'. Suppose there are three distinct sequences s_1, s_2, and s_3 in $M(C)$, such that the removal of any of these sequences breaks all the conflicts in $M(C)$. Then one of those three rows must contain the state-pair $(0, 1)$ for sites c, d. Applying lemma 8.9.2 leads to a contradiction, so when C has at least three sites, there can be at most *two* arrangements of the sites of C on Q_C.

Now assume that each side of $G_0(M)$ has at least two sites, and suppose there are two sequences, s_1 and s_2, where the removal of either one breaks all conflicts in $M(C)$. Since $G_0(M)$ is connected, there must be a node c which is adjacent to at least two nodes. Let $N(c)$ be the set of nodes that c is adjacent to in $G_0(M)$. If none of the nodes in $N(c)$ is adjacent to a node $c' \neq c$, then $N(c) \cup \{c\}$ would be disconnected from the rest of G_0. Hence, there must be an edge (c, d), such that c is also adjacent to a node $d' \neq d$, and d' is adjacent to a node $c' \neq c$ (on the same side as c).

Applying lemma 8.9.2 to the (c, d) pair, the (c, d) state-pair cannot be $(0, 1)$ (and it can't be $(0, 0)$) in either sequence s_1 or s_2. Further, the (c, d) state-pair cannot be the same in s_1 and s_2. So, in one of those two sequences, the (c, d) state-pair must be $(1, 0)$. But node d is also adjacent to node c', so (after relabeling), we can apply lemma 8.9.2 to obtain a contradiction to the assumption that the removal of either s_1 and s_2 breaks all conflicts in $M(C)$. Hence, there is only one sequence whose removal from $M(C)$ breaks all conflicts in $M(C)$, and so the arrangement of sites on Q_C is unique. ∎

What we have shown is that "typically" there will be a *unique* permitted arrangement of the sites on a gall. When not, there can only be a small number of arrangements on any gall. Moreover, since the arrangements on galls are *independent*, and all galled trees for M have the same backbone tree, all the galled trees for M can be compactly represented and easily generated. This is another *invariant* or *near-invariant* property of galled trees.

We note that what was proved in this section for galls in a galled tree, also applies to galls in *any* ARG. This point is made more explicitly in [146].

8.10   Advanced Topic: A Concise Necessary and Sufficient Condition for a Galled Tree

In sections 8.3.1 and 8.5 we developed polynomial-time algorithms to determine whether a set of sequences M can be derived on a galled tree (in both the root-known and the root-unknown cases). Those time bounds were improved in section 8.8. But an algorithm, no matter how efficient, is not a concise mathematical characterization of the conditions that allow or disallow a galled tree. Having such a concise mathematical characterization, in the form of a *necessary and sufficient condition* for the existence of a galled tree for M, establishes additional combinatorial structure, and may facilitate the development of deeper insights into the galled-tree problem. In this section we develop a concise necessary and sufficient condition for the existence of a galled tree. These results were established by Yun Song in [393]. As a byproduct, we will also prove theorem 8.3.4 from section 8.3.1.

We first consider the case of a reduced galled tree T for M that only contains a *single* gall. It must be that $G(M)$ only contains a single nontrivial connected component C. Let (c, d) be an incompatible pair of sites in C, and let x be the recombination node on Q_C. We claim the following:

Lemma 8.10.1 *The (c, d) state-pair at node x is the (c, d) state-pair at each leaf in T that is reachable (by a directed path) from node x, and no other leaf in T has that (c, d) state-pair.*

Proof: By lemma 7.3.5 the state of any site in C at a node reachable from x must be the same as it is at node x. Therefore, the (c, d) state-pair at each leaf reachable from x will be the same as it is at x.

Next we claim that the (c, d) state-pair at x is different from the (c, d) state-pair at any other node on Q_C. Note first that other than at node x, the (c, d) state-pair can only change on the two edges where c and d mutate. Therefore, the (c, d) state-pair at any node on Q_C, other than x, is either equal to the (c, d) state-pair at the root of Q_C, or at the node at the head of an edge where c or d mutates. So, suppose for contradiction, that the (c, d) state-pair at x is the same as the (c, d) state-pair at some other node y of Q_C. But then there would only be *three* distinct (c, d) state-pairs at all the nodes of Q_C, and by lemma 7.3.5, there would only be three distinct (c, d) state-pairs at the leaves of T reachable from the nodes on Q_C. The (c, d) state-pair of any leaf not reachable from a node on Q_C must be the (c, d) state-pair at the root of T, which is also the (c, d) state-pair at the root of Q_C. Therefore, the sequences in M would contain at most three distinct (c, d) state-pairs, contradicting the assumption that c and d are incompatible. Hence it must be that the (c, d) state-pair at x is different from the (c, d) state-pair at any other node on Q_C. The lemma then follows from lemma 7.3.5. ∎

Lemma 8.10.1 leads directly to a *necessary* condition for the existence of a galled tree for M, with a single gall.

Definition Let C be the single nontrivial connected component of $G(M)$. A subset of taxa, $\mathcal{S}_C$, of M is called a *C-covering subset*, if for every incompatible pair (c, d) in C, all the taxa in $\mathcal{S}_C$ have the same (c, d) state-pair, and no taxon outside of $\mathcal{S}_C$ has that (c, d) state-pair.

Theorem 8.10.1 *Suppose M can be generated on a galled tree with a single gall Q_C. Then there exists a C-covering subset $\mathcal{S}_C$ of the taxa of M. Also, there is a single integer k, such that for every incompatible pair of sites (c, d) in M, where $c < d$, it must be that $c < k \leq d$.*

Proof: Let T be a reduced galled tree for M with a single gall Q_C, with recombination node x and crossover index b_x. Let $\mathcal{S}_C$ be the taxa labeling the leaves of T reachable from node x. By lemma 8.10.1, all the taxa in $\mathcal{S}_C$ have the same (c, d) state-pair, and no taxon outside of $\mathcal{S}_C$ has that state-pair. So $\mathcal{S}_C$ is a C-covering subset. Further, the crossover index b_x is an integer with the properties required of k.   ∎

Note that theorem 8.10.1 holds in both the case that there is a known, specified ancestral sequence, and in the case that no ancestral sequence is specified.

The surprise
The surprising structural result established in [393] is that the *necessary* condition established in theorem 8.10.1 is also *sufficient*, as shown next.

Theorem 8.10.2 *Suppose that $G(M)$ contains a single nontrivial connected component C. Suppose also that there exists a C-covering subset $\mathcal{S}_C$ of the taxa of M, and that there is an integer k, where $c < k \leq d$ for every incompatible pair of sites $c < d$ in M. Then M can be generated on a galled tree T with a single gall Q_C, where the sequences in $\mathcal{S}_C$ are exactly the sequences labeling the leaves of T reachable from the recombination node of Q_C.*

Note that there may be more than one C-covering subset of M, but since the statement of theorem 8.10.2 does not specify any particular one, it must hold for *every* C-covering subset.

To prove theorem 8.10.2, we follow the ideas in [393], starting by first establishing several needed lemmas.

Recall that $M(C)$ is matrix M restricted to the sites of C. By theorem 8.5.3, the correctness of algorithm *Root-Unknown-Galled-Tree*, M can be derived on a galled tree if and only if $M(C)$ can be derived on a single gall. So, we can ignore any sites not in C, and we will assume (for ease of exposition and without loss of generality) that all sites in M are in C.

Lemma 8.10.2 *Suppose that the conditions in theorem 8.10.2 are satisfied for matrix M, and that $\mathcal{S}_C$ is a C-covering subset for the single nontrivial connected component C of $G(M)$. All taxa in $\mathcal{S}_C$ have the identical sequence, denoted s_f, in M, and no taxon outside of $\mathcal{S}_C$ has sequence s_f in M.*

Proof: Let (c, d) be an incompatible pair. By the definition of a C-covering subset, the (c, d) state-pair must be the same for each taxon in $\mathcal{S}_C$, so the state of c (and of d) must be the same for each taxon in $\mathcal{S}_C$. Every site in M is in some incompatible pair, so this conclusion applies to all sites in M, and hence all taxa in $\mathcal{S}_C$ have the same sequence in M. We call that sequence s_f. No sequence outside of $\mathcal{S}_C$ has the (c, d) state-pair in s_f, so no taxon outside of $\mathcal{S}_C$ has the sequence s_f in M. $\blacksquare$

We will assume (again without loss of generality) that all rows (taxa) of M are distinct. Clearly, removing any duplicate rows has no effect on the validity of theorem 8.10.2. With that assumption, lemma 8.10.2 implies that $\mathcal{S}_C$ consists of just a *single* taxon of M, denoted f. For ease of exposition, we assume (again without loss of generality) that s_f is the *all-one* sequence.[3]

Let m be the number of sites in M, and recall the assumption that all sites in M are in C. Since there is an integer k where $c < k \leq d$ for every incompatible pair of sites $c < d$, the ordered set of sites in M can be divided into two ordered lists $L = \{c_1 < c_2 < ... < c_{k-1}\}$ and $R = \{c_k < ... < c_m\}$, where $c_{k-1} < c_k$, such that no pair of sites in L (respectively R) are incompatible, and every site in L (or R) is incompatible with some site in R (or L).

Every site c in M defines a bipartition of the taxa into two *non-empty* subsets: those with state 1 for site c and those with state 0. Recall from chapter 2 that the first set is denoted O_c and the second set is denoted $\overline{O_c}$. Since every site in M is incompatible with some other site, by theorem 2.3.2 (the four-gametes theorem), both O_c and $\overline{O_c}$ are non-empty for every site c in M. The following is the main technical lemma from [393] required to prove theorem 8.10.2.

Lemma 8.10.3 *Suppose that M satisfies the conditions of theorem 8.10.2, and let $\mathcal{S}_C$ be a C-covering subset. If (c, d) is a pair of sites in list L, then either*

$$O_c \subseteq O_d \text{ or } O_d \subseteq O_c.$$

---

3 In [393], s_f is assumed to be the *all-zero* sequence, but we have reversed that for uniformity of exposition in the book, since the all-zero sequence is the assumed ancestral sequence when an ancestral sequence is known.

		c			d		
		0			1		
		0			1		
$\overline{O_c}$		0			1		O_d
		0			1		
		0			1		
	s_f	1	1	1	1	1	1
		1			1		
O_c		1			0		
		1			0		$\overline{O_d}$
		1			0		

Figure 8.13 Sites c and d, sequence s_f, set $\overline{O_c} \subset O_d$, and set $\overline{O_d} \subset O_c$.

Similarly, If (c, d) is a pair of sites in list R, then either

$O_c \subseteq O_d$ or $O_d \subseteq O_c$.

Proof: There is nothing to prove if $O_c = O_d$, so suppose $O_c \neq O_d$. Clearly, $O_c \cap O_d \neq \emptyset$, since taxon f is in both sets. Now, suppose for contradiction that $O_c \not\subset O_d$ and $O_d \not\subset O_c$. Then, $O_c \cap \overline{O_d} \neq \emptyset$ and $O_d \cap \overline{O_c} \neq \emptyset$. Therefore, $\overline{O_c} \cap \overline{O_d} = \emptyset$, for otherwise all four binary pairs 0,0; 0,1; 1,0; 1,1 would appear at site-pair (c, d), and (c, d) would then be an incompatible pair. This implies that $\overline{O_c} \subset O_d$ and $\overline{O_d} \subset O_c$. See figure 8.13.

Since $\overline{O_d} \subset O_c$ and the sets are both non-empty, there exists a row r_0 of M with the (c, d) state-pair $(1, 0)$. Now every site in L is incompatible with some site in R, so site $c \in L$ is incompatible with some site $c' \in R$, and therefore there are rows r_2 and r_3 whose (c, c') state-pairs are $(0, 0)$ and $(0, 1)$, respectively. Further, since c and c' are incompatible, and f is the only taxon in $\mathcal{S}_C$, s_f is the only sequence that can have ones at both c and c'. So r_0 must have state zero at site c'. See figure 8.14.

Now, since r_2 and r_3 are both in $\overline{O_c}$, and $\overline{O_c} \subset O_d$, taxa r_2 and r_3 must both have state one at site d. See figure 8.15.

At this point, we have established that the (d, c') state-pair for taxon r_3 is $(1, 1)$ which is the (d, c') state-pair in s_f. But since r_3 is not f, it is not in $\mathcal{S}_C$, so it must be that d and c' are a compatible pair. Therefore, d must be incompatible with some other site $d' \neq c'$ in R. Now, r_2 and r_3 have state one at site d, and since (d, d') is an incompatible pair and neither r_2 nor r_3 are in $\mathcal{S}_C$, it must be that the state of both r_2 and r_3 at site d' is zero.

	c	d	c'
r_0	1	0	0
r_2	0		0
r_3	0		1
s_f	1	1	1

Figure 8.14 Initial entries in matrix M deduced in the proof of lemma 8.10.3.

	c	d	c'
r_0	1	0	0
r_2	0	1	0
r_3	0	1	1
s_f	1	1	1

Figure 8.15 Entries at site d deduced in the proof of lemma 8.10.3.

	c	d	c'	d'
r_0	1	0	0	0
r_2	0	1	0	0
r_3	0	1	1	0
s_f	1	1	1	1

Figure 8.16 Site d' deduced in the proof of lemma 8.10.3.

This establishes that (c', d') has all three state-pairs $(0, 0)$; $(1, 0)$; and $(1, 1)$. But c' and d' are both in R, so they are a compatible pair and therefore, the state of r_0 at site d' must be zero. We now have completely deduced the r_0, r_2, r_3, s_f states at sites c, d, c', d'. See figure 8.16.

Since d is incompatible with d', and state-pair $(0, 1)$ does not appear at sites d, d' in the part of M deduced so far, M must contain another taxon r_5 whose (d, d') state-pair is $(0, 1)$. See figure 8.17.

Since c' and d' are a compatible pair, the state of c' for taxon r_5 must be one. But c is incompatible with c', and $r_5 \notin \mathcal{S}_C$, so the (c, c') state-pair at r_5 cannot be $(1, 1)$ so the state of c at r_5 must zero. See figure 8.18. But this establishes that all four binary pairs appear in the state pair (c, d), which contradicts the assumption that sites c and d are compatible in M. Therefore, the assumption that $O_c \not\subset O_d$ and $O_d \not\subset O_c$ must be false.

	c	d	c'	d'
r_0	1	0	0	0
r_2	0	1	0	0
r_3	0	1	1	0
s_f	1	1	1	1
r_5		0		1

Figure 8.17   Row r_5 and some of its entries deducted in the proof of lemma 8.10.3.

	c	d	c'	d'
r_0	1	0	0	0
r_2	0	1	0	0
r_3	0	1	1	0
s_f	1	1	1	1
r_5	0	0	1	1

Figure 8.18   The final deduced submatrix of M, showing all four gametes at c, d.

We conclude that $O_c \subseteq O_d$ or $O_d \subseteq O_c$. The second claim of the lemma follows by symmetry.   ■

We can now obtain the following lemma:

Lemma 8.10.4 *Suppose that M satisfies the conditions of theorem 8.10.2, and let S_C, s_f, L, and R be as described above. Then there exists a sequence s in M, other than s_f, that is identical to s_f at sites $c_1 \ldots c_{k-1}$. Similarly, there exists a sequence s' in M, other than s_f, that is identical to s_f at sites $c_k \ldots c_m$.*

Proof: Let c_L be the site in L containing the minimum number of ones. Then by lemma 8.10.3, $O_{c_L} \subseteq O_c$ for every site $c \in L$. So the state of every $c \in L$ is 1 for every taxon in O_{c_L}. Site c_L is incompatible with some site in R, so the state of c_L must be 1 for at least two taxa, and therefore $|O_{c_L}| \geq 2$. Therefore, there is some taxon $f' \neq f$ with state 1 at each site in L. By a symmetric argument, there is some taxon $f'' \neq f$ that has state 1 at each site in R. Taxon f' cannot be the same taxon as f'', for if they were the same, then the sequence at f' (f'') would be the all-one sequence, contradicting the assumption that s_f is the only all-one sequence in M.   ■

Corollary 8.10.1 *A single-crossover recombination with crossover index k, (i.e., between sites $k-1$ and k), with parental sequences $s_{f'}$ and $s_{f''}$, creates sequence s_f.*

We can now prove the main result.

Proof (Of theorem 8.10.2): Recall that all taxa in M are assumed to be distinct, and that all sites in M are in C. For every incompatible pair of sites (c, d), sequence s_f is the only sequence in M with (c, d) state-pair of $(1, 1)$. Therefore, the set of sequences $M - s_f$ has no incompatible pair of sites and can be generated on a perfect phylogeny T. Note that both sequences $s_{f'}$ and $s_{f''}$ are in $M - s_f$, and so there are two leaves v and v' of T labeled with those sequences. Let $\mathcal{N}$ be the ARG created from T by adding a recombination node x, labeled with s_f, with parent nodes v and v'. Corollary 8.10.1 ensures that such an ARG $\mathcal{N}$ exists. Clearly $\mathcal{N}$ has only a single recombination node and a single recombination cycle, hence $\mathcal{N}$ is a galled tree for M with a single gall Q_C.

Until now, we assumed (for ease of exposition) that all rows of M are distinct and that there are no fully-compatible sites in M. But when there are duplicate taxa or there are fully-compatible sites in M, lemma 8.10.2 established that all sequences in any C-covering set $\mathcal{S}_C$ will be identical in $M(C)$, and so they can all be generated at the recombination node of Q_C. Therefore, in the galled tree for M, all sequences in $\mathcal{S}_C$ will label leaves reachable from the recombination node of Q_C. ∎

8.10.1 The Root-Known Case

The statement of theorem 8.10.2 does not cover the case that a specific ancestral sequence is required. When there is a known ancestral sequence, s_r, theorem 8.10.2 is modified as follows:

Theorem 8.10.3 *Suppose that $G(M)$ contains a single nontrivial connected component C. Suppose also that there exists a C-covering subset $\mathcal{S}_C$ of the taxa of $M_r = M \cup \{s_r\}$ where s_r is not in $\mathcal{S}_C$, and suppose there is an integer k, where $c < k \leq d$ for every incompatible pair of sites $c < d$ in M_r. Then M can be generated on a galled tree T, with ancestral sequence s_r, and a single gall Q_C, where the sequences in $\mathcal{S}_C$ are exactly the sequences labeling the leaves of T reachable from the recombination node of Q_C.*

The essential difference between theorems 8.10.2 and 8.10.3 is that in the latter theorem, s_r must *not* be in $\mathcal{S}_C$.

Proof (Of theorem 8.10.3): The conditions of theorem 8.10.3 are more restrictive than those in theorem 8.10.2, so when the conditions of theorem 8.10.3 are satisfied, there will be a galled tree T for M_r with a single gall Q_C, where $\mathcal{S}_C$ is the set of sequences labeling the leaves reachable from the recombination node of Q_C. Sequence s_r might not be the ancestral sequence of T, but since $s_r \notin \mathcal{S}_C$, the leaf for s_r is not on a directed path from the recombination node of Q_C. Hence, T can be rerooted at that leaf (as in the proof of theorem 8.5.1), creating a galled tree for M with ancestral sequence s_r. ∎

8.10.1.1   Proof of Theorem 8.3.4

Given theorem 8.10.3, we can now prove a result stated in section 8.3.1 (page 248).

Proof (Of theorem 8.3.4): There is at least one incompatible pair of sites in $M(C)$, but by the statement of theorem 8.3.4, there is a perfect phylogeny T for $M(C) - s_f(C)$, so the removal of $s_f(C)$ from $M(C)$ breaks all the incompatibilities in $M(C)$. Therefore, for any incompatible pair (c, d) in $M(C)$, the (c, d) state-pair in $s_f(C)$ must be different from the (c, d) state-pair in every sequence in $M(C) - s_f(C)$. So, $s_f(C)$ is a C-covering subset $\mathcal{S}_C$. Perfect phylogeny T has ancestral sequence $s_r(C) \neq s_f(C)$, so $s_r(C) \notin \mathcal{S}_C$. Also, by the statement of theorem 8.3.4, there is an integer k such that for every incompatible pair of sites $c < d$ of $M(C)$, $c < k \leq d$. All the conditions required to apply theorem 8.10.3 to $M(C)$ and $s_r(C)$, are now satisfied. So by theorem 8.10.3, there is a galled tree for $M(C)$ with ancestral sequence $s_r(C)$. ∎

8.10.2   Extension to Multiple Galls

In this section we address the case that $G(M)$ contains more than one nontrivial connected component, so the galled tree contains more than one gall. Let s_r be a specified ancestral sequence, and recall that M_r is defined as $M \cup \{s_r\}$. By the correctness of algorithm *Root-Known-Galled-Tree*, there is a galled tree for M with ancestral sequence s_r if and only if, for each nontrivial connected component C of $G(M_r)$, the sequences in $M(C)$ can be generated on a gall with ancestral sequence $s_r(C)$. That means that in the *root-known* case we can simply apply theorems 8.10.1 and 8.10.3 to each nontrivial connected component C. So we have:

Theorem 8.10.4 *There is a galled tree for M, with ancestral sequence s_r, if and only if, for each nontrivial connected component C of $G(M_r)$, there is a C-covering subset $\mathcal{S}_C$ of the taxa of M_r where s_r is not in $\mathcal{S}_C$, and there is an integer k, such that for every incompatible pair of sites (c, d) in C, with $c < d$, it must be that $c < k \leq d$.*

In the *root-unknown* case, theorem 8.5.1 (that if there is a galled tree for M, there is one where the ancestral sequence is in M) together with theorems 8.10.1 and 8.10.3 imply:

Theorem 8.10.5 *There is a galled tree for M if and only if, for each nontrivial connected component C of $G(M)$, there is a C-covering subset $\mathcal{S}_C$ of the taxa of M where $\cup_{C \in G(M)} \mathcal{S}_C \neq M$, and there is an integer k, with $c < k \leq d$, for every incompatible pair of sites (c, d) in C, where $c < d$.*

8.11 Advanced Topic: The Character-Removal, Site-Compatibility Problem on Galled Trees

It is a common practice in phylogenetics that when data M cannot be generated on a perfect phylogeny, some characters are removed from M so that the remaining data fits the perfect-phylogeny model [87, 112, 114, 281, 373, 383]. This was discussed earlier in section 1.4.1.4. When characters are removed some data are lost, but the expectation is that if a perfect phylogeny can be constructed using a large percentage of the original characters then the resulting tree will give valid evolutionary information about the taxa. In that view, it is the evolutionary relationship of the taxa that is important, rather than the history of any given character, and so the primary use of characters is to provide enough *discriminating* information to be able to create a tree showing the evolution of the taxa.[4] Of course, in order to get the most informative tree, we want to remove as few characters or sites as possible, motivating the following:

> **Character-removal, site-compatibility problem** Given data M, remove the *minimum* number of characters (columns) from M so that the remaining data can be generated on a perfect phylogeny.

Definition A *node cover* of a graph is a set of nodes NC such that every edge in the graph touches at least one node in NC. A *minimum* node cover is a node cover with the fewest nodes. An *independent set* of a graph is a set of nodes IS such that no pair of nodes in IS is connected by an edge. A *maximum* independent set is an independent set with the largest number of nodes.

---

4 A more technically informative way of viewing this is that for each pair of taxa (f, g), there must be at least one character c that discriminates f from g, that is, exactly one of the two taxa possess character c. Thus, we need a set of characters that provides such a level of discrimination, and all other characters can be safely removed.

For a binary matrix M, it is easy to see that the character-removal problem can be solved by finding a minimum node cover of the incompatibility graph $G(M)$. After removing the sites associated with the nodes in any node cover of $G(M)$, no incompatible pair of sites remains, and conversely any set of sites whose removal breaks all incompatible pairs of sites defines a node cover of $G(M)$. Equivalently, a maximum independent set of $G(M)$ identifies the largest set of sites to retain in any subset of sites allowing a perfect phylogeny for M.[5]

The problem is NP-hard in general

Unfortunately, for general graphs the problems of finding a minimum node cover or of finding a maximum independent set, are NP-hard. So no polynomial-time algorithm is known to solve either of these problems in general. The solution to the character-removal problem, based on using node cover (or independent set), is therefore not a polynomial-time solution. In fact, it is easy to reduce (in polynomial time) the node cover problem to the character-removal problem, and so the character-removal problem is also NP-hard [87].

But linear time solvable for galled trees

When M can be derived on a galled tree, the incompatibility graph $G(M)$ has structure that allows the node cover problem on $G(M)$ to be solved in polynomial-time, even linear-time. We established in lemma 8.9.1 that if there is a galled tree for M, then $G(M)$ must be bipartite. It is well-known that the minimum node cover problem can be solved in polynomial time (by network flow) on any bipartite graph [6]. To obtain a linear-time solution, we exploit additional structure.

Definition Let L and R be the node sets on the two sides a bipartite graph G. G is said to be *convex* for L if the nodes of L can be ordered so that for each node $v \in R$, the set of nodes in L that v is adjacent to form a *closed interval*. That is, v is adjacent to u and $u' > u$ in L if and only if v is adjacent to all nodes in the closed interval $[u, u']$.

Definition A bipartite graph G is called *biconvex* if L and R can be ordered so that G is simultaneously convex for A, and convex for B.

---

5 In [112] the character-removal problem in binary matrices is cast as the problem of finding a *maximum clique* in a graph that is the complement of $G(M)$. These three approaches to the character removal problem are essentially the same, reflecting the well-known relationships between node covers, independent sets, and cliques. [39]

It was shown in [146] that when M can be derived on a galled tree, $G(M)$ must be convex, and in fact biconvex. A linear-time algorithm for the maximum independent set problem, restricted to convex bipartite graphs, was shown in [258] and a different linear-time solution was given in [384]. This leads to:

Theorem 8.11.1 *The character-removal problem can be solved in linear time for any data M that can be derived on a galled tree.*

Theorem 8.11.1 was conjectured in [146, 147], and first established, by a different approach, in [135]. Theorem 8.11.1 is another reflection of the significant combinatorial structure possessed by galled trees and the sequences that can be derived on galled trees.

8.12 Advanced Topic: Relation of Galled Trees to the Back-Mutation Model

Another deviation from the perfect-phylogeny model that is of interest is to allow a *limited* number of back mutations, but no recombinations. A *back mutation* is a mutation at a site from its derived state *back* to its ancestral state, that is not due to recombination. Hence the back mutation occurs on an edge of a genealogical network. Allowing only one mutation per site, but also allowing a limited number of back mutations, specializes the well-known *Dollo-parsimony* model [92, 112, 356]. In the Dollo-parsimony model, as in the infinite-sites model, only one mutation per site is allowed where the state changes from ancestral state to derived state. This occurs at a *founding* individual for that mutation. But thereafter, any occurrence of the derived state can mutate back to its ancestral state at any descendant of the founder. The authors of [356] argue that Dollo parsimony is appropriate "for reconstructing evolution of the gene repertoire of eukaryotic organisms because although multiple, independent losses of a gene in different lineages are common, multiple gains of the same gene are improbable."

In [41], a variant of Dollo parsimony is defined where at most one mutation and one back mutation per site are allowed. A tree that derives the input M, obeying these constraints, is called a *persistent phylogeny* for M. See [345] for a justification of this model.

For this discussion, we assume that the ancestral sequence is the all-zero sequence, so the derived state is 1 and a back mutation for a site is a change on an edge from state 1 to state 0.

Theorem 8.12.1 *Any set of sequences M that can be derived on a galled tree, can be derived on a true tree (no recombinations and hence no underlying undirected cycles) with* *at most one mutation and one back mutation per site. That is, M can be derived on a persistent phylogeny.*

Proof: We take a galled tree T for M and transform each gall Q separately, so that no cycles remain, but all the node labels are preserved. Since a gall has only one recombination node x, and the two recombination edges into x are identified as the S-edge and P-edge, we can identify the two paths from the root node of Q to the recombination node of Q, as the S-path and the P-path.

The simplest case of the theorem is when one path, say the S-path, has no mutations (sites), so it just consists of the S-edge into node x. Let v denote the P-edge parent of x. In this case, we remove the S-edge, and then for any site c which has state 1 at node v, but has state 0 at node x, we write a back mutation for site c on the (v, x) edge. Note that Q is no longer a cycle, but all the node labels on Q remain unchanged.

The more complex case is that both the S-path and the P-path contain at least one mutation. In this case, we remove the first edge on Q out of the root node of Q, on either the P or the S-path, say the S-path, and then reverse the direction of all the remaining edges on the S-path. Next, for every site c that has state 1 at node v, but state 0 at x, write a back mutation for c on the (v, x) edge; for every site c that has state 0 at v, but state 1 at x, write the mutation c on edge (v, x). Now let u denote the parent of x on the S-path of Q. For every site c that has state 1 at u, but state 0 at x, write the mutation c on the (u, x) edge (which now runs from x to u). For every site c that has a state 1 at x, but state 0 at u, write the back mutation for c on the (x, u) edge. Finally, convert each original mutation on the remaining edges of the S-path to a back mutation. The result is that Q is no longer a gall, but all the node labels are preserved. Processing each gall in this way creates a true tree, i.e., without cycles, that derives M using at most one mutation and one back mutation per site. See figure 8.19.   ∎

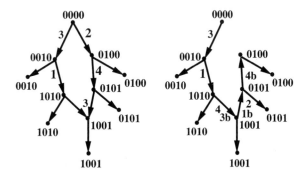

Figure 8.19 Gall Q is shown on the left and the result of the transformation is shown on the right. The crossover index for Q is 3. A number followed by the letter b denotes a back mutation.

9   General ARG Construction Methods

In the previous chapter, we considered the problem of constructing a MinARG, but only for the special case that there is a galled tree for the input M. In this chapter we consider the problem of constructing good ARGs and MinARGs for arbitrary input.

The problem of constructing a MinARG for a set of sequences M, or even of computing $Rmin(M)$, is NP-hard. Thus, we do not have, nor do we expect to have, a worst-case *polynomial-time* algorithm for those problems. Instead, we have heuristic algorithms that empirically run fast (and sometimes can be made to run in worst-case polynomial time) and that produce ARGs with a number of recombination nodes "close" to $Rmin(M)$, on meaningful data. When comparing the number of recombination nodes in those ARGs to the highest available lower bound on $Rmin(M)$, we see that those methods often, but not always, produce MinARGs. We also have *worst-case super-exponential-time* algorithms that compute $Rmin(M)$ exactly, and build MinARGs; those methods are practical for different, but generally small-sized, ranges of biological data. In this chapter, we will discuss two examples of the first kind of algorithm, implemented into the programs *SHRUB* [390] and *mARGarita* [286]; and three examples of the second kind of algorithm, two of which have been implemented into the programs *Beagle* [265] and *RecMinPath* [388, 394]. We will also mention two related methods and will discuss an extension of *SHRUB* to handle the case of gene-conversion [387].

9.1   ARG Construction Methods that Destroy M

The urge to destroy is also a creative urge.
— Pablo Picasso

In this section we develop an approach to ARG construction that also introduces a viewpoint commonly used in thinking about ARGs. The approach is used in several practical methods for finding good ARGs (but ones that are not guaranteed to be MinARGs), and in one program that is guaranteed to construct MinARGs.

Many ARG construction methods build ARGs *backward in time*, from the leaves of the ARG up to the root.[1] The primitive operations in those methods act on the input sequences, M, by removing (or "destroying") rows and columns of the matrix M, until M consists of only a single row with no sites. An ARG $\mathcal{N}$ for M can be built in *parallel* with the destruction of M, by associating each *destructive* step on M with a *constructive* step that adds to a growing ARG $\mathcal{N}$. This is a common approach to ARG construction (and in reasoning about ARGs), so it is important to clearly understand how destructive operations on M translate into constructive operations for building an ARG. In general, there are *three* types of destructive operations, and we will describe each of them in this chapter. However, it is instructive to start with the simplest case, when M has a perfect phylogeny with all-zero ancestral sequence. In that case, only two of the destructive operations are needed. So we begin with that case.

9.1.1   The Perfect-Phylogeny Case with the All-Zero Ancestral Sequence

For this exposition, we assume that the ancestral sequence is the all-zero sequence, and assume that it is in M, adding it if needed. As the matrix M, initially with n rows m columns, is destroyed, we let $\widetilde{M}$ denote the remaining submatrix of M; so initially, $\widetilde{M} = M$. At any point in the destructive process, each row in the current $\widetilde{M}$ will be *named by*, and *associated with*, a subset of taxa, and those sets will *partition* the original n taxa of M. Initially, each row f in M is associated with the singleton set $\{f\}$. The two destructive rules remove columns and rows as follows:

> **Rule Dc:** If a column c of $\widetilde{M}$ contains at most one entry with value 1, then remove column c from $\widetilde{M}$.

> **Rule Dr:** If two rows in $\widetilde{M}$ are identical, then (arbitrarily) remove one. Let $\mathcal{U}$ be the union of the two sets of taxa associated with those two rows. Then rename the remaining row $\mathcal{U}$, that is, *associate* that row with $\mathcal{U}$.

Assuming there is a perfect phylogeny for M, with all-zero ancestral sequence, algorithm *Clean*, shown in figure 9.1, reduces M to a matrix containing a single row with no sites, that is, no entries.

---

1 This bottom-up approach follows the major and original way that trees and ARGs were thought of, analyzed, and constructed in *coalescent theory* [130, 131, 168, 180, 225, 274, 315, 316, 421].

Algorithm Clean (M)

   Set $\widetilde{M}$ to M.

   Execute rules **Dc** and **Dr** on $\widetilde{M}$ in any order until neither rule applies.

Figure 9.1 Algorithm *Clean*.

Algorithm *Clean* is a simplified version of the data-reduction method from [394]. Note that the execution of rule **Dc** may create the conditions where rule **Dr** applies, and the converse is also true. See figures 9.3–9.11. Since rules **Dc** and **Dr** can be applied in any order, and to different columns and rows, it is conceivable that different executions of algorithm *Clean* could produce different results. However, that is essentially not true.

Lemma 9.1.1 *The resulting submatrix $\widetilde{M}$ of M created by running algorithm Clean on M (to completion), is invariant (ignoring any permutation of the rows) over all executions of algorithm Clean. Further, the partition of the taxa is invariant, and the association of subsets of taxa with rows of $\widetilde{M}$ is also invariant.*

Proof: We will first prove that $\widetilde{M}$ is invariant (ignoring permutations of rows). For this part of the argument, it is easier if the original row indices are maintained during an execution of algorithm *Clean*, so that each **Dr** operation removes a row in $\widetilde{M}$ that originates with an identified row in M. But for clarity, note that "$\widetilde{M}$" refers only to a (possible row-permuted) submatrix of M, and does not include the row indices from M.

Let c be a column removed in some execution of algorithm *Clean*, and let P be the series of operations just *before* column c is removed. M is changed in P by removal of specific rows and columns from M, so any permutation of the operations in P that can be executed by algorithm *Clean* would lead to the conditions where rule **Dc** applies to c. Moreover, the insertion of any additional operations into P (or a permitted permutation of P) that removes additional columns or rows, still leads to the conditions where rule **Dc** applies to column c. So, in any execution of algorithm *Clean* where all of the operations in P are applied, column c will eventually be removed. A similar argument can be made for any row removal.

We will prove, by induction on the length of P, that every execution of algorithm *Clean* will apply all the operations in P. Clearly, the first operation in P must be an application of rule **Dc** or **Dr** that applies to a full column c' or full row in M. Two full rows in M that are identical remain identical no matter what columns are removed, and a column in M with at most one entry of value one continues to have that property no matter what rows are removed. So, the first rule applied in P will also be applied somewhere in any other execution of algorithm *Clean*. Thus, the basis of the inductive claim is established.

Next, assume that the inductive claim is true for P of length $k > 1$, and consider P of length $k + 1$. Operation $k + 1$ in P is either a row removal, or a column removal. So, in P, those k operations either create a column with at most one entry of value 1, or create two rows that are identical (or both). By the inductive hypothesis, in any other execution of algorithm *Clean*, the first k operations of P will be applied. As argued above, no matter what permutation of P occurs, or what other operations are inserted into a permuted P, the specific row and column removals from M, specified by P, will create the conditions required for the application of operation $k + 1$ of P. So, in any execution of algorithm *Clean*, operation $k + 1$ of P will eventually be applied, and hence the submatrix $\widetilde{M}$ is invariant over all executions of algorithm *Clean*.

Now we prove that the partition of the taxa of M is invariant, and that the association of subsets of taxa with rows of $\widetilde{M}$ is the same over all executions of algorithm *Clean* on M. By the fact that $\widetilde{M}$ is invariant, we can select a set of columns $\widetilde{C}$, and a set of rows $\widetilde{R}$ in M, such that (up to permutation of rows) $\widetilde{M}$ is created by removing columns in $\widetilde{C}$ and rows in $\widetilde{R}$, from M. Let M' denote M after removing the columns of $\widetilde{C}$, but leaving all rows of M. By the workings of algorithm *Clean*, each maximal subset S of identical rows in M' labels one of the rows of $\widetilde{M}$, and it is identical to each row in S. Hence, the row labels, and partition of the taxa, are also invariant over all executions of algorithm *Clean*.   ∎

The parallel construction of a perfect phylogeny

> In a time of destruction, create something.
> — Maxine Hong Kingston

Here we develop an algorithm that constructs a forest $\widetilde{\mathcal{F}}$ in parallel with the execution of the destructive rules **Dc** and **Dr** in algorithm *Clean*. If M has a perfect phylogeny, T, with all-zero ancestral sequence, then the final forest will be tree T.

To construct the parallel forest (which will be a perfect phylogeny, if one exists) we start with a forest $\widetilde{\mathcal{F}}$ containing one node v_f for each taxon f in M, and one unlabeled edge directed into v_f. The forest $\widetilde{\mathcal{F}}$ will grow and coalesce as algorithm *Clean(M)* is executed; each time a *destructive* rule **Dc** or **Dr** is executed, a parallel *constructive* rule will be executed that changes $\widetilde{\mathcal{F}}$.

Definition At any point in construction, we say that a node v in the current forest $\widetilde{\mathcal{F}}$ is *associated with* the set of taxa that label the leaves in the subtree rooted at v.

Two nodes might be associated with the same subset of taxa, but if this happens then one of the nodes will be an ancestor of the other. Therefore, for any subset of taxa that is associated with some node(s) in $\widetilde{\mathcal{F}}$, there is a well-defined *most ancestral* node that is associated with that subset. Recall that algorithm *Clean(M)* also associates a subset of taxa with each row in the current $\widetilde{M}$. The constructive rules (to be described) will imply (inductively) that if any row in the current $\widetilde{M}$ is associated with a set of taxa $\mathcal{U}$, then some node in the current $\widetilde{\mathcal{F}}$ will also be associated with $\mathcal{U}$.

Each execution of destructive rule **Dc** or **Dr** triggers a parallel execution of constructive rule **Cc** or **Cr**, respectively. Those constructive rules are:

Rule Cc (mutation-event rule): If column c has *exactly* one entry with value 1, let $\tilde{f}$ denote the row in $\widetilde{M}$ that contains the entry of value 1 in column c. Let $\mathcal{U}$ be the subset of taxa associated with row $\tilde{f}$ in $\widetilde{M}$.

If $|\mathcal{U}| = 1$, let e be the (existing) edge in $\widetilde{\mathcal{F}}$ directed into the leaf for $\tilde{f}$. Then add the label c to edge e.

{Note that if there are multiple columns in the original M where each has only a single entry of one, and each is in row $\tilde{f}$, then e will ultimately be labeled with all of those columns.}

If $|\mathcal{U}| > 1$, create a new node v and an edge directed from v into the most ancestral node u in $\widetilde{\mathcal{F}}$ associated with $\mathcal{U}$; then label edge (v, u) with c.

Rule Cr (coalescent-event rule): Suppose the two identical rows in the current $\widetilde{M}$ are associated with the subsets of taxa $\mathcal{U}$ and $\mathcal{U}'$, and let v and v' be the most ancestral nodes in $\widetilde{\mathcal{F}}$ associated with $\mathcal{U}$ and $\mathcal{U}'$, respectively. Then merge nodes v and v' into a single node.

Rule Cc corresponds to creating a *mutation event* and rule **Cr** corresponds to creating a *coalescent event*, in the growing $\widetilde{\mathcal{F}}$. Neither of these rules corresponds to creating a recombination event. Later, we will add a third rule that will do that. Note that if rule **Dc** removes a column that only contains zeros, then rule **Cc** does not apply and $\widetilde{\mathcal{F}}$ will be unchanged.

Algorithm *Clean-Build*, shown in figure 9.2, completely destroys M and builds a perfect phylogeny, T, for M, with all-zero ancestral sequence, if one exists. Otherwise, it constructs a forest of two or more trees. A complete example of the execution of algorithm *Clean-Build*, in the case that M has a perfect phylogeny with all-zero ancestral sequence, is shown in figures 9.3–9.12.

Algorithm Clean-Build (M)

    while (rule **Dc** or **Dr** applies) **do**
       if (rule **Dc** applies) **then**
            execute rules **Dc** and **Cc**.
       endif

       if (rule **Dr** applies) **then**
            execute rules **Dr** and **Cr**.
       endif
    endwhile

    return the resulting matrix $\widetilde{M}$ and the forest (possibly a single tree) $\widetilde{\mathcal{F}}$.

Figure 9.2   Algorithm *Clean-Build*.

	1	2	3	4
r_1	0	0	1	0
r_2	0	0	1	0
r_3	1	1	0	1
r_4	1	1	0	0
r_5	1	0	0	0

(a) The original matrix M (b) The initial forest $\widetilde{\mathcal{F}}$

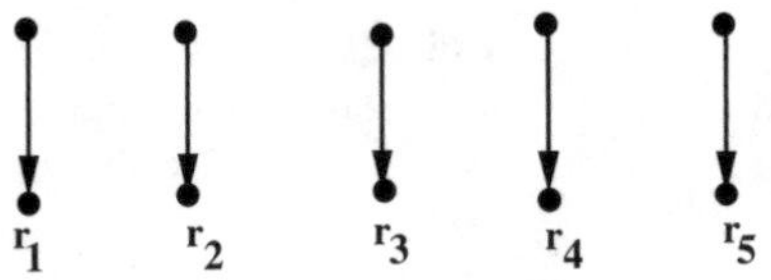

Figure 9.3   Matrix M has a perfect phylogeny. Column 4 has only one entry of 1 (in row r_3), so rules **Dc** and **Cc** can be applied. See figure 9.4 for the results.

9.1.1.1   *Another View of Algorithm* Clean-Build

There is another helpful way to view algorithm *Clean-Build*. Assume M has a perfect phylogeny with all-zero ancestral sequence. Then any submatrix $\widetilde{M}$ of M, and in particular any submatrix created by algorithm *Clean-Build*, will also have a perfect phylogeny with all-zero ancestral sequence.

	1	2	3
r_1	0	0	1
r_2	0	0	1
r_3	1	1	0
r_4	1	1	0
r_5	1	0	0

(a) $\widetilde{M}$ after removal of column 4.

(b) The modified $\widetilde{\mathcal{F}}$

Figure 9.4 Rule **Dc** removed column 4 from the matrix in figure 9.3. Rule **Cc** added label 4 to the edge directed into leaf r_3. Now rows r_1 and r_2 are identical, so rule **Dr** can be applied. See figure 9.5 for the results.

	1	2	3
$\{r_1, r_2\}$	0	0	1
r_3	1	1	0
r_4	1	1	0
r_5	1	0	0

(a) The modified $\widetilde{M}$

(b) The modified $\widetilde{\mathcal{F}}$

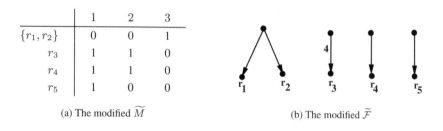

Figure 9.5 Rule **Dr** merged rows r_1 and r_2 of the matrix in figure 9.4 into one row labeled $\{r_1, r_2\}$. Rule **Cr** merged the parents of leaves r_1 and r_2 into single node associated with the set $\{r_1, r_2\}$. As a result, column 3 now has only a single entry with value one (in the row for $\{r_1, r_2\}$), so rules **Dc** and **Cc** can be applied. See figure 9.6 for the results.

In general, let $\widetilde{T}$ denote the unique perfect phylogeny for $\widetilde{M}$ with all-zero ancestral sequence. Since any row $\tilde{f}$ of $\widetilde{M}$ is represented by a leaf v in $\widetilde{T}$, if row $\tilde{f}$ is associated with a subset $\mathcal{U}$ of taxa of M, we also associate $\mathcal{U}$ with leaf v. The following two lemmas are easy to establish by inducting on the number of rules applied in an execution of algorithm *Clean-Build*.

Lemma 9.1.2 *In algorithm Clean-Build, when rule* **Dc** *applies to a site* c *in* $\widetilde{M}$, *the edge labeled* c *in* $\widetilde{T}$ *must be directed into a leaf of* $\widetilde{T}$.

Lemma 9.1.3 *In algorithm Clean-Build, when rule* **Dr** *merges two rows of* $\widetilde{M}$ *that are associated with subsets of taxa* $\mathcal{U}$ *and* $\mathcal{U}'$, *there must be two sibling leaves in* $\widetilde{T}$ *that are*

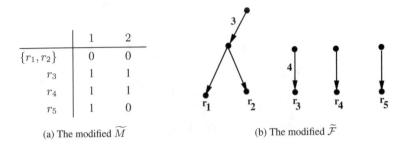

	1	2
$\{r_1, r_2\}$	0	0
r_3	1	1
r_4	1	1
r_5	1	0

(a) The modified $\widetilde{M}$ 　　　　　　　　(b) The modified $\widetilde{\mathcal{F}}$

Figure 9.6　Rule **Dc** removed column 3 from the matrix in figure 9.5. Rule **Cc** created an edge directed into the most ancestral node associated with set $\{r_1, r_2\}$, and labeled that edge with column 3. Note that the new node is now the most ancestral node associated with $\{r_1, r_2\}$. Rows r_3 and r_4 are now identical, so rules **Dr** and **Cr** can be applied. See figure 9.7 for the results.

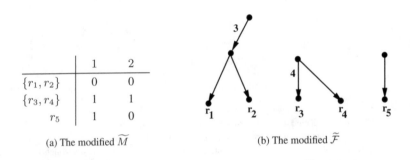

	1	2
$\{r_1, r_2\}$	0	0
$\{r_3, r_4\}$	1	1
r_5	1	0

(a) The modified $\widetilde{M}$ 　　　　　　　　(b) The modified $\widetilde{\mathcal{F}}$

Figure 9.7　Rule **Dr** merged rows r_3 and r_4 in the matrix from figure 9.6 into one row associated with $\{r_3, r_4\}$. Rule **Cr** merged the parents of r_3 and r_4 into a single node associated with the set $\{r_3, r_4\}$. Column 2 now only contains one entry with value 1, so rules **Dc** and **Cc** can be applied. See figure 9.8 for the results.

associated with $\mathcal{U}$ and $\mathcal{U}'$, respectively. *Further, the directed edges into those two sibling leaves must be unlabeled.*

Given lemmas 9.1.2 and 9.1.3, an insightful way to view the action of algorithm *Clean-Build* is with the following thought experiment. Imagine that we know the unique perfect phylogeny T for M, with all-zero ancestral sequence. Imagine also that for every destructive operation on M, we execute a parallel *destructive* operation on T. At the start of an execution of algorithm *Clean-Build*, each leaf in T will be associated with the taxon that labels it. Then, in the thought experiment, whenever rule **Dc** applies to a site c in $\widetilde{M}$,

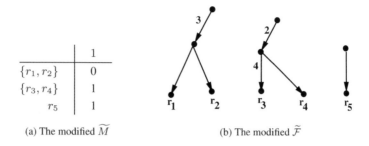

	1
$\{r_1, r_2\}$	0
$\{r_3, r_4\}$	1
r_5	1

(a) The modified $\widetilde{M}$ (b) The modified $\widetilde{\mathcal{F}}$

Figure 9.8   Rule **Dc** removed column 2 from the matrix in figure 9.7. Rule **Cc** created a new node and an edge directed into the most ancestral node associated with the set $\{r_3, r_4\}$, and labeled that edge with column 2. The new node is now the most ancestral node associated with $\{r_3, r_4\}$. Rows $\{r_3, r_4\}$ and r_5 are now identical, so rules **Dr** and **Cr** can be applied. See figure 9.9 for the results.

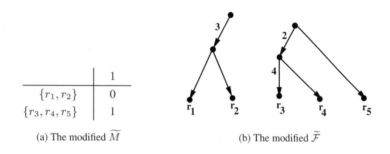

	1
$\{r_1, r_2\}$	0
$\{r_3, r_4, r_5\}$	1

(a) The modified $\widetilde{M}$ (b) The modified $\widetilde{\mathcal{F}}$

Figure 9.9   Rule **Cr** merged rows associated with $\{r_3, r_4\}$ and $\{r_5\}$ from figure 9.8 into one row labeled $\{r_3, r_4, r_5\}$. Rule **Cr** merged the parents of leaf r_5 and the most ancestral node associated with $\{r_3, r_4\}$. Column 1 now has only a single entry with value one, so rules **Dc** and **Cc** can be applied. See figure 9.10.

remove the label c from the edge (into a leaf) in $\widetilde{T}$ labeled by c. That edge exists by lemma 9.1.2. Also, whenever rule **Dr** applies to two rows in $\widetilde{M}$, let v and v' be the sibling leaves in $\widetilde{T}$ associated with sets $\mathcal{U}$ and $\mathcal{U}'$. Those sibling leaves exist by lemma 9.1.3. Then, in the thought experiment, associate the parent node of leaves v and v' with $\mathcal{U} \cup \mathcal{U}'$; and remove from $\widetilde{T}$ the nodes v and v' and the edges into them.

Figure 9.13 shows the seven steps of this thought experiment corresponding to the first seven steps in the destruction of M shown in figures 9.3–9.10. The (not shown) eighth step would remove the two leaves and label the one remaining node with the full set of taxa of M. Lemmas 9.1.2 and 9.1.3 also lead to the following theorem:

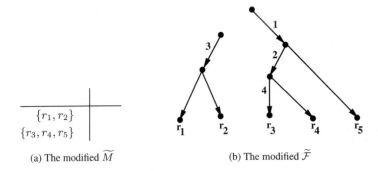

(a) The modified $\widetilde{M}$ (b) The modified $\widetilde{\mathcal{F}}$

Figure 9.10   Rule **Dc** removed column 1 from the matrix in figure 9.10. Rule **Cc** created a new node and an edge directed into the most ancestral node associated with the $\{r_3, r_4, r_5\}$, and labeled the edge with column 1. All entries in M have now been removed, but two rows remain, labeled $\{r_1, r_2\}$ and $\{r_3, r_4, r_5\}$. These rows are identical, so rules **Dr** and **Cr** can be applied. See figure 9.11 for the results.

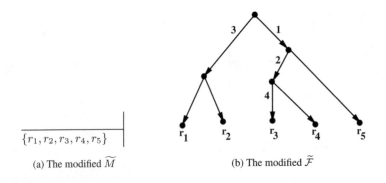

(a) The modified $\widetilde{M}$ (b) The modified $\widetilde{\mathcal{F}}$

Figure 9.11   Rule **Dr** merged the two rows of figure 9.10 into a single row labeled $\{r_1, r_2, r_3, r_4, r_5\}$. Rule **Cr** merged the most ancestral node associated with $\{r_3, r_4, r_5\}$ and the most ancestral node associated with $\{r_1, r_2\}$. Matrix M has now been reduced to a single row, associated with all of the taxa, and with no sites. The construction now consists of a single tree whose root node is associated with all of the taxa. The tree is a perfect phylogeny for M. See figure 9.12.

Theorem 9.1.1 *Let T be the unique perfect phylogeny for M with all-zero ancestral sequence. Let $\widetilde{M}$ be a submatrix of M created during an execution of algorithm Clean-Build. Then the perfect phylogeny $\widetilde{T}$ for $\widetilde{M}$ is a subtree of T, rooted at the root node of T.*

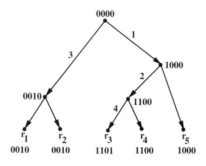

Figure 9.12   The perfect phylogeny T for M with all of the node and edge labels.

The point of the thought experiment is that the (assumed) known T is destroyed from the leaves upward, in parallel with the destruction of M, and in parallel with the construction of unknown T, by algorithm *Clean-Build*. When M is reduced to a single row with no sites, T is reduced to a single node with no edges, and that node is associated with the set of all the taxa in M. Essentially, the forest construction part of algorithm *Clean-Build* constructs edges and labels of the perfect phylogeny corresponding to edges and labels of T, which the destructive thought experiment removes. The actual correspondence is left to the reader, but the idea is simple: as columns and rows of are *removed* from M, edge labels and sibling edges are *removed* from the original T, and corresponding labels, sibling edges, and labeled edges are *added* to the growing $\widetilde{\mathcal{F}}$. Extension of the thought experiment to the case when M does not have a perfect phylogeny, so that the final $\widetilde{\mathcal{F}}$ is a forest, is straightforward and left to the reader.

Formal correctness
Now we can formally prove the correctness of algorithm *Clean-Build*.

Theorem 9.1.2 *The set of sequences M can be derived on a perfect phylogeny T with all-zero ancestral sequence if and only if algorithm Clean-Build reduces M to a single row containing no sites; and if and only if the forest constructed by the algorithm is the perfect phylogeny T for M.*

Proof: We first prove the "only if" side of the theorem, so suppose that M can be derived on a perfect phylogeny T with all-zero ancestral sequence. Clearly, the theorem holds if T contains only a single node, so that the corresponding M contains a single row with no entries; or if T contains only a single labeled edge, so that M contains a single row and a single column with an entry of value one.

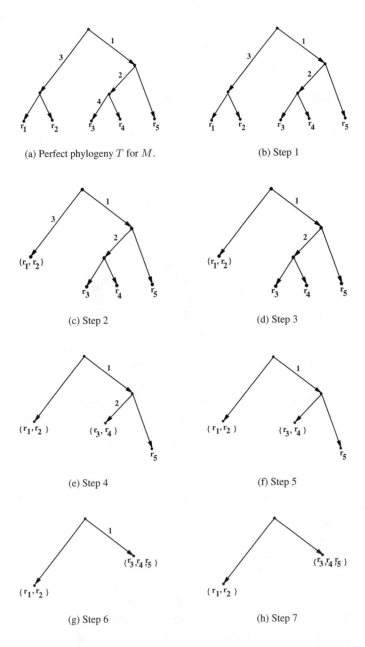

(a) Perfect phylogeny T for M.

(b) Step 1

(c) Step 2

(d) Step 3

(e) Step 4

(f) Step 5

(g) Step 6

(h) Step 7

Figure 9.13   Seven steps of the thought experiment, corresponding to the first seven destructive operations on M shown in figures 9.3–9.11. The last step would merge the remaining two leaves into a single node associated with all the taxa in M.

More generally, note that if M has been reduced to a single row, but there are sites remaining, those sites will be removed by subsequent applications of rule **Dc**. So, if M is reduced to a single row, it can be reduced to a row containing no sites. Therefore, for contradiction, assume that there is a counterexample to the theorem, that is, a matrix M that can be derived on a perfect phylogeny T with all-zero ancestral sequence, but algorithm *Clean-Build* does not reduce M to a single row. Then, there is a counterexample with the minimum number of sites, and among such counterexamples, there is one with a minimum number of taxa. Let M be such a minimal counterexample. Now consider the deepest leaf node v in the perfect phylogeny T for M and suppose it is labeled by taxon f. Recall that all interior nodes of a perfect phylogeny must have degree at least three, so the parent of v must have at least two children. Therefore, v has a sibling v', which is also a leaf node. Let f' be the taxon labeling v'.

If the edge into v or v' (say v) is labeled by a site c, then site c in M has only a single one entry and rule **Dc** applies. Removing site c from M creates a matrix $\widetilde{M}$, and removing label c from the edge in T into v creates a perfect phylogeny $\widetilde{T}$ for $\widetilde{M}$. $\widetilde{M}$ has fewer sites than M, so the theorem applies to $\widetilde{M}$. Therefore, algorithm *Clean-Build* reduces $\widetilde{M}$ to a single row, showing that M also reduces to a single row.

If neither edge into v or v' is labeled by a site, then the two rows for taxa f and f' are identical and rule **Dr** applies. Merging the rows for f and f' creates a matrix $\widetilde{M}$; removing v' from T and the edge into v', and labeling v with $\{f, f'\}$ creates a perfect phylogeny for $\widetilde{M}$. $\widetilde{M}$ has fewer rows than M, so the theorem again applies to $\widetilde{M}$, and so algorithm *Clean-Build* reduces $\widetilde{M}$, and M, to a single row. We have therefore proved that if M can be derived on a perfect phylogeny T then algorithm *Clean-Build* reduces M to a single row with no sites.

In a similar way, we can prove by contradiction that if there is a perfect phylogeny T for M with all-zero ancestral sequence, then algorithm *Clean-Build* will construct T. But it is more insightful to understand constructively that as M is destroyed, the algorithm builds a set of trees (a forest) that generate the sequences defined by submatrices of M of increasing size. In particular, at any point in the algorithm the submatrix consists of those rows and those columns that have been involved in an application of rule **Dc** or **Dr**. Therefore, when M is reduced to a single row with no sites, the set of trees consists of a single tree, that is, the unique perfect phylogeny T for M.

In more detail, assuming that there is a perfect phylogeny T for M with all-zero ancestral sequence, each application of rule **Cc** or **Cr** identifies a *forced* feature (a new edge label, a new labeled edge, or the merging of two nodes) of the unique perfect phylogeny T for M. That is what is established by lemmas 9.1.2 and 9.1.3. For example, if site c in the current $\widetilde{M}$ has only a single one (so rule **Dc** applies) and the one is in row $\tilde{f}$ associated with the subset of taxa $\mathcal{U}$, then in T there must be an edge into a node associated with $\mathcal{U}$, and that edge must be labeled with site c. Hence, rule **Cc** is forced. Similarly, when rule **Dr** applies, rule **Cr** is forced. So, as M is destroyed, the algorithm constructs features of T that are forced, finishing the proof of the "only-if" side of the theorem.

To prove the "if" side of the theorem, suppose that algorithm *Clean-Build* reduces M to a single row with no sites. As above, as M is destroyed, a perfect phylogeny T for M is constructed, so certainly a perfect phylogeny for M exists.   ∎

9.1.2   The Case of the Root-Unknown Perfect Phylogeny

Earlier, we assumed that M can be derived on a perfect phylogeny with all-zero ancestral sequence. It is easy to extend that to the case of a different *known* ancestral sequence in the same way that the perfect phylogeny problem with any known ancestral sequence can be reduced to the case of the all-zero ancestral sequence (see section 2.2). But if no ancestral sequence is known, then we modify the approach as follows: First assume that no column in M contains only zeros or only ones. Next, modify rule **Dc** so that column c is removed if it contains only a single one or only a single zero. Such a column is called *uninformative*. The parallel rule **Cc** is not changed; the edge created is labeled with site c. However, the ancestral state of c is set whenever rule **Cc** is applied: If column c contains only a single one, the ancestral state of c is set to zero; if column c only contains a single zero, the ancestral state of c is set to one. Rules **Dr** and **Cr** are not changed. When these variants of rules **Dc** and **Cc** are used in algorithms *Clean* and *Clean-Build*, the resulting algorithms will be called *RU-Clean* and *RU-Clean-Build*.

9.1.3   The General ARG Case

We now move from the case of perfect phylogeny to ARGs that must contain recombination nodes. We again assume that the ancestral sequence is required to be the all-zero sequence. By theorem 9.1.2, if algorithm *Clean-Build* reduces M to a single row with no sites, then M can be derived on a perfect phylogeny with all-zero ancestral sequence. Hence, if there is no perfect phylogeny for M, then any execution of algorithm *Clean-Build must* reach a point where $\widetilde{M}$ still contains entries, but neither rules **Dc** nor **Dr** apply. Let $\widetilde{\mathcal{F}}$ be the forest constructed by algorithm *Clean-Build* at that point; let D be the set of sites *removed* from M; and now let $\widetilde{R}$ be the set of taxa involved in any application of rule **Dr**, that is, that set of rows in M that were part of some merge(s). $M(D)$ represents the submatrix of M restricted to the sites in D. Application of lemmas 9.1.2 and 9.1.3 implies the following generalization of the perfect-phylogeny theorem (page 40):

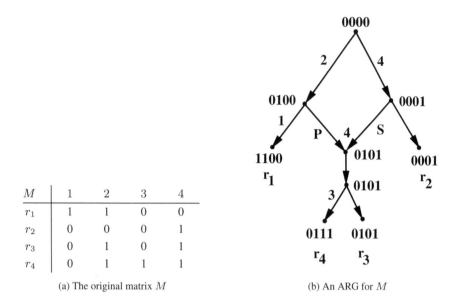

M	1	2	3	4
r_1	1	1	0	0
r_2	0	0	0	1
r_3	0	1	0	1
r_4	0	1	1	1

(a) The original matrix M (b) An ARG for M

Figure 9.14 Matrix M does not have a perfect phylogeny when the required ancestral sequence is the all-zero sequence.

Lemma 9.1.4 *The set of sequences in $M(D)$ can be derived on the unique forest $\widetilde{\mathcal{F}}$ of perfect phylogenies, each with all-zero ancestral sequence.*

As an example, consider the matrix M and the ARG for it shown in figure 9.14. Since sites r_2 and r_4 are incompatible (after adding the all-zero ancestral sequence to M), there is no perfect phylogeny for M with all-zero ancestral sequence. Algorithm *Clean-Build* removes columns 1 and 3 and merges rows r_3 and r_4. The resulting matrix $\widetilde{M}$ is shown in figure 9.15a; the forest $\widetilde{\mathcal{F}}$ is shown in figure 9.15b; and the matrix $M(D)$ is shown in figure 9.15c. Consistent with lemma 9.1.4, $\widetilde{\mathcal{F}}$ is a forest of three perfect phylogenies, each with all-zero ancestral sequence, and these generate the set of sequences $M(D)$. Also, as in the thought experiment done for the case of perfect phylogenies, $\widetilde{\mathcal{F}}$ is a forest of trees that *hangs off* the periphery of the ARG for the sequences in $\widetilde{M}$, shown in figure 9.14. That observation will be formalized below in theorem 9.1.3.

So, at the first point in an execution of algorithm *Clean-Build* where neither rule **Dr** nor **Dc** applies, algorithm *Clean-Build* will have constructed a forest $\widetilde{\mathcal{F}}$ of perfect phylogenies whose leaves are labeled by the taxa in M, and whose edges are either unlabeled or are

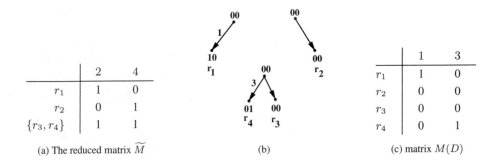

(a) The reduced matrix $\widetilde{M}$

	2	4
r_1	1	0
r_2	0	1
$\{r_3, r_4\}$	1	1

(b)

(c) matrix $M(D)$

	1	3
r_1	1	0
r_2	0	0
r_3	0	0
r_4	0	1

Figure 9.15 Sites 1 and 3 have been removed from the matrix M shown in figure 9.14, and rows r_3 and r_4 have been merged. The set D of removed columns is $\{1, 3\}$. The forest $\widetilde{\mathcal{F}}$, restricted to the sites in D, is shown in (b), and the matrix $M(D)$ is shown in (c). As stated in lemma 9.1.4, forest $\widetilde{\mathcal{F}}$ generates the sequences in $M(D)$.

labeled by the sites in D. Intuitively, this forest $\widetilde{\mathcal{F}}$ should be part of a larger ARG for M, and it should form part of the "bottom" of the ARG. We formalize and prove that next.

Theorem 9.1.3 *There exists an ARG $\mathcal{N}$ for the original M that contains the forest $\widetilde{\mathcal{F}}$. More exactly, if $\widetilde{\mathcal{F}}$ consists of k trees, then there are k nodes in $\mathcal{N}$ that root the trees in $\widetilde{\mathcal{F}}$.*

Proof: Let $\widetilde{M}$, $\widetilde{\mathcal{F}}$, D, and $\widetilde{R}$ be as defined above. By construction, $\widetilde{M}$ does not contain any of the sites in D, but for each tree in $\widetilde{\mathcal{F}}$, $\widetilde{M}$ does contain one row $\tilde{f}$ associated with the taxa labeling the leaves of that tree. Let $T_{\tilde{f}}$ denote that tree, and let $\mathcal{T}(\tilde{f})$ be the set of taxa of M labeling the leaves of $T_{\tilde{f}}$ (and hence associated with row $\tilde{f}$ of $\widetilde{M}$). Let $s_{\tilde{f}}$ be the sequence in $\widetilde{M}$ in row $\tilde{f}$. Two rows are merged in an execution of algorithm *Clean-Build* only when the two rows are identical in $\widetilde{M}$, so in the original M all of the taxa in $\mathcal{T}(\tilde{f})$ must be identical at all of the sites in $\widetilde{M}$. More exactly, when restricted to the sites in $\widetilde{M}$, each sequence in M, for the taxa in $\mathcal{T}(\tilde{f})$, must be identical to the sequence $s_{\tilde{f}}$.

Now let $\widetilde{\mathcal{N}}$ be an ARG for $\widetilde{M}$. Since $\tilde{f}$ is a taxon in $\widetilde{M}$, there must be a node $\tilde{v}$ in $\widetilde{\mathcal{N}}$ labeled by $s_{\tilde{f}}$. If we attach the root of tree $T_{\tilde{f}}$ at node $\tilde{v}$, the result is an ARG that generates the sequences in M for all the taxa in $\mathcal{T}(\tilde{f})$. Let $\mathcal{N}$ be the ARG with all-zero ancestral sequence created by repeating this operation for each tree in $\widetilde{\mathcal{F}}$, that is, attaching each tree to the appropriate node in $\widetilde{\mathcal{N}}$. Clearly, the ARG $\mathcal{N}$ correctly generates all the sequences in $\widetilde{R}$. We need to prove that it also correctly generates the sequences *not* in $\widetilde{R}$. By assumption, $\widetilde{\mathcal{N}}$ correctly generates those sequences at all the sites not in D. We next show that each sequence not in $\widetilde{R}$ has a value of zero at any site in D.

Every site in D must label an edge in some tree in $\widetilde{\mathcal{F}}$. Consider a site c in D that labels an edge of tree $T_{\tilde{f}}$. Clearly, only taxa in $\widetilde{R}$ can label leaves of $T_{\tilde{f}}$, so we need to prove that any taxon not in $\widetilde{R}$ has a value of zero at site c. That is true because at the point in the execution of algorithm *Clean-Build* where site c was removed, column c had only one entry with a value of one, so in M, no taxon outside of $\mathcal{T}(\tilde{f})$ could have a value of one at site c. ∎

Given theorem 9.1.3, an ARG for M can be built by first applying algorithm *Clean-Build* until no further application of rule **Dc** or **Dr** is possible, creating the forest $\widetilde{\mathcal{F}}$ and the resulting matrix $\widetilde{M}$. Then, recursively, an ARG $\mathcal{N}$ for $\widetilde{M}$ can be constructed; after that, $\widetilde{\mathcal{F}}$ can be attached to $\mathcal{N}$ as in the proof of theorem 9.1.3. Since there are no recombination nodes in $\widetilde{\mathcal{F}}$, the number of recombination nodes for the resulting ARG $\mathcal{N}$ will be the number of recombination nodes in $\widetilde{\mathcal{N}}$.

But how can a good ARG for $\widetilde{M}$ be constructed? We know that no application of rule **Dc** or **Dr** on $\widetilde{M}$ is possible. Instead, we introduce the *third* destructive rule that operates on $\widetilde{M}$, and the third constructive rule that builds part of $\widetilde{\mathcal{N}}$, again bottom up.

9.1.3.1   The Third Destructive and Constructive Rules

> **Rule Dt:** If neither rule **Dc** nor **Dr** can be applied, pick a row $\tilde{f}$ in the current $\widetilde{M}$ (other than the all-zero row that corresponds to the ancestral sequence) and remove row $\tilde{f}$ from $\widetilde{M}$.

For clarity of the discussion below, we will use $\widetilde{M}_1$ to refer to the matrix $\widetilde{M}$ before the first application of rule **Dt**. $\widetilde{M}$ will always refer to the current matrix incorporating all of the removals of columns and rows due to applications of rules **Dc**, **Dr**, and **Dt**. We use $s_{\tilde{f}}$ to denote the sequence removed from $\widetilde{M}_1$.

The third constructive rule

After the application of rule **Dt**, we use the third *constructive* rule to begin the construction of $\widetilde{\mathcal{N}}$:

> **Rule Ct:** Find a series of recombinations (without mutations) of sequences in $\widetilde{M} = \widetilde{M}_1 - s_{\tilde{f}}$ that derive sequence $s_{\tilde{f}}$. (This will always be possible, as explained below.) Then construct a DAG containing one node for each sequence in $\widetilde{M}$, and one node for $s_{\tilde{f}}$, and all the needed edges and additional recombination nodes specified by the recombinations that derive $s_{\tilde{f}}$ from $\widetilde{M}$.

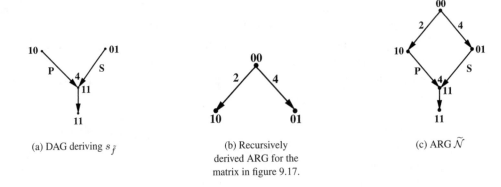

(a) DAG deriving $s_{\tilde{f}}$

(b) Recursively derived ARG for the matrix in figure 9.17.

(c) ARG $\widetilde{\mathcal{N}}$

Figure 9.16   ARG $\widetilde{\mathcal{N}}$ for $\widetilde{M}_1$ is constructed by adding the DAG deriving $s_{\tilde{f}}$ to the ARG for the matrix in figure 9.17.

The DAG constructed by applications of rule **Ct** forms a bottom or peripheral part of the desired ARG $\widetilde{\mathcal{N}}$ for $\widetilde{M}_1$. The full ARG $\widetilde{\mathcal{N}}$ for $\widetilde{M}_1$ is formed by adding that DAG to an ARG that derives $\widetilde{M}$. Of course, the ARG for $\widetilde{M}$ is found by recursive application of the entire method.

For example, consider the reduced matrix in figure 9.15, and denote it $\widetilde{M}_1$. If we pick $\tilde{f}$ to be the row labeled $\{r_3, r_4\}$, sequence $s_{\tilde{f}} = 11$ can be generated by recombining the two sequences in rows r_1 and r_2. The DAG representing the generation of $s_{\tilde{f}}$ is shown in figure 9.16a. Then, the algorithm must recursively find an ARG for the matrix $\widetilde{M}$ shown in figure 9.17. That ARG is the tree shown in figure 9.16b. The ARG $\widetilde{\mathcal{N}}$ for $\widetilde{M}_1$, shown in figure 9.16c, is formed by combining the DAG in figure 9.16a with the ARG in figure 9.16b. The full ARG $\mathcal{N}$ for M, shown in figure 9.14, is formed by adding the forest $\widetilde{\mathcal{F}}$ shown in figure 9.15b to ARG $\widetilde{\mathcal{N}}$. If instead of picking row $\{r_3, r_4\}$ to be $\tilde{f}$, we pick row r_2, the resulting ARG for M is shown in figure 9.18. In this example, both choices for $\tilde{f}$ result in an ARG with only one recombination node. In general, however, different choices for $\tilde{f}$ can result in different numbers of recombination nodes in the ARG for M. In this example, the DAG generating $s_{\tilde{f}}$ only has one recombination node, but that need not be true in general; also the recursively found ARG is a tree, but that also need not be true in general. For a more involved example, See figure 1 in [390].

	2	4
r_1	1	0
r_2	0	1

Figure 9.17 The reduced matrix $\widetilde{M}$ after rule **Dt** removes row $\tilde{f}$, labeled $\{r_3, r_4\}$.

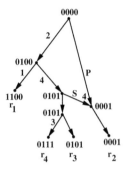

Figure 9.18 The ARG constructed by algorithm *Clean-Build-with-Recombination* when row r_2 is selected for row $\tilde{f}$ in rule **Dt**.

9.1.3.2   Summarizing the Full Algorithm

Algorithm *Clean-Build-with-Recombination*, shown in figure 9.19, reduces M to a single row with no sites, and builds an ARG $\mathcal{N}$ for M. As before, $\widetilde{M}$ denotes the current matrix created during an execution of the algorithm, changing as the algorithm proceeds. The algorithm is recursive, and input to it is denoted $\widetilde{M}_0$. To build an ARG for M, the algorithm is called with $\widetilde{M}$ set to M.

It should be clear that algorithm *Clean-Build-with-Recombination* does build an ARG for M, but we have not yet given any reason to believe that it will be an ARG where the number of recombination nodes is close to $Rmin_0(M)$. To achieve that goal, we will have to add criteria for the way we select the row $\tilde{f}$ in rule **Dt**, and we will have to be explicit about how recombinations are found to derive $s_{\tilde{f}}$. We do that next.

Algorithm Clean-Build-with-Recombination $(\widetilde{M}_0)$

{The algorithm will return an ARG $\widetilde{\mathcal{N}}$ for input matrix $\widetilde{M}_0$.}

if $\widetilde{M}_0$ contains more than one row or contains some sites **then**

    Set $\widetilde{M}$ to $\widetilde{M}_0$ and run algorithm *Clean-Build* on $\widetilde{M}$.

    Let $\widetilde{\mathcal{F}}$ denote the forest created by algorithm *Clean-Build*,
    and let $\widetilde{M}_1$ denote $\widetilde{M}$ at this point, i.e., modified by *Clean-Build*.

    Apply rule **Dt** to $\widetilde{M}$
    {so $\widetilde{M}$ now is $\widetilde{M}_1 - \{s_{\bar{f}}\}$}

    Apply rule **Ct** to $\widetilde{M}$ and $s_{\bar{f}}$, creating a DAG G_t that derives the sequence $s_{\bar{f}}$
    by recombinations of sequences in $\widetilde{M}$.

    (Recursively) call algorithm *Clean-Build-with-Recombination* with input $\widetilde{M}$.

endif

Add the DAG G_t to the ARG returned from the above recursive call,
creating an ARG $\widetilde{\mathcal{N}}_1$ for $\widetilde{M}_1$.

Add the forest $\widetilde{\mathcal{F}}$ to $\widetilde{\mathcal{N}}_1$, creating the ARG $\widetilde{\mathcal{N}}$ for $\widetilde{M}_0$.

return $\widetilde{\mathcal{N}}$

Figure 9.19   Algorithm *Clean-Build-with-Recombination*

9.1.3.3 *Specifying and Reducing Recombinations*

To reduce the number of recombination nodes created by algorithm *Clean-Build-with-Recombination*, we first examine how to derive $s_{\bar{f}}$ from a set of sequences $\widetilde{M}$, using the fewest number of (single-crossover) recombination events.

Definition Given a set of sequences M, each of length m, and an m-length sequence s not in M, $Rmin(M; s)$ denotes the *minimum* number of *single-crossover* recombinations needed to create s from the sequences in M, without using any mutations. $Rmin(M; s)$ is not defined if it is not possible to create s from M without mutations.

In the context of algorithm *Clean-Build-with-Recombination*, we want to compute $Rmin(\widetilde{M}; s_{\tilde{f}})$, or determine that it is not defined.

Minimizing recombinations

Algorithm *Min-Crossover*, shown in figure 9.20, efficiently computes $Rmin(\widetilde{M}; s_{\tilde{f}})$, or determines that it is not defined, for an arbitrary matrix $\widetilde{M}$ and a sequence $s_{\tilde{f}}$. This algorithm is a straightforward *greedy* algorithm that generalizes algorithm *Multiple-Crossover-Test* (page 261), discussed in section 8.6.

Algorithm Min-Crossover (M, s)

    Set $\mathcal{C}$ to the list containing the number 1.
    Set c to 1, and cr to 0.

    repeat

        Over all the sequences in M, find the longest substring starting at site c
        that matches the substring in s starting at site c.
        Let k denote the length of that longest matching substring.

        if $(k == 0)$ **then**
            report that $Rmin(M; s)$ is undefined, and exit.

        else
            set cr to $cr + 1$, and set c to $c + k$.
            Add c to the end of $\mathcal{C}$.
        endif
    until $(c > m)$

    Set $cr^* = cr$, and add $m + 1$ to the end of $\mathcal{C}$.
    return cr^* and $\mathcal{C}$.

Figure 9.20 Algorithm *Min-Crossover*, which generalizes algorithm *Multiple-Crossover-Test*.

Clearly, when called with $M = \widetilde{M}$ and $s = s_{\tilde{f}}$, if $Rmin(\widetilde{M}; s_{\tilde{f}})$ is defined, algorithm *Min-Crossover* will return some value for cr^*, and so $s_{\tilde{f}}$ can be created from $\widetilde{M}$ using

cr^* single-crossover recombinations (whose crossover indexes are recorded in $\mathcal{C}$) and no mutations. More completely, we have the following:

Theorem 9.1.4 *Algorithm Min-Crossover correctly determines whether* $Rmin(\widetilde{M}; s_{\tilde{f}})$ *is defined, and when defined,* $Rmin(\widetilde{M}; s_{\tilde{f}})$ *equals* cr^*. *The running time is polynomial in the size of* M.

Proof: In each iteration, the algorithm greedily attempts to extend the length of the prefix of $s_{\tilde{f}}$ that can be generated from $\widetilde{M}$ by single-crossover recombinations. In particular, the length of the prefix starts at zero and increases by exactly k characters in any iteration where $k > 0$. Therefore, unless $k = 0$ in some iteration, the algorithm constructively shows that $Rmin(\widetilde{M}; s_{\tilde{f}})$ is defined, and $s_{\tilde{f}}$ can be created from exactly $cr^* + 1$ substrings of sequences in $\widetilde{M}$.

Conversely, if $k = 0$ at some iteration, it means that there is a site c where all sequences in $\widetilde{M}$ have a state that is unequal to the state of c in $s_{\tilde{f}}$. Therefore, the algorithm is correct when it reports that $Rmin(\widetilde{M}; s_{\tilde{f}})$ is not defined.

To show that $cr^* = Rmin(\widetilde{M}; s_{\tilde{f}})$ (when it is defined), assume that the equality does not hold. Since $s_{\tilde{f}}$ can be created from $cr^* + 1$ substrings in $\widetilde{M}$, it must be that $cr^* > Rmin(\widetilde{M}; s_{\tilde{f}})$. Let $\mathcal{C}_{min}$ be the list consisting of 1, followed by all the crossover indexes in some scenario that creates $s_{\tilde{f}}$ from $\widetilde{M}$, using $Rmin(\widetilde{M}; s_{\tilde{f}})$ single-crossover recombinations, followed by $m + 1$. If cr^* is strictly greater than $Rmin(\widetilde{M}; s_{\tilde{f}})$, then there must be a *first* index i, such that $\mathcal{C}[i]$, the ith entry in $\mathcal{C}$, is *strictly* less than $\mathcal{C}_{min}[i]$. Note that by the choice of i, $\mathcal{C}[i-1] \geq \mathcal{C}_{min}[i-1]$. But then, the substring of $s_{\tilde{f}}$ from site $\mathcal{C}_{min}[i-1]$ to site $\mathcal{C}_{min}[i] - 1$ must match the substring of some sequence s in $\widetilde{M}$, in that same range of sites. And, since $\mathcal{C}_{min}[i-1] \leq \mathcal{C}[i-1]$, the substring of $s_{\tilde{f}}$ from site $\mathcal{C}[i-1]$ to site $\mathcal{C}_{min}[i] - 1$ must match s in that range of sites. But that would contradict the fact that the algorithm reported a crossover at site $\mathcal{C}[i] < \mathcal{C}_{min}[i]$. Hence there is no index i where $\mathcal{C}[i] < \mathcal{C}_{min}[i]$ and therefore it follows that $cr^* = Rmin(\widetilde{M}; s_{\tilde{f}})$. ∎

Although the proof is written for $\widetilde{M}$ and $s_{\tilde{f}}$, it clearly also proves the correctness of algorithm *Min-Crossover* for any binary matrix M and binary sequence f with length equal to the number of sites in M. Further, although algorithm *Min-Crossover* was written for binary sequences, it works without modification for sequences over any alphabet. Problems of computing $Rmin(M, s)$ in nonbinary alphabets (usually DNA) arise frequently in the computational biology literature. The solutions are often said to be examples of *dynamic programming*, but it is really more appropriate to call the solutions *greedy* methods.

Given theorem 9.1.4, the obvious modification of algorithm *Clean-Build-with-Recombination* is to use algorithm *Min-Crossover* to determine how the chosen $s_{\tilde{f}}$ should be derived from $\widetilde{M}$, during an application of rule **Ct**. Further, we have:

Lemma 9.1.5 *When algorithm Min-Crossover is used in rule* **Ct***,* $Rmin(\widetilde{M}; s_{\tilde{f}})$ *will be defined.*

Proof: Clearly, $Rmin(\widetilde{M}; s_{\tilde{f}})$ will be defined if and only if, for every site c in $s_{\tilde{f}}$, the state of c in $s_{\tilde{f}}$ equals the state of c for some sequence in $\widetilde{M}$. Rule **Ct** is executed only after rule **Dt** is executed, which is executed only when no application of either rule **Dc** or **Dr** is possible. At that point, every column of $\widetilde{M}$ will contain some entry with value 1, and the all-zero ancestral sequence will contain a 0 in column c. So, the state of c in $s_{\tilde{f}}$ will be equal to the state of c for some sequence in $\widetilde{M}$. Hence, $Rmin(\widetilde{M}; s_{\tilde{f}})$ will be defined. ∎

We will assume that algorithm *Min-Crossover* is used in rule **Ct**. A more effective change is to modify destructive rule **Dt**. Recall that $\widetilde{M}_1$ is the matrix $\widetilde{M}$ before the application of rule **Dt**.

> **Modified rule Dt:** In rule **Dt**, choose the row $\tilde{f}$ in $\widetilde{M}_1$ that minimizes $Rmin(\widetilde{M}_1 - s_{\tilde{f}}; s_{\tilde{f}})$.

Modified rule **Dt** and algorithm *Min-Crossover* are clearly good heuristics to *locally* (in each application of rules **Dt** and **Ct**) reduce the number of recombination nodes in the resulting ARG $\mathcal{N}$ for M. But their use does *not* guarantee that the resulting ARG will minimize the number of recombination nodes over all possible executions of algorithm *Clean-Build-with-Recombination*. We consider that goal next.

Definition For a matrix M, let $\mathcal{N}^*(M)$ be the ARG with the *minimum* number of recombination nodes over all possible executions of algorithm *Clean-Build-with-Recombination* on input M, and let $RN^*(M)$ denote the number of recombination nodes in $\mathcal{N}^*(M)$.

It is *not* true that $\mathcal{N}^*(M)$ is necessarily found by applying the modified rule **Dt** in every iteration of algorithm *Clean-Build-with-Recombination*. It is also *not* true that $\mathcal{N}^*(M)$ is necessarily a MinARG for M, or an ARG with all-zero ancestral sequence that uses $Rmin_0(M)$ recombination nodes. Still, it is desirable to build $\mathcal{N}^*(M)$ and to compute $RN^*(M)$.

9.1.3.4   *Branching to Compute* $RN^*(M)$ *and Build* $\mathcal{N}^*(M)$

The conceptually simplest way to compute $RN^*(M)$, and to build $\mathcal{N}^*(M)$, is to *branch* on all choices of $\tilde{f}$ in each application of rule **Dt**, building a search tree of choices.[2] Each path in that search tree defines an ARG for M, and the ARG with the fewest recombination nodes defines $RN^*(M)$. The algorithm that conducts such an exhaustive branching is called algorithm *CBR-Branch*.

Exhaustive branching in algorithm *CBR-Branch* makes it computationally prohibitive, except for small matrices.[3] However, there is considerable redundancy in any execution of algorithm *CBR-Branch*, because the same submatrix $\widetilde{M}$ of M can be formed many times along many different search paths. We will, in section 9.1.3.6, speed up the branching by avoiding such redundancy. There, the addition of *bounding* ideas will make the branching approach practical for a much wider range of data. Moreover, understanding the redundancy in algorithm *CBR-Branch* leads to a *dynamic programming* algorithm to compute $RN^*(M)$, achieving a worst-case running time that is significantly less than the $O(n!)$ worst-case time for algorithm *CBR-Branch*.

9.1.3.5   *Dynamic Programming Computation of* $RN^*(M)$ *and* $\mathcal{N}^*(M)$

Definition Given a matrix M, and a subset $\mathcal{U}$ of rows of M, let $M_{\mathcal{U}}$ denote the submatrix of M consisting of the rows in $\mathcal{U}$ and *all* of the columns of M.

Algorithm $DP\text{-}RN^*$, shown in figure 9.21, uses dynamic programming to compute $RN^*(M_{\mathcal{U}})$, for each subset $\mathcal{U}$ of the rows of M. The algorithm mimics the action of algorithm *CBR-Branch* on matrix $M_{\mathcal{U}}$, but avoids redundancy by enumerating each required matrix $\widetilde{M}$ only once, and computing each $RN^*(\widetilde{M})$ only once.

In the dynamic program, the values will be computed in order of increasing size of $\mathcal{U}$. $RN^*(M)$ is the extreme case that $\mathcal{U}$ is the entire set of rows of M. In the algorithm, we use the variable $rn^*(M)$, rather than $RN^*(M)$, but later prove that $rn^*(M) = RN^*(M)$. After $RN^*(M)$ is computed, the ARG $\mathcal{N}^*(M)$ can be constructed during a standard dynamic programming traceback. That traceback will specify a series of applications of

---

2 Note that the original rule **Dt** is used here, not the modified rule **Dt**.

3 The only established upper bound on the running time of this version of algorithm *CBR-Branch* is $O(n!)$.

Algorithm *DP-RN** (M)

    if (M has fewer than three rows) **then**
        set $rn^*(M)$ to 0
        return $rn^*(M)$
    endif

    for ($k = 2, \ldots, n$) **do**
        {where n is the number of rows in M}

        for (each subset $\mathcal{U}$ of k rows of M) **do**
            Form the submatrix $M_{\mathcal{U}}$ of M and run algorithm *Clean* on $M_{\mathcal{U}}$.
            Let $\widetilde{M}$ denote the resulting matrix, and let $\widetilde{\mathcal{U}}$ denote
            the set of rows of $\widetilde{M}$.

            Set $rn^*(M_{\mathcal{U}}) = \min_{s_{\tilde{f}} \in \widetilde{M}} [Rmin(\widetilde{M} - s_{\tilde{f}}; s_{\tilde{f}}) + rn^*(M_{(\widetilde{\mathcal{U}} - \tilde{f})})]$
        endfor
    endfor

    return $rn^*(M)$

Figure 9.21   Algorithm *DP-RN*$^*(M)$.

rules **Dc**, **Dr**, and **Dt**, and $\mathcal{N}^*(M)$ can therefore be built by applying the corresponding rules **Cc**, **Cr**, and **Ct**, as discussed earlier.

The most effective part of the dynamic program is that it avoids enumeration of subsets of *columns*. The recurrences in the dynamic program require $rn^*(M_{(\widetilde{\mathcal{U}} - \tilde{f})})$, where $M_{(\widetilde{\mathcal{U}} - \tilde{f})}$ is the submatrix of M containing the *rows* in $\widetilde{\mathcal{U}} - \tilde{f}$, and *all* of the columns of M. In that way, the recurrences enumerate only 2^n submatrices of M, each specified by a choice of *rows*, rather than enumerating $2^n 2^m$ submatrices that would be specified by choices of both rows *and* columns. The correctness of this idea essentially follows from lemma 9.1.1 (page 287), and is formally proved next.

Theorem 9.1.5 *The value of $rn^*(M)$ computed by algorithm DP-RN* on input M is $RN^*(M)$.*

Proof: The proof is by induction on $|\mathcal{U}|$. When $|\mathcal{U}| < 3$, matrix $M_{\mathcal{U}}$ cannot have an incompatible pair of sites, so it can be derived on a perfect phylogeny with all-zero ancestral sequence. Therefore, by theorem 9.1.2, algorithm *Clean* reduces $M_{\mathcal{U}}$ to a matrix with one row and no sites. Then, $RN^*(M_\mathcal{U}) = rn^*(M_\mathcal{U}) = 0$, and the theorem holds when $|\mathcal{U}| < 3$. Now suppose the theorem holds up to some $k \geq 3$, and that $|\mathcal{U}| = k + 1$.

Consider the action of algorithm *CBR-Branch* on $M_\mathcal{U}$. It would first run algorithm *Clean* on $M_\mathcal{U}$, resulting in the same matrix $\widetilde{M}$ defined in algorithm *DP-RN**. At that point, it would branch on all choices of $s_{\tilde{f}} \in \widetilde{M}$. Each path from that branching point would compute $Rmin(\widetilde{M} - s_{\tilde{f}}; s_{\tilde{f}}) + RN^*(\widetilde{M} - s_{\tilde{f}})$ for one of those $s_{\tilde{f}}$. So clearly,

$$RN^*(M_\mathcal{U}) = \min_{s_{\tilde{f}} \in \widetilde{M}} [Rmin(\widetilde{M} - s_{\tilde{f}}; s_{\tilde{f}}) + RN^*(\widetilde{M} - s_{\tilde{f}})].$$

The right-hand side of that recurrence differs from the right-hand side of the recurrence given in algorithm *DP-RN** in that the second term in the recurrence is $RN^*(\widetilde{M} - s_{\tilde{f}})$, rather than $rn^*(M_{(\tilde{\mathcal{U}}-\tilde{f})})$. So, if we can prove that those two terms are equal, the recurrence for rn^* will be the same as for RN^*, and the theorem will be proved. Now $rn^*(M_{(\tilde{\mathcal{U}}-\tilde{f})}) = RN^*(M_{(\tilde{\mathcal{U}}-\tilde{f})})$, by the induction hypothesis, so we want to prove that $RN^*(M_{(\tilde{\mathcal{U}}-\tilde{f})}) = RN^*(\widetilde{M} - s_{\tilde{f}})$.

Note that $\tilde{\mathcal{U}}$ is the set of rows in $\widetilde{M}$, so $\widetilde{M} - s_{\tilde{f}}$ and $M_{(\tilde{\mathcal{U}}-\tilde{f})}$ have the same set of rows but a different set of columns. Recall that the computation of $RN^*(M_{(\tilde{\mathcal{U}}-\tilde{f})})$ begins with the application of algorithm *Clean* to $M_{(\tilde{\mathcal{U}}-\tilde{f})}$. Consider a column c in $M_\mathcal{U}$ that is not in $\widetilde{M}$. In reducing $M_\mathcal{U}$ to $\widetilde{M}$, column c was removed at a point where it had at most one entry with value 1. Since the rows in $\tilde{\mathcal{U}} - \tilde{f}$ are a subset of the rows at that point, column c in $M_{(\tilde{\mathcal{U}}-\tilde{f})}$ will also have at most one entry with value 1. If we run algorithm *Clean* on $M_{(\tilde{\mathcal{U}}-\tilde{f})}$ by first removing all the columns in $M_\mathcal{U}$ not in $\widetilde{M}$ (which is permitted by lemma 9.1.1), the result will be $\widetilde{M} - s_{\tilde{f}}$. Therefore, the full execution of algorithm *Clean* on $M_{(\tilde{\mathcal{U}}-\tilde{f})}$ will be the same as the result of running algorithm *Clean* on $\widetilde{M} - s_{\tilde{f}}$. Further, the destructive operations in algorithm *Clean* induce constructive application of rules **Cc** and **Cr** that build a forest containing *zero* recombination nodes. Therefore, $rn^*(M_{(\tilde{\mathcal{U}}-\tilde{f})}) = RN^*(M_{(\tilde{\mathcal{U}}-\tilde{f})}) = RN^*(\widetilde{M} - s_{\tilde{f}})$, and we conclude that the value $rn^*(M)$ returned by algorithm *DP-RN** is $RN^*(M)$. ∎

Time analysis

There are 2^n subsets of rows of M, and for each one we examine at most n individual sequences $s_{\tilde{f}}$, running algorithm *Min-Crossover* on each. With the appropriate data structures and string algorithms, each execution of algorithm *Min-Crossover* takes $O(nm)$ time, so: we have:

Theorem 9.1.6 *Algorithm DP-RN** *can be implemented to run in* $O(n^2 m 2^n)$ *time.*

9.1.3.6 Major Speedups of Algorithm CBR-Branch

In addition to the dynamic programming approach to computing $RN^*(M)$, and then $\mathcal{N}^*(M)$, there are three ways to speed up algorithm *CBR-Branch*.

Speedup 1
The first speedup is through the standard, well-known way that *memoization* converts a top-down, recursive branching algorithm into an algorithm whose worst-case running time is of the same order as a bottom-up dynamic programming solution [80]. In that approach, the search tree used by algorithm *CBR-Branch* would be expanded *depth-first*, and a table would be built that stores $RN^*(\widetilde{M})$ for every submatrix of M formed during the search. When the depth-first search backs up to a node in the search tree, where a recursive call to compute $RN^*(\widetilde{M})$ was made, the value of $RN^*(\widetilde{M})$ will be known and can be put in the table. Then, when another search path requires $RN^*(\widetilde{M})$, it can retrieve the value from the table, rather than expanding the search tree to (re)compute it. This standard memoizing idea has the consequence that the number of internal nodes in the search tree is bounded by the number of *distinct* submatrices of M. Further, by lemma 9.1.1, the number of submatrices where any branching occurs is bounded by 2^n, and so we have:

Theorem 9.1.7 *Algorithm CBR-Branch can be implemented using memoization to run in $O(n^2 m 2^n)$ worst-case time.*

Speedup 2
The second speedup occurs if the search-tree in algorithm *CBR-Branch* is *not* expanded in a strictly depth-first manner. In that case there can be two or more nodes in the expanded tree that require the computation of $RN^*(\widetilde{M})$, which has not yet been computed. Because the order that rows are removed by rule **Dt** differs along different search paths that create submatrix $\widetilde{M}$, the total number of recombination nodes (in the implied ARGs that could be created) specified along those paths can differ. Clearly, among all the paths in the search tree that create the same $\widetilde{M}$, only the one with the smallest total number of specified recombination nodes should be expanded. The other paths can be terminated.

Speedup 3
The third speedup is through application of the classic idea of using *lower bounds* to cut off paths in the search tree. Suppose that some of the paths in the search tree have reached leaves, and let w be the minimum number of recombination nodes in the ARGs for M specified along those paths. Then, consider an incomplete path in the search tree ending at a node v, creating submatrix $\widetilde{M}$. Let $w(\widetilde{M})$ be the total number of recombination nodes

specified by that path, and let $L(\widetilde{M})$ be a *lower bound* on $Rmin_0(\widetilde{M})$. If $w \leq w(\widetilde{M}) + L(\widetilde{M})$, then there is no need to expand the search tree at node v.

9.1.4   Program *SHRUB*

Algorithm *CBR-Branch* with the second and third speedups discussed above has been implemented in a program called *SHRUB*, which stands for *"simulated history recombination upper bound"* [390]. *SHRUB* does not necessarily assume that the ancestral sequence is the all-zero sequence, but we will continue to use the notation $\mathcal{N}^*(M)$ for the best possible ARG produced by *SHRUB*, and use $RN^*(M)$ for the number of recombination nodes in $\mathcal{N}^*(M)$.

With no additional modifications, when given an input matrix M, *SHRUB* will compute the *full upper bound* $RN^*(M)$ and build the ARG $\mathcal{N}^*(M)$. The lower bound used in the branch-and-bound is either the *HK bound* or the *haplotype bound* discussed in chapter 5. $RN^*(M)$ is called an "upper bound" because $Rmin(M) \leq RN^*(M)$. This inequality is due to the fact that $\mathcal{N}^*(M)$ is an ARG with all-zero ancestral sequence that derives M and uses exactly $RN^*(M)$ recombination nodes.

The user can also select a number k, and specify that the search tree should branch at most k ways. In that case, when *SHRUB* is at a branch node of the search tree, it *randomly* selects k rows (or all rows if there are k or fewer rows) in the current $\widetilde{M}$, and then successively chooses each of those rows as $\tilde{f}$. Since the rows are selected randomly, different executions of the computation can give different results, and it may be advantageous to repeat the computation several times to select the best result. In experiments reported in [390], the numbers of recombination nodes in the ARGs found with k between 3 and 5 were very close to $RN^*(M)$.

SHRUB can also be used to compute a *fast upper bound* which is obtained by an implementation of algorithm *Clean-Build-with-Recombination* using the modified rule **Dt**, that is, always choosing the row $\tilde{f}$ to minimize $Rmin(\widetilde{M}_1 - s_{\tilde{f}}; s_{\tilde{f}})$. The computation time for this approach is polynomially bounded since there is no branching when the modified rule **Dt** is applied. The resulting number of recombination nodes might not be close to $RN^*(M)$, but it can be used as a first value w in the branch-and-bound version of algorithm *CBR-Branch*.

The program *SHRUB* not only computes $RN^*(M)$, but outputs the specifications for $\mathcal{N}^*(M)$ in a format that can then be processed by a graphical display program and drawn

Figure 9.22 SHRUB was used in the location of a single gene influencing body size in dogs. The larger dog has fifty times the volume of the smaller dog, a variation larger than in any other mammalian species. (From *Science*, April 6, 2007. Reprinted with permission from AAAS. Photograph by Deanne Fitzmaurice.)

on the plane. *SHRUB*, and the ARG drawing that it specified, were used to help identify a single gene that greatly influences the body size of dogs [403][4]. See figure 9.22.

Experimental studies of the accuracy of *SHRUB* are reported in [390] for both real SNP data and for simulated data. In each dataset, $RN^*(M)$ is compared to the lower bound on $Rmin(M)$ computed by program *HapBound*. Table 5.1 (page 161) shows both of the numbers for real population data. The results there show that the observed upper and lower bounds are close, often equal, in which case both *HapBound* and *SHRUB* have computed

---

4 The use of *SHRUB* is discussed in the supplemental material for the paper.

$Rmin(M)$ exactly. Simulated data was produced by the coalescent simulation program *ms* [182] which is commonly used to generate test data for questions about ARGs. The critical parameters in that simulation are the *sample size* (number of taxa in the terminology of this book), the *mutation rate* θ (which affects the number of polymorphic sites in the dataset), and the *recombination rate* ρ. In the simulations, the parameters were chosen to reflect a range of realistic biological datasets. The results show that when θ and ρ are modest, and the sample size is 25, the upper and lower bounds agreed more than 95 percent of the time; as the sample size increases to 100, the agreement falls slowly to about 85 percent. In the cases where the upper and lower bounds were not equal, the lower bound was on average about 80 percent of the value of the upper bound. More details on these experiments can be found in [390].

SHRUB also efficiently, almost instantly, builds an ARG with 7 recombination nodes for Kreitman's classic SNP data (See figure 5.7 on page 144). In section 5.2.8 we stated that program *HapBound* produces a lower bound of 7 for this data, so we can now conclude that $Rmin(M)$ for Kreitman's data is in fact 7. This will be additionally confirmed in section 9.1.5.3, when we discuss the program *Beagle*.

9.1.4.1   *Extending SHRUB to Handle Gene-Conversion*

When gene-conversion is allowed, the problem is to construct an ARG for M with a small *total* number of single-crossover and gene-conversion events. Program *SHRUB* can be easily extended to do that by use of the dynamic programming method in [235] that finds the *minimum* number of single-crossover and two-crossover recombination events to derive a new sequence $s_{\tilde{f}}$ from a set of sequences $\widetilde{M}$. Program *SHRUB* is modified by replacing every use of algorithm *Min-Crossover* with the dynamic program from [235]. The resulting program is called *SHRUB-GC* and is described in more detail in [387].

9.1.5   Building MinARGs with Program *Beagle*: A Coalescent Inspired Approach

Program *SHRUB* constructs ARGs which empirically have been observed to use a number of recombination nodes "close" to $Rmin(M)$ for many datasets examined. *SHRUB* is reasonably efficient and can handle many datasets of interest in biology (but not genomic-scale datasets). However, *SHRUB* is not guaranteed to produce a MinARG for M, nor is there any guaranteed bound on how many more recombination nodes *SHRUB* will create, compared to a MinARG for M. To date, three algorithms, and two *programs* have been developed that are *guaranteed* to produce a MinARG for any input M. Of course, since the problem of building a MinARG is NP-hard, these programs cannot handle data as large as

SHRUB can handle. In this section we will discuss the ideas behind the program *Beagle*,[5] the faster of the two MinARG-building programs. The other program that is guaranteed to find a MinARG is called *RecMinPath*, and is based on an idea called *tree-scanning*, which will be discussed in section 9.2.

Why does SHRUB fail?

Before explaining the ideas behind program *Beagle*, it is worthwhile considering why *SHRUB* might fail to construct a MinARG. *SHRUB* builds a branching search tree that (without cutting off any of the search) considers all ordered choices of rows to remove during the applications of rule **Dt**; and at every removal of a row $\tilde{f}$, it uses algorithm *Min-Crossover* to determine the specific recombinations. This is *locally* optimal, but it is not necessarily *globally* optimal. That is, *SHRUB* does not consider or coordinate the recombinations used to derive two or more successive choices in **Dt** operations. Therefore, *SHRUB* doesn't have any mechanism for *sharing* or minimizing recombination nodes in the successive derivation of two or more sequences. Moreover, it is not obvious how such sharing could be reasonably explored in a program like *SHRUB*. In order to do that, we need a more fine-grained approach to ARG building, again from the bottom up. The approach leads to program *Beagle*. First we need the following:

Definition An ARG, where there is a *total ordering* on the occurrence of its events (mutations, coalescences, recombinations), is called *time-ordered ARG*.

For example, figure 9.23 shows a time-ordered ARG for the ARG shown in figure 3.8.

A change in display

Because of the total-ordering of the events (that is, the assumption that only a single event occurs at one time), we will depict a *time-ordered* ARG a bit differently than the way we depict a standard ARG. A mutation on directed edge (u, v) occurs at the time associated with node u; a coalescent event occurs at the time associated with the coalescent node; and a recombination event occurs at the time associated with the recombination node. So, to avoid any ambiguity about the time when mutations occur relative to recombinations and coalescences, when we depict a *time-ordered* ARG we use the convention that two edges directed out of (down from) a coalescent node must be unlabeled, as are the two edges into a recombination node. Also, a mutation occurs on a directed edge (u, v), where u has out-degree one. Hence, the standard assumption for ARGs (the way we have treated

---

5 Unfortunately, another program that is now widely used in population genetics, for haplotyping (see section 12.1), is also called *Beagle* [51]. We will only briefly mention the other program *Beagle*, and only in section 12.1.

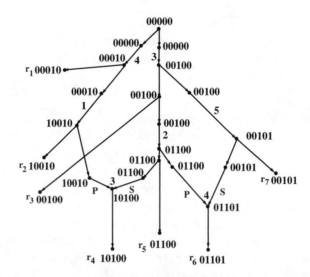

Figure 9.23   One possible *time-ordered* ARG derived from the ARG in figure 3.8 (page 75). Time-ordered ARGs have the constraint that only one event occurs at a time. As a consequence, each coalescent node has out-degree two, those two edges must be unlabeled, and each mutation occurs on a tree edge (u, v) where u has out-degree one. For example, observe how the two edges out of the root of the ARG in figure 3.8 have become four edges in the time-ordered ARG. Also, the two parents of a recombination node must be displayed at the same height, since they are part of the representation of one recombination event, which occurs at a particular point in time. Except for the parents of a recombination node, no other pair of nodes in the ARG should be displayed at the same height. (Lack of precision in this figure may suggest otherwise.)

them in this book), that each internal node has degree *at least* three, does not hold for time-ordered ARGs—tree nodes with out-degree one are permitted, and a coalescent event is represented by a coalescent node with out-degree *exactly* two. All the other assumptions about nodes and degrees are unchanged: for example, mutations only occur on tree edges; and a recombination is represented at a node with in-degree two and out-degree one. See figure 9.23.

In a *time-ordered* ARG, there is a total order on the occurrences of events, but no absolute times need be specified. Thus if two ARGs are isomorphic as labeled DAGs and the total orders on the events are the same, then the two *time-ordered* ARGs are considered to be identical. However, two ARGs that are isomorphic as labeled DAGs, but have different total orderings on the events, are not considered to be identical *time-ordered* ARGs.

MinARG from time-ordered ARGs

The following theorem is immediate, since any ARG can be converted to a *time-ordered* ARG without changing the number of recombination nodes.

Theorem 9.1.8 *A time-ordered ARG that derives a set of sequences M and uses the fewest recombination nodes over all time-ordered ARGs for M will be a MinARG for M.*

Beagle

Program *Beagle* [265] searches over orderings of events, to build a time-ordered ARG with the fewest recombination nodes, over all time-ordered ARGs for M. Hence, by theorem 9.1.8, the result is a MinARG for M. *Beagle* builds on algorithm *Clean-Build-with-Recombination* (as does *SHRUB*), constructing a time-ordered ARG for M, beginning with M and working backward in time (building the ARG bottom up). It successively chooses which of the three (permitted) events should next occur; which of the current (one or two) sequences should be involved; and what the (permitted) effect of the event should be. The process also keeps track of what is known about the derivation of the sequences after each new event. Moreover, the process only keeps track of segments of the sequences that can be known *with certainty*. In the terminology of coalescent theory, these segments are said to be *ancestral to the sample*.[6] We use the symbol $*$ for the state of a site that is unknown or cannot be determined, that is, is not ancestral to the sample.

For example, in figure 9.24, consider the recombination event creating the sequence 10100. Since the chosen crossover index is 3, we know that one parent contributes the prefix 10 and one parent contributes the suffix 100. But we cannot know the suffix of the first parent's sequence after the prefix 10, and similarly we cannot know the prefix of the second parent's sequence before the suffix 100. So what we can deduce about the sequence for the first parent (at the time that the recombination event occurred) is only that it is 10***, and we only deduce that the sequence for the second parent is **100. A subsequent event depicted in figure 9.24 is a coalescence of sequence 10*** and 10010. In this case we deduce that the parent of those two sequences must be labeled with 10010, so certainty about the sequence has been regained. That happens because there is no site where both sequences have an unknown state. See figure 9.25.

---

6 A site c in a sequence labeling a node v, is "ancestral to the sample" if and only if there is a path P in the ARG from v to some leaf, where for every node $u \neq v$ on P, the sequence at u receives its value for c from its parent on P. Analogously, a physical descendant of the molecule at site c in the sequence at node v appears at some leaf of the ARG. This issue is related to the issue of a "detectable recombination," discussed in section 3.6.

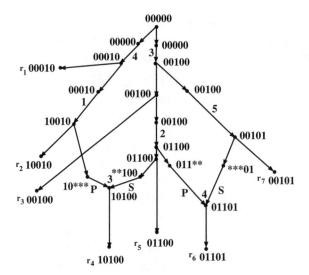

Figure 9.24   A time-ordered ARG with the deduced sequences. Note that the sequences are deduced backward in time (following the coalescent-theory viewpoint), and so some of the derived sequences contain sites whose state is unknown. In contrast, a top-down (forward in time) deduction of sequences completely specifies the state of each site in each sequence.

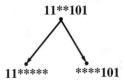

Figure 9.25   A coalescent event where parts of the coalescent sequence remain unknown.

Maximal deduced sequences

During *Beagle's* (backward-in-time) construction of an ARG for M, each event creates one or two new nodes, and labels the new node(s) with a sequence that might contain * symbols, each indicating an unknown state.

Definition At any point in the execution of *Beagle*, a node v is called a *Maximal Deduced Node (MDN)*, if no ancestor of v has yet been created. The sequence labeling an *MDN* is called a *maximal deduced sequence MDS*. The set of maximal deduced sequences is denoted *MDSet*.

At the start of an execution of *Beagle*, *MDSet* consists of M; and at the end of the algorithm, *MDSet* consists of the single all-zero sequence (the assumed ancestral sequence). The size of *MDSet* increases by one after each recombination event, decreases by one after each coalescent event, and remains the same after each mutation event.

Permitted events and MDS rules

As *Beagle* constructs an ARG, it successively chooses the next event (mutation, coalescence, or recombination), to extend the ARG, and create one or two new sequences in *MDSet*. There are often multiple events possible, but choices are constrained. The rules for the choice of a *permitted* next (backward-in-time) event, and for a change in *MDSet* are:

Permitted ARG-event rules

Mutation rule If *exactly* one sequence s in *MDSet* has an entry of value one, at site c, then a mutation at site c is permitted. Let s' be sequence s where the one at site c is changed to a zero. If this mutation is the next event chosen, replace s with s' in *MDSet*. In the growing ARG, add an edge $e = (u, v)$ into the node v labeled s, label e with c, and label u with s'.

Coalescence rule Let s and s' be two sequences in the current *MDSet*. If there is a site c where one of the sequences has value zero and the other has value one, then s and s' cannot coalesce. Otherwise, they can coalesce to sequence s'', where $s''[c] = s[c]$ for every site c such that $s[c] \neq *$, and $s''[c] = s'[c]$ otherwise. For example, if $s = 1**01$ and $s' = 11**1$, then $s'' = 11*01$. If the coalescent event is the next event chosen, replace s and s' with s'' in *MDSet*; extend the growing ARG by adding a node labeled s'' with a directed edge to the node labeled s, and a directed edge to the node labeled s'.

Recombination rule Let s be any sequence in *MDSet*. A recombination event creating s is always permitted as the next event, but a crossover index must be selected. If crossover index b_x is selected, let s' be the sequence matching s up through site $b_x - 1$, followed by $*$ at all remaining sites; and let s'' be the sequence containing $*$ at sites 1 through $b_x - 1$ followed by the sequence matching s at all remaining sites. If that recombination event is the next event chosen, replace s in *MDSet* with s' and s''; extend the growing ARG by adding two nodes labeled by s' and s'', respectively, each with an edge directed to the node labeled s.

The following lemma is easy to establish by induction on the number of events.

Lemma 9.1.6 *After each event, M can be derived from the current set of maximal deduced sequences, MDSet, using the part of the ARG constructed to that point.*

At each event, an additional part of the ARG is constructed and the problem of completing the full ARG for M reduces to the problem of building an ARG to derive the current *MDSet*.

The fact that no sequence in *MDSet* is an ancestor of another sequence in *MaxDS* explains why the *mutation rule* requires that *exactly* one sequence s in *MDSet*, must have an entry of value one at site c. If two or more sequences in *MDSet* have a value of one at a site c, then more than a single mutation at c would be required to derive *MDSet* from the all-zero sequence, violating the requirement that in an ARG, no site mutates more than once. The following theorem follows immediately from lemma 9.1.6.

Theorem 9.1.9 *Given a set of sequences M, any series of permitted ARG events that starts with M and ends with the all-zero sequence creates a time-ordered ARG for M.*

Conversely:

Theorem 9.1.10 *Let N_t be a time-ordered ARG for M with $Rmin_0(M)$ (single-crossover) recombination nodes. Then, there is a series of permitted ARG events that creates N_t.*

The approach handles missing data naturally

It is common in biological data that some of the entries in a sequence in M are missing or unknown. As before, we use $*$ to indicate such missing values. Since the ARG-event rules address unresolved sites, they can also handle data where the input sequences in M have some missing entries. This greatly expands the range of applications of this approach. In the case of missing data, theorem 9.1.9 must be modified to:

Theorem 9.1.11 *Given a set of sequences M, with some missing entries, any series of permitted ARG events that starts with M and ends with the all-zero sequence, creates a time-ordered ARG for M.*

9.1.5.1   Using the Permitted ARG-Event Rules to Find a MinARG

Given theorems 9.1.9 and 9.1.10, a MinARG for M can be found by building a *branching search tree* to explore all possible series of permitted ARG events, finding a series that uses the fewest recombination events to build a time-ordered ARG for M.

The branching search approach is clearly correct, but as described it is practical for only very small problem instances. To extend the range of practicality, we must use ideas that reduce the size of M; or add constraints to the search that still allows it to find a MinARG; or add in bounding rules to cut off large parts of the search tree. Program *Beagle* follows this general approach. We next discuss the major ideas that *Beagle* uses to make the branching-search more efficient.

9.1.5.2 Improving the Efficiency of the Search

Branch-and-bound
The most important idea is to use lower bounds to cut off branches of the search tree in the well-known *branch-and-bound* manner: Let v be a node in the search tree, and let $\mathcal{S}_v$ denote the *MDSset* for the partial ARG constructed along the search path to v. It follows that:

> If the number of recombination events on the search path to v, plus a lower bound on $Rmin_0(\mathcal{S}_v)$, is larger than or equal to the number of recombination events in a known ARG for M, then there is no need to expand the search from node v.

Beagle follows this branch-and-bound approach.[7]

The lower bound on $Rmin_0(\mathcal{S}_v)$ used in *Beagle*, is a heuristic simplification of the *optimal RecMin bound*, based on the *haplotype bound* (section 5.2.4.3). However, since a sequence in the search may contain sites with unknown value (each denoted by a $*$), we must modify the basic *haplotype bound* so that it is a valid lower bound. For that we consider the following abstract problem.

Haplotype bound with missing data
Let $\mathcal{S}_v$ be a set of sequences, where some of the entries are $*$, representing missing data. Any $*$ can be converted to a zero or a one. We use $\mathcal{S}'_v$ to denote a set of sequences obtained from $\mathcal{S}_v$ by converting each $*$ to zero or one.

Definition Given a set of sequences $\mathcal{S}_v$ with missing data, $R'min_0(\mathcal{S}_v)$ is defined to be the *smallest* value $Rmin_0(\mathcal{S}'_v)$, over all $\mathcal{S}'_v$ obtained from $\mathcal{S}_v$ by converting each $*$ to a zero or a one.

Clearly, when the branching search algorithm is at a node v and the current set of sequences is $\mathcal{S}_v$, $R'min_0(\mathcal{S}_v)$ is a valid lower bound on the number of recombination events needed on any search path extending from v. Therefore, any lower bound on

---

7 More accurately, *Beagle* organizes its computation in phases, where in each phase it has a target number of recombination events, k, and it determines whether k recombination events are sufficient to build an ARG for the input M. If k is not sufficient, then it begins a new phase with target $k + 1$. Therefore, in the phase with target k, if the number of recombination events on the path to v, plus a lower bound on $Rmin_0(\mathcal{S}_v)$, is greater than k, then the search is cut at node v.

$R'min_0(\mathcal{S}_v)$ can also be used to bound the number of recombination events on the paths extending from v. One way to get a valid lower bound on $R'min_0(\mathcal{S}_v)$ is to examine each pair of sequences s, s' in $\mathcal{S}_v$ to see if they can be made identical (by converting each $*$ to zero or one). If they can, then put them into the same class of an equivalence relation $\mathcal{R}$. After all pairs of sequences have been compared, the number of classes of $\mathcal{R}$ is less than or equal to the number of distinct sequences in $\mathcal{S}'_v$, for any $\mathcal{S}'_v$. Also, the number of distinct sites in the sequences in $\mathcal{S}_v$ (considering $*$ unequal to zero or one) is larger or equal to the number of distinct sites in $\mathcal{S}'_v$, for any $\mathcal{S}'_v$. This idea was first developed in [19]. In summary, we have:

Lemma 9.1.7 *The number of classes of the equivalence relation* $\mathcal{R}$, *minus the number of distinct sites in* $\mathcal{S}_v$, *minus one, is a lower bound on* $R'min_0(\mathcal{S}_v)$.

Lemma 9.1.7 shows how a valid lower bound can be obtained for use in the branching search algorithm, and it can be extended to subsets of sites along with the use of the composite method to obtain a better lower bound, as discussed in section 5.2.4. In actuality, program *Beagle* does not use this lower bound, but rather uses a heuristic simplification of the *haplotype bound*. It is reported to generally be as good as the *haplotype bound*, although examples can be found where it is strictly smaller.

Cleaning the sequences

The next idea that *Beagle* uses is to run algorithm *Clean* at each node in the search tree before expanding the node (by branching on all possible next events). An additional rule was also suggested in [265]: if two *neighboring* columns are identical, or one is the perfect complement of the other, then one of the neighboring columns can be removed (see section 3.2.3.4).

The use of algorithm *Clean* and the additional rule can considerably reduce the problem size and the size of the search tree. An example is given in [265], where the size of *Drosophila melanogaster* alcohol dehydrogenase data is reduced by such cleaning from 11 rows and 44 columns to 9 rows and 16 columns.

Since the sequences in the search can have unresolved sites (each represented by a $*$), the removal rules in algorithm *Clean* can be extended in another way. Let s and s' be two sequences that match at every site in s' whose value is zero or one. Then if s contains a site c whose value is zero or one while the value of c in s' is $*$, then s' is *subsumed by s* and can be removed. That is, if some $*$ entries of s' can be changed to zero or one so that s and s' become identical, then s' can be removed. Similarly, if a site c' is subsumed by a *neighboring* site c, then c' can be removed.

Restricted event ordering

The use of algorithm *Clean* has the effect of giving priority to the mutation and coalescent rules over the recombination rule. So, the search will always create a mutation or a coalescence event when possible, and only use a recombination event when neither of the other events is possible. This constrains the set of ARGs that *Beagle* can produce, making the program more efficient, while still guaranteeing that a MinARG will be produced.

Additional restrictions on the set of achievable ARGs are possible, further improving the efficiency of the search. The main restriction is in the relationship of recombination events to coalescent events. It is shown in [265] that a MinARG will still be found if the set of permitted ARGs are constrained so that no directed path in the ARG contains more than two successive recombination nodes. Further, looking backward up a directed path with two successive recombination nodes, one of the parents of the second recombination node must be a coalescent node. These constraints limit the set of permitted ARGs, and reduce the size of the search tree. The search is additionally restricted by ensuring that each recombination event creates a "maximal segment," constraining the allowed crossover indexes used in a recombination. The reader is directed to [265] for details on these two restrictions.

9.1.5.3   Empirical Results for Program *Beagle*

Empirical results of applying *Beagle* to the LPL data, introduced in section 5.2.8.1, are discussed in [265] and shown in table 9.1. The tests are broken into three populations and three genomic regions. Each cell of the table reports the results from programs *HapBound*, *Beagle*, and *SHRUB*, in that order. As required (and providing empirical validation of the correctness of the three programs), the values reported from *Beagle* are always in the closed interval defined by the other two values. Generally, *HapBound* produced its results in seconds, *SHRUB* in minutes, and *Beagle* in a range of minutes to hours.[8]

For Kreitman's classic data (See figure 5.7 on page 144) with 11 sequences and 43 sites, *Beagle* computes $Rmin(M) = 7$ almost instantly. That agrees with the matching lower and upper bounds on $Rmin(M)$, claimed in section 9.1.4, computed by programs *HapBound* and *SHRUB*.

---

8 See the preface for comments on empirical testing.

	Beagle compared to HapBound and SHRUB		
Population	reg 1	reg 2	reg 3
Jackson	11, 13, 13	10, 10, 10	13, 15, 16
N. Karelia	2, 2, 2	15, 16, 17	8, 8, 10
Rochester	1, 1, 1	14, 14, 14	8, 8, 8

Table 9.1   The result of programs *HapBound*, *Beagle*, and *SHRUB* on the partitioned LPL data from [311].

9.1.6   Program *KwARG*

The program *kwARG*, developed by Rune Lyngsø, implements heuristic speedups to *Beagle* in order to handle larger datasets. It does not guarantee finding a MinARG, but it has been successfully used to address ARG reconstruction and recombination problems on realistic biological datasets [65, 209]. In [209], *kwARG* was used to identify gene-flow in yeast; and in [65] *kwARG* was used to identify coevolution in fungi. There is no publication describing *kwARG*, but it is partially described on two webpages [263, 264]. Empirical comparisons of programs *Beagle*, *kwARG*, and *Margarita* are reported in [56].

In [65], *KwARG* was used to study the evolution of double-stranded RNA in fungi. The data had seven taxa and 229 sites. Program *KwARG* constructed an ARG for the data using 32 recombination nodes. See figure 9.26. This is the largest ARG we know of that has appeared in the published biological literature. Program *SHRUB* constructed an ARG with 31 recombination nodes, essentially instantly, and both programs *RecMin* and *HapBound* produced a lower bound value of 24. The *HK bound* was 23, which is surprisingly high.

9.1.7   Program *mARGarita*

In a paper on ARG-based *association mapping*, Minichiello and Durbin [286] developed an ARG construction method that is also modeled on the backward-in-time, coalescent-theory viewpoint. The association mapping program, called *Margarita*, a.k.a. *mARGarita*, implements their ARG construction method. We will discuss association mapping in chapter 13 and see how ARGs are used in that application; here we only discuss the ARG construction method from [286].

As in *Beagle*, *Margarita* builds an ARG for a set of sequences M backward-in-time by successively choosing the next event, either a mutation, coalescence, or recombination. However, unlike *Beagle*, it does not systematically explore different combinations of events in a search tree. Instead, in a single execution, it uses heuristic rules to select a *single*

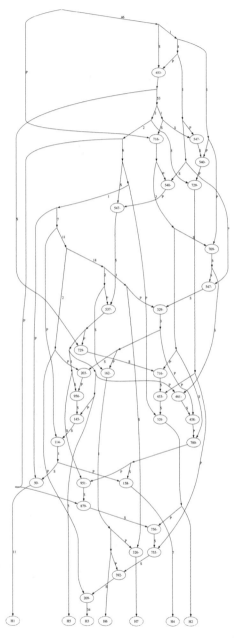

Figure 9.26 The fungi ARG published in [65]. Each crossover index is displayed inside a recombination node, and may be too small to read. The point of the figure is to demonstrate the capabilities of *KwARG*, and document its utility in analyzing population data. (Figure reprinted with copyright permission from Mycologia.)

series of events, building a single ARG. The program also incorporates some randomized selections, so that different ARGs can be constructed in different executions with the same input. The program can then generate a set of ARGs for M and output the one with the fewest recombination nodes or, in the case of association mapping, use the set of ARGs in further analysis.

Margarita uses the same concepts of *maximal deduced nodes* and *maximal deduced sequences* that *Beagle* uses. As in *Beagle*, *Margarita* begins the construction of an ARG for M with *MDSet* equal to M, and *MDSet* is modified after each chosen event. Also, as in *Beagle*, *Margarita* always chooses a mutation or a coalescent event if possible, and only chooses a recombination event when forced. The *permitted ARG-event rules*, specifying when each of the three types of events are possible, are the same as in *Beagle*. But additional heuristics are added, in order to *coordinate* recombination and coalescence events. If no mutation or coalescent event is possible, *Margarita* chooses one or two recombination events, which create the appropriate sequences for a subsequent "maximum length" coalescent event. To be more precise, we need the following:

Definition Two sequences, s and s', in *MDSet* have a *maximal shared tract* in the closed interval $[c_1, c_2]$, if (1) for every site c in $[c_1, c_2]$, $s[c] = s'[c]$ when neither $s[c]$ or $s'[c]$ is a $*$; (2) $s[c] = s'[c] \neq *$ for at least one site c in $[c_1, c_2]$; and (3) the interval $[c_1, c_2]$ cannot be extended on either side without violating condition (1).

When a recombination event is required, with probability 0.9 *Margarita* searches *MDSet* for a pair of sequences (s, s') whose maximal shared tract $[c_1, c_2]$ is the *longest* over all pairs of sequences in *MDSet*. And, with probability 0.1 it randomly chooses two sequences, s and s', in *MDSet* that have a maximal shared tract $[c_1, c_2]$ of length at least two. It then selects one or two recombination events involving s or s' or both, using crossover indices at one or both ends of $[c_1, c_2]$, so that one parent of s, and one parent of s', can coalesce. That coalescent event is immediately executed after the recombination event(s). Two recombination events will be used if $c_1 > 1$ and $c_2 < m$, and only one recombination event will be used otherwise.[9]

For example, suppose $s = *10001011*0$ and $s' = 1100110*1*1$. Then $m = 11$, and s, s' have two maximal shared tracts in the closed intervals: $[1, 4]$ and $[6, 10]$. If *Margarita* chooses $[6, 10]$ for $[c_1, c_2]$, then for sequence s, creating one recombination node with crossover index 6, creates the P-parent sequence $*1000******$, and creates the S-parent

---

9 Note that if $c_1 = 1$ and $c_2 = m$, then s and s' can themselves coalesce, violating the assumption that no coalescence event is possible.

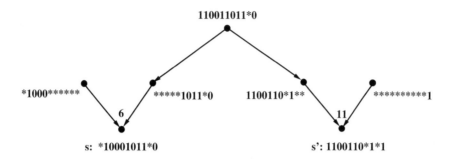

Figure 9.27 *Margarita* creates two recombination events, one for *s* and one for *s'*, designed so that one created parental sequence of *s*, and one created parental sequence of *s'*, obey the conditions of the coalescent-event rule (**Cr**). Those parents then immediately coalesce.

sequence *****1011*0. These decisions are made looking backward in time. Looking forward in time, the construction has created a *P*-parent sequence and an *S*-parent sequence that can recombine with crossover index 6, to create sequence *s*. Similarly, for sequence *s'*, one recombination node with crossover index 11, creates the two parent sequences **********1 and 1100110*1**. Those two recombination events create two sequences that can coalesce. So, finally, *****1011*0, a parent of *s*, and 1100110*1**, a parent of *s'*, coalesce to create the sequence 110011011*0. See figure 9.27.

When multiple coalescent events are possible, *Margarita* chooses randomly between them. This source of randomness, along with the randomness used in choosing maximal shared tracts, means that different ARGs can be created for the same input *M*. When used for association mapping (discussed in section 13.4), *Margarita* creates one hundred ARGs for *M*. The ARGs created are called *plausible* ARGs for *M*, but no argument is given in [286] for their biological validity. However, the inferred ARGs are successfully used in association mapping to find the locations of mutations contributing to genetic-influenced disease. So, the strongest argument for the plausibility of those ARGs is "that the inferred ARGs work for disease mapping" [286]. The method is efficient enough that it can handle thousands of sequences with hundreds of sites. However, in a different application domain (detecting SNPs from low-coverage sequencing data, discussed in section 13.9), the ability of *Margarita* to scale to a large number of sequences ("sample" in the terminology of [286]) was less clear, "since we have noticed that for 400 samples, *Margarita*, which implements a greedy algorithm, gets locked into incorrect structures" [246].

9.1.7.1   A Related Method

We mention here another method that is based on ideas similar to those presented in this chapter, but differing in some significant ways. The method of Parida et al. [328] constructs an ARG, for SNP data, that is not guaranteed to be a MinARG. At the high level, it shares some of the ideas of *Margarita* in that it exploits long segments (haplotypes) shared between individuals, but it differs in detail and to some extent in the underlying genealogical model. The model in [328] also shares some features with a model of recombination (the mosaic model) that will be discussed in chapter 14. The interested reader is referred to [328] for a more complete discussion.

9.2   ARG Building by Tree-Scanning

The next method we will discuss for general ARG reconstruction is based on the idea of *tree-scanning*. This is the central idea in the first developed method and program guaranteed to construct a MinARG for any set of sequences. The theoretical basis for the method was first developed by Hein [165] in 1990, and a heuristic program based on these ideas was discussed in [166]. Implementation details and practical improvements for the *exact*, non-heuristic method were developed in [388, 394]. The method has been implemented as a working program called *RecMinPath*, but the program is very computationally expensive and works only on relatively small problem instances.

From a *practical* standpoint *RecMinPath* is not competitive with *Beagle*, the only other implemented program guaranteed to find a MinARG. However, from a *conceptual* standpoint, the idea of tree-scanning is important and gives a different way to define and construct ARGs, developed in [436]. Further, tree-scanning has lead to the *sequential Markov coalescent* approach for simulating the coalescent-with-recombination [283, 275], computing probabilities related to recombination [331, 377, 445]), and inferring feature of population history. In fact, the sequential Markov coalescent approach, developed in [283], is dominant way that statistical inference over ARGs is implemented today. Similarly, it was recently claimed to be the most effective approach to simulate sequence evolution with complex recombination [453].

Tree-scanning is also used in *mathematical* results in coalescent theory (for example, see [283, 436]). Furthermore, the key objects in tree-scanning, the *marginal trees* of an ARG, are widely used, for example in association mapping. This connection will be discussed in chapter 13. For those reasons, and because this book emphasizes fundamental ideas (which almost surely will be of repeated use), we discuss the *general idea* of tree-scanning, and uses marginal trees, in this book.

9.2.1 Sequence-Labeled Marginal Trees

The central insight underlying the tree-scanning method is that in any ARG $\mathcal{N}$ for M, the *transmission history* of a *single* site c, forms a *directed subtree* of $\mathcal{N}$ rooted at the root of $\mathcal{N}$, and with the same leaves as in $\mathcal{N}$. By "transmission history of c" we mean the history of the transmissions of the *state* of site c, from the root down to the leaves of $\mathcal{N}$.

The top-down view

To visualize the transmission history of a single site, imagine the *physical* DNA molecule at the root of $\mathcal{N}$, and imagine how the physical molecule is successively replicated, mutated, and recombined to create a molecule at each node of $\mathcal{N}$. In such a physical process, at any node v, the molecular material containing site c that v receives comes from exactly one parent of v. This is obvious if v is a tree node, since a tree node has only one parent; but it is also true when v is a recombination node. So, for any site c, the subgraph of $\mathcal{N}$ that shows, at each node v, which parent transmits the molecular material containing c, must form a directed tree rooted at the root of $\mathcal{N}$. An example of the transmission history of site 4 is shown in figure 9.28*e*.

The bottom-up view

An alternative way to see that the transmission history of a site c forms a directed subtree of $\mathcal{N}$ is to imagine a *backward* traversal of $\mathcal{N}$ starting at the leaves. Such a traversal can trace out in reverse the transmission history of any specific site c. In more detail, first note that since each leaf and each tree node in $\mathcal{N}$ has only one parent the backward traversal from a leaf or a tree node proceeds on a *unique* edge. But a recombination node x has two edges into it, so there is a choice for which edge the backward traversal should use. Suppose the crossover index at node x is b_x. If $c < b_x$ then the state of c at x was given to it by its P-parent. In that case, the backward traversal from x should proceed up the edge to the P-parent of x. But if $c \geq b_x$, then the state of c at x was given to it by its S-parent and the backward traversal from x should proceed up the edge to its S-parent. In either case, only one edge into x is used in the traversal. It follows that the complete backward traversal traces out a directed subtree of $\mathcal{N}$, rooted at the root of $\mathcal{N}$ and containing the leaves of $\mathcal{N}$. This directed subtree of $\mathcal{N}$ identifies the transmission history of c. See figure 9.28 for an example of the bottom-up traversal to find the transmission history of a site.

Definition The subtree T of $\mathcal{N}$ that displays the transmission history of a site c in $\mathcal{N}$, and where every node in T has the same label it has in $\mathcal{N}$, is called the *sequence-labeled marginal tree* of $\mathcal{N}$ at c.

For simplicity, we will often use the term "SL-marginal tree" in place of "sequence-labeled marginal tree." Anticipating the method to come, where we do not know $\mathcal{N}$, and hence do not know any internal-node labels from $\mathcal{N}$, we have the following:

Definition Let T be an SL-marginal tree of $\mathcal{N}$ at site c. After contracting any internal node of degree two, and replacing each internal node label from $\mathcal{N}$ with a distinct label unrelated to the sequences in $\mathcal{N}$, the resulting tree is called the *marginal tree* of $\mathcal{N}$ at site c. Marginal trees are sometimes also called "local trees." See figure 9.28f for an example.

Conceptually, it will be convenient to think about SL-marginal trees. However, often only the marginal trees are known, and the algorithms will have to work with marginal trees, making the algorithms more complex. Note that a marginal tree, or SL-marginal tree of $\mathcal{N}$, at a site c must have a distinct leaf for each taxon in M, even though a single site c only defines a *bipartition* of the taxa. Note also that the SL-marginal trees, at two adjacent sites in M, might be identical, and so all the sites in a contiguous *interval* of sites might have the same SL-marginal tree T. In that case, we can say that a tree T is the SL-marginal tree for an interval of sites, I_T. We also can also use a single marginal tree for all sites in I_T.

A simpler (middle) way to identify the (SL-) marginal tree

We have developed top-down and bottom-up ways to think about the (SL-) marginal tree of $\mathcal{N}$ at c; now we develop a *middle way*: The (SL-) marginal tree of $\mathcal{N}$ at site c can be obtained from $\mathcal{N}$ by removing exactly one edge into each recombination node x. In particular, if $c < b_x$, then remove the S-edge into x, and if $c \geq b_x$, then remove the P edge into x. In some cases these edge removals will transform a recombination node into a leaf node, and if this happens, all such leaves should be deleted, so a leaf remains only if it is labeled by a taxon in M. See figure 9.29 for an example.

9.2.2   The Key Relationship of SL-Marginal Trees to Their Originating ARG

The relationship of the SL-marginal trees of $\mathcal{N}$ to the full ARG $\mathcal{N}$ is seen by *unioning* the SL-marginal trees of $\mathcal{N}$, as follows.

Definition Let $G_1 = (V, E_1)$ and $G_2 = (V, E_2)$ be two directed graphs over the same set of labeled nodes V, where each edge is specified by an ordered pair of nodes. The *union* of G_1 and G_2 is the graph $G_{1,2} = (V, E_1 \cup E_2)$, that is, where G_1, G_2, and $G_{1,2}$ all have the same node set, and the edge set of $G_{1,2}$ is the union of the edge sets in G_1 and G_2.

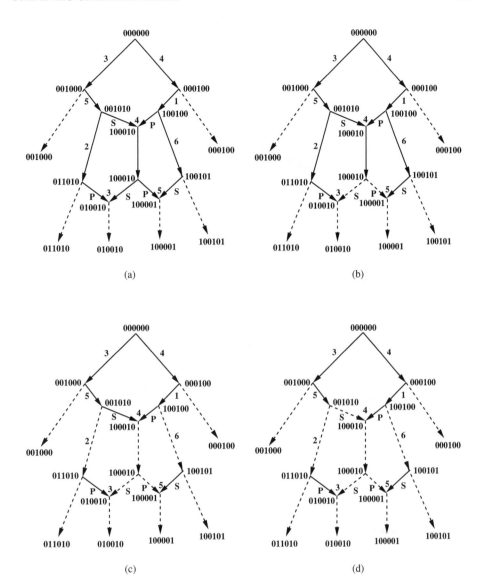

(a)

(b)

(c)

(d)

(Continued from previous page)

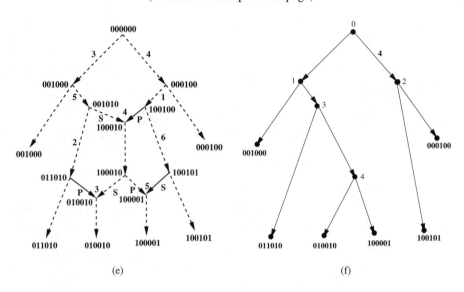

(e)   (f)

Figure 9.28   The backward traversal in ARG $\mathcal{N}$ constructs the transmission history of site 4. The dashed edges in panels (a)–(e) show the progress of the backward traversal. The SL-marginal tree of $\mathcal{N}$ at site 4 is shown in (e), and the marginal tree of $\mathcal{N}$ at site 4 is shown in (f). The internal node labels in that marginal tree are arbitrary and have no connection to the sequences in $\mathcal{N}$. Note that the (SL-)marginal tree at site 4 has a distinct leaf for each sequence in M, even though 4 is the only site of M labeling an edge in the (SL-) marginal tree, and a single site only creates a *bi*partition of the taxa.

The union of G_1 and G_2 superimposes ("identifies" in graph-theory terminology) nodes of G_1 and G_2 that have the same node label, and superimposes any pair of edges from E_1 and E_2 that are specified by the same ordered pair of nodes.

Definition The union of a *set* of graphs is obtained by first unioning two graphs in the set, and then unioning the remaining set of graphs, with the result of the first merge. The

following theorem is the capstone theorem of this section.

Theorem 9.2.1 *Any ARG $\mathcal{N}$ can be created by the union of the set of SL-marginal trees of $\mathcal{N}$. So any ARG $\mathcal{N}$ is defined by the set of SL-marginal trees it contains.*

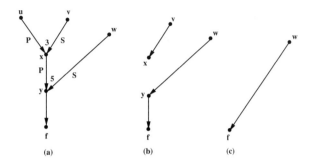

Figure 9.29 Panel (a) shows a fragment of an ARG containing two recombination nodes x and y and a leaf labeled by taxon $f \in M$. To create the marginal tree for site 7, as an example, the P-edges into node x and y are deleted, giving the fragment shown in panel (b). Now node x is a leaf node, but it is not labeled by any taxon in M. So node x should be deleted, and the node y should be contracted, creating the ARG fragment shown in panel (c).

Theorem 9.2.1 shows the relationship of $\mathcal{N}$ with its SL-marginal trees. We will see later (section 9.2.3.2) that a time-ordered MinARG for a set of sequences M can be constructed by reversing the logic behind theorem 9.2.1. However, we will only have *marginal* trees available, rather than *SL-marginal* trees. The lack of sequence labels on the internal nodes will require that a "union" of marginal trees be done differently than a union of SL-marginal trees. That will be addressed in section 9.2.5.2.

9.2.2.1 A Digression: Return to Undetectable Recombination

Now that we have defined marginal trees, we can expand on the issues discussed in section 3.6. To start, we state a condition that identifies some *locally inconsequential* recombinations.

Lemma 9.2.1 *A recombination event at a recombination node x in an ARG $\mathcal{N}$ is locally inconsequential if for at least one edge e into x, there is no marginal tree that contains edge e.*

For example, in figure 3.14 (page 96) there is a marginal tree that uses the P-edge into the recombination node labeled 1001, but no marginal tree using the S-edge into that node. Hence that recombination edge can be deleted, which removes the recombination event.

Note that the condition in lemma 9.2.1 only concerns the ARG topology and the crossover-indices—it is independent of any mutations that may occur on the ARG. Hence, it does not characterize all conditions that make a recombination event locally inconsequential. For example, it does not identify the recombinations in figures 3.13b and 3.15 as locally inconsequential.

We can also now be more precise about the approach to undetectable recombinations developed in [168, 435]. Consider a recombination at node x with crossover-index b_x, and consider the marginal trees at sites $b_x - 1$ and b_x. Then, the recombination at x is detectable if and only if the two marginal trees have "different topologies" [435]. We interpret a "topology" of a rooted tree to be determined by its clusters, i.e., its rooted splits.

9.2.3 Tree-Scanning

Conceptually, each of the SL-marginal trees, for the m sites in M, can be found separately and in any order. However, the SL-marginal trees at two *neighboring* sites are similar. In fact, the SL-marginal tree at site c will be identical to the SL-marginal tree, T, at site $c - 1$, unless c is the crossover index for some recombination node x in $\mathcal{N}$. In that case, the P-edge into x is removed from the current SL-marginal tree T, and replaced with the S-edge into x. If there is more than one recombination node with crossover index c, then an edge exchange must be made at each of those recombination nodes. The result is the SL-marginal tree at site c. Moreover, that tree is the SL-marginal tree at all the sites in the interval $[c, c' - 1]$, where c' is the smallest crossover index larger than c at any recombination node in $\mathcal{N}$.

So, all of the SL-marginal trees in $\mathcal{N}$ can be found by sorting the crossover indexes in $\mathcal{N}$, smallest first, then constructing the SL-marginal tree for $c = 1$; and then successively modifying the SL-marginal tree at each crossover index in the sorted list. An example is shown in figure 9.30. The ARG $\mathcal{N}$ has three distinct crossover indexes, at sites 3, 4, and 5, and so it has four distinct SL-marginal trees. Another example is shown in figure 13.1 (page 441).

Definition The process of successively constructing all the SL-marginal trees of $\mathcal{N}$, as described in the previous paragraph, is called *tree-scanning*, and the result is called the *tree-scan* of $\mathcal{N}$.

The idea of a tree-scan (under different terminology) was first introduced in [436], where it was used to stochastically generate ARGs. In that view, an ARG is the result of a stochastic tree-scan, or a "walk through tree space."

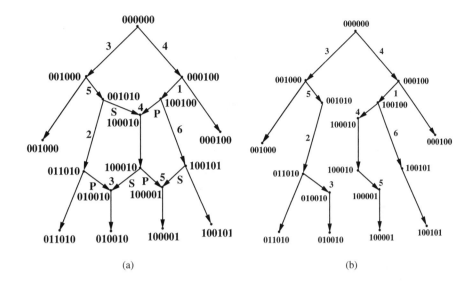

(a)   (b)

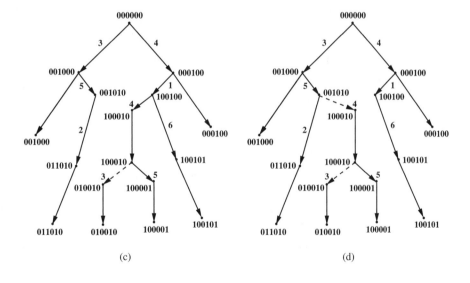

(c)   (d)

(Continued from previous page)

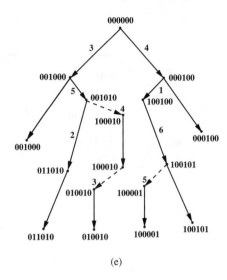

(e)

Figure 9.30   An ARG $\mathcal{N}$ with three recombination nodes and four SL-marginal trees. We have left in all of the mutations and crossover indexes to help see how the SL-marginal trees are embedded in $\mathcal{N}$. In $\mathcal{N}$, the recombination nodes have crossover indexes 3, 4, and 5. The first SL-marginal tree, shown in (b), is for sites 1 and 2. The second and third SL-marginal trees, shown in (c) and (d), are for sites 3 and 4, respectively. The fourth SL-marginal tree, shown in (e), is for sites 5 and 6. The dashed edges show the new edge that was added into the SL-marginal tree. The removed edge is in the previous tree. The process of successively constructing these SL-marginal trees, in order, is called *tree-scanning*.

Exposing relationships

Our interest in obtaining all the SL-marginal trees in an ARG $\mathcal{N}$, by tree-scanning, rather than by finding each one separately, is not for improved efficiency, but for the way it exposes the *relationship* of neighboring SL-marginal trees and neighboring marginal trees. That relationship is the key to how an ARG $\mathcal{N}$ for M can be *constructed* from a set of *hypothesized* marginal trees. This is developed next, starting with a description of the key operation called *rooted subtree-prune-and-regraft (rSPR)*.

9.2.3.1   Rooted Subtree-Prune-and-Regraft (rSPR)

Definition Let T be a rooted, leaf-labeled tree. Informally, a *rooted subtree-prune-and-regraft (rSPR)* operation modifies tree T by *cutting* or breaking away one edge e and the

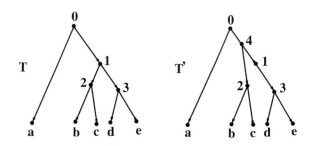

Figure 9.31 An rSPR operation removes the edge $(1, 2)$ from the rooted tree T and reattaches it in the interior of edge $(0, 1)$, creating the new node 4 and the new tree T'. Essentially, the $(1, 2)$ edge has been converted to the $(4, 2)$ edge. As drawn, we can picture the rSPR operation as a "pivot" of edge $(1, 2)$.

subtree below e, and then *reattaching* e and its subtree to some point (a node or a point in the interior of an edge) in the remaining part of T.

If edge e is reattached to the interior of an edge, then a new node is created and it is given a node label distinct from the other node labels in T. All the other node labels remain the same in the two trees, in this informal description of an rSPR operation. Later this definition will be somewhat expanded.

Essentially, an rSPR operation changes an edge (u, v) to an edge (u', v), where u' might be a newly created node. In relation to the change of neighboring marginal trees, it is visually useful to note that v remains the same, making an rSPR operation look like a "pivot" of edge (u, v) around node v. For example, the successive changes to the trees in figure 9.30 can be viewed a pivots, created by rSPR operations. See figure 9.31.

The *cutting* act is the "prune" in an rSPR operation, and the *reattaching* act is the "regraft" in an rSPR operation. For a fully formal definition of an rSPR operation, we will have to be more precise in order to take care of the case that e is pruned from the root of T, or that e is regrafted so that it creates a new root of T. We will do that in section 9.2.3.3.

Relating rSPR operations to tree-scanning

It should be easy to see the relationship of rSPR operations to tree-scanning. In a tree-scan of an ARG $\mathcal{N}$, if c is the crossover index at a recombination node x, then when the tree-scan reaches site c, the P-edge into x is removed from the current SL-marginal tree T, and replaced with the S-edge into x. That modification of T is an rSPR operation. Further, if c is the crossover index at k_c different recombination nodes, then the new SL-marginal tree at c can be derived from T by a sequence, σ_c, of exactly k_c different rSPR operations. For example, in the tree-scan shown in figure 9.30, each of the trees in panels (c), (d), and (e) can be obtained from the previous tree, by one rSPR operation.

Example from retrotransposons

In a quite different biological context, an rSPR operation models a horizontal gene transfer from one organism to another. An example is a study of horizontal transfer of retrotransposons [425]. In that paper they show two phylogenetic trees for sequences of the BovB retrotransposon and display nine rSPR operations that are required to convert one tree to the other.

Definition A sequence σ of rSPR operations that transforms a tree T into a tree T', is said to be *simple* if the rSPR operations can be done in any order.

A trivial case of a simple sequence of rSPR operations is when there is only one rSPR operation.

Lemma 9.2.2 *In a simple sequence, σ, of rSPR operations, for any node v, at most one edge (u, v) is removed, and at most one edge (u', v) is added. Also, if $e = (u', v)$ is an edge created by an rSPR operation in σ, then no subsequent rSPR operation in σ will insert a node into e.*

Proof: Let the first rSPR operation involving v be the removal of edge (u, v), and the addition of edge (u', v). Since a tree can have only one edge into node v, any subsequent rSPR operation involving v must remove edge (u', v), and add an edge (u'', v). But edge (u', v) cannot be removed before it is created, so these two rSPR operations cannot be done in the opposite order, violating the definition of a simple sequence of rSPR operations. Similarly, $e = (u', v)$ is created in one rSPR operation, and a subsequent rSPR operation inserts a new node u'' into e, then those two rSPR operations must be done in that order. ∎

Simple sets

So, a simple sequence σ of rSPR operations can be viewed as a *set* of independent operations on T, since the order of these operations does not matter. Therefore, we will also refer to a simple sequence as a "simple set." For example, the three rSPR operations that transform the tree in panel (b) in figure 9.30 into the tree in panel (e), is a simple set of rSPR operations. See figure 9.32 for a sequence of rSPR operations that is not simple.

We think of the SL-marginal trees of $\mathcal{N}$, in the order they are found by the tree-scan of $\mathcal{N}$, as being produced by successive simple sets of rSPR operations on SL-marginal trees. While the order of the rSPR operations inside any simple set σ_c does not matter, the order of the sets themselves is fully determined by the order of the sites in M. Similarly, we

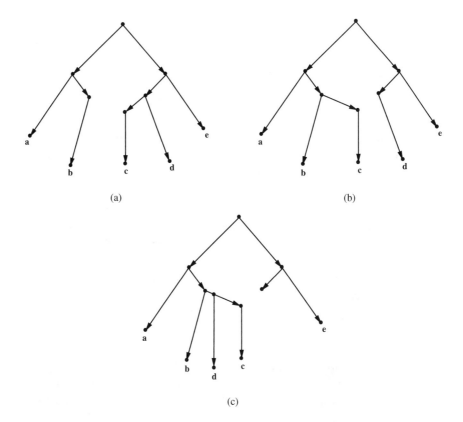

(a)   (b)

(c)

Figure 9.32 A sequence of two rSPR operations that is not a simple sequence. The second operation cannot be done before the first.

can think of the marginal trees of M as being produced by successive simple sets of rSPR operations on marginal trees. This view is reflected in the following quotation from Joseph Felsenstein [112]:

> As one moves along the genome, one will have a series of sites that all have the same coalescent tree, with no recombination anywhere in that tree. But suddenly, there will be a recombination somewhere in the tree. It will cause a particular kind of rearrangement of the tree (breaking of one lineage and its attachment elsewhere). That tree will then hold for a while as one moves along the genome, and then there will be another breakage and reattachment. After some distance along the genome, the tree will have changed to be totally different.

The "coalescent tree" for a site refers to the tree we call "the marginal tree" for that site. "Recombination in the tree" means an rSPR operation on the tree, corresponding to a recombination in the chromosome. The word "lineage" in this context can be interpreted as an edge in a tree.

The following lemma summarizes the relationship between SL-marginal trees and rSPR operations.

Lemma 9.2.3 *Let T_1 be the SL-marginal tree of an ARG $\mathcal{N}$ at site 1. If $\mathcal{N}$ has k recombination nodes, then starting from T_1, all the successive SL-marginal trees of $\mathcal{N}$ can be created by k rSPR operations.*

Note that k is the total number of rSPR operations, summing over all the sets of rSPR operations, so the number of SL-marginal trees of $\mathcal{N}$ is at most k, and could be strictly less than k. Note also that within a single simple set σ_c, for any node v at most one rSPR operation in σ_c prunes an edge into v, but this property does not extend beyond a single simple set: for each site c, one rSPR operation in σ_c is permitted to prune an edge into v. Conversely:

Lemma 9.2.4 *An ARG $\mathcal{N}$ can be reconstructed if we know T_1, and a series of simple sets of rSPR operations that create the successive SL-marginal trees of $\mathcal{N}$.*

So given the SL-marginal trees for $\mathcal{N}$, and simple sets of rSPR operations that construct the SL-marginal trees in order of the sites, ARG $\mathcal{N}$ can be fully reconstructed. This is intuitively the key to the method we will develop for constructing a MinARG for M. However, when we only have the sequences M, we don't know any SL-marginal *or* marginal trees of M. Further, although lemma 9.2.3 continues to hold if we replace "SL-marginal" with "marginal," lemma 9.2.4 does not. We next address these issues.

9.2.3.2   *Reversing the Tree-Scanning Viewpoint to Construct ARGs*

In this section we will see how to construct an unknown MinARG for a set of sequences M, by reversing the relationship of rSPR operations to tree-scanning. In that reversal, we will *select* a *feasible* tree for each site c, and consider it as the *marginal* tree at c in some (yet) unknown MinARG for M. Then we will find successive simple sets of rSPR operations that construct the selected marginal trees in order, and use them to construct a MinARG for M. In order to do this we will have to be more precise about which rSPR operations are allowed, and we have to define exactly what a "feasible" tree is.

We assume, in the rest of the discussion of tree-scanning, that the ancestral sequence is required to be the all-zero sequence. Recall that for any site c in M, the *split* defined by c

is the bipartition of the taxa of M into the taxa with state zero at site c and the taxa with state one at site c.

Definition Given M, a rooted tree T with one leaf for each taxon in M is *feasible* for site c in M, if there is an edge in T whose removal creates two subtrees, one that contains all the leaves for the taxa with state zero at c, and one that contains all the leaves for taxa with state one at c. That is, there is an edge in T whose removal creates the split defined by c.

Note that given an ARG $\mathcal{N}$ for M, the marginal tree for a site c is a feasible tree for c.

A plausible approach

Having specified what it means for a tree to be feasible, we now have the outlines of a *plausible* approach to construct a (time-ordered) ARG for M, reversing the logic of tree-scanning. Further, given lemmas 9.2.3 and 9.2.4 it should also seem plausible that the problem of computing $Rmin(M)$ (and building a MinARG for M) can be solved as follows:

> Find a feasible tree for each site c such that those trees can be generated in order, starting from $c = 1$, using the *minimum* number of rSPR operations. Then create ARG $\mathcal{N}$ from the chosen trees and the successive rSPR operations that generate them.

This approach seems plausible, although not efficient. However, it does *not* work. But, it can be made to work with a few technical modifications that we discuss next.

9.2.3.3 A More Complete Description of rSPR and Ordered-SPR Operations

In order to correctly reverse the logic of tree-scanning, to construct a *time-ordered* MinARG, we have to give a more complete definition of an rSPR operation, and also add a restriction to the allowed rSPR operations. The informal definition of an rSPR operation given earlier (in section 9.2.3.1, page 336) covers the typical case, when the removed edge is not incident to the root of T; and conversely, when its reattachment does not *create* a new root. Also, the informal definition of an rSPR operation allows the creation of internal nodes of degree two, caused by an edge pruning. We need an expanded rSPR operation to handle those issues.

An expanded rSPR definition

Suppose that the pruned edge is $e = (r, v)$, where r is the root of tree T, with degree two in T. After the pruning of e, r will have only one child, denoted w. In that case, the expanded rSPR operation will contract the edge (r, w), making w the new root. Conversely, suppose that the pruned edge is (u, v), where u is not the root of T. The expanded rSPR operation allows u to become the new root. To achieve this, add a new directed edge, (u, r), into the root r of T, and attach edge $e = (u, v)$ to node u, making u the new root of the new tree.

We will also require that after an rSPR operation, if any internal node v has degree two, then v must be contracted. These modifications of the informal definition of an rSPR operation completely define an rSPR operation.

Ordered rSPR operations

In addition to the expanded definition of an rSPR operation, in order to use rSPR operations to create a proper *time-ordered* ARG for M, we need to restrict where the removed edge e can be reattached in an rSPR operation.

Recall from section 1.6.1.2 (page 30) that the nodes of any DAG G can be topologically sorted into a list Π, where for any pair of nodes (u, v) with a directed path in G from u to v, node u appears before v in Π. Clearly, any rooted tree is a DAG, so its nodes can be topologically sorted.

Definition An *ordered tree* T is a rooted tree, along with some topological sorting, Π, of its non-leaf nodes.

The interpretation of Π is that it specifies the relative ages of the nodes, so if node v precedes node v' in Π, v is considered older than v'. In this view, every non-leaf node occurs at a *distinct* time. In contrast, the leaves correspond to observed sequences that can all be observed at the same point in time.

Definition Let T and T' be two ordered trees with the same leaf labels, and some common non-leaf labels. Let Π and Π' be topological sorts of the non-leaf nodes of T and T', respectively. Π and Π' are *consistent orders* if for every pair of non-leaf nodes (u, v) in *both* T and T', if u precedes v in Π, then u precedes v in Π'.

Definition Let $e = (u, v)$ be the directed edge that is pruned from ordered tree T (with topological sort Π) in an rSPR operation; and let u' be the node where e is reattached to T, creating the rooted tree T'. Note that u' is either in T or is created when e is reattached. We say that the rSPR operation is an *ordered-SPR operation* if there exists a topological sort, Π', of the non-leaf nodes of T' so that Π and Π' are consistent orders.

See figure 9.33 for an example of an rSPR operation that is not an ordered-SPR operation, and examples of ordered-SPR operations. As shown there, the number of ordered rSPR operations that are needed to transform one ordered tree to another may be larger than the number of regular rSPR operations needed for the transformation. The importance of consistent orders and of ordered-SPR operations is that no time-relationship in

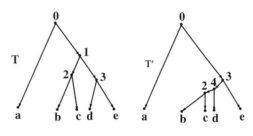

(a) An rSPR operation that cannot be an ordered-SPR
operation.

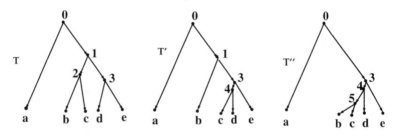

(b) Two ordered-SPR operations that create an ordered tree T'' with the same leaf-labeled
topology as T', but with some different interior node labels.

Figure 9.33 In panel (a) the rSPR operation prunes the edge $(1, 2)$ from the ordered tree T and reattaches it in the interior of edge $(3, d)$. This creates the new node 4. Node 1 is removed because it only has degree two after edge $(1, 2)$ is removed. Tree T' is the resulting tree, shown in panel (b). Assuming that the total order Π of the non-leaf nodes in T is $0, 1, 2, 3$, this rSPR operation cannot be an ordered-SPR operation, because in any topological sorting Π' of the non-leaf nodes of T', node 3 must precede node 2. However, both nodes 2 and 3 are also in T, and 2 must precede 3 in any topological sort Π of the nodes of T. Panel (b) shows that it is possible to transform T to T'', a tree with the same labeled leaves and the same topology as T', using two ordered-SPR operations. Note that node 2 was removed in the first ordered-SPR operation, resolving the ancestry conflict with node 3. In this example, $\Pi' = 0, 1, 3, 4$ and $\Pi'' = 0, 3, 4, 5$, and the two ordered-SPR operations do *not* form a simple set of ordered-SPR operations.

Π' contradicts a time-relationship already established in T by Π. Hence, the ordered-SPR operations form a possible scenario for how T was transformed into T'.

Definition For two ordered trees T_c and T_{c+1}, which are respectively feasible for sites c and $c + 1$ of M, $oSPR(T_c, T_{c+1})$ denotes the *minimum-length* sequence of ordered-SPR operations needed to transform T_c to T_{c+1}; similarly, $sSPR(T_c, T_{c+1})$ denotes the minimum-size *simple set* of ordered-SPR operations needed to transform T_c to T_{c+1}.

If T_c cannot be transformed to T_{c+1} by a simple set of ordered-SPR operations, then $sSPR(T_c, T_{c+1})$ is defined to be infinity. It is an open question whether any such pair of trees exists, but the definition allows for them. It is known [386] that $oSPR(T_c, T_{c+1}) \leq n - 2$, where n is the number of leaves of T_c. The following two lemmas are immediate:

Lemma 9.2.5 *Let $\mathcal{N}$ be an ARG for M, and let Π be a topological order of its nodes. Let T_c and T_{c+1} be any two neighboring marginal trees of $\mathcal{N}$, and let Π_c and Π_{c+1} be topological sorts of the nodes in T_c and T_{c+1}, respectively, induced by Π.[10] Then, Π_c and Π_{c+1} are consistent orders.*

Lemma 9.2.6 *Let σ_c be the rSPR operations determined by the tree-scan at site c, that is, that transform T_c to T_{c+1}. With the topological orderings Π_c and Π_{c+1} induced by Π, trees T_c and T_{c+1} are ordered trees; each rSPR operation in σ_c is an ordered-SPR operation; and σ_c is a simple set of rSPR operations.*

Lemmas 9.2.5 and 9.2.6 lead the following conclusion:

Theorem 9.2.2 *For any input M with m sites and n taxa, let $\mathcal{N}$ be a MinARG for M. Then, there are ordered trees $T_1, T_2, \ldots, T_m$ with the following properties:*

(1) Each tree T_c is a feasible tree for site c.

(2) For each $c \leq m - 1$, tree T_c can be transformed to T_{c+1} with a simple set of ordered-SPR operations, σ_c.

(3) $\sum_{c \leq m-1} |\sigma_c| = Rmin(M)$.

(4) For each $c \leq m - 1$, $|\sigma_c| = sSPR(T_c, T_{c+1}) \leq n - 2$.

Proof: Let $T_1, T_2, \ldots, T_m$ be the marginal trees of MinARG $\mathcal{N}$ for M. Properties (1) and (2) follow from the fact that any marginal tree in $\mathcal{N}$ at site c is a feasible tree in M at site c; and from lemma 9.2.5. Property (3) follows from lemmas 9.2.3 and 9.2.5.[11] The fact that $|\sigma_c| = sSPR(T_c, T_{c+1}) = oSPR(T_c, T_{c+1})$ then follows from property (3), and the assumption that $\mathcal{N}$ is a MinARG. The fact that $sSPR(T_c, T_{c+1}) \leq n - 2$ is based on a related result in [386], which is based on a proof of the same fact for unordered rSPR operations in [385]. To see a weaker bound (sufficient for our purposes), note that a simple

---

10 "Induced" here means that Π_c (or Π_{c+1}) is obtained from Π (or Π'), by restricting to the nodes in T_c (or T_{c+1}).

11 Also, it was stated (without an explicit proof) in [165, 388, 394], that the number of recombination nodes in any ARG (in particular a MinARG $\mathcal{N}$) equals the number of ordered-SPR operations needed to generate, in order, all of the marginal trees of the ARG.

set of rSPR operations, σ_c, prunes an edge into a node v at most once, and since (after the removal of any internal node of degree two) $|T_c| < 2n$, it follows that $|\sigma_c| < 2n$. ∎

9.2.3.4   The Capstone Theorem

We can now state the capstone theorem of this section.

Theorem 9.2.3 *Over all sequences of m ordered trees $\{T_1, T_2, \ldots, T_m\}$, where each tree T_c is feasible for site c in M, let $\mathcal{T}$ be a sequence of ordered trees that minimizes the total number of ordered-SPR operations in a sequence Σ of $m - 1$ simple sets of ordered-SPR operations, that create, in order, the trees in $\mathcal{T}$. Then the total number of rSPR operations in Σ equals $Rmin(M)$.*

Proof: Let $\mathcal{N}^*$ be a time-ordered MinARG for M. Then the tree-scan of $\mathcal{N}^*$ identifies a sequence of ordered-trees $\mathcal{T}$ that satisfy the conditions of the theorem. The conclusion of the theorem follows by theorem 9.2.2. ∎

9.2.4   A Tree-Scanning Method to Compute $Rmin(M)$

Given theorem 9.2.3 we can finally describe the general approach, inspired by tree-scanning, which is *guaranteed* to compute $Rmin(M)$, for any M:

1. For each site c in M, enumerate the set Δ_c of all of the leaf-labeled, ordered trees that are feasible for site c, where the leaves are labeled by the n taxa in M.
2. Next, build a directed graph $G(\Delta)$ with one node for each enumerated tree, and a source node with a directed edge to each node for a tree in Δ_1; and a sink node with a directed edge from each node for a tree in Δ_m. Assign a distance of zero to each of those edges. Then, for c from 1 to $m - 1$, add a directed edge from each node in Δ_c to each node in $\Delta c + 1$ (associated with ordered trees T and T', respectively), and assign that edge the distance of $oSPR(T_c, T_{c+1})$.
3. Finally, find a shortest path from the source node to the sink node in $G(\Delta)$. That shortest path must include exactly one node from each Δ_c for $c = 1...m$, and therefore the path identifies one feasible ordered tree for each site c. The length of the shortest path is $Rmin(M)$.

Implementation details and heuristics to compute $oSPR$ values
The ideas described above are at the conceptual heart of the program *RecMinPath* [388, 394], that computes $Rmin(M)$. However, we have not explained how to compute the

needed $oSPR$ values, and without some heuristic speedups, the method would only be practical for very small problem instances.

The two major bottlenecks in the method are the time to enumerate all the feasible ordered trees in each set Δ_c, and the time needed to find $oSPR(T_c, T_{c+1})$ for each pair of ordered trees in $\Delta_c \times \Delta_{c+1}$. The time for both of those tasks grows rapidly as a function of n, the number of taxa in M. In fact, the number of ordered trees grows *super-exponentially*[12] with n [386, 394], and the problem of computing the minimum number of rSPR operations between rooted trees is NP-hard [44]. That problem is not the same as computing $oSPR$ or $sSPR$ values (and the complexity of those problems seems to still be open), but the available techniques for computing $oSPR(T, T')$ or $sSPR(T, T')$ all require worst-case exponential time.

Program *RecMinPath* incorporates several heuristics and implementation ideas, in addition to applying algorithm *Clean* to reduce the size of the problem instance. What we describe next is an adaptation of the central heuristic, which can be used to find a simple set of ordered-SPR operations that transform a given feasible tree for a site c to a given feasible tree for site $c + 1$. Program *RecMinPath* only computes $Rmin(M)$, and does not construct a MinARG for M. It therefore only needs to compute $oSPR$ values, and the heuristic it uses is somewhat simpler than what we present here.

The most important heuristic is to avoid the explicit computation of $oSPR(T_c, T_{c+1})$ for each *separate* pair of trees (T, T') in $\Delta_c \times \Delta_{c+1}$. Computing each value separately would involve a huge amount of redundant computation that can be avoided with the right implementation.

Definition For a tree $T \in \Delta_c$ let $\mathcal{T}^d(T)$ be the set of ordered trees in Δ_{c+1} that can be reached from T by a simple set of at most d ordered-SPR operations.

Clearly, $\mathcal{T}^d(T)$ can be found by building a search tree, where each node represents a tree derived from T by a simple set of d or fewer ordered-SPR operations; the root of the search tree represents tree T. To make sure that the sequence of rSPR operations is a simple sequence, each search path must avoid pruning an edge that is into a node v in T, if an edge into v has already been pruned by an ordered-SPR operation on that search path. Similarly, it must regraft any pruned edge either to a node in T, or into an edge in T.

The speedup comes from the following: At a node in the search tree, the search can expand along all next permitted ordered-SPR operations, but if the paths to two or more nodes in the search tree each contain the same set of ordered-SPR operations, then only

---

12 A function f is considered super-exponential if for all $x, y > 0$, $f(x)f(y) < f(x + y)$. So the function 2^n is exponential, but 2^{n^2} is super-exponential.

one of those nodes needs to be expanded. This follows from the fact that the path contains a *simple* set of ordered-SPR operations. If a tree T' is enumerated along some search path, then $sSPR(T, T') \leq d$. So, for a given site c, and tree T that is feasible for c, the method first computes $\mathcal{T}^1(T)$ and notes which trees are in Δ_{c+1}. Then, it continues expanding search trees to find $\mathcal{T}^2(T)$, etc. This approach clearly avoids redundant computation that would occur if $sSPR(T, T')$ were computed separately for each $T' \in \Delta_{c+1}$. However, one such search tree is needed for each $T \in \Delta_c$. Another speedup is obtained by using the fact that the maximum needed value for d is $n - 2$.

We will not provide a formal time analysis, but note that the time for the original version of this method was estimated in [17] to be $\Omega(n!!)$.

Empirical results for RecMinPath

For Kreitman's benchmark dataset (shown in figure 5.7) with 11 sequences and 43 sites, *RecMinPath* was efficient enough to compute that $Rmin(M)$ is 7. Previous lower bounds were below 7 (although later lower bound methods also computed a lower bound of 7), and no method before *RecMinPath* had been able to compute $Rmin(M)$ for this dataset. Recall that *Beagle* (developed later) almost instantly finds that $Rmin(M)$ is 7.

Additional empirical results on the practical behavior of *RecMinPath* are discussed in [394]. One question examined in detail there is how accurately *RecMinPath* reconstructs the marginal trees. In that investigation, a dataset M and an ARG $\mathcal{N}$ for M, and the marginal trees for $\mathcal{N}$ were constructed using the coalescent simulation program *ms* [182]. Then, program *RecMinPath* was executed with dataset M, and the m feasible trees selected by that execution were compared to the marginal trees of $\mathcal{N}$. In particular, for each site c in M, the feasible tree for site c was compared to the marginal at site c using a variant of the *Robinson-Foulds* [355] measure of tree similarity. That measure is the number of nontrivial splits[13] the two trees have in common, divided by $n - 3$, which is the largest possible number of nontrivial splits in a tree with n nodes. The conclusion is that even though the selected feasible trees might not be usable to perfectly reconstruct the unknown ARG $\mathcal{N}$, the feasible trees that *RecMinPath* selects did capture a substantial amount of the structure of the marginal trees underlying $\mathcal{N}$. Moreover, in some applications, the structure of the selected feasible trees is sufficient to address relevant biological questions. See [394] for more details.

---

13 Nontrivial splits are splits where both sides of the bipartition contain at least two taxa. Trivial splits correspond to the split created by removing an edge into a leaf.

Modifying RecMinPath to compute lower bounds

If, instead of computing $sSPR(T, T')$ or $oSPR(T, T')$ for each pair (T, T') in $\Delta_c \times \Delta_{c+1}$, the method computes the minimum number of rooted, but unordered, rSPR operations to transform T to T', denoted $rSPR(T, T')$, then the result would be a *lower bound* on $Rmin(M)$. Since the number of rooted trees with n leaves is vastly smaller than the number of ordered trees with n leaves (and there is a more developed literature on practical computation of rSPR transformations), this lower bound can be computed much faster than *RecMinPath* can compute $Rmin(M)$. A practical approach to computing $rSPR(T, T')$ is discussed in section 14.5. It is reported in [394] that for the data tested, the lower bounds were mostly equal to $Rmin(M)$, and never much lower. Generally, the lower bounds were comparable to the bounds obtained by the *History* method (to be discussed in chapter 10). See [394] for details.

9.2.5   How a Simple Set of Ordered-SPR Operations Is Used to Construct a MinARG

Program *RecMinPath* computes $Rmin(M)$ and finds a sequence of feasible marginal trees used to obtain that value. However, it does not explicitly construct a MinARG for M, although it is stated in [394] that "we can explicitly construct evolutionary histories that are consistent with the data by combining the trees which appear in the sequence of trees." With the increased use of the *sequential Markov coalescent* approach to generate ARGs, the issue of how to assemble an ARG from a sequences of marginal trees has become even more significant. We are not aware of any part of the literature that explicitly addresses this issue, and there seem to be some subtleties in how to assemble an ARG from feasible marginal trees. The subtleties are related to the comment made after lemma 9.2.4, that lemma 9.2.4 requires SL-marginal trees, but here we only have trees that are not sequence labeled.

In this section, we detail how a simple set of ordered-SPR operations can be used to construct a MinARG for the data M. The construction method relies on the fact that the sequence of rSPR operations is *simple*. In fact, simple sequences were not discussed in [394], but they are introduced in this chapter in order to more easily construct an ARG from a sequence of ordered-SPR operations. We start by considering how to combine two trees related by a simple sequence of ordered SPR operations.

9.2.5.1 The SPR-Union of Two Trees

Given input sequences M, let T_c and T_{c+1} be feasible trees for sites c and $c + 1$, and suppose that a *simple* set, σ_c, of k_c ordered-SPR operations transforms T_c to T_{c+1}. By theorem 9.2.2, such feasible trees T_c and T_{c+1} exist.

Definition The SPR-union of T_c and T_{c+1}, given a simple set σ_c, is the ARG $\mathcal{N}(c, c + 1)$ created by algorithm *SPR-Union*, shown in figure 9.34.

The correctness of the algorithm *SPR-Union* depends critically on lemma 9.2.2, which depends on the fact that σ_c is a *simple* set of SPR operations.

Note that every recombination node created by algorithm *SPR-Union* has out-degree one. Note also that the two marginal trees of $N(c, c + 1)$ at sites c and $c + 1$ *contract* to T_c and T_{c+1}, respectively. That is, after removing any internal node with degree two from the two marginal trees in $N(c, c + 1)$, the resulting trees are T_c and T_{c+1}. So the marginal trees of $N(c, c + 1)$ describe the same transmission history of sites c and $c + 1$, as do T_c and T_{c+1}. See figure 9.35 for an example.

Why so complex?

It may seem that algorithm *SPR-Union* is more complex than need be. In particular, it may not be clear why a new node x is needed, and why it is inserted into edge (u, v). Why not just add the edge (u', v) to G_k, making v a recombination node? By the fact that σ_c is a simple set of ordered-SPR operations, this should cause no problem. In fact, that would work to find the SPR-union of two trees. However, we will want the SPR-union of a *sequence* of m neighboring feasible trees. Intuitively, the SPR-union of a sequence of $m - 1$ trees, given the sequence σ_c, for each c from 1 to $m - 1$, is just the ARG created by the *accumulated* result of the SPR-unions of each successive pair of trees in $\mathcal{T}$. However, there is a technical problem that arises because of the possibility of two or more sequences σ_i and σ_j, for $i \neq j$, which each prune and regraft an edge into the same node v. A straightforward accumulation of those edges into the growing ARG would result in node v having in-degree greater than two. Inserting a new node x into the pruned edge (u, v) resolves this problem.

9.2.5.2 The SPR-Union of a Sequence of Trees

Next we extend the definition of the SPR-union of two neighboring feasible trees to the SPR-union of a *sequence* of more than two neighboring feasible trees.

Algorithm SPR-Union (T_c, T_{c+1}, σ_c)

if (T_{c+1} is identical to T_c, except for edge labels) **then**
    Set $\mathcal{N}(c, c+1)$ to T_{c+1}.
else

    Set G_0 to T_c.

    for $(k = 1, \ldots, k_c)$ **do**
        Let (u, v) be the edge in T_c that is pruned, and let (u', v) be the edge
        that is regrafted in the kth ordered-SPR operation of σ_c.

        if (node u' is created in this operation) **then**
            let $e' = (w, v')$ be the directed edge into which u' is inserted.
            {In this case, edge $e' = (w, v')$ will also be in G_k.}
            {This can be can be proved by induction on k.}
            Insert node u' into edge e' in G_k.
        endif

        Add node x to the (u, v) edge of G_k, creating directed edges (u, x) and (x, v).
        {Theorem 9.2.2, statement (2), guarantees that (u, v) is also in G_k.}
        Add in the directed edge (u', x) to G_k.
        Label edge (u, x) with P and edge (u', x) with S
        and assign node x the crossover index $b_x = c + 1$, making x a
        recombination node.
        Set G_{k+1} to G_k.
    endfor

    Let e be the edge in T_{c+1} that is labeled with site $c + 1$.
    Label edge e in G_{k_c} with site $c + 1$.
    Set $\mathcal{N}(c, c+1)$ to G_{k_c}.
endif

Figure 9.34 The procedure to create the SPR-union of two neighboring feasible trees, T_c and T_{c+1}, given the simple sequence, σ_c, of ordered-SPR operations that transforms T_c to T_{c+1}.

Given the input M of m sites, let $\mathcal{T} = T_1, T_2, \ldots, T_m$ be an ordered sequence of trees such that T_c is a feasible tree for site c of M, for c from 1 to m. Also, for each c from 1 to $m - 1$, let σ_c be a simple set of ordered-SPR operations that transforms T_c into T_{c+1}. Let $\Sigma = (\sigma_1, \ldots, \sigma_{m-1})$. We define the SPR-union of $\mathcal{T}$ procedurally.

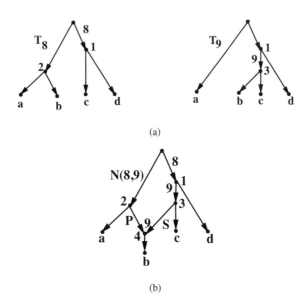

(a)

(b)

Figure 9.35 (*a*) Feasible trees T_8 and T_9 for sites 8 and 9, respectively, of some unshown dataset M. T_9 is created from T_8 by a single ordered-SPR move that cuts edge $(2, b)$ and creates edge $(3, b)$. In this example, $k_c = 1$, so $N(8, 9) = G_1$. Node 2 is u, leaf b is v, node 3 is u', and node 4 is x.
(*b*) ARG $N(8, 9)$ created by the SPR-union of T_8 and T_9. The marginal trees of $N(8, 9)$ at sites 8 and 9 *contract* to T_8 and T_9: after contracting any internal node with degree two, from the marginal trees of $N(8, 9)$ (at sites 8 and 9), the resulting trees are T_8 and T_9.

Definition The *SPR-union* of $\mathcal{T}$, given Σ, is the ARG $\mathcal{N}$ created by the *accumulated* result of running algorithm *SPR-Union* of each successive pair of trees in $\mathcal{T}$. That is, edges, nodes, and labels are added to the growing ARG, but never removed (although some edges are subdivided).

See figures 9.36 and 9.37 for an example of four steps, creating the SPR-union of five trees. Figure 9.36 shows the first rSPR operation, and the DAG N_2. Figure 9.37 successively shows the three additional rSPR operations, and the resulting DAGs. Note that trees T_3 and T_4 are identical, except for edge labels. Also, the first and second rSPR operations remove and add an edge into node v, illustrating the need for the more complex details in algorithm *SPR-Union*. The ARG N_5 is the SPR-union of the five trees.

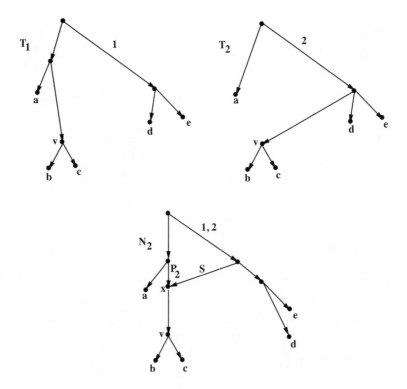

Figure 9.36 The first SPR-union of two of five neighboring feasible trees. The remaining SPR-unions are shown in figure 9.37.

Note that the SPR-union of a sequence of trees must be formed in successive steps because an edge in one step may be cut and reattached to an edge, or a node, only created in an earlier step. We state the following theorem that summarizes the discussion.

Theorem 9.2.4 *Given M, $\mathcal{T}$, and Σ as above, the SPR-union of $\mathcal{T}$ creates an ARG $\mathcal{N}$ where the marginal tree in $\mathcal{N}$, at any c, contracts to tree $T_c \in \mathcal{T}$.*

Why ordered-SPR operations?

Now that we have seen how to obtain the SPR-union of a sequence $\mathcal{T}$ of feasible trees, we can more fully explain why *ordered*-SPR operations are needed. Ordered-SPR operations ensure that the SPR-union of the trees in $\mathcal{T}$ creates a DAG, and moreover that the parents of any recombination node x are *older* than x. Therefore, the resulting DAG is a proper time-ordered ARG. When regular (unordered) rSPR operations are allowed, directed cycles can

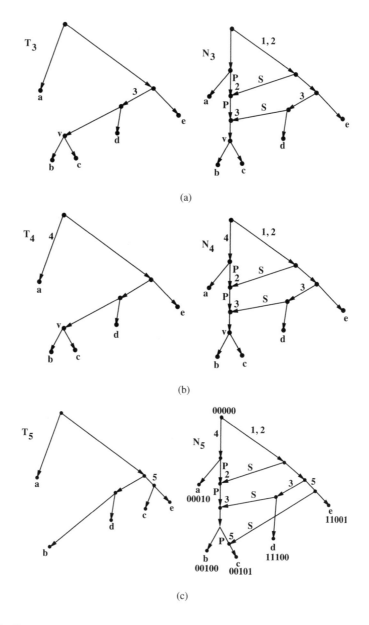

Figure 9.37 The last three steps of the SPR-union of five neighboring feasible trees. The first SPR-union was shown in figure 9.36.

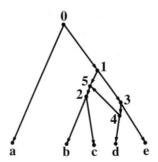

Figure 9.38   The union of the two ordered trees shown in figure 9.33a creates a DAG where the node labeled 5 is older than its parent labeled 4. This is because node 5 is an ancestor of node 2, node 2 is older than node 3 in the ordering Π in figure 9.33, and node 3 is an ancestor of node 4.

be created in the union, or it can happen that one or both of the parents of x are younger than x, in which case the DAG is not a proper time-ordered ARG. For example, the union of the two ordered trees in figure 9.33 is shown in figure 9.38, where the recombinant node labeled 5 is older than one of its parents.

Missing data

It is easy to extend the basic method to handle *missing entries* in M. For each c, Δ_c should be the set of all ordered trees that are feasible for the known data at site c. The size of Δ_c increases as the fraction of missing entries at site c increases, but otherwise all aspects of the method remain the same. The ARG that is produced is an ARG that minimizes the number of recombination nodes over all ARGs that generate the known data. After an ARG $\mathcal{N}$ is produced, it can be used to generate the leaf sequences, which then give imputations for the missing data.

9.3   Advanced Material: The Fastest (in Worst-Case) MinARG Algorithm

We have presented two programs, *Beagle* and *RecMinPath*, that are guaranteed to find a MinARG for any input M. Although *Beagle* is practical for many small, but realistic problem instances, neither of these methods has a worst-case running time that is bounded by a purely *exponential* function of input size, rather than a *super-exponential* function.

9.3.1 The Problem

In 2004, Bafna and Bansal [17] wrote the following about the problem of finding a MinARG: "This problem is computationally challenging and has resisted efforts for even an exponential time algorithm." For most natural computational problems that we encounter, finding a worst-case exponential-time method is straightforward, and not a significant accomplishment. But the MinARG problem belies that experience, and the question of whether an exponential-time MinARG method is possible, is *still* open.

In the hopes of facilitating the solution of this open problem, we present here a different idea for guaranteeing a MinARG. In terms of worst-case behavior it is the fastest of the three guaranteed methods, but its worst-case running time is still super-exponential. The method we present here is a corrected version of one developed in [447]. It has not been implemented as a program.

The return of self-derivability

To begin, we return to the issue of the *self-derivability* of a set of sequences M, discussed earlier in section 5.2.7.4 (page 155). Recall that M is *self-derivable* if there is an ARG $\mathcal{N}$ that derives M, where every node in $\mathcal{N}$ is labeled with a sequence in M. In that case, we also say that $\mathcal{N}$ *self-derives* M. Earlier, we also discussed whether M is *unary*-self-derivable, but now the only issue is self-derivability, so an edge can be labeled with more than one site. To solve the self-derivability question, we modify the dynamic program used by algorithm *DP-USD*, in section 5.2.7.3 (page 156).

The DP for self-derivability

We let S represents a set of sequences, (generally a superset of the input M), and let K be a subset of S.

Definition $SD(K)$ is a Boolean variable that is set to **true** if the sequences in K are self-derivable, and **false** otherwise. If the set of sequences, K, is self-derivable, let $R(K)$ denote the *minimum* number of recombination nodes in any ARG $\mathcal{N}$ that self-derives K.

Note that $R(K)$ has value infinity (meaning that it is undefined) if $SD(K)$ is **false**. Note also, that even when $R(K)$ is defined, $Rmin(K) < R(K)$ is possible, since $Rmin(K)$ is the minimum over the set of ARGs that generate K, including those that do not self-derive K.

Algorithm SD $(\mathcal{S})$

Set $SD(K)$ to **true**, and $R(K)$ to zero, for every singleton subset K of $\mathcal{S}$.

for each subset $K \subseteq \mathcal{S}$, where $\|K\| > 1$, (in order of the size of K) **do**
{This block determines if K is self-derivable.}

Set $SD(K)$ to **false** and $R(K)$ to ∞.
for each sequence $s \in K$ **do**

for each pair of sequence (s', s'') in $K - \{s\}$ **do**
Test if s can be created by a recombination of s' and s''.
If so, set $SD(K)$ to **true** and set $R(K)$ to
the *minimum* of $R(K)$ and $R(K - \{s\}) + 1$.
endfor

for each sequence $s' \in K - \{s\}$ **do**
Test if s differs from s' at a set of sites C, where for each site $c \in C$,
all sequences in $K - \{s\}$ have the *same* state at site c.

{i.e., Test if s can be derived from s' by unused mutations.
The condition that for any $c \in C$, each sequence in $K - \{s\}$ must
have the same state at site c, ensures that none of the sites in C
mutated in the generation of $K - \{s\}$.}

If so, set $SD(K)$ to **true** and set $R(K)$ to
the *minimum* of $R(K)$ and $R(K - \{s\})$.
endfor

endfor
endfor
Output $SD(\mathcal{S})$ and $R(\mathcal{S})$.

Figure 9.39 Algorithm *SD* determines if $\mathcal{S}$ is self-derivable, and if so, computes the *minimum* number of recombination nodes needed in any ARG that self-derives $\mathcal{S}$.

9.3.2 Algorithm SD($\mathcal{S}$)

We can compute $SD(\mathcal{S})$ and $R(\mathcal{S})$, if defined, by the dynamic programming approach shown in figure 9.39.

Note that if $\mathcal{N}$ is a MinARG for M, and $\mathcal{S}$ is the set of node labels in $\mathcal{N}$, then $\mathcal{S}$ is self-derivable. Further, $R(\mathcal{S})$ will be equal to $Rmin(M)$. So:

Theorem 9.3.1 *Algorithm SD correctly computes each value $SD(K)$ and $R(K)$, and so $SD(\mathcal{S})$ and $R(\mathcal{S})$ will be correctly computed after the algorithm completes the case of $K = \mathcal{S}$. The total time to compute $SD(\mathcal{S})$ and $R(\mathcal{S})$ is $O(q^2 m^2 2^q)$, where $q = |\mathcal{S}|$.*

Theorem 9.3.2 *A MinARG for any input M of n taxa and m sites can be constructed by algorithm MinARG, shown in figure 9.40. The running time of the algorithm is bounded by $O(n^3 m^3 2^{(nm)^2})$.*

Proof: Let $\mathcal{N}$ be a MinARG for M. We can assume that every row in M is distinct, and that every node in $\mathcal{N}$ has a distinct label. $\mathcal{N}$ has at most m tree nodes since each is the head of an edge labeled by some distinct site(s) of M. Also, $R(\mathcal{N})$, the number of recombination nodes in $\mathcal{N}$, is at most nm since by theorem 3.2.1 (page 81) M can be generated on an ARG with only nm recombinations. Therefore, the number of interior nodes in $\mathcal{N}$ and the number of distinct labels on those nodes is bounded by $nm + m$. Clearly then, the enumerative method in algorithm *MinARG* considers all possibilities and finds a MinARG for M.

Now we analyze the running time of algorithm *MinARG*. The node labels in $\mathcal{N}$ are selected from the set of all 2^m binary sequences of length m. It follows that the set of distinct node labels in $\mathcal{N}$ is one of $\binom{2^m}{R(M)}$ sets. Now $\binom{2^m}{R(M)} \leq \binom{2^m}{nm} \leq 2^{nm^2}$. Also, $\sum_{k=0}^{k=R(M)} \binom{2^m}{k} \leq \sum_{k=0}^{k=R(M)} 2^{mk} \leq R(M) 2^{mR(M)} \leq nm 2^{nm^2}$. So algorithm *MinARG* examines at most $nm 2^{nm^2}$ subsets of labels. When a superset, $\mathcal{S}$, of M, is examined, algorithm *MinARG* computes $SD(\mathcal{S})$ in $O(q^2 m^2 2^q)$ time, and since $q \leq nm$, a MinARG for M can be found in $O(n^3 m^3 2^{n^2 m^2}) = O(n^3 m^3 2^{(nm)^2})$ time. Note that nm is the size of the input. Using theorem 5.3.1, this bound can be reduced by a factor of $\log n$, but that is not significant. ∎

Minimizing Steiner nodes

Note that if algorithm *MinARG* is changed so that subsets of $P(m)$ are tested only *until* a subset $\mathcal{S}$ is found where $SD(\mathcal{S})$ is *true*, then the algorithm will compute the minimum number of *Steiner* nodes needed in any ARG to generate M. However, that might not be the minimum total number of recombinations needed. That is possible when more than

---

Algorithm MinARG (M)

    Set R^* to nm
    Let P(m) be the set of all binary sequences of length m
    for $(k = n \ldots nm + m)$ **do**
        for (each subset $\mathcal{S}$ of P(m) that has size k and includes M) **do**
            Compute $SD(\mathcal{S})$ and $R(\mathcal{S})$
            if (SD($\mathcal{S}$) is *true*, and R($\mathcal{S}$) $< R^*$) **then**
                Set R^* to $R(\mathcal{S})$ and record the ARG $\mathcal{N}$ constructed
                during the computation of SD($\mathcal{S}$) and R($\mathcal{S}$)
            endif
        endfor
    endfor
    Output $R(M) = R^*$; and ARG $\mathcal{N}$

Figure 9.40   Algorithm *MinARG* is guaranteed to find a MinARG for M.

one site labels a tree-edge of an ARG for set $\mathcal{S}_1$ of size q_1. In that case, there will be fewer than m edge-labeled tree-edges, and so there will be strictly more than $q_1 - m - 1$ recombination nodes. Then, for a larger set $\mathcal{S}_2$ of size $q_2 > q_1$, an ARG for $\mathcal{S}_2$ where every edge is labeled with at most one site, will have exactly $q_2 - m - 1$ recombination nodes, which could be smaller than the number of recombination nodes in the first ARG.

9.3.3 Some Improvements

Algorithm *MinARG* can be made more practical with the use of lower bounds. Throughout the algorithm, the value of R^* is an *upper bound* on $Rmin(M)$. As the algorithm proceeds, R^* is reduced until it ultimately equals $Rmin(M)$ (although the algorithm cannot recognize the equality until the end of the execution). Still, it seems plausible that as the algorithm proceeds, and particularly after it has set R^* to $Rmin(M)$, lower bounds applied to each enumerated set S will find that S needs more than R^* recombinations. In that case, there is no need to compute $R(\mathcal{S})$, and that time can be saved.

    A further improvement occurs when no edge in the MinARG is permitted to be labeled with more than one site. In that case, the number of distinct sequences that label a leaf or tree node or the root of the ARG is exactly $m + 1$. So if an ARG $\mathcal{N}$ unary-self-derives set $\mathcal{S}$, then the number of recombination nodes $\mathcal{N}$ is exactly $|\mathcal{S}| - m - 1$. Therefore, algorithm *MinARG* only needs to test each enumerated $\mathcal{S}$ to see if it is unary-self-derivable, using algorithm *USD* (page 156), and it can stop the first time it finds a unary-self-derivable set $\mathcal{S}$.

Finally, if we bound the number of Steiner sequences allowed by some fixed value, independent of n and m, then the algorithm will run in worst-case exponential-time, as a function of n and m. This kind of fixed-parameter bound is analogous to fixed-parameter tractability results often studied in computer science, but with *exponential* instead of polynomial time bounds.

Theorem 9.3.3 *The problem of computing a MinARG is fixed-parameter exponential for the number of allowed recombinations.*

10 The *History* and *Forest Lower Bounds*

10.1 The *History Lower Bound*

If you don't know history, then you don't know anything. You are a leaf that doesn't know it is part of a tree.
— Michael Crichton

10.1.1 Introduction to the *History Bound*

In chapters 5, 7, 8, and 9, we examined several lower bounds on $Rmin(M)$. Those bounds differed in the times needed for their computation and in their level of accuracy (how close they are to $Rmin(M)$ in practice). We stated earlier that two additional lower bounds would be discussed after methods for general ARG construction were presented. Here we discuss the *history lower bound* on $Rmin(M)$ developed by Myers and Griffiths [299, 302], and the related *forest lower bound* developed in [449]. These two bounds, usually called the *history bound* and the *forest bound*, were deferred until after the discussion of methods for building general ARGs, because the methods for general ARG construction and for computing the *history bound* are very similar. In particular, the destructive rules **Dc**, **Dr**, and **Dt** that were central in the general construction of ARGs are also central in defining and computing the *history bound*. The *forest bound* is related to the *history bound* and so was deferred until after discussing the *history bound*.

Empirically, the *history bound* is usually (but not always, as we will see) the most accurate (i.e., highest) lower bound of all the practical lower bounds[1] on $Rmin(M)$. However,

---

1 This excludes the lower bound discussed in the last chapter, based on Tree-Scanning.

its computation requires exponential time[2] in best and worst case, and no empirically efficient method to compute it is known. Therefore, it can only be computed on modest-sized problem instances. The worst-case time bound for the *forest bound* is also exponential, but it can be computed using integer linear programming, and is empirically efficient on a much larger range of problem instances than is the *history bound*.

We will discuss in detail the *history bound*, specialized to the case when the ancestral sequence is known and assumed to be the all-zero sequence. Hence, the *history bound* discussed here will be a lower bound on $Rmin_0(M)$. It will also be a lower bound on $R^m min_0(M)$. Other small changes from the original method in [302] are made for simplicity of exposition. Modification of the *history bound* described here, to cover the case when no ancestral sequence is known, is easy and left to the reader.

Destructive rules

The definition and computation of the *history bound* uses the three destructive rules **Dc**, **Dr**, and **Dt** (discussed in sections 9.1.1 and 9.1.3) to reduce M to a single row with no sites. As before, we use $\widetilde{M}$ to denote the matrix derived from M, at any point in the algorithm; initially $\widetilde{M}$ is set to M.

10.1.2   The Candidate History Bound (CHB)

Algorithm *CHB* (candidate-history-bound), shown in figure 10.1, is analogous to algorithm *Clean-Build-with-Recombination*, but it does not do any construction, and it is organized as an iterative algorithm rather than a recursive algorithm. It computes a value $CLB(M)$ which is a *candidate* for the *history bound*, but not all executions of algorithm *CHB* will result in a $CLB(M)$ value that is a lower bound on $Rmin_0(M)$. However, we will show later that the *minimum* computed $CLB(M)$ value is a valid lower bound on $Rmin_0(M)$.

Note that the algorithm is *nondeterministic* in that it allows choices for which row $\tilde{f}$ to remove when rule **Dt** is applied. Hence, different executions of the algorithm can return different $CLB(M)$ values. The term CLB stands for candidate lower bound, emphasizing the fact that not every computed $CLB(M)$ value is a lower bound on $Rmin_0(M)$. Only the *minimum* $CLB(M)$ value, over all possible executions of algorithm *CHB*, is guaranteed to be a lower bound on $Rmin_0(M)$. Figures 10.2–10.10 show an execution of the algorithm that returns a *CLB* value of three, while the *history bound* for this data is two.

---

2 In its original development, it required $\theta(mn!)$ time, where M has n sequences with m sites each.

Algorithm CHB (M)

   Set $CLB(M) = 0$.

   Set $\widetilde{M}$ to M.

   while ($\widetilde{M}$ contains more than one row or contains some sites) **do**

      Run algorithm *Clean* on $\widetilde{M}$.

      if $\widetilde{M}$ contains a non-empty row **then**
        Select a row $\tilde{f}$ in $\widetilde{M}$ and remove it
        {i.e., apply rule **Dt** to $\widetilde{M}$}
        Set $CLB(M) = CLB(M) + 1$.
      endif
   endwhile

   return $CLB(M)$

Figure 10.1   Algorithm *CHB (candidate-history-bound)* computes and returns a candidate lower bound for M, $CLB(M)$.

	1	2	3	4	5	6
r_1	0	0	1	0	0	0
r_2	0	1	1	0	1	0
r_3	0	1	0	0	1	0
r_4	1	0	0	0	0	1
r_5	1	0	0	1	0	1
r_6	0	0	0	1	0	0

Figure 10.2   Example input M. No application of rule **Dc** or **Dr** is possible. So pick a row, say r_6, for rule **Dt**.

Definition An execution of algorithm *CHB* is called a *minimum execution* if it computes the minimum $CLB(M)$ value over all possible executions of the algorithm. We use $MLB(M)$ to denote the minimum $CLB(M)$ value.

Definition The *history lower bound* on $Rmin_0(M)$ is the value $MLB(M)$.

	1	2	3	4	5	6
r_1	0	0	1	0	0	0
r_2	0	1	1	0	1	0
r_3	0	1	0	0	1	0
r_4	1	0	0	0	0	1
r_5	1	0	0	1	0	1

Figure 10.3   Now apply rule **Dc** to column 4.

	1	2	3	5	6
r_1	0	0	1	0	0
r_2	0	1	1	1	0
r_3	0	1	0	1	0
r_4	1	0	0	0	1
r_5	1	0	0	0	1

Figure 10.4   Now apply rule **Dr**, removing row r_5.

	1	2	3	5	6
r_1	0	0	1	0	0
r_2	0	1	1	1	0
r_3	0	1	0	1	0
r_4	1	0	0	0	1

Figure 10.5   Now apply rule **Dc** twice to remove columns 1 and 6.

	2	3	5
r_1	0	1	0
r_2	1	1	1
r_3	1	0	1
r_4	0	0	0

Figure 10.6   Now apply rule **Dt** to remove row 4.

	2	3	5
r_1	0	1	0
r_2	1	1	1
r_3	1	0	1

Figure 10.7 Now apply rule **Dt** again to remove row 3.

	2	3	5
r_1	0	1	0
r_2	1	1	1

Figure 10.8 Now apply rule **Dc** twice to remove columns 2 and 5.

	3
r_1	1
r_2	1

Figure 10.9 Now apply rule **Dr** to remove row 2.

	3
r_1	1

Figure 10.10 Now apply rule **Dc** to remove column 3 to obtain a single row with no entries. In this execution of algorithm *CHB*, rule **Dt** is applied three times, so $CLB(M)$ has ending value of three. In other executions, $CLB(M)$ is two, which is the actual *history bound* for M.

Algorithm CHB-Branch

As in the computation of $RN^*(M)$, we can compute the *history bound* using an algorithm that branches over all possible choices of $\tilde{f}$. That branching algorithm is called algorithm *CHB-Branch* in this book, and its worst-case running time is $\theta(mn!)$, the original time bound stated in [299, 302]. We will later see that the *history bound* can be computed in $O(nm2^n)$ time using dynamic programming.

Use with the composite method

As with other lower bounds, the *history bound* can be used with the composite bound method, computing *history bounds* in intervals of M; then using each one as a local bound for input to the composite bound method. Note that unlike the *haplotype bound*, when computing a local bound for an interval of sites I, there is no advantage to computing the *history bound* on *subsets* of sites inside I. The reason is that the *history bound*, computed for a set of sites S in M, will be greater or equal to the *history bound* for any subset of sites in S. That fact is easy to prove. Therefore, when the composite bound method uses *history bounds* as local lower bounds, only $\binom{m}{2}$ such lower bounds are needed. That is a great reduction in the number of bounds needed, as compared to using *haplotype bounds* for local lower bounds, but this does not affect the time needed to compute any of those local *history bounds*.

Similarities

The reader should see the similarity of algorithms *CHB* and *Clean-Build-with-Recombination*; and the similarity of algorithms *CBR-Branch* and *CHB-Branch*. In particular, if we remove all the constructive rules from algorithm *Clean-Build-with-Recombination*, leaving the destructive rules, and *count* the number of times rule **Dt** is applied, the resulting algorithm would essentially be algorithm *CHB*. Also, if that modified algorithm were used in *CBR-Branch*, the resulting algorithm would compute the *history bound*. Because the computation of algorithm *CHB* does not construct an ARG, it can be expressed in an iterative form rather than the recursive form of algorithm *Clean-Build-with-Recombination*. Myers and Griffiths [302] proved:

Theorem 10.1.1 $MLB(M) \leq Rmin_0(M)$, *i.e., the history bound is a valid lower bound on* $Rmin_0(M)$.

Although not stated in [302], the proof of theorem 10.1.1 can also be used to establish a more general statement that includes a bound on $R^m min_0(M)$, that is, when multiple-crossover recombination is allowed.

Theorem 10.1.2 $MLB(M) \leq R^m min_0(M) \leq Rmin_0(M)$.

We first present a synthetic proof of theorem 10.1.2, and then a more elementary proof.

Proof (synthetic): As suggested in the comments above, for any execution of algorithm *Clean-Build-with-Recombination* on M, there is an execution of algorithm *CHB* on M, using the same sequence of rules **Dr**, **Dc**, and **Dt**. In those executions, algorithm *CHB* counts the number of applications of rule **Dt**, while algorithm *Clean-Build-with-Recombination* constructs an ARG $\mathcal{N}$ for M, where for each application of rule Dt, it creates one or more recombination nodes in $\mathcal{N}$. Hence, for every execution of *Clean-Build-with-Recombination(M)*, returning an ARG $\mathcal{N}$ with $R(\mathcal{N})$ recombination nodes, there is an execution of algorithm *CHB* returning a value $CLB(M) \leq R(\mathcal{N})$. By letting $\mathcal{N}$ be a MinARG for M, the theorem is proved. Note that the type of recombination used in $\mathcal{N}$ has no affect on the argument.   ∎

Proof (more elementary): We prove that $MLB(M) \leq Rmin_0(M)$, and leave the extension to the case of multiple-crossover recombination to the reader. The proof is by induction on $Rmin_0(M)$.

When $Rmin_0(M)$ is zero, M can be derived on a perfect phylogeny with all-zero ancestral sequence, so by theorem 9.1.2, algorithm *Clean* reduces M to a single row with no entries: no application of rule **Dt** is needed and so $MLB(M)$ is zero. This proves the theorem when $Rmin_0(M) = 0$.

Now assume that the theorem holds for $Rmin_0(M) < k$ and consider a matrix M where $Rmin_0(M) = k$; let $\mathcal{N}$ be a MinARG for M. Consider the effect of running algorithm *CHB* until the end of the first execution of algorithm *Clean*. By theorem 9.1.3, the actions of algorithm *Clean-Build* to that point would build a unique forest $\mathcal{F}$ that is part of $\mathcal{N}$, hanging off the "bottom" of $\mathcal{N}$. So the actions of algorithm *Clean* to that point correspond (recalling the thought experiment of section 9.1.1) to *removing* $\mathcal{F}$ from $\mathcal{N}$.

Let $\widetilde{\mathcal{N}}$ be the remaining part of $\mathcal{N}$, i.e., $\mathcal{N} - \mathcal{F}$, where the edge and node labels are restricted to the sites in $\widetilde{M}$, the matrix at the end of the first execution of algorithm *Clean* on M. Clearly, $\tilde{N}$ is an ARG for $\widetilde{M}$, and every non-leaf node in $\tilde{N}$ is in some recombination cycle. Therefore, there is a row $\tilde{f}$ in $\widetilde{M}$ such that the sequence $s_{\tilde{f}}$ labels a leaf L_x whose parent, x, is a recombination node labeled $s_{\tilde{f}}$. If we remove nodes L_x and x from $\tilde{N}$, the result is an ARG that derives all the sequences in $\widetilde{M} - s_{\tilde{f}}$, using one fewer recombination node than is in $\mathcal{N}$. So if the first application of rule **Dt** in algorithm *CHB* removes row $\tilde{f}$ from $\widetilde{M}$, the sequences in the updated $\widetilde{M}$ can be derived on an ARG with $Rmin_0(M) - 1$ recombination nodes. After rule **Dt** is applied to row $\tilde{f}$, $Rmin_0(\widetilde{M}) \leq Rmin_0(M) - 1 = k - 1$, and by the induction hypothesis, $MLB(\widetilde{M}) \leq k - 1$. Then the partial execution of algorithm *CHB* described above, followed by whatever execution of algorithm *CHB* on $\widetilde{M}$ produces $MLB(\widetilde{M})$, defines an execution of algorithm *CHB* that computes the value $CLB(M) \leq k$. Therefore, $MLB(M) \leq k$ and the induction is proved.   ∎

Theorem 10.1.1 in [302] was actually proved for the case when *no* ancestral sequence is specified. The only change needed is in algorithm *CHB* where rule **Dc** is extended to:

> **Extended Rule Dc** If a column c of $\widetilde{M}$ contains at most one entry with value one, *or at most one entry with value zero*, then remove column c from $\widetilde{M}$.

The proof of correctness is left to the reader.

10.1.3   A Graphical View of *CHB* and the *History Bound*

It is useful to have a graphical interpretation of the effect of rules **Dc**, **Dr**, and **Dt** in an execution algorithm *CHB*. This graphical interpretation extends the "thought experiment" discussed in section 9.1.1.1, in the context of algorithm *Clean*. Consider an arbitrary ARG $\mathcal{N}$ for M. Similar to the earlier thought experiment, for rules **Dc** and **Dr**, and only for a tree, each application of a destructive rule in algorithm *CHB* removes a column or row from M, and we will define a parallel rule that removes a part of $\mathcal{N}$. We let $\widetilde{\mathcal{N}}$ denote the remaining portion of $\mathcal{N}$, and use $\widetilde{M}$, as before, to denote the current matrix derived from M. The rules that remove parts of $\mathcal{N}$ will be such that each $\widetilde{\mathcal{N}}$ will be an ARG that generates the corresponding $\widetilde{M}$.

Cases
The simplest case is when rule **Dc** applies, removing a column of $\widetilde{M}$. To modify $\tilde{N}$, we simply remove label c from the edge it labels.

When rule **Dr** applies, removing a row in $\widetilde{M}$ with sequence s, there must be two leaves of $\widetilde{\mathcal{N}}$ labeled s; we first remove one of those leaves and the edge into it.

Now consider an application of rule **Dr** or **Dt** that removes a row $\tilde{f}$ in $\widetilde{M}$ with sequence $s_{\tilde{f}}$. We have assumed inductively that ARG $\widetilde{\mathcal{N}}$ generates $\widetilde{M}$, so it must have a leaf $\tilde{f}$ labeled $s_{\tilde{f}}$. When either rule **Dr** or **Dt** applies, removing row f, we first remove leaf $\tilde{f}$ from $\widetilde{\mathcal{N}}$. Then, we successively remove *all* nodes and edges that are ancestral to $\tilde{f}$ in $\widetilde{\mathcal{N}}$, following any such path backward *until* the path reaches a node v with out-degree more than one. The key point is that there is a path in $\widetilde{\mathcal{N}}$ from v to a leaf other than $\tilde{f}$.

Each time we modify $\widetilde{\mathcal{N}}$ as above, we also do some "general housekeeping" on $\widetilde{\mathcal{N}}$: contract any internal node of degree two; contract any edge that has no label, other than an edge into a leaf or into a recombination node; merge any parallel edges directed in the same way; remove the edge (v, x) into a recombination node x if the label on x is identical to the label on the other parent (not v) of x.

We distinguish the modifications to $\widetilde{\mathcal{N}}$ that are made when a rule **Dc**, **Dr**, or **Dt** is applied to $\widetilde{M}$, and the general housekeeping modifications. The critical distinction is that the modifications of the first type change the set of sequences generated by the ARG, while the modifications of the second type do not; they change the ARG, but leave the set of sequences generated the same.

For example, consider the MinARG in figure 7.6, reproduced in figure 10.11a. Initially, no applications of rules **Dc** or **Dr** are possible. Suppose the first application of rule **Dt** removes sequence 100001, which labels a recombination node and a leaf. The effect on $\mathcal{N}$ is shown in figure 10.11b. Next, rule **Dc** can be applied to remove the edge label 6, as shown in figure 10.11c. Then, rule **Dt** can remove the sequence 01001 that labels a recombination node and a leaf. The resulting ARG $\widetilde{\mathcal{N}}$ is shown in figure 10.11d. Note that in this step of the thought experiment, the recombination node labeled 10001 was removed *along with* the recombination node labeled 01001. So a single application of rule **Dt** can lead to the removal of *more than* a single recombination node. After the second application of rule **Dt**, ARG $\widetilde{\mathcal{N}}$ is a tree and applications of rules **Dc** and **Dr** reduce the matrix to a single row with no sites; and the corresponding thought experiment reduces the ARG to a single node.

Having defined the way that each ARG $\widetilde{\mathcal{N}}$ is modified when rules **Dc**, **Dr**, and **Dt** are applied, we can easily prove the following lemma, by induction:

Lemma 10.1.1 *During the execution of algorithm CHB, each $\widetilde{\mathcal{N}}$ will be an ARG that generates the sequences in the corresponding matrix $\widetilde{M}$.*

Hence, at every step of the execution, the sequences in $\widetilde{M}$ are in one-to-one correspondence with the leaf sequences in $\widetilde{\mathcal{N}}$. It follows that when $\widetilde{M}$ is a single row with no entries, $\widetilde{\mathcal{N}}$ will contain only the root node of $\mathcal{N}$. Note that since the initial ARG $\mathcal{N}$ was an arbitrary ARG for M, lemma 10.1.1 holds simultaneously for *all* ARGs that generate M.

10.1.3.1   *Why the History Bound Might Not Be Tight*

The removal of two recombination nodes in one application of rule **Dt**, as shown in figure 10.11d, illustrates one reason why the *history bound* is not always tight: the CLB count is increased by one, but two recombination nodes are removed. In this section, we will more precisely analyze this type of gap.

Definition A recombination node x in an ARG $\mathcal{N}$ is called *rec-visible* if there is some path from v to a leaf of $\mathcal{N}$ that does not contain a recombination node other than x. Otherwise it is called *rec-invisible*, that is, every path from v to a leaf encounters another recombination node.

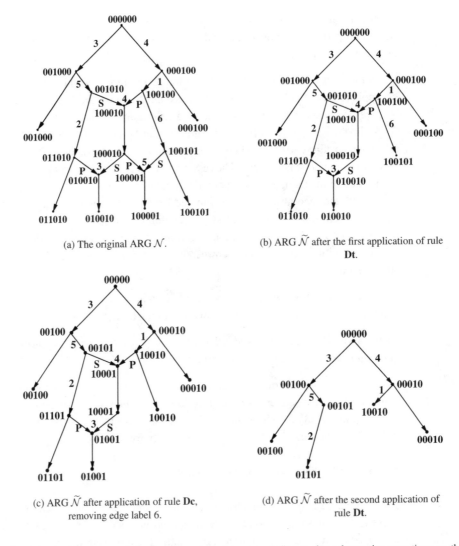

(a) The original ARG $\mathcal{N}$.

(b) ARG $\widetilde{\mathcal{N}}$ after the first application of rule **Dt**.

(c) ARG $\widetilde{\mathcal{N}}$ after application of rule **Dc**, removing edge label 6.

(d) ARG $\widetilde{\mathcal{N}}$ after the second application of rule **Dt**.

Figure 10.11 Three steps of the thought experiment corresponding to three destructive operations on the sequences shown at the leaves of the ARG in panel (a). Those sequences are also shown in figure 10.2. Note that the second application of rule **Dt**, creating the tree in panel (d), also removes the recombination node labeled 10001 (labeled 100010 in $\mathcal{N}$). That node is called a *rec-invisible* recombination node.

The concept of a of rec-visible recombination node may seem related to the concept of a *detectable* recombination (discussed in section 3.6), but the two concepts are actually independent.

Theorem 10.1.3 *The history bound for M will be strictly less than $Rmin_0(M)$ if there is any MinARG $\mathcal{N}$ for M containing a rec-invisible recombination node x in $\tilde{\mathcal{N}}$.*

Proof: For any M and any ARG $\mathcal{N}$ for M, there is an execution of algorithm *CHB*, where every application of rule **Dt** removes a sequence s from $\widetilde{M}$ which labels a recombination node in $\mathcal{N}$. Call such an execution *rec-intensive*. So when $\mathcal{N}$ is a MinARG for M, a rec-intensive execution returns a value $CLB(M)$ which is less than or equal to $Rmin_0(M)$. (This is the crux of the synthetic proof of theorem 10.1.2.)

We will show that if x is a rec-invisible recombination node in $\mathcal{N}$, then in no execution of algorithm *CHB* will the sequence labeling x, in an ARG $\tilde{\mathcal{N}}$, be chosen for removal by rule **Dt**. This is actually stronger than necessary, since it suffices to show this in a rec-intensive execution, and only when $\mathcal{N}$ is a MinARG for M.

Note that in the graphical interpretation, no recombination node x is removed from $\tilde{\mathcal{N}}$ by an application of rule **Dc** or **Dr**; only an application of rule **Dt** can result in the removal of a recombination node. Further, if rule **Dt** removes a sequence s in $\widetilde{M}$ that labels a leaf v in $\tilde{\mathcal{N}}$ whose parent is not a recombination node, then only leaf v and the edge into it is removed from $\tilde{\mathcal{N}}$. This is because the parent of v must be tree node, and hence has out-degree at least two. So, until node x is removed, it remains rec-invisible in every $\tilde{\mathcal{N}}$ created in an execution of algorithm *CHB*. Also, as noted earlier, when an application of rule **Dt** removes a row $\tilde{f}$ in $\widetilde{M}$ whose sequence $s_{\tilde{f}}$ labels a recombination node x in $\tilde{\mathcal{N}}$, x must have a leaf-child in $\tilde{\mathcal{N}}$ also labeled $s_{\tilde{f}}$. Therefore, if sequence $s_{\tilde{f}}$ is chosen for removal from $\widetilde{M}$ in an application of rule **Dt**, all of the descendants of x in $\mathcal{N}$, other than the single node that is then its leaf-child, must have already been removed.

Now, consider the point in an execution of algorithm *CHB* where the last recombination node x' that is a descendant of x is removed. Node x' is removed from $\tilde{\mathcal{N}}$ because the sequence association with x', or associated with some recombination node that is a descendant of x', has been chosen for removal by rule **Dt**. But because node x is rec-invisible in that $\tilde{\mathcal{N}}$, and x' is the last recombination node removed that is a descendant of x, all paths from x to a leaf must go through x'. Therefore, by the graphical interpretation of rule **Dt**, the removal of x' will result in the removal of x. Hence the number of applications of rule **Dt** will be strictly less than the number of recombination nodes in $\mathcal{N}$. The theorem is then proved by letting $\mathcal{N}$ be a MinARG for M. ∎

Stating theorem 10.1.3 differently, in order for the *history bound* to have a *chance* to be equal to $Rmin_0(M)$, every recombination node must be rec-visible in every MinARG $\mathcal{N}$ for M.

For example, the recombination node labeled 100010 is rec-invisible in the MinARG shown in figure 10.11a. Consistent with theorem 10.1.3, the execution of algorithm *CHB* described earlier shows that the *history bound* is at most two, while the MinARG actually has three recombination nodes. A single step of the thought experiment removed both a rec-visible recombination node and a rec-invisible one. This is also an example where program *HapBound -M* produces a lower bound that is higher than the *history bound*.

In corollary 10.2.3 (page 378), we will show that the *history bound* is always greater than or equal to the *optimal haplotype bound*, and so also to the *haplotype bound*. Hence, the *optimal haplotype bound*, or the *haplotype bound*, can be tight *only if* every recombination node is rec-visible in every MinARG for M. The results here are the first examples of the importance of the concept of *node visibility*. We will later see several additional ways that node visibility or invisibility is the critical distinction in certain ARG properties.

Corollary 10.1.1 *The difference between $Rmin_0(M)$ and the history bound for M is at least as large as the maximum number of rec-invisible recombination nodes in any MinARG for M.*

Algorithm CHB constructs a forest

The graphical view of the actions of algorithm *CHB* presented earlier focused on the destructive rules **Dc**, **Dr** and **Dt** and their use in the thought experiment that decomposes a MinARG for M. However, if we consider the parallel actions of rules **Cc** and **Cr** (but not **Ct**), we can also view the actions of algorithm *CHB* as constructing a forest of trees. In particular, consider the execution of algorithm *CHB* that computes the *history bound*. In that execution, if we also execute rules **Cc** and **Cr** whenever rules **Dc** and **Dr**, respectively, are executed, we get a forest of perfect phylogenies that derive the sequences in M. Moreover, each taxon is a leaf in exactly one of the perfect phylogenies, and each site labels exactly one of the perfect phylogenies. Note, however, that the perfect phylogenies might have *different* ancestral sequences. This set of disjoint perfect phylogenies (which is called a *perfect phylogenetic forest*), will be important in our later discussion of the *forest lower bound* in section 10.2.

Algorithm DP-History-Bound (M)

    for (k= $2, \ldots, n$) **do**
        {where n is the number of rows of M}
        for (each subset K of k rows of M) **do**
            Form the submatrix M_K of M and run algorithm *Clean* on M_K.
            Let $\widetilde{M}$ denote the resulting matrix, and
            let $\widetilde{K}$ denote the set of rows of $\widetilde{M}$.

            Set $mlb(M_K) = \min_{s_{\tilde{f}} \in \widetilde{M}} [1 + mlb(M_{(\widetilde{K} - \tilde{f})})]$
        endfor
    endfor

    return $mlb(M)$

Figure 10.12   Algorithm *DP-History-Bound*.

10.1.4   Computing the *History Bound* by Dynamic Programming

> Let me recite what history teaches. History Teaches.
> — Gertrude Stein

The original method for computing the *history bound* in [302] required $\theta(n!)$ executions of algorithm *candidate-history-bound* in worst case, corresponding to the $n!$ ways to order the choice of rows to remove in rule **Dt**. Such an exhaustive approach is practical only for small problem instances. However, Bafna and Bansal [18, 19] showed that the *history bound* can be computed much more efficiently using dynamic programming, very similar[3] to the dynamic programming approach developed in section 9.1.3.4 to compute $RN^*(M)$. The detailed dynamic programming algorithm to compute the *history bound* is shown in figure 10.12.

Theorem 10.1.4 *The value of $mlb(M)$ computed by algorithm DP-History-Bound on input M is $MLB(M)$, the history lower bound on $Rmin_0(M)$.*

---

3 Actually, the work of Bafna and Bansal [18, 19] came first and is the model for the dynamic program for $RN^*(M)$.

The proof of theorem 10.1.4 is essentially the same as the proof of theorem 9.1.5 and is left to the reader. The proof of theorem 10.1.2, that the *history bound* is a valid lower bound on $Rmin_0(M)$, also helps explain the recursive logic of algorithm *DP-History-Bound*.

Theorem 10.1.5 *The history bound can be computed in $O(nm2^n)$ worst-case time.*

10.1.5   Final Comments on the *History Bound*

We make three additional comments on the *history bound*.

Comment 1 The *history bound* can be increased in some cases by an idea developed in [19]. The main insight behind this idea is that the *history bound* never reflects needed rec-invisible nodes in MinARGs for M. The improved *history bound* identifies situations where the MinARG must have such recombination nodes, and increases the lower bound to reflect them. The reader is directed to [19] for details. The improvement leads to a lower bound of 7 for Kreitman's ADH data (see figure 5.7, page 144), while the original *history bound*, implemented in *RecMin*, only gave a lower bound of 6. Other reported improvements in real datasets show typical increases in lower bounds of between five and 50 percent. An extension of this idea was developed in [259].

Comment 2 An example is given in [19] showing a dataset M where $Rmin_0(M)$ is 6, but the *history bound* (applied to the whole data, that is, without use of the composite method) is only one. That construction can be generalized to construct data where $Rmin_0(M)$ is any specified value but the *history bound* is always one. Therefore, the difference between $Rmin_0(M)$ and the *history bound* can be made arbitrarily large.

Comment 3 Note that the *history bound* is defined *procedurally*. That is, it is defined as the output of algorithm *CHB-Branch*. Therefore, to prove that algorithm *DP-History-Bound* actually computes the *history bound*, one shows that it produces the same result as algorithm *CHB-Branch*. Until recently, we did not know a nonprocedural definition of the *history bound*; we will return to that point in the next section. But surprisingly, despite the lack of a nonprocedural definition of the *history bound*, Bafna and Bansal [18, 19] were able to establish the following:

Theorem 10.1.6 *The problem of computing the history bound is NP-hard.*

The original proof is very involved and we will not discuss its details in this book.

10.2   The *Forest Bound*

In this section, we present a new lower bound on $Rmin(M)$ which has a static and intuitive meaning. This lower bound, called the *forest bound*, was developed in [449]. It is related to the *history bound* and to an extension of the perfect-phylogeny problem, called the *minimum perfect phylogenetic forest problem*.

We will prove that the *forest bound* on $Rmin(M)$ is always greater than or equal to the *optimal haplotype bound* on $Rmin(M)$ (defined in section 5.2.5), and this relation holds for any interval of sites I in M. Therefore, when the *forest bound* on $Rmin(M(I))$ is used as the *local* lower bound in each interval I, the resulting global lower bound obtained from the composite method will be greater than or equal to the *optimal RecMin bound* (discussed in section 5.2.5).

We will also show that the *forest bound* is always less than or equal to the *history bound*, showing that the *history bound* is always greater than or equal to the *optimal haplotype bound*. We then give an integer linear programming formulation, whose solution computes the *forest bound* exactly. We show empirically that this formulation can be solved in practice for data with a small number of sites.

10.2.1   Definition of the *Forest Bound*

Given an ARG $\mathcal{N}$ for M, suppose we remove *all* of the recombination edges (edges directed into recombination nodes). The resulting DAG consists of two or more connected components, each of which is a rooted tree.

Definition $\mathcal{F}(\mathcal{N})$ denotes the *forest* of directed trees obtained by removing *all* the recombination edges in $\mathcal{N}$, and let $|\mathcal{F}(\mathcal{N})|$ denote the number of trees in $\mathcal{F}(\mathcal{N})$.

Since $\mathcal{F}(\mathcal{N})$ is derived from an ARG $\mathcal{N}$ for M, each site in M labels exactly one edge in $\mathcal{F}(\mathcal{N})$, and each taxon in M labels exactly one leaf in $\mathcal{F}(\mathcal{N})$. Therefore, $\mathcal{F}(\mathcal{N})$ defines a *partition* of the taxa and sites of M into $|\mathcal{F}(\mathcal{N})|$ distinct subsets, with a perfect phylogeny for each subset. The ancestral sequences of those perfect phylogenies do not have to be the same.

An alternative way to express this is that $\mathcal{F}(\mathcal{N})$ defines a partition of M into $|\mathcal{F}(\mathcal{N})|$ *submatrices*, each defined by a distinct subset of taxa and a distinct subset of sites of M, such that there is a perfect phylogeny for each of these submatrices, and each site labels an edge in one of them. It is necessary to add this last condition, since the ancestral sequences of the perfect phylogenies can be different, allowing the possibility that some site might never need to mutate.

Definition A partition P of M into submatrices is called a *perfect partition* if there is a perfect phylogeny for each submatrix, and each site labels an edge of one of them. The ancestral sequences of the perfect phylogenies need not be the same.

Definition The set of perfect phylogenies induced by a perfect partition of M is called a *perfect phylogenetic forest* for M.

It is easy to see that a perfect partition (and a perfect phylogenetic forest) always exists, for any M: One trivial perfect phylogenetic forest consists of a single directed edge, with the leaf labeled by sequence s_f, where f is an arbitrary taxon in M, and the root labeled with ancestral sequence obtained from f by reversing all of the states. So, all of the sites label that single edge. Then each of the other trees in the forest consists of a single node for one of the taxa in $M - \{f\}$. This is a trivial perfect phylogenetic forest with n trees, and a perfect partition of M with n submatrices. However, we are interested in a perfect partition with a *small* number of submatrices, motivating the following optimization problem.

> **The minimum perfect phylogenetic forest (MPPF) problem:** Given a binary matrix M, find a perfect partition of M containing the *minimum* number of submatrices, and hence the minimum number of trees in the associated perfect phylogenetic forest for M.

Definition $\mathcal{F}_{min}(M)$ denotes the perfect phylogenetic forest given by a solution to the MPPF problem for M; and $f_{min}(M)$ denotes the number of trees in $\mathcal{F}_{min}(M)$.

The forest bound
Definition The *forest bound* on M is defined as $f_{min}(M) - 1$.

Lemma 10.2.1 *The forest bound on M is a valid lower bound on $Rmin(M)$.*

Proof: Let $\mathcal{N}$ be a MinARG for M, and let $\mathcal{F}(\mathcal{N})$ be as defined earlier. By definition of $f_{min}(M)$, $|\mathcal{F}(\mathcal{N})| \geq f_{min}(M)$. By construction of $F(\mathcal{N})$, all but one of the trees in $\mathcal{F}(\mathcal{N})$ contains at least one node that is a recombination node in $\mathcal{N}$. Hence $\mathcal{N}$ contains at least $|\mathcal{F}(\mathcal{N})| - 1$ recombination nodes, and $Rmin(M) \geq |\mathcal{F}(\mathcal{N})| - 1 \geq f_{min}(M) - 1$. ∎

Recall that a node in an ARG $\mathcal{N}$ for M that is not labeled with a sequence in M is called a *Steiner* node.

Lemma 10.2.2 *Let $\mathcal{F}(M)$ be a perfect phylogenetic forest for M containing n_s Steiner nodes. Then the number of trees in $\mathcal{F}(M)$ is at least $n + n_s - m$, where M has n taxa and m sites.*

Proof: Let $|\mathcal{F}(\mathcal{N})| = k$. For each tree $T_i \in \mathcal{F}(M)$, let n_i denote the number of distinct sequences labeling nodes in T_i, and note that not all of these sequences need be in M. Also let m_i denote the number of sites in M that label edges of T_i, and note that an edge might be labeled by more than a single site. Then, $\sum_{i=1}^{k} n_i = n + n_s$. Since there are n_i distinct sequence labels in T_i, there must be exactly $n_i - 1$ edges labeled with one (or more) sites of M, so $m_i \geq n_i - 1$. We require each site in M to appear in $\mathcal{F}(M)$ exactly once, so we have $m = \sum_{i=1}^{k} m_i \geq \sum_{i=1}^{k} (n_i - 1) = n + n_s - k$, and hence $|(\mathcal{F}(\mathcal{N})| = k \leq n + n_s - m$. ∎

Corollary 10.2.1 *A solution to the MPPF problem generates a perfect phylogenetic forest using the minimum number of Steiner nodes, and conversely a perfect phylogenetic forest using the minimum number of Steiner nodes specifies a solution to the MPPF problem.*

Theorem 10.2.1 *The forest bound is always greater than or equal to the haplotype bound, but lower than or equal to the history bound, and for both of these relations, examples exist where the inequality is strict.*

Proof: Suppose a perfect phylogenetic forest has $k = f_{min}(M)$ trees. From lemma 10.2.2, $k \geq n + n_s - m \geq n - m$. So $k - 1 \geq n - m - 1$, which is the *haplotype bound* for M. It is simple to find examples where the inequality is strict.

Recall from the discussion in section 10.1.3 that each execution of algorithm *CHB* identifies a perfect phylogenetic forest, where the number of trees in the forest is equal to the value $CLB(M)$ returned by the algorithm. The *forest bound* must therefore be less than or equal to any value returned by algorithm *CHB*, including the smallest of those values, which is equal to the *history bound*. It is again simple to construct examples where this inequality is strict. ∎

There is another way to understand the relationship of the *history* and the *forest* bounds. The forest produced by algorithm *CHB*, used in the computation of the *history bound*, has an additional *time ordered* property that is not required in the definition of a *perfect phylogenetic forest*. In the perfect phylogenetic forest created from an execution of algorithm *CHB*, the trees in that forest can be ordered so that if site c appears in a tree T_i, the state of site c in all sequences labeling nodes in earlier trees must be the ancestral state of c. So when the perfect phylogenetic forest contains more than one tree, the *history bound* corresponds to a perfect phylogenetic forest selected from a *strict subset* of the forests used in the definition of the *forest bound*. Since both the *history* and the *forest bounds* are obtained from the *minimum* value in their respective sets of forests, the *forest bound* will always be less than or equal to the *history bound*, and it is strictly less in some cases.

We now relate the *forest bound* to the *optimal haplotype bound*.

Theorem 10.2.2 *The forest bound is always greater than or equal to the optimal haplotype bound, and is strictly greater in some cases.*

Proof: By theorem 10.2.1, the *forest bound* computed for a subset of sites S will be greater than or equal to the *haplotype bound* computed for S. In particular, if S^* is the subset of sites of M that yields the *optimal haplotype bound* for M, the *forest bound* computed for S^* will be greater than or equal to the *optimal haplotype bound*. Hence, to prove the theorem we only need to show that the *forest bound* computed on all of M is greater than or equal to the *forest bound* computed on any subset of sites s of M.

Let $\mathcal{F}_{min}(M)$ be a *minimum perfect phylogenetic forest* for M, with $f_{min}(M)$ trees. For a subset of sites S, recall that $M(S)$ denotes the sequences in M restricted to the sites in S. We can create a *perfect phylogenetic forest* for $M(S)$ by removing the sites not in S from $\mathcal{F}(\mathcal{N})$, and then contracting any edge that is not labeled by any site. The number of trees in the resulting forest is still $f_{min}(M)$, so $f_{min}(M(S)) \leq f_{min}(M)$, proving that the *forest bound* is greater than or equal to the *optimal haplotype bound*. It is easy to construct examples where this inequality is strict.   ■

Corollary 10.2.2 *If the forest bound is used in the composite bound method as the local lower bound in each interval of sites, then the resulting global lower bound will be greater than or equal to the optimal RecMin bound.*

Unfortunately, the *MPPF* problem is NP-hard [449], as are the problems of computing the *history bound* and the *optimal haplotype bound*.

Combining theorem 10.2.2 with theorem 10.2.1 we obtain:

Corollary 10.2.3 *The optimal haplotype bound is less than or equal to the history bound, and examples exist where that inequality is strict.*

Corollary 10.2.3 was stated in [299], but no proof of it was given there. We can also conclude that if the *history bound* is used in the composite method as the local lower bound in each interval of sites, then the resulting global lower bound will be greater than or equal to the *optimal RecMin bound*.

10.2.2   The *Forest Bound* **Is Statically Defined**

Unlike the *forest bound*, the *history bound* was only defined procedurally (that is, through an algorithm we call *CHG-Branch*) when it was introduced in [302]. Later, algorithm *DP-History-Bound* was developed to compute the *history bound*, but still the *history bound* was implicitly defined as "what those two algorithms produce." A nonprocedural, *static* definition of the *history bound* was not known until recently. The lack of a static definition makes it difficult to understand the semantics of the *history bound*, or to reason about it, or to find alternative methods to compute it. For example, the lack of a static definition for the *history bound* has made it difficult to develop an integer programming formulation for it, and no such formulation is yet known. Having an ILP formulation is desirable because integer programming has been empirically observed to allow effective solution of many NP-hard problems, in datasets of current interest.

In contrast to the *history bound*, the *forest bound* has a natural static definition through the *MPPF* problem. In the next section, we exploit the static definition to describe an integer programming formulation for the *forest bound*.

But the *history bound* **now has a static definition**

Recently, a static definition for the *history bound* was stated [220], and now a full proof has been written [279]. The static definition will be discussed in section 14.6.2, since it relies on material developed in chapter 14.

10.2.3 Computing the *Forest Bound* by Integer Programming

Since computing the *forest bound* is NP-hard, an integer programming formulation was developed in [449] to effectively compute it for data within certain useful ranges. The ILP formulation we present has a size that grows exponentially in the number of sites. Our experience shows that the formulation can be solved within reasonable time for data with up to eight sites, using the ILP solution package CPLEX.

Recall that M has n taxa and m sites. There are 2^m possible sequences that could label nodes in $\mathcal{F}_{min}(M)$, the forest induced by a solution to the *MPPF* problem for M. Of course, the n input sequences in M must label leaves in $\mathcal{F}_{min}(M)$. Using corollary 10.2.1, we can solve the *MPPF* problem by building a perfect phylogenetic forest for M, using the minimum number of Steiner nodes. Equivalently, since each sequence in M labels a leaf in the forest, we want a perfect phylogenetic forest using the minimum number of node labels. Here we develop an integer programming formulation that follows that approach.

ILP formulation

The ILP formulation has one binary integer programming variable v_i for each sequence s_i in the set of 2^m possible sequences labeling Steiner nodes. Setting v_i to 1 indicates that sequence s_i labels a node in the forest. The ILP formulation also has one binary variable, $e_{i,j}$, for every pair of sequences (s_i, s_j) that differ at exactly one site of M. Setting $e_{i,j}$ to 1 indicates that sequences s_i, s_j label nodes at the two ends of an edge e, and that e is labeled by the site c where s_i and s_j differ. The ILP formulation includes the constraints

$e(i, j) \leq v_i$, and $e(i, j) \leq v_j$, which ensure that when $e_{i,j} = 1$, both v_i and v_j are also set to 1.

We define E_c as the set of variables $e_{i,j}$, where s_i and s_j differ at the single site c. The infinite-sites model requires that exactly one variable in E_c be set to 1, and that the other variables in E_c be set to 0. Then, the following ILP formulation computes the *forest bound* exactly:

Objective function Minimize $\left(\sum_{i=1}^{2m} v_i\right) - m - 1$

Subject to

$v_i = 1$, for each sequence $s_i \in M$.

$e_{i,j} \leq v_i$, and $e_{i,j} \leq v_j$, for each variable $e_{i,j}$.

$\sum_{(s_i, s_j) \in E_c} e_{i,j} = 1$, for each site c.

All variables are binary

Missing data

The ILP formulation can be easily extended to handle the case of missing values in M, which is a common occurrence in real biological data. For a sequence s_i with missing values, we change the constraint $v_i = 1$ to the constraint $\sum_j v_j \geq 1$, for each sequence s_j that matches the values of s_i at all sites where values are not missing. Our experience is that this formulation can be solved in practical time for data with up to eight sites, using CPLEX. The interested reader is referred to [449] for more details. Note that the dynamic programming method for computing the *history bound* is not easily extended to the problem of computing the bound when there is missing data. This distinction is another indication of the advantage of having a *static* definition of a computational problem.

11 Conditions to Guarantee a Fully Decomposed MinARG

> Success is a science; if you have the conditions, you get the result.
> — Oscar Wilde

As discussed in chapter 7, the task of constructing ARGs is simplified (both computationally and conceptually) if we restrict attention to fully decomposed ARGs. We showed in chapter 7 that there is always a fully decomposed ARG for any input M, but for some M there is no MinARG for M that is fully decomposed. So, although it is attractive to restrict attention to fully decomposed ARGs, if we do so we may sometimes fail to construct a MinARG. It is therefore desirable to identify conditions that *guarantee* the existence of a fully decomposed MinARG. We have already developed one such condition in the case of *galled trees*: If M can be generated on a galled tree, then any reduced galled tree for M is a MinARG for M, and any reduced galled tree is a fully decomposed ARG. In this chapter we develop several additional *sufficient* conditions that guarantee a fully decomposed MinARG for input M, and then develop a general *necessary and sufficient* condition for there to be a fully decomposed MinARG for M. The results were first published in [144, 145], but are somewhat modified or corrected here.

11.1 Introduction

Recall (from section 3.3, page 82), the definitions for $R^m min(M)$, $R^1 min(M)$, $Rmin_s(M)$, $R^1 min_s(M)$, $R^m min_s(M)$, and $R(\mathcal{N})$. These will be central in this chapter.

Definition A set of binary sequences M is "min-m decomposable" or "min-1 decomposable" if there is a fully decomposed ARG for M allowing multiple-crossovers, or only allowing single-crossovers, with exactly $R^m min(M)$ or $R^1 min(M)$ recombination nodes, respectively. Similarly, a set of binary sequences M is

"s_r-min-m decomposable" or "s_r-min-1 decomposable" if there is a fully decomposed ARG for M with ancestral sequence s_r, allowing multiple-crossovers or only allowing single-crossovers, using exactly $Rmin_{s_r}(M)$ or $R^1min_{s_r}(M)$ recombination nodes, respectively.

When M is min-1-decomposable (or is min-decomposable in any of the other senses) we can find a MinARG for M by focusing separately on each connected component of the incompatibility graph or of a conflict graph. In addition, we can compute lower bounds on $R^1min(M)$, (and $R^mmin(M)$, $Rmin_{s_r}(M)$, and $R^1min_{s_r}(M)$) by computing corresponding lower bounds separately for each nontrivial connected component, and then adding these bounds together for a valid overall lower bound.

11.2   Sufficient Conditions for a Fully Decomposed MinARG

Here we develop sufficient conditions for the existence of a fully decomposed MinARG. In the process, we establish combinatorial properties that any M must possess, if there is no fully decomposed MinARG for M. We will focus on the case when the ancestral sequence is known. Analogous results for the root-unknown case are left to the reader. We begin with the following definitions and central technical lemma.

Definition Let $\mathcal{N}$ be an ARG for M with ancestral sequence s_r, and let s_x be a recombinant sequence created in $\mathcal{N}$ by the recombination of parent sequences s_1 and s_2. The recombination event (at node x) creating s_x is defined to be a *0-component recombination* if for *every* component $C \in G_{s_r}(M)$, $s_x(C) = s_1(C)$ or $s_x(C) = s_2(C)$. Similarly, the recombination event is defined to be a *1-component recombination* if there is exactly one component $C \in G_{s_r}(M)$ such that $s_x(C) \neq s_1(C)$ and $s_x(C) \neq s_2(C)$.

In other words, in a 0-component recombination, for every connected component C in $G_{s_r}(M)$, when restricted to the sites of C, the recombinant sequence s_x must be identical to at least one of its parent sequences. In a 1-component recombination, this property fails for *exactly* one connected component of $G_{s_r}(M)$. Note however that in a 0-component recombination, the full sequence s_x can differ (and must differ in any MinARG) from both of its parent sequences. This will happen for example when $s_x(C) = s_1(C) \neq s_2(C)$ and $s_x(C') = s_2(C') \neq s_1(C')$ for two connected components C and C'. Note also, that the definitions of 0-component and 1-component recombination are independent of any constraints on the number of crossovers allowed.

Lemma 11.2.1 *Let $\mathcal{N}$ be an arbitrary ARG for M, with ancestral sequence s_r. Suppose that every recombination in $\mathcal{N}$ is either a 0-component or 1-component recombination. If* *multiple-crossover recombinations are allowed in $\mathcal{N}$, then*

$$R(\mathcal{N}) \geq \sum_{C \in G_{s_r}(M)} R^1 min_{s_r(C)}(M(C)) \geq \sum_{C \in G_{s_r}(M)} R^m min_{s_r(C)}(M(C)).$$

Proof: We define a map h from the 1-component recombination nodes in $\mathcal{N}$, to the connected components of $G_{s_r}(M)$. Let x be any 1-component recombination node in $\mathcal{N}$, and let s_x, s_1, s_2 be the recombinant and parent sequences, respectively, at x. Since there is exactly one connected component $C \in G_{s_r}(M)$ for which $s_x(C) \neq s_1(C)$ and $s_x(C) \neq s_2(C)$, we define the map $h(x)$ of node x to component C.

Now, consider any recombination node x' in $\mathcal{N}$ that is *not* mapped to component C. Restricted to the sites in C, the recombinant sequence $s_{x'}$ at node x' must be identical to the sequence of one of its parents (say at node v'), because either x' is a 0-component recombination node or it is a 1-component recombination node that is mapped to a component different from C. Therefore, the recombination event at x' does not create a distinct sequence of $M(C)$.

Suppose we now remove the edge into node x' from the parent of x' that is not v'. The result is that x' is no longer a recombination node. Now remove any node, other than an original leaf in $\mathcal{N}$, that has out-degree zero, remove any edge label not in C, and contract any node of degree two. The result is an ARG, $\mathcal{N}_C$, that creates all the sequences in $M(C)$ and that has ancestral sequence $s_r(C)$. Further, every recombination node in $\mathcal{N}_C$ is a 1-component recombination that maps to C. Certainly then, the number of recombination nodes in $\mathcal{N}_C$, that is, $R(\mathcal{N}_C)$, is at least $R^1 min_{s_r(C)}(M(C))$. Since each recombination node in $\mathcal{N}$ is mapped to at most one component,

$$
\begin{aligned}
R(\mathcal{N}) &\geq \sum_{C \in G_{s_r}(M)} R(\mathcal{N}_C) \geq \sum_{C \in G_{s_r}(M)} R^1 min_{s_r(C)}(M(C)). \\
&\geq \sum_{C \in G_{s_r}(M)} R^m min_{s_r(C)}(M(C)).
\end{aligned}
$$

∎

Note that lemma 11.2.1 holds whether or not the ancestral sequence s was specified in the problem input.

Theorem 11.2.1 *Let $\mathcal{N}$ be an ARG for M with ancestral sequence s such that every recombination node in $\mathcal{N}$ is either a 0-component or a 1-component recombination node. Then there is fully decomposed ARG for M with ancestral sequence s, using at most $R(\mathcal{N})$ recombination nodes, with the same type of recombination (single- or multiple-crossover) as in $\mathcal{N}$.*

Proof: We consider the case of single-crossover recombination. By lemma 11.2.1, $\mathcal{N}$ has at least

$$\sum_{C \in G_{s_r}(M)} R^1 min_{s_r}(C)(M(C))$$

recombination nodes. By theorem 7.3.4 (page 226), there is a fully decomposed ARG $\mathcal{N}$ for M with ancestral sequence s_r, that has

$$\sum_{C \in G_{s_r}(M)} R^1 min_{s_r(C)}(M(C))$$

recombination nodes, so it is a MinARG for M. The same argument holds for multiple-crossover recombination.   ∎

We will next use theorem 11.2.1 to establish several sufficient conditions that guarantee a fully decomposed MinARG.

11.2.1   Spatial Disjointness

Definition Remember that the sites in M are ordered, $\{1, 2 \ldots m\}$. Let s be a specified binary sequence of length m. M is said to be *spatially disjoint*, if for every connected component C of $G_s(M)$, the sites in C form a *contiguous interval* in M.

Theorem 11.2.2 *Given s, if M is spatially disjoint, then M is s-min-1 decomposable.*

Proof: Let $\mathcal{N}$ be an ARG for M with ancestral sequence s, with $R^1(M \cup \{s\})$ single-crossover recombination nodes. Since M is spatially disjoint, a single-crossover falls either within the interval of sites for a single connected component of $G_s(M)$, or between two such intervals. In the former case, the recombination must be either a 0-component or a 1-component recombination (most likely a 1-component recombination), and in the latter case, the recombination must be a 0-component recombination. The theorem then follows from theorem 11.2.1 by letting $\mathcal{N}$ be a MinARG for M with ancestral sequence s.   ∎

Note that theorem 11.2.2 is proved only for the case of *single-crossover* recombination.

11.2.2   Component Respect

Definition For an ARG $\mathcal{N}$ for M, $L_{\mathcal{N}}$ is the set of sequences that label the nodes of $\mathcal{N}$. Clearly, $M \subseteq L_{\mathcal{N}}$, and $M \subset L_{\mathcal{N}}$ is possible.

Definition If the addition of the sequences $L_{\mathcal{N}} - M$ to $M \cup \{s\}$ does not create any incompatibilities between sites in *different* components of $G_s(M)$, then we say that $L_{\mathcal{N}}$ *respects* the component structure of $G_s(M)$.

Note that since $M \cup \{s\} \subseteq L_{\mathcal{N}}$, any pair of sites that is incompatible in $M \cup \{s\}$, is incompatible in $L_{\mathcal{N}}$, so $L_{\mathcal{N}}$ respects $G_s(M)$, if and only if $G_s(L_{\mathcal{N}})$, and $G_s(M)$ have the same *number* of components, although they need not be identical graphs.

Theorem 11.2.3 *Let $\mathcal{N}$ be an ARG (not necessarily a MinARG) for M with ancestral sequence s_r, and suppose that $L_{\mathcal{N}}$ respects $G_{s_r}(M)$. Then there is a fully decomposed ARG $\mathcal{N}'$ for M with ancestral sequence s_r, using at most $R(\mathcal{N})$ recombination nodes, where $\mathcal{N}$ and $\mathcal{N}'$ allow the same type of recombinations (single- or multiple-crossover).*

Proof: Consider a recombination in $\mathcal{N}$ between sequences s_1 and s_2 resulting in recombinant sequence s_x. We will show that this recombination is a 0-component or a 1-component recombination. If not, then there must be (at least) *two* connected components C_a and C_b in $G_{s_r}(M)$ such that $s_x(C_a) \neq s_1(C_a)$, $s_x(C_a) \neq s_2(C_a)$, $s_x(C_b) \neq s_1(C_b)$, and $s_x(C_b) \neq s_2(C_b)$. We will show that no such pair of connected components exists.

Consider a trivial connected component C in $G_{s_r}(M)$. Since C consists of just one site, if $s_1(C) \neq s_2(C)$, then one of those values is 0, and the other is 1. Similarly, $s_x(C)$ is either 0 or 1. Hence $s_x(C) = s_1(C)$ or $s_x(C) = s_2(C)$. Otherwise, if $s_1(C) = s_2(C)$, then $s_x(C) = s_1(C)$. In either case, neither C_a nor C_b can be component C. So if C_a and C_b exist, they must both be nontrivial connected components of $G_s(M)$.

Recall the concept of a *constant sequence* developed in chapter 6 (page 184). Consider two nontrivial connected components C and C' in $G_s(M)$, and let s_C be the constant sequence in $M(C)$, with respect to $< C, C' >$; let $s_{C'}$ be the constant sequence in $M(C')$ with respect to $< C', C >$. By assumption, $L_{\mathcal{N}}$ respects $G_{s_r}(M)$, so the subsets of sites in C and C' are also the subsets of sites in two connected components of $G_{s_r}(L_{\mathcal{N}})$, although the edges inside the components might be different. Therefore, by lemma 6.2.5 (page 185), the constant sequence, s_C, of C in $M(C)$ with respect to $< C, C' >$ is also the constant sequence of C in $L_{\mathcal{N}}(C)$ with respect to $< C, C' >$. Similarly, $s_{C'}$ is the constant sequence of C' in $L_{\mathcal{N}}(C')$ with respect to $< C', C >$. By the definition of $L_{\mathcal{N}}$, the recombinant sequence s_x is in $L_{\mathcal{N}}$. Theorem 6.2.4 (page 184) then implies that either $s_x(C) = s_C$, or $s_x(C') = s_{C'}$, or both. We will examine the case that $s_x(C) = s_C$ (the other case is symmetric and left to the reader).

If either $s_1(C) = s_C$ or $s_2(C) = s_C$, then neither C_a nor C_b can be component C. Conversely, if neither $s_1(C)$ nor $s_2(C)$ is s_C (which is equal to $s_x(C)$ by assumption), then by theorem 6.2.4, $s_1(C') = s_2(C') = s_{C'}$, and so $s_x(C') = s_{C'}$ no matter where any crossovers occur in the recombination of s_1 and s_2. In that case, neither C_a nor C_b is C'.

From the assumptions that $L_{\mathcal{N}}$ respects $G_{s_r}(M)$, and that C and C' are nontrivial components of $G_{s_r}(M)$, we have established that the pair (C, C') cannot be the (unordered) pair (C_a, C_b). Therefore, the pair (C_a, C_b) cannot exist, and so the recombination must be a 0-component or a 1-component recombination in $\mathcal{N}$. The theorem then follows by applying theorem 11.2.1.   ∎

Constructively then, in order to show that M is s-min-1 (or s-min-m) decomposable, it is sufficient to find one MinARG for M (using single- or multiple-crossovers, respectively) where $G_s(M)$ and $G_s(L_{\mathcal{N}})$ have the same number of connected components. Consistent with this, note that in figure 7.6 (page 207), the sequence 100010 at the internal recombination node of the MinARG is not in M, and sites 1 and 5 are incompatible in $L_{\mathcal{N}}$ but are compatible in M. Since sites 1 and 5 are in different connected components of $G_s(M)$, the number of connected components of $G_s(L_{\mathcal{N}})$ is one less than the number in $G_s(M)$.

11.2.3   Visibility and Tight Lower Bounds

Definition In an ARG $\mathcal{N}$ for M, a node is called *visible* if it is labeled with a sequence in M.

When a node is not visible, it is a *Steiner* node, but it is sometimes also called an *invisible* node. When all nodes in $\mathcal{N}$ are visible, $L_{\mathcal{N}} = M$, so theorem 11.2.3 leads immediately to the following conclusions.

Theorem 11.2.4 *Let $\mathcal{N}$ be an ARG for M where all the nodes in $\mathcal{N}$ are visible. Then there is a fully decomposed ARG for M with $R(\mathcal{N})$ recombination nodes.*

Corollary 11.2.1 *Let $\mathcal{N}$ be a MinARG for M with ancestral sequence s_r, allowing multiple-crossover recombination (respectively allowing only single-crossover recombination). Then M is s-min (respectively s-min-1) decomposable, if every node in $\mathcal{N}$ is visible.*

Corollary 11.2.1 was proved with a more complex proof in [141].

Corollary 11.2.2 *There is no fully decomposed MinARG for M, only if every MinARG for M has at least one Steiner node.*

The example shown in figure 7.6 has a Steiner node and is therefore consistent with corollary 11.2.2.

Next we observe a relationship between the $H(M)$, the *haplotype bound* for M, and full-decomposition.

Theorem 11.2.5 *Assume that the set of sequences M contains no duplicate sites. For a sequence s_r, if $H(M \cup \{s_r\}) = R^1 min(M \cup \{s_r\})$, then M is s-min-1 decomposable; and if $H(M \cup \{s_r\}) = R^m min(M \cup \{s_r\})$, then M is s_r-min-m decomposable. That is, if the haplotype bound is tight (for a particular type of recombination), then there is a fully decomposed MinARG (with that particular type of recombination) for M.*

Proof: The theorem follows immediately from theorem 5.2.10 (page 155) and corollary 11.2.1. ∎

11.3   The Most General Result

Here we state the most general result established for the existence of a fully decomposed MinARG.

Theorem 11.3.1 *M is s-min-m decomposable (or s-min-1 decomposable) if and only if there is a MinARG $\mathcal{N}$ for M with ancestral sequence s (using multiple- or single-crossover recombinations, respectively) such that conflict graphs $G_s(M)$ and $G_s(L_\mathcal{N})$ have the same number of connected components.*

Proof: The sufficient condition follows immediately from theorem 11.2.3.

Conversely, suppose M is s-min-m (or s-min-1) decomposable, and let $\mathcal{N}$ be a fully decomposed MinARG for M with ancestral sequence s. Since $\mathcal{N}$ is a MinARG for M, every blob in $\mathcal{N}$ contains at least two sites of M. Now add an edge from any Steiner node v in $\mathcal{N}$, to a new leaf labeled s_v. This creates an ARG $\mathcal{N}'$, with ancestral sequence s, for the set of sequences $L_\mathcal{N}$. Note that the blobs of $\mathcal{N}$ and $\mathcal{N}'$ are identical, so $\mathcal{N}$ and $\mathcal{N}'$ have the same number of blobs, say k.

$L_\mathcal{N}$ differs from M by the addition of some sequences, so every pair of sites that is in conflict in M, is in conflict in $L_\mathcal{N}$. Hence, the number of connected components in $G_{s_r}(L_\mathcal{N})$ must be less than or equal to the number in $G_{s_r}(M)$, which is k, since $\mathcal{N}$ is a fully decomposed ARG. But, if the number of components in $G_{s_r}(L_\mathcal{N})$ is strictly less than k, then by lemma 7.2.2 (page 206), there is no ARG for $L_\mathcal{N}$ with k blobs, where each blob contains some site(s) of M. That contradicts the existence of ARG $\mathcal{N}'$. Hence, the conflict graphs $G_s(M)$ and $G_s(L_\mathcal{N})$ must have the same number of connected components. ∎

11.4   Additional Applications

As discussed before, many mathematical and algorithmic results that we establish out of an interest in ARGs and population genetics, also apply in other biological contexts and to other problems. Here we note the following such extension. Recall the *maximum parsimony problem* introduced in section 7.5 (page 231).

Theorem 11.4.1 *The maximum parsimony problem for* M, *with specified ancestral sequence* s_r, *can be solved by separately solving the maximum parsimony problem for the sites in each connected component of* $G_{s_r}(M)$.

Proof: Recall that the maximum parsimony problem can be modeled and solved by using multiple-crossover recombination, as detailed in section 7.5.2. Now consider the tree T with ancestral sequence s_r, which solves the maximum parsimony problem, minimizing the number of mutations when back and recurrent mutations are allowed. Let $\mathcal{N}$ be the ARG that implements the same derivation from s_r as in T, using recombinations to model each back or recurrent mutation. A recombination event at node x is a two-crossover recombination, where the two crossovers occur just before and just after a single site c. By construction of $G_{s_r}(M)$, c is only in one connected component $C \in G_s(M)$. Therefore, for every other connected component $C' \neq C$, the recombinant sequence s_x is identical to one (or two) of its parent sequences, at the sites in C'. Therefore, the recombination event is either a 0-component or a 1-component recombination, and theorem 11.2.1 applies.   ∎

Because of theorem 11.4.1, when incompatibilities are caused by recurrent and/or back mutation, we can solve the maximum parsimony problem exactly, for each set $M(C)$ separately, and then connect the individual trees as specified by the backbone tree $\overline{T}(M)$. Since the maximum parsimony problem is NP-hard, and the only exact methods to solve it take exponential time in worst-case, decomposing the problem into several smaller problems may increase the practicality of the exponential-time methods.

Theorem 11.4.1 was first implicitly established using results on median networks [23, 373].

12 Tree and ARG-Based Haplotyping

There are two types of people in this world: Those who believe there are two types of people in this world and those who are smart enough to know better.
— Modified quote from Tom Robbins

12.1 Introduction

Recall that a *haplotype* is a sequence obtained from individuals in a diploid population, and that for each individual, a haplotype is obtained from only *one* of the two homologs of some chromosome. Recall also that a *genotype* is a mixture of the data from the two haplotypes, and that generally, a genotype does not unambiguously determine the two originating haplotypes. The concept of a *haplotype* was extensively discussed in section 4.1 (page 98); the reader should review that section. The concept of a *genotype* was also introduced in section 4.1, but will be more deeply discussed in this chapter. To start, see figure 12.1 below.

Implicit assumption
Until now, we have implicitly assumed that the binary sequences in input M are derived from experimentally determined, or observed, haplotype data, that is, from only one of a homologous pair of chromosomes. However, that is generally not the case. Rather, it is *genotypes* that are experimentally obtained, and haplotypes must be *computationally inferred* from the genotype data. So, in this chapter, we consider problems where the input data is based on *genotypes* rather than on haplotypes.

	1	2	3	4	5	6
individual 1: haplotype 1	0	0	0	0	0	0
individual 1: haplotype 2	0	1	0	0	0	1
individual 1: genotype	0	2	0	0	0	2
individual 2: haplotype 1	0	0	0	0	0	1
individual 2: haplotype 2	0	0	0	0	0	0
individual 2: genotype	0	0	0	0	0	2
individual 3: haplotype 1	0	0	1	1	1	0
individual 3: haplotype 2	0	0	1	1	1	0
individual 3: genotype	0	0	1	1	1	0
individual 4: haplotype 1	1	0	1	1	0	0
individual 4: haplotype 2	0	0	1	1	1	0
individual 4: genotype	2	0	1	1	2	0

Figure 12.1   The four hypothetical pairs of SNP haplotypes from four individuals shown in figure 4.2 (page 99), have been recoded so that the ancestral state at any site is taken from the ancestral sequence AGGCCA shown in figure 2.2 (page 38). For each site, the ancestral state is also the most frequent state in the haplotype frequencies shown in figure 4.1 (page 99). This is common in haplotype data, and in fact, when the true ancestral haplotype is not known, it is often assumed that the most frequent state of a site is the ancestral state. The four genotypes derived from each pair of haplotypes are shown for each individual.

12.1.1   Genotypes and Haplotype Inference

The combined description of the alleles in two homologous chromosomes is called a *genotype* (see figure 12.1). The definition of a genotype at a single site was given in chapter 4 (page 98). We will elaborate on that definition below.

Haplotype inference

For technological reasons, it has been (and still continues to be) much cheaper and easier to obtain genotypic data than haplotypic data, while generally, haplotypic data is more biologically informative: biological and chemical activity tends to take place on each of the two haplotypes separately. So, a critical technical problem that has received great attention in the last two decades, is the problem of inferring the underlying haplotypes (two per individual), or some haplotypic information, from genotypes (one per individual). This problem comes in different variants, but is generally referred to as *the haplotype inference*

(HI) problem. In this chapter, we will discuss some of the *combinatorial* variants of this problem and detail some of the most notable results.

12.2 The Haplotype Inference (HI) Problem

We assume that in each haplotype there are m SNP sites, where each site can have one of two states (alleles), 0 and 1. The sampled population consists of n individuals. For each individual, we know the $2m$ states (the genotype data) possessed by an individual, and do not know the two haplotypes of that individual. This leads to information that is *ternary* rather than binary.

12.2.1 Genotypes from Haplotypes

Definition Two haplotypes of length m define a single *genotype* of length m under the following rules:

> A site in the genotype has a value of 0 or 1, if the two corresponding sites in the underlying haplotypes both have that value. In that case, the site is called *homozygous*. A site in the genotype has value of 2 if the corresponding values in the haplotypes do not agree with each other, i.e., one is 0 and one is 1. In this case, the site is called *heterozygous*. See figure 12.1.

Definition A set of genotypes is denoted $\mathcal{G}$.

It would be mathematically cleaner to encode a site in a genotype as the *sum* of the values at that site in the two haplotypes. In that approach, a heterozygous site would be encoded as 1, and a homozygous site would be encoded as 0 or 2, respectively, depending on whether the haplotypes both have value 0 or 1. Some tasks and concepts are simplified by this encoding. For example, with this encoding, the leaf-count (to be defined below) for a site c would just be the sum of the values in column c of $\mathcal{G}$. A few papers do use this alternative encoding, and it is becoming more popular, but the majority of the literature uses the encoding defined above. For that reason, we have chosen to use it here.

The HI problem

Abstractly, input to the haplotype inference (HI) problem consists of a set of n *genotypes* $\mathcal{G}$, each of length m, where each site in the genotype has value either 0, 1, or 2. Each site in a genotype corresponds to the same site in two (unknown) haplotypes.

Definition Given an input set of n genotypes $\mathcal{G}$, a *solution* to the HI problem is a set of n pairs of binary sequences, $M(\mathcal{G})$, one pair for each genotype in $\mathcal{G}$. For any genotype $g \in \mathcal{G}$, the associated binary sequences h_1, h_2 in $M(\mathcal{G})$ must both have value 0 (or 1) at any site where g has value 0 (or 1); but for any site where g has value 2, exactly one of h_1, h_2 must have value 0, while the other has value 1.

That is, in a solution to the HI problem, h_1 and h_2 must be a feasible explanation for the true (but unknown) haplotype pair that gave rise to the observed genotype g. We will sometimes say that a solution $M(\mathcal{G})$ to the HI problem "explains" $\mathcal{G}$. See figure 12.2 for an instance of the HI problem and two solutions. See also figure 12.4 for another instance of the HI problem and a particular type of solution (a PPH solution) that will be explained in section 12.3.

Definition When a solution to the HI problem $\mathcal{G}$, with n genotypes, is shown as a binary matrix $M(\mathcal{G})$ with $2n$ rows, then the binary values in a column c represent the way that c has been *phased*, and the binary values in a pair of columns (c, c') represent the way that the sites c and c' have been *phased* (relative to each other).

Definition A solution to the HI problem $\mathcal{G}$ is also called a *phasing* of the genotypes.

For an individual with k heterozygous sites there are 2^{k-1} haplotype pairs that could appear in a solution to the HI problem. That is, there are 2^{k-1} ways to phase the genotypes. For example, if the observed genotype g is 0212, then the pair of binary sequences 0110, 0011 is one feasible phasing (or explanation), out of two feasible phasings. See figure 12.2 and 12.4. Of course, we want to find the explanation that actually gave rise to genotype g, and to all of $\mathcal{G}$. However, without additional biological insight, one cannot know which of the exponential number of phasings is the correct one.

Many solution approaches
The seminal paper that introduced the HI problem in 1990 and proposed a combinatorial algorithm for its solution was by Andrew Clark [71]. Since then, many methods have been explored, and some are intensively used for the general task of computationally inferring haplotype information from genotype information. These methods are divided into *statistical-based* methods and *optimization or combinatorial-based* methods. See [52] for a review of some of the currently used statistical methods, and also some laboratory-based methods. See [151] for a review of some of the combinatorial-based methods. Some of the methods give impressively accurate results in some circumstances, particularly for identifying common haplotypes of moderate length (a few hundred SNPs). However, research into haplotyping methods continues, because no single method is considered fully adequate in all applications, particularly for long haplotypes or rare haplotypes (of low frequency in a population).

12.2.2 The Need for a Genetic Model

Algorithm-based haplotype inference would be impossible without the implicit or explicit use of some genetic model of haplotype evolution, either to assess the biological fidelity of proposed solutions to the HI problem, or to guide the algorithm in constructing a solution. Many of the underlying models that have been articulated are stochastically based (often based on a coalescent model of haplotype evolution), but several are more combinatorially based, and one can sometimes view a model in either light.

Combinatorial methods
Combinatorial (or optimization) methods often state an explicit objective function that indirectly reflects the evolutionary model of the underlying haplotypes; one then tries to optimize the objective in order to solve the HI problem. There are several combinatorial models that have been proposed and extensively studied for the HI problem. The best studied ones are Clark's model [71, 138, 321], the perfect-phylogeny haplotyping (PPH) model [139], and the *pure-parsimony* model [140]. The latter two problems will be discussed in sections 12.3 and 12.4.7, respectively. The literature on these methods has grown rapidly in the last several years, particularly for the pure-parsimony model [127]. See [42, 151, 156, 157] for other reviews of combinatorial models and HI problems.

Statistical methods
Statistical methods are usually based on an explicit stochastic model of haplotype evolution and of the creation of genotypes from haplotypes; the HI problem is then cast as a maximum-likelihood or a Bayesian inference problem, and attacked by statistical methods such as the EM algorithm, Markov chain Monte Carlo, Gibbs sampling, importance sampling, or other related methods. The choice of the underlying genetic model often influences, or dictates, the type of algorithm used to solve the associated HI problem.

The seminal statistical paper
The paper that most clearly articulates a stochastic model of haplotype evolution (in this case the coalescent model), and the relationship of the model to the solution method is [398], written by Stephens, Smith, and Donnelly. The methods discussed in that paper were implemented into a surprisingly accurate and fast program called *PHASE* that solves the HI problem on data that is of small to moderate size by today's standards. Still, *PHASE* is considered the gold standard for phasing accuracy on the data that it can efficiently handle. Extensions of that methodology for genome-scale data (for example in [370]) have been somewhat less effective and there are now other contenders for the most effective stochastic method to solve large-scale instances of the HI problem [272]. One newer method that is widely used is developed in [51]. Unfortunately, their program is called *Beagle*, the same

name given to the MinARG construction program in [265], discussed in section 9.1.5. We will not further discuss the haplotyping *Beagle*, so outside of this paragraph, all reference to program *Beagle* are to the MinARG construction program of [265].

Our focus

In this chapter, we focus on *combinatorial* methods related to the HI problem, since these are the problems and methods that are the most related to the combinatorial treatment of trees and ARGs in this book. Also, combinatorial formulations and approaches to the HI problem have received considerable attention in the computer science community, although statistical methods have also been addressed there and are more commonly used by biologists. The combinatorial HI problem that we will first discuss in detail is the *perfect-phylogeny haplotyping* (PPH) problem.

12.3   Perfect-Phylogeny Haplotyping

We start by defining the problem, and then explain the biological basis for it.

The Perfect-Phylogeny Haplotype (PPH) Problem:

> Given a set of n genotypes, $\mathcal{G}$, determine if there is a set of $2n$ haplotypes, $M(\mathcal{G})$, that explain $\mathcal{G}$, such that $M(\mathcal{G})$ can be generated by a perfect phylogeny $T(\mathcal{G})$.

A solution must also specify a partition of the leaves of $T(\mathcal{G})$ into pairs, to explicitly explain $\mathcal{G}$. Note that a PPH problem instance might not have a solution, i.e., when no such set of $2n$ haplotypes is possible.

A solution to the PPH problem (if there is one) is a solution to HI problem. But it is highly constrained by the requirement that the haplotypes in $M(\mathcal{G})$ be derived on a perfect phylogeny. It is this constraint that makes the PPH problem nontrivial to solve, and, in the right context, makes it more likely to capture the historically correct biological solution. See figure 12.2 for a simple example. In that example, there are two solutions to the HI problem, but only one solves the PPH problem.

Similar, but not the same

The PPH problem is similar to the *incomplete directed perfect-phylogeny problem* considered in [332], where a ternary matrix $\mathcal{G}$ is also given. In that problem, one must change each 2 to either a 0 or a 1, so that the resulting binary matrix has a perfect phylogeny with the all-zero ancestral sequence. An elegant polynomial-time algorithm is presented in [332] for that problem. The *undirected* variant of that problem in NP-hard, but an effective heuristic method is developed in [368], and an integer programming approach is developed

	1	2
person 1	2	2
person 2	0	2
person 3	1	0

(a) Matrix G with three genotypes, each with two sites

	1	2
1	0	0
1'	1	1
2	0	0
2'	0	1
3	1	0
3'	1	0

(b) First HI solution (phasing)

	1	2
1	0	1
1'	1	0
2	0	0
2'	0	1
3	1	0
3'	1	0

(c) Second HI solution (phasing)

Figure 12.2   A set of genotypes, G, and two HI solutions, i.e., two ways to *phase* the genotypes. There are three individuals and two sites, shown in panel (a). The first HI solution, shown in panel (b), cannot be derived on a perfect phylogeny with all zero ancestral sequence since it violates the perfect-phylogeny theorem. The second HI solution, shown in panel (c), can be derived on a perfect phylogeny, and is a solution to the PPH problem. The tree $T(G)$ corresponding to the PPH solution is shown in figure 12.3.

in [148]. However, neither of those problems is the same as the PPH problem, because a solution to an instance of the PPH problem has twice as many rows, and the content of pairs of rows is constrained.

12.3.1   The Biological Basis for the PPH Problem

As mentioned earlier, finding the historically correct solution to the HI problem would be impossible without some implicit or explicit genetic model of haplotype evolution, to guide a method to the most promising solutions. The most powerful such genetic model is the population-genetic *coalescent*, previously discussed in several sections of the book.

The coalescent model for haplotype evolution without recombination

"In the absence of recombination, each sequence has a single ancestor in the previous generation" [181]. So, if we follow, backward in time, the history of a single haplotype h from a given individual f, that haplotype h is a copy of one of the haplotypes in one of the parents of individual f. It doesn't matter that f had two parents, or that each parent

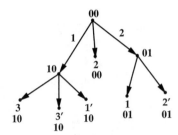

Figure 12.3   The perfect phylogeny corresponding to the second HI solution shown in figure 12.2. These six haplotypes explain the three genotypes shown in figure 12.2: haplotypes f and f' pair to form genotype f, for $f = 1, 2, 3$.

had two haplotypes. The backward history of a single haplotype in a single individual is a simple path, assuming no recombination. It follows that the history of a set of haplotypes (in the same genomic region) from a set of individuals forms a rooted tree. The histories of two sampled haplotypes (looking backward in time), coalesce at the most recent common ancestor haplotype of the two sampled haplotypes. The history forms a *coalescent* tree.

There is one additional element of the coalescent model: the *infinite-sites* assumption; that the m sites in the sequence (SNP sites in our case) are so sparse, relative to the mutation rate, that in the time frame of interest, the probability of two mutations at the same site is so low that we can assume at most one mutation will have occurred at any site.

12.3.1.1   The Coalescent Implies the Perfect-Phylogeny Model

The coalescent model of haplotype evolution (without recombination) implies that the derivation of the $2n$ historically correct haplotypes underlying the observed set of genotypes, $\mathcal{G}$, can be represented by a *perfect phylogeny*. In more biological detail, each observed genotype (from a single individual in a sample of n individuals) was obtained by *pairing* two of the $2n$ haplotypes derived on an unknown perfect phylogeny $T(\mathcal{G})$. So, $T(\mathcal{G})$ displays the history of haplotypes one generation above the sampled individuals. This is a subtle point:

> The $2n$ leaves of the perfect phylogeny, $T(\mathcal{G})$, are labeled with the haplotypes of the $2n$ *parents* of the n individuals whose genotypes have been sampled. Those $2n$ haplotypes can be partitioned into pairs, each of which gives rise to one of the n observed genotypes in $\mathcal{G}$. See figure 12.3.

This explains how the haplotypes were historically generated and how the genotypes were generated from the haplotypes. But the problem is that we observe only the genotypes and want to learn the historically correct haplotypes that generated them. This leads to the *perfect-phylogeny haplotyping problem* that we have already defined.

The converse

Here we have argued that the coalescent model supports the validity of the PPH problem. Conversely, the arguments for the validity of perfect characters and of the perfect-phylogeny model, given in chapters 1 and 2, support the coalescent model.

12.3.1.2 Additional Support for the Coalescent Model and the PPH Problem

Haplotype blocks

Most of the support for the coalescent model of haplotype evolution (in the appropriate contexts) comes from population genetics, for example see [231]. Additional strong support comes from recurrent empirical observations of *little or no recombination* in long segments of human DNA. Rather, much of the recombination tends to occur in *recombination hotspots* separated by long intervals of recombination deserts. The long intervals lacking recombination are called "haplotype blocks" (sometimes "recombination coldspots" or "recombination deserts"). The observations and descriptions of haplotype blocks were initially made in [85, 118, 170, 396], and those observations are reviewed in [424]. Confirming observations in humans have continued since then, for example, recently in [175]. Quoting from [9]: stretches of human DNA " ... were observed to form a block-like structure consisting of regions characterized by little evidence for historical recombination and limited haplotype diversity."

Similar phenomena have been observed in other species [16, 382], but not all species, and the locations of the hotspots can vary dramatically between species; for example, there is almost no agreement in the location of hotspots in humans and chimpanzees [301, 433]. Also, the location of hotspots in a species seems to be unstable and can change over evolutionary time [204]. We also note that there still remains uncertainty about the causes of observed recombination hotspots [212], although for our purposes, we only need to exploit their existence, without needing to know the underlying cause.

Validity in long segments

The existence of observed recombination hotspots and haplotype blocks, coupled with the general acceptance of the infinite-sites model—that many mutations will have only occurred once in recent[1] human history [176]—implies that the PPH model is valid in many long segments of the human genome. It also follows that the number of distinct haplotypes observed in a human population should be dramatically smaller than is combinatorially possible, and much less than had been expected before large-scale genomic data became available. This has in fact been observed:

> We observed strikingly limited haplotype diversity across long distances punctuated by sites of multiple historical recombination events. Essentially, this long genomic region can be parsed into blocks of low diversity in which recombination plays little or no role. [84]

The observation of no recombination in long intervals imposes severe combinatorial constraints on solutions to the HI problem, leading precisely to the PPH problem in those intervals. The observation provides a biological justification for what would otherwise be an attractive algorithmic problem, but one that might have seemed contrived—made up by computer scientists and mathematicians just for the fun of solving it.[2]

Even further justification for the coalescent model

In addition to the "no-recombination in long blocks" observation which leads to the coalescent model and the PPH problem, there are some less-direct justifications for the model.

First, the program *PHASE* [398] that solves the HI problem for genotypes of moderate length (up to several hundred SNP sites), is based on the assumption that the historically correct haplotypes evolved according to a coalescent model. Those assumptions are used to guide the program, and *PHASE* has been widely observed to be extremely accurate (in simulations and in cases when the true haplotypes are known). The effectiveness of *PHASE*, compared to programs that do not incorporate any coalescent assumptions, is an indirect validation of the underlying model, suggesting that much of haplotype evolution does indeed follow the coalescent model.

Second, in the biological literature, haplotypes that have been found (by whatever means), are often displayed in a tree to demonstrate the evolutionary (or mutational) relationship of the haplotypes. Those trees are not always perfect phylogenies, but many are. An example is shown in figure 2.2 (page 38). For another example, see [211]. Note that the approach in that literature is to first determine the haplotypes (solve the HI problem) by

---

1 "Recent" here means since the founding of populations that can be studied today.

2 Actually, there is some truth to this supposition—the work in [139] began slightly before the announcements of the discovery of haplotype blocks.

some means, and then to lay out a tree that explains the haplotypes; it often happens that the tree is a perfect phylogeny. The approach we take here is the *reverse*: we try to solve the HI problem by finding explaining haplotypes that can be derived on a perfect phylogeny, that is, by solving the PPH problem.

12.3.2 Algorithms and Programs for the PPH Problem

The PPH problem was introduced and first solved in [139]. There it was also shown that after one PPH solution is obtained (by whatever method), one can get an implicit representation, and a count, of the set of all PPH solutions in $O(m)$ time. The algorithm given in [139] is based on *reducing* the PPH problem to a well-studied problem in graph theory, called the *graph-realization* problem. The time for the reduction itself is $O(nm)$, and the graph-realization problem can be solved by several polynomial-time methods [34, 124, 410]. The best theoretical running time [34] for solving the graph-realization problem is $O(nm\alpha(nm))$, where α is the inverse Ackerman function, usually taken to be a constant in practice. Hence, the theoretical worst case time for the method is nearly linear in the size of the input, $n \times m$. The reduction of the PPH problem to the graph-realization problem is fairly easy to understand, and we will detail that reduction in section 12.4. However, the fastest (worst-case) algorithm that solves the graph-realization problem (in [34]) is complex and was too difficult to implement. Instead, the first implementation [69] of the reduction approach used the solution to the graph-realization problem in [124], and has a worst-case running time of $O(nm^2)$. In this chapter we will discuss the reduction of the PPH problem to the graph-realization problem, but we will not detail any solution to the graph-realization problem itself.

After the publication of [139], several additional methods for solving the PPH problem were developed [21, 106], and further exploited in [105, 159]. These methods also have worst-case running time of $O(nm^2)$ and can also be used to count and implicitly represent the set of all solutions. The method in [21] is not based (explicitly or implicitly) on a graph-realization algorithm; rather, it is based on deeper insights into the combinatorial structure of the PPH problem and its solution. Another paper [434] has insights similar to those in [21], but does not exploit them to develop an explicit algorithm for solving the PPH problem. Although the worst case running times for these methods are slower than for the fastest theoretical solution in [139], they are self-contained and easier to understand and implement. The results of empirical testing of the $O(nm^2)$ time methods can be found in [69].

Subsequently, $O(nm)$-time (i.e., truly linear time) worst-case algorithms were developed to solve the PPH problem [40, 89, 90, 366, 367], and another solution was outlined

[260]. Some comparisons of the first of these algorithms to the fastest of the earlier algorithms are reported in [89]. The linear-time method is significantly faster when the number of sites is large. For example, in tests with $n = 1000$ and $m = 2000$, the fastest of the $O(nm^2)$-time methods ran for an average of 467 seconds, while linear-time method ran for an average of only 1.89 seconds. All four of the PPH programs are available on the web, through the link given in the Preface.

The perfect phylogeny approach to haplotyping was additionally investigated in [100], where the data consists of short reads obtained from *next generation* DNA sequencing platforms. Compared to genotype data, in short read data the SNP sites are denser, some actual haplotypes are provided for small intervals, the types of errors are different, and the frequency of errors is greater. The haplotyping method presented in [100], based on perfect phylogeny and small deviations from perfect phylogeny, exploits this data and is reported to perform well in certain ranges of data.

12.4 Solving the PPH Problem

We now discuss the original polynomial-time solution to the PPH problem, developed in [139]. The solution *reduces* the PPH problem to the graph-realization problem. To begin, we address the case that the ancestral haplotype (ancestral sequence) at the root of the unknown perfect phylogeny is specified, and assume without loss of generality that it is the all-zero sequence. We will later show how to handle the case when the ancestral haplotype is not known.

12.4.1 Restating the PPH Problem

Given a (genotype) matrix $\mathcal{G}$, duplicate each row f in $\mathcal{G}$, creating a pair of rows (f, f'). The resulting matrix is denoted by χ. Then for each such pair of rows (f, f'), and every column c where $\chi(f, c) = \chi(f', c) = 2$, we are required to set *exactly* one of those two cells to 0, and set the other to 1, so that the resulting binary matrix, $M(\mathcal{G})$, has a perfect phylogeny $T(\mathcal{G})$. By theorem 2.1.1, the perfect-phylogeny theorem, the setting of values in χ must avoid creating two columns in $M(\mathcal{G})$, where three rows contain the binary pairs 0,1; 1,0; and 1,1.

12.4.2 Simple Methods First

We first show that parts of any PPH solution can be found by simple ideas and that if these ideas find a solution, it must be the *unique* PPH solution. Then we show that if these ideas are not sufficient to deduce the entire solution, the remaining part of the PPH problem can be solved by *reduction* to the problem of *graph realization*. Initially, we will assume that for any $\mathcal{G}$, there is a solution to the PPH problem; later we will show how to remove this assumption.

12.4.2.1 Simple Tools for Simple Paths

Definition We use $M^*(\mathcal{G})$ to denote the unknown, historically correct set of $2n$ haplotypes that define the n genotypes $\mathcal{G}$, and use $T^*(\mathcal{G})$) to denote the unique perfect phylogeny for $M^*(\mathcal{G})$.

It may seem that the mating of pairs of haplotypes in $M^*(\mathcal{G})$ would so disguise the underlying tree structure of $T^*(\mathcal{G})$, that it would be impossible to learn much about $T^*(\mathcal{G})$ from $\mathcal{G}$. However, the genotypes in $\mathcal{G}$ actually encode a significant amount of information about paths that *must* be in *any* PPH solution, in particular $T^*(\mathcal{G})$. This information is contained in three observations that are easy to establish:

Three observations

Observation 1 For any individual f in $\mathcal{G}$, the set of 1 entries in row f of $\mathcal{G}$ specify (without order), the *exact set* of edge labels on the path from the root to the *least common ancestor* of the leaves for f and f', in *every* PPH solution $T(\mathcal{G})$ for $\mathcal{G}$.

For example, in figure 12.4, the entries with value 1 in row r_2 are in columns 1 and 3, which are the sites labeling edges on the path from the root to the least common ancestor of leaves r_2 and r_2'.

Observation 2 For any individual f in $\mathcal{G}$, and any column c, if $\mathcal{G}(f, c)$ is 2, then the edge labeled with c must be on the path from the root to *exactly one* of the leaves f and f' in *every* perfect phylogeny $T(\mathcal{G})$ for $\mathcal{G}$. Another way to say that is that the path between f and f' must contain the edge with label c. Conversely, it is also easy to see that the only labels on that path are from columns where individual f has a value of 2 in $\mathcal{G}$.

For example, in figure 12.4, the entries with value 2 in row r_3 are in columns $1, 2, 3, 4, 6$, which are the sites labeling edges on the path between leaves r_3 and r_3'.

Observation 3 For any individual f in $\mathcal{G}$, and any PPH solution $T(\mathcal{G})$ for $\mathcal{G}$, the path from the root to the leaf labeled f first contains the edges labeled by sites with value 1 in row f, and then contains some edges (perhaps all or none) labeled by sites with value 2 in row f. The same is true for f'.

For example, see the paths to r_2 and r_2' in figure 12.4. Observation 3 actually follows from the two others but is stated explicitly for emphasis.

These observations imply that for any individual f in $\mathcal{G}$, and any column c, if $\mathcal{G}(f, c)$ is 0, then the edge labeled with site c must *not* be on the path from the root to either leaf f or f', in *any* PPH solution $T(\mathcal{G})$ for $\mathcal{G}$. It also follows that c cannot label an edge on the path between leaves f and f'. For example, see row r_4, and leaves r_4, r_4' in figure 12.4.

Extending and exploiting the observations

Observation 1 allows us to deduce the complete *set* of labels of the edges on the path from the root to the least common ancestor of leaves f and f' in *any* PPH solution $T(\mathcal{G})$, but without deducing the *order* of those labels. However, we will show that the order of the labels on that path is also invariant over all solutions to the PPH problem, and we can efficiently deduce that order. We will then use the other observations to detail more of the structure of any solution.

12.4.2.2 Deducing the Order of the Labels

Definition We define a *leaf-count*, denoted $t(c)$, for any column c in $\mathcal{G}$ as the sum of counts contributed by each entry in column c, as follows: Each 1 in column c contributes a count of *two* to the sum, and each 2 in column c contributes a count of *one*. Let $t(c)$ denote the total leaf-count for c.

Clearly, all the leaf-counts can be found in $O(nm)$ time, and since the largest leaf-count possible is $2n$, the leaf-counts can be *sorted* (using bucket-sort for example) in $O(nm)$ time, and the columns of $\mathcal{G}$ can be ordered by leaf-count, largest first. We will assume that this ordering has been done.

From the three observations above, we see that in *every* PPH solution $T(\mathcal{G})$ for $\mathcal{G}$, the edge labeled with site c must have exactly $t(c)$ leaves in its subtree. For example, the leaf-count of site 1 in figure 12.4a is four, and there are four leaves in the subtree below the edge labeled one.

It also follows that all labels that appear together on the same edge of any perfect phylogeny $T(\mathcal{G})$, must correspond to columns with the same leaf-count. In fact, all edge labels that appear together on the same edge must have exactly the same entries in their respective columns of $\mathcal{G}$ (this is a necessary but not sufficient condition for being on the same edge).

Leaf-counts tell all

More insightfully, the leaf-counts of sites labeling *successive* edges on any directed path from the root must *strictly decrease*, since the number of leaves in the subtree below an edge must strictly decrease along the path from the root (recall that no perfect phylogeny has an interior node with out-degree one). Hence, by using leaf counts, if we know the *set* of labels that appear on a directed path from the root, we can deduce the *unique* order of the edges on the path. This is called the *monotonic property* of leaf-counts. In summary:

Lemma 12.4.1 *For any individual f in $\mathcal{G}$ the path from the root to the least common ancestor of leaves for f and f', including the exact order of the edges and all the edge labels, must be the same in every perfect phylogeny $T(\mathcal{G})$ for $\mathcal{G}$. Further, that path can be efficiently determined.*

Lemma 12.4.1 is illustrated by figure 12.4.

Paths must superimpose

On the assumption that there is a PPH solution for $\mathcal{G}$, for any two individuals, f and g, the paths described in lemma 12.4.1 for f and for g, must be superimposable. That is, the two paths must agree on edge labels and edge order for some initial subpath (possibly empty) starting from the root, and after that have no edge labels in common. This can be proved more formally from observation 3, and the *shared-prefix* property of perfect phylogenies (page 42). The details are left to the reader.

Definition Let $\widehat{T}(\mathcal{G})$ denote the superposition of the paths described in lemma 12.4.1.

By the fact that the paths are superimposable, $\widehat{T}(\mathcal{G})$ must be a tree, and by lemma 12.4.1, $\widehat{T}(\mathcal{G})$ is a unique, well-defined tree. Further, $\widehat{T}(\mathcal{G})$ can be determined easily in $O(nm)$ time.

Lemma 12.4.2 *Every perfect phylogeny $T(\mathcal{G})$ for $\mathcal{G}$ must contain the invariant subtree $\widehat{T}(\mathcal{G})$, rooted at the root node of $T(\mathcal{G})$.*

Note that the leaves of $\widehat{T}(\mathcal{G})$ are not labeled. However, any character c with an entry of value 1 in column c of $\mathcal{G}$ must label one of the edges in $\widehat{T}(\mathcal{G})$. Thus, we have:

Theorem 12.4.1 *If every column in $\mathcal{G}$ contains at least one 1, then there is only one PPH solution $T(\mathcal{G})$ for $\mathcal{G}$ (up to relabeling each pair of leaves (f, f')), and it can be obtained from $\widehat{T}(\mathcal{G})$ by possibly adding new leaves, and by adding leaf labels. This can be done in $O(nm)$ time. So when every column in $\mathcal{G}$ contains at least one 1, the PPH problem can be solved in $O(nm)$ time.*

Proof: Since every column in $\mathcal{G}$ contains at least one 1, every character in $\mathcal{G}$ labels an edge of $\widehat{T}(\mathcal{G})$, and since no column labels two edges, $T(\mathcal{G})$ cannot have any additional *labeled* edges not already in $\widehat{T}(\mathcal{G})$. However, for each individual f, we have to label one leaf of $T(\mathcal{G})$ with f and one leaf with f', and this may require adding new leaves to $\widehat{T}(\mathcal{G})$. For

	1	2	3	4	5	6
r_1	2	2	0	0	2	0
r_2	1	0	1	2	0	0
r_3	2	2	2	2	0	2
r_4	0	1	0	0	2	2
leaf-count	4	4	3	2	2	2

(a) Genotypes $\mathcal{G}$ for four individuals at six sites

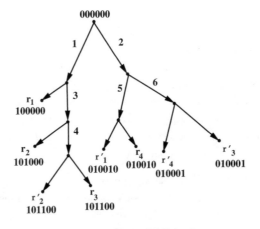

(b) A PPH solution $T(\mathcal{G})$ for $\mathcal{G}$

Figure 12.4   (a) Genotypes $\mathcal{G}$ and the leaf-counts of the columns.
(b) A PPH solution for $\mathcal{G}$. Note that $T(\mathcal{G})$ conforms to observations 1 through 3, and that along any path from the root, the leaf-counts of the columns labeling the edges strictly fall. The table of haplotypes determined by $T(\mathcal{G})$ is shown in figure 12.5.

any individual f, tree $\widehat{T}(\mathcal{G})$ contains the path from the root to the least common ancestor, call it v_f, of the leaves for f and f'. If the row in $\mathcal{G}$ for f does not contain any 2, then the path to v_f contains all of the characters that f and f' possess and the leaves for f and f' should hang directly off of v_f. This is the case when f is completely homozygous.

Now suppose that the row for f does contain one or more entries of value 2. On the assumption that there is a perfect phylogeny $T(\mathcal{G})$ for $\mathcal{G}$, and the fact that all the characters in $\mathcal{G}$ label edges in $\widehat{T}(\mathcal{G})$, it follows that there is a unique path P in $\widehat{T}(\mathcal{G})$ that contains all and only the characters of value 2 in row f. Further, path P must contain node v_f since, by

	1	2	3	4	5	6
r_1	1	0	0	0	0	0
r_1'	0	1	0	0	1	0
r_2	1	0	1	0	0	0
r_2'	1	0	1	1	0	0
r_3	1	0	1	1	0	0
r_3'	0	1	0	0	0	1
r_4	0	1	0	0	1	0
r_4'	0	1	0	0	0	1
leaf-count	4	4	3	2	2	2

Figure 12.5 The haplotype pairs derived from $T(\mathcal{G})$ in figure 12.4.

observation 2, in any perfect phylogeny for $\mathcal{G}$, the path from the root to f (or f') contains all and only the sites that f (or f') possesses. It also follows that the leaves for f and f' must hang off of, and label, the two ends of path P. If either end of P is an internal node of $\widehat{T}(\mathcal{G})$, extend P by attaching a new edge and leaf to that end of P. Then, label the two ends of (the extended) P with f and f'. Doing this for every row f creates a PPH solution $T(\mathcal{G})$ for $\mathcal{G}$.

By lemma 12.4.2, $\widehat{T}(\mathcal{G})$ is unique, and since for each individual f, P is uniquely defined in $\widehat{T}(\mathcal{G})$, the only variable element in the above procedure is which end of P is labeled f and which is labeled f'. In either case, for each individual f, the same pair of rows are created for f and f' in the PPH solution, although which row is labeled f and which is f' is not determined.

The only remaining issue is how to find all the paths in $O(nm)$ total time. Recall that the columns of $\mathcal{G}$ are sorted by leaf-count, largest first. Therefore, for any individual f, the path to v_f can be determined in $O(m)$ time by a walk from the root, guided by a scan of row f to find the sites that contain a 1. Next, we must identify path P. The two parts of P both extend from node v_f in $\widehat{T}(\mathcal{G})$. Each part of P contains, in order, edges labeled by some subsequence of sites that contain a value of 2 in row f of $\mathcal{G}$. These two subsequences can be interleaved, but together they contain all the sites of value 2 in row f. So one part of P can be determined in $O(m)$ time in a walk from v_f guided by scanning row f: When a 2 is encountered at a site c, the walk is advanced one edge if there is an edge labeled c extending from the current node in the walk; otherwise, the walk is not extended, but the scan of row f is continued. When the scan of row f is finished, the walk will be at one end of P. Then, the walk returns to v_f and a second walk is taken, using the sites with value 2 that did not label edges on the first part of P. This finds the other end of P. These three walks can be done in $O(m)$ time, so all of the needed paths can be found in $O(nm)$ time, and so $T(\mathcal{G})$ can be determined in $O(nm)$ time.   ∎

12.4.3   Graph Realization

Given theorem 12.4.1, the PPH problem is only difficult if there is a column c in $\mathcal{G}$ that contains no entries of value 1, so no edge in $\widehat{T}(\mathcal{G})$ will be labeled by c. In that case, we reduce the PPH problem to the *graph-realization* problem. We start with a definition of the graph-realization problem, although specialized to the case when the graph to be realized is a tree.[3]

Definition Let E_q be a set of q distinct integers. A *path-set* is an *unordered* subset π of E_q. A path-set is *realized* in an *undirected*, edge-labeled tree T consisting of q edges, if each edge of T is labeled by a distinct integer from E_q, and there is a contiguous path in T whose labels consist only of the integers in π.

For example, the path-set $\pi_3 = \{1, 2, 5, 6\}$ is realized by the path 2,1,5,6 in tree T shown in figure 12.6. Note that the order of the integers in a path-set need not agree with the order of the integers in a path that realizes the path-set. This is illustrated by path-set π_3 in figure 12.6.

In quite different terms from the treatment here, from the 1930s to the 1960s Hassler Whitney [430, 431] and William T. Tutte [410] and others studied and solved the following problem:

> **The graph-realization problem** Given a set of distinct integers, E_q, and a family $\Pi = \{\pi_1, \pi_2, ..., \pi_k\}$ of path-sets, find a tree T in which each path-set is realized, or determine that no such tree exists. Further, determine whether there is only one such T, and if there is more than one, characterize the relationship between the realizing trees.

Note that not every path in T needs to be in Π. Note also that a solution to an instance of the graph-realization problem is an *undirected tree*. Since we will use graph realization algorithms to solve the PPH problem, which requires a *directed* tree, we will need to generate a family of path-sets that forces the graph-realization algorithm to create an undirected graph that can be directed to become a PPH solution.

---

3 The general graph problem and the specialized tree version are essentially the same problem, which can be seen through the concept of "fundamental cycles" of a graph. However, the general problem is not needed here.

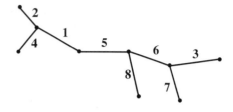

π_1	1	5	8	
π_2	2	4		
π_3	1	2	5	6
π_4	3	6	8	
π_5	1	5	6	7

(a) A family Π of five path-sets   (b) A tree T realizing the family of path-sets Π

Figure 12.6   (a) An instance Π of the graph-realization problem.
(b) A tree T realizing the path-sets in Π. Note that T is undirected, as are the realizing paths. The order of the integers on the path realizing π_3 does not agree with the order that the integers are given in π_3. By coincidence only, the order of the integers in each of the other three path-sets does agree with the order in the realizing path.

There are elegant mathematical results for each of the three problems included in the graph-realization problem. Building on the mathematical results, algorithms for the graph-realization problem were studied by several people in the 1950s and work continued until the 1980s. One of the first algorithms [261] finds a realizing tree T by an algorithm whose naive implementation runs in exponential time. Later, Bixby, and Wagner [34] showed how it could be implemented in *almost linear* time. Hence, the problem of finding a single realizing tree T for a given Π can be solved efficiently. We will not discuss any method of solving the graph-realization problem but will rely on the fact that there are efficient solutions.

12.4.4   Reducing the PPH Problem to the Graph-Realization Problem

In this section we show how to efficiently reduce any instance $\mathcal{G}$ of the PPH problem, to an instance Π of the graph-realization problem, so that any realizing tree T for Π specifies a PPH solution $T(\mathcal{G})$ for $\mathcal{G}$.

12.4.4.1   *The Reduction*

The reduction given here is a corrected version of the reduction in [139]. To begin, given $\mathcal{G}$, the set of integers E_q will consist of the set of column labels from $\mathcal{G}$, plus one additional integer, e_0, representing a "glue edge" (explained below).

    We assume that the leaf-counts in $\mathcal{G}$ have been calculated and that the columns of $\mathcal{G}$ have been sorted in order of the leaf-counts, largest first. The reduction has three steps.

Reduction step 1 (analysis of $\mathcal{G}$) In this step, we will create a set of paths that will be used to define path-sets in step 2. This is for exposition purposes, and explicit construction of these paths can be avoided in a more efficient implementation.

For every row f that contains either a 1 or a 2, let $[c_1, c_2, ...c_q]$ be columns in $\mathcal{G}$ (possibly none) where row f contains an entry of value 1. As established earlier, if there is a PPH solution $T(\mathcal{G})$, then these columns must label, in order, the edges of a directed path, $P_1(f)$, starting from the root, in any PPH solution $T(\mathcal{G})$ for $\mathcal{G}$.

If row f contains a 2, let $c^2(f)$ be the first column where row f has a 2, i.e., the column with the largest leaf count of the columns that have a 2 in row f. By the monotonic property of leaf-counts, and by observation 3, the edge labeled with $c^2(f)$ must be adjacent to the last node on path $P_1(f)$, in any PPH solution $T(\mathcal{G})$. So we extend $P_1(f)$ by adding an edge e labeled $c^2(f)$ to the end of $P_1(f)$. We also add a "glue edge" labeled e_0 to the start of $P_1(f)$. The resulting path is denoted $P'_1(f)$. Note that if row f contains no 1 entries, then e_0 is added to the start of an empty path, and if row f contains no 2 entries, then no edge e is added to the end of $P_1(f)$.

Reduction step 2 (generation of the first path-sets) For each row f, add to the family, Π, a path-set for each *adjacent pair* of edges on $P'_1(f)$, and create a path-set for each *adjacent triple* of edges on $P'_1(f)$.

It is easy to prove inductively on its length that the only tree that realizes these path-sets for a row f must be exactly the ordered path $P'_1(f)$ described in step 1. By creating similar path-sets for each row (each time with the same glue edge e_0), we create a family of path-sets Π, whose only realizing tree T consists of the P'_1 paths, along with the common glue edge e_0.

Reduction step 3 (generation of the second path-sets) For each row f in $\mathcal{G}$ that contains more than one 2, add a path-set π_f to Π consisting of the column numbers where row f has a value of 2.

Note that the size of the family of path-sets Π is at most $O(nm)$.

Theorem 12.4.2 *Assuming that there is a PPH solution for $\mathcal{G}$, the family of path-sets Π created in the reduction will have a realizing tree T. Further, T can be extended and labeled to become a PPH solution $T(\mathcal{G})$ for $\mathcal{G}$. The total time to find a PPH solution in this approach is $O(nm)$, plus the time to solve the instance of the graph-realization problem with $O(n)$ path-sets, each of maximum size m.*

Proof: We will show that there is a realizing tree for family Π created from $\mathcal{G}$, if there is a PPH solution for $\mathcal{G}$. We have already observed that any PPH solution $T(\mathcal{G})$ for $\mathcal{G}$ must contain the path $P_1(f)$, for each row f in $\mathcal{G}$. By observation (3) and the fact that the leaf-counts of sites labeling successive edges on any directed path from the root in $T(\mathcal{G})$ must strictly decrease (i.e., the monotonicity property), it follows that path $P_1(f)$ must be extended by the edge labeled $c^2(f)$. Each such path $P_1'(f)$ must be in any PPH solution $T(\mathcal{G})$ for $\mathcal{G}$. Clearly, path $P_1'(f)$ realizes any path-set specified for f in step (2) of the reduction, if that path-set does not contain integer e_0. Now add an edge labeled e_0 to the root of $T(\mathcal{G})$. Then, $T(\mathcal{G})$ realizes all of the path-sets specified in step (2).

For each f in $\mathcal{G}$, the path-set π_f created in step (3), will be realized in the extended $T(\mathcal{G})$, by observation (2). Hence, the family Π of path-sets created in the reduction has a realizing tree, namely $T(\mathcal{G})$ extended with edge labeled e_0.

Now we prove the second statement in the theorem. Let T be a realizing tree for Π. We claim that after contracting the glue edge e_0, the resulting tree T can be extended to become a PPH solution for $\mathcal{G}$. The path-sets created in step (2) ensure that for each f in $\mathcal{G}$, T contains the required path $P_1(f)$. Path $P_1(f)$ in T is followed with an edge labeled $c^2(f)$, if row f contains any entries of value 2. Notice that for any f in $\mathcal{G}$, there is at most one integer in both the path-sets created in steps (2) and (3), and it is $c^2(f)$. Hence the path-set π_f created in step 3 will connect to the realized path $P_1(f)$ via the edge labeled $c^2(f)$. By observation (3), this is required of any PPH solution for Π.

The only remaining issue is to explain exactly how to add the required leaf labels, and possibly create some new leaves. Clearly, we should place the leaves for f and f' at opposite endpoints of the path required by the path-set π_f, although if either of those endpoints is an internal node in T, we will need to extend T with an additional leaf node and edge, so that each row in χ labels a leaf. Finally, if row f in $\mathcal{G}$ has no 2, then f is homozygous, and the leaves for f and f' should branch off the end node of path $P_1(f)$.

After the leaves and leaf labels have been added, the resulting tree will have a path from the root to the least common ancestor of the leaves for f and f', labeled by all and only the columns where row f has a 1. It will also have a path between the leaves for f and f', labeled by all and only the columns where row f has a 2. And, the two paths will meet at the node at the end of the first path. Therefore, the resulting tree explains genotype f, and in general explains all of $\mathcal{G}$. Since no integer labels more than one edge in the tree, it is a PPH solution $T(\mathcal{G})$ for $\mathcal{G}$.

We leave it to the reader to verify that the solution of the PPH problem takes $O(nm)$-time plus the time to solve the instance of the graph-realization problem with $O(n)$ path-sets of size $O(m)$ each. ∎

Once the tree $T(\mathcal{G})$ is constructed, for each genotype f in $\mathcal{G}$, we can extract the two haplotypes that explain f, obtaining an HI solution that can be derived on a perfect phylogeny, namely $T(\mathcal{G})$.

12.4.4.2   The Case of a Non-Zero or Unknown Ancestral Haplotype

For convenience, we assumed that the ancestral haplotype at the root of the perfect phylogeny was specified, and that it was the all-zero sequence. Here we address the question of what to do when this is not the case: either a non-zero ancestral haplotype is known, or no ancestral haplotype has been specified.

The case when there is a known ancestral haplotype that is not the all-zero sequence can be handled easily by reduction to the all-zero case. Similar to the reduction discussed in section 2.2, in any column c of $\mathcal{G}$ where the ancestral state is 1, we interchange the 0 and 1 entries in column c, leaving all the 2 entries unchanged. It is easy to see that this new problem instance will have a PPH solution if and only if there is a PPH solution for $\mathcal{G}$, and that a PPH solution for $\mathcal{G}$ is obtained by the reverse reduction from the solution to the reduced PPH problem.

The case of the unknown ancestor

The case that there is no known ancestral haplotype is more interesting, but has a computationally easy solution. First, if $\mathcal{G}$ has even one row f that has at most one 2, then we know exactly what the haplotypes for f and f' are, and so we know two leaf labels that will be in any PPH solution $T(\mathcal{G})$. Therefore, we can "declare" the leaf for one of these, f say, to be the root of $T(\mathcal{G})$, and then have a problem instance with a known ancestral haplotype. The tree built from this instance may not be historically correct, but it will solve the PPH problem correctly. Tree $T(\mathcal{G})$ can then be rerooted at a more biologically valid root.

If the above approach is not possible, we can still find a PPH solution by using the following theorem, established in [282]:

Theorem 12.4.3 *Let M be a binary matrix. If there is a perfect phylogeny for M (with some unspecified root sequence), then there is a perfect phylogeny for M where the root sequence s, called the* majority *sequence, is formed by setting the state of each site c in s to be the most frequent state in column c of M. If there is a tie for the most frequent state, then either 0 or 1 will work.*

Theorem 12.4.3 implies that the PPH problem, when no ancestral haplotype is specified, can be reduced to the PPH problem where the ancestral haplotype is the majority sequence s described in the theorem. This is correct because the state, 0 or 1, that is most frequent in column c in $\mathcal{G}$, will also be the most frequent in column c in any PPH solution $M(\mathcal{G})$. This follows since each 2 in $\mathcal{G}$ will be create exactly one 0 and one 1 in $M(\mathcal{G})$. So, even without knowing any PPH solution, we know which binary state, 0 or 1, will be most frequent (or know that there will be a tie) in each column of any PPH solution for $\mathcal{G}$. Theorem 12.4.3

then allows a simple reduction of the root-unknown PPH problem to the root-known PPH problem.[4]

12.4.5 What Happens When There Is No PPH Solution?

Until now, we have assumed that there is a PPH solution for in input $\mathcal{G}$. To handle the possibility that there is no solution, we should always check that any tree $T(\mathcal{G})$ created by the above method is actually a PPH solution for $\mathcal{G}$. That can clearly be done in $O(nm)$ time. Additionally, if the method cannot be fully executed, for example because the P_1 paths cannot be superimposed, or because there is no solution to the instance of the graph-realization problem Π created by the reduction, then we can conclude that there is no PPH solution for $\mathcal{G}$.

12.4.6 Uniqueness and Multiplicity of PPH Solutions

We know that when there is a perfect phylogeny for a set of binary sequences M, it is unique, so an additional check on the biological fidelity of a solution to the PPH problem is whether that solution is unique. Conversely, having many different solutions to a problem reduces our confidence that any particular solution is biologically valid. So, we would like to know if a given PPH solution for $\mathcal{G}$ is the only PPH solution. If it is not unique, we would like to know the number of solutions, or characterize the set of solutions to see how much they differ, and to identify parts that are common to the solutions.

The uniqueness question for the general graph-realization problem was resolved (but in terms quite different from our treatment here) by Whitney [430] in 1932: Given a tree T

---

4 As a side note about the utility of mathematics, when I first learned theorem 12.4.3, my reaction that it was an example of interesting, but unneeded and useless, mathematics. The reason is that every sequence in M must label a leaf of any perfect phylogeny for M, so when no ancestral sequence is specified in a perfect-phylogeny problem, we can declare any sequence in M to be the ancestral sequence, reducing the root-unknown problem to the root-known case of the perfect-phylogeny problem: There will be a perfect phylogeny for M with some ancestral sequence, if and only if there is a perfect phylogeny for M where the ancestral sequence is any sequence in M. So, it seemed that there would never be a need to use the majority sequence as the ancestral sequence in the perfect-phylogeny problem; and theorem 12.4.3 looked to be useless in that "provable" sense. However, the theorem is very useful and needed: in an instance of the PPH problem (rather than an instance of the perfect-phylogeny problem), where no leaf sequence of any PPH solution $T(\mathcal{G})$ is known because every row of $\mathcal{G}$ has two or more 2s, theorem 12.4.3 is exactly the right tool to apply to solve the problem. This illustrates that one can never know whether a mathematical result will be useful or needed. Probably there is no "provably" useless mathematics.

in which all the path-sets of input set Π are realized, add a new edge between the two endpoints of path P_i in T, for each path-set P_i in Π. Call these new edges *scaffold edges*. Contract any edge in T that is not labeled, and call the new graph $G(T)$. Then T is the only labeled tree that realizes the path-sets in Π if and only if $G(T)$ is a *three-connected* graph. A connected graph is three-connected if there is no pair of nodes whose removal disconnects the graph.

Clearly, determining whether a graph is three-connected can be done trivially by testing all pairs of nodes, in $O(n^3)$ total time, and in fact there are much more efficient, nontrivial algorithms for this task. Whitney [431] also solved the problem of counting and implicitly representing the set of solutions to the graph-realization problem. We can use Whitney's results to determine whether a PPH solution $T(\mathcal{G})$ is unique, or to count, represent, or enumerate the set of solutions. The application of Whitney's results to the PPH problem is not completely straightforward, because of the glue-edge e_0. The interested reader should see [139] for details. Also, the PPH algorithms developed in [21, 89, 90, 106, 366, 367] all lead to different, computationally simple methods to determine if a PPH solution is unique, and to count and characterize the set of solutions if not, after the PPH algorithm has found one solution. An interesting result that follows from these methods is:

Theorem 12.4.4 *The number of PPH solutions for any $\mathcal{G}$ is always a power of two. If the problem instance has a PPH solution, the exact number of solutions can be determined in linear time, and the solutions can be enumerated in linear time per solution, after the first PPH solution has been found.*

The interested reader can find a proof and explanation for this in [21, 139].

12.4.7   The PPH Problem with Additional Criteria

When the PPH solution is not unique, it is desirable to use additional criteria to select one of the PPH solutions from among the multiple solutions. An attractive secondary criterion is to find a PPH solution that uses the *fewest* number of *distinct* haplotypes, over all PPH solutions. This criterion mixes the PPH problem with the *pure-parsimony* haplotype inference problem.

Pure parsimony

The combinatorial model for the HI problem that has been most widely studied and supported is the *pure-parsimony* model (sometimes called the *maximum parsimony* model), which suggests that a biologically valid solution to the HI problem should use a *small* number of *distinct* haplotypes. The pure-parsimony model is supported by the empirical observation that the number of distinct haplotypes seen in a population is generally small, and vastly smaller than the number of haplotypes possible, given the observed genotypes. For example, in a study of haplotypes associated with asthma [99], only 10 distinct haplotypes were found in a region with 13 SNP sites (which have the potential for $2^{13} = 8192$ distinct haplotypes). The standard explanation for the small number of distinct haplotypes seen in humans is the rapid expansion of the human population, which has not allowed enough time for the establishment of highly diverse haplotypes.

The pure-parsimony model leads to the *pure-parsimony problem* of finding an HI solution that *minimizes* the number of *distinct* haplotypes used. Note that this HI solution might not be a PPH solution. The pure-parsimony model was first suggested by Earl Hubbel [178], who showed that the pure-parsimony problem is NP-hard. A practical ILP solution for the pure-parsimony problem was suggested and studied in [140], and a very large body of literature on the pure-parsimony problem has developed since then (see [127]).

The problem of finding a PPH solution that minimizes the number of distinct haplotypes (i.e., mixing PPH with pure-parsimony) was shown to be NP-hard [20], and yet was also shown to be solvable in practice (on problem instances of moderate size) by an ILP formulation [148]. In addition to its biological utility, this solution is of combinatorial interest because it builds on a particular ILP formulation for the pure-parsimony problem [48, 50, 243] that is not empirically practical for problems of the same size. Apparently, the requirement that the HI solution minimize the number of distinct haplotypes over all *PPH solutions*, rather than over *all HI solutions*, constrains the ILP sufficiently so that it is solved efficiently in practice. The paper [214] establishes a lower bound, based on the rank of $\mathcal{G}$, on the number of distinct haplotypes needed in any HI solution for input $\mathcal{G}$. That bound is further discussed and studied in [49].

Other extensions of the PPH model and problem
Several other extensions of the basic PPH model and problem have been considered in the literature. The papers [128, 224] show that the PPH problem is NP-complete when there is (sufficient) missing data. Paper [20] addresses the question of whether the matrix-based PPH methods in [21, 106], which all run in $O(nm^2)$ time, can be implemented to run in $O(nm)$ time. It is shown that if either method could be implemented to run in $O(nm+m^2)$ time, then they could be implemented to run in $O(nm)$ time. The methods in [21, 106] run in $O(nm)$ time, except for initial computations that run in $O(nm^2)$ time, but only produce output of size $O(m^2)$. So it seemed attractive to see if that initial computation could be implemented to run in $O(m^2)$ time. However, the main result in [20] is that the initial computational tasks are equivalent to the problem of *Boolean matrix multiplication*, and hence if the initial tasks could be implemented to run in $O(nm)$ time, then two n-by-n Boolean matrices could be multiplied in $O(n^2)$ time, which is significantly faster than is currently possible. This provides evidence that the matrix-based approaches to the PPH

problem will not be made to run in linear time, in contrast to the graph-based methods in [40, 89, 90, 366, 367].

An interesting paper [162] explores the linear algebraic aspects of haplotype inference, to find redundancies which can be removed to reduce the number of sites in a problem instance. While this approach is not guaranteed to preserve the set of solutions, and can behave differently when used with different haplotyping methods, when it was tested in the context of the PPH problem, the results in [162] showed little loss of accuracy, and a large increase in speed.

Different genotype models

Several papers discuss variants of the original PPH problem that arise in different biological contexts, and that generate genotypic data differently than in the pure PPH problem. The papers [128, 129] consider the PPH problem where the input is assumed to have a row consisting of all-2 entries. An immediate consequence is that there is a pair of haplotypes whose state differs at every site. Such a pair is called a *Yin-Yang* haplotype, and it is the common occurrence of such haplotype pairs [456] that motivates the assumption that the input has an all-2 row. The paper [28] examines the *exclusive-OR (XOR) PPH problem*, which is based on a different way that genotype data is generated and collected. In the *XOR PPH* problem, the input genotype sequence for an individual is obtained from the XOR of the two binary haplotypes for the individual. That means that the genotype for an individual only indicates whether a site is *heterozygous* or *homozygous*, but unlike the original PPH problem, when the site is homozygous, the genotype does not indicate the specific state of the site. The main result in [28] is that this problem can also be efficiently reduced to an instance of the graph-realization problem, as in the original PPH problem. Further elaboration and extension of the XOR problem and results appear in [29, 30].

Near-perfect phylogeny

One modification of the PPH problem, called "imperfect" or "near-perfect phylogeny haplotyping" deserves particular attention. This approach is developed in three papers by Eleazar Eskin, Eran Halperin, and Richard M. Karp [105, 106, 158]. They address the issue that the pure PPH model may sometimes be an overly-idealized model of reality, or too brittle to handle errors in real data, and yet it might almost correctly reflect the historically correct evolution of the haplotypes. In those papers, they develop a haplotyping program called *HAP*, which is applied to data, M, for which there is no PPH solution. *HAP* tries to modify a *small* amount of the data in M, in order to make the modified data fit the PPH model. If it can, then the PPH problem is solved on the modified data, giving a set of haplotypes that do not perfectly explain the original M, but might actually be more accurate than an HI solution for M. They report that *HAP* is very effective in solving the HI problem on appropriate datasets.

We consider the effectiveness of *HAP* to be a validation of the PPH model and approach. Its success illustrates the utility of developing a clean, formal model and problem definition with an efficient solution, that can then be used in the core of a program that handles messier data or data produced by less precise models. There are many other such successful examples, which together provide encouragement and justification for clean methods and solutions of the type developed and discussed in this book. To some, these methods look too clean to be usable in biology (a field with messy data), but to others, clean methods form the *algorithmic core* of more robust and practical *programs*.

Galled trees

Another extension of the PPH problem, in [133, 134], considers the case that a set of genotypes cannot be derived on a perfect phylogeny, but might be derivable on a galled tree, or specialized variants of a galled tree. This is a natural extension of the PPH problem, since galled trees model small to modest deviations from the perfect-phylogeny model.

12.5 Dense Haplotyping with ARGs and Sparse Haplotypes

Here we discuss the HI problem in a particular context that allows an HI solution in the spirit of the PPH model, although much simpler. This comes from a paper by Li and Durbin [246], which primarily concerns the problem of SNP discovery, but also discusses the particular HI problem that we present here. We will examine their approach to SNP discovery in section 13.9.

The problem setup

The situation considered in [246] is that a set of $2n$ SNP haplotypes, M, are *known* for a set of n individuals (two haplotypes per individual), but at a relatively *sparse* set of sites. These are called *sparse haplotypes*. In particular, the haplotypes used in [246] are from the *HapMap 3* Project [200]. Later, in a study of the same individuals, SNP *genotypes*, $\mathcal{G}$, are obtained at a *dense* set of sites in the same genomic region as M, the known haplotypes. These are called *dense genotypes*. In particular, the genotypes used are from the *1000 genomes project* [1].

The goal then is to solve the HI problem for the dense genotypes, $\mathcal{G}$, making use of the available sparse haplotypes, M.

12.5.1   The Idealized Logic Behind the Method

The approach suggested in [246] is another example of exploiting the *shared-haplotype* structure captured in an ARG. We explain here a simplified variant of the full method in [246]. We think it exposes the underlying logic and spirit of the method, without developing the full maximum-likelihood formalities of the actual method.

To explain the underlying *idealized logic* of the approach, *suppose* we somehow know the historically correct ARG $\mathcal{N}$, with $2n$ leaves, that generated the sparse haplotypes M. Each individual is associated with exactly two leaves of $\mathcal{N}$, and each leaf is associated with exactly one individual.[5] This induces a pairing of the leaves of $\mathcal{N}$. That pairing is retained by, and shown in, the marginal tree T_c of $\mathcal{N}$, at any site c in M. See figures 12.7 and 12.8, showing an ARG with leaf pairings, and the marginal tree at site 3, with the same leaf pairings.

Key observation

Let $\mathcal{N}^*$ denote the historically correct (but unknown) ARG for the sites in $M \cup \mathcal{G}$. ARG $\mathcal{N}$ is ARG $\mathcal{N}^*$ restricted to the sites in M, and so the marginal tree T_c of $\mathcal{N}$ at site c, is also the marginal tree of $\mathcal{N}^*$ at c. Let $\tilde{c}$ be the first site in $\mathcal{G}$ to the right of site c in M. Since we assume we know $\mathcal{N}$, and hence T_c, *if* we also know the marginal tree $T_{\tilde{c}}$ of $\mathcal{N}^*$ at $\tilde{c}$, we would know the correct *phase* of c relative to $\tilde{c}$. This is because both T_c and $T_{\tilde{c}}$ inherit their $2n$ leaf labels from $\mathcal{N}^*$, and those leaf labels specify the two haplotypes of each individual. So, for any individual, say r_1, we would explicitly see the states of haplotypes r_1 and r'_1 at sites c and $\tilde{c}$, establishing how the two sites are phased relative to each other.

Nice, but ...

This is a nice observation, but we don't know $T_{\tilde{c}}$ (although we assume, without justification yet, that we know T_c), so how is this observation useful? The answer is that even though we don't know $T_{\tilde{c}}$, it is natural to expect that it is highly "similar" to the (assumed) known marginal tree T_c. That similarly might be high enough to allow T_c to *substitute* for the unknown $T_{\tilde{c}}$. Further, if we have some relevant, and efficiently computable, measure of the similarity of those two marginal trees, then we can decide if the similarity is sufficient to allow that substitution, and hence to reliably use T_c, to infer the phase of $\tilde{c}$ relative to c. The measure of similarity used in [246] is:

---

5 Here we relax the general assumption that duplicate rows have been removed, in order to explicitly represent a pair of haplotypes for each individual.

The number of edges of T_c that would have to be labeled with site $\tilde{c}$ (each label representing a mutation at site $\tilde{c}$), in order for the labeled tree T_c to explain the n genotypes in $\mathcal{G}$, at site $\tilde{c}$.

In the best case, when T_c and $T_{\tilde{c}}$ are identical, only one mutation at $\tilde{c}$ must be added to T_c, but when T_c and $T_{\tilde{c}}$ are very different (caused by recombination events between c and $\tilde{c}$), it may be necessary to add many mutations of $\tilde{c}$ to T_c. We will not discuss *how* to determine where to place the mutations, but certainly this measure is well defined.

An example

As a running example, assume site c in M is 3, and consider table 12.1, specifying three genotypes, $\mathcal{G}$, at four new sites between sites 3 and 4 in M. The first site, Γ_1 of $\mathcal{G}$, only needs to be added once to the marginal tree T_3 (from figure 12.8), in order to create haplotypes that explain the genotype of 2, for individuals r_1 and r_2, and the genotype of 0 for individual r_3. Similarly, the next two sites, Γ_2 and Γ_3, only need to be added once each to T_3 to explain the genotypes given in table 12.1, at those two sites. See figure 12.9. This suggests that tree T_3 is highly similar to the marginal trees in $\mathcal{N}^*$ at the first three sites of $\mathcal{G}$. (We will discuss site Γ_4 later.)

The most important consequence of this similarity is that we can use marginal tree T_3 in place of the three unknown marginal trees, to infer that the haplotypes of individuals r_1, r_2, and r_3, at sites c, Γ_1, Γ_2, and Γ_3, are as shown in tree T_3 in figure 12.9. Those haplotypes are shown in table 12.2. Additionally, since we already know the haplotypes for M, we can infer the haplotypes at the sites of $M \cup \{\Gamma_1, \Gamma_2, \Gamma_3\}$. Hence T_3 can be used to determine the phase of the first three sites in $\mathcal{G}$, relative to *all* the sites in M.

The justification for these inferences is that it seems highly unlikely that a single addition of each of the sites Γ_1, Γ_2, and Γ_3 to T_3 would be sufficient to generate haplotypes that explain $\mathcal{G}$ at those sites, unless T_3 is topologically similar to the marginal trees in $\mathcal{N}^*$ at those three sites. Further, if T_3 is topologically similar to the marginal tree at Γ_1 (and Γ_2, and Γ_3), then T_3 can be used in place of the true marginal trees, in order to establish the correct phase of c relative to Γ_1 (or Γ_2 or Γ_3). Moreover, the fact that T_3 works for several *consecutive* sites in $\mathcal{G}$ provides greater confidence in these inferences.

More generally, it seems highly unlikely that a small number of mutations at $\tilde{c}$, labeling edges in T_c, would suffice to explain the n genotypes of $\mathcal{G}$ at $\tilde{c}$, if T_c and $T_{\tilde{c}}$ were topologically very different. It follows that if a few mutations do suffice, then T_c probably can be used as a reliable substitute for $T_{\tilde{c}}$, to phase $\tilde{c}$ relative to c.

What about site Γ_4?

As sites in $\mathcal{G}$ become more distant from site c in M (site 3 in the running example), it becomes less likely that they will be consistent with T_c. For example, site Γ_4 in table 12.1

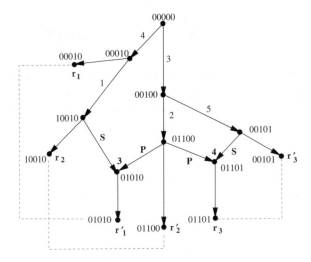

Figure 12.7 The presumed correct ARG $\mathcal{N}$ for six haplotypes in M. The ancestral sequence is the all-zero sequence. The haplotype pairs are indicated by dashed lines connecting leaf pairs. Individual r_1 has the haplotypes r_1 and r'_1; individual r_2 has the haplotypes r_2, r'_2; and individual r_3 has haplotypes r_3, r'_3.

	Γ_1	Γ_2	Γ_3	Γ_4
r_1	2	0	1	2
r_2	2	2	2	0
r_3	0	2	0	2

Table 12.1 Genotypes for individuals r_1, r_2, and r_3, seen at the four sites $\Gamma_1, \Gamma_2, \Gamma_3, \Gamma_4$ in $\mathcal{G}$, just to the right of site 3 in M. The marginal tree for site 3 is shown in figure 12.8.

must be added twice to T_3 to generate haplotypes that explain the genotypes in $\mathcal{G}$, at site Γ_4. Therefore, we are less confident of any phase inference for site Γ_4, than we are for inferences about the first three sites in $\mathcal{G}$. However, if site Γ_4 is consistent with the marginal tree T_{c+1} at site $c + 1$ of M (site 4 in the running example), we can use the marginal tree T_{c+1} in $\mathcal{N}$ to infer the phase of Γ_4 relative to site $c + 1$, and hence relative to all of M. In the running example, site Γ_4 still must be added twice to the marginal tree T_4 of $\mathcal{N}$, so we are left unsure about the proper phasing of site Γ_4.

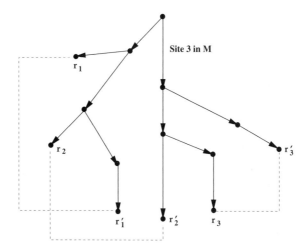

Figure 12.8 The marginal tree in $\mathcal{N}$ for site 3, with the same leaf pairs from $\mathcal{N}$ as in figure 12.7.

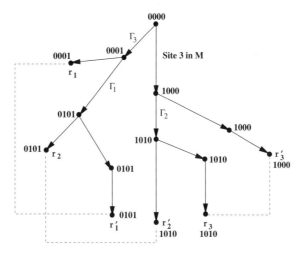

Figure 12.9 The marginal tree T_3 in $\mathcal{N}$ for site 3, with mutations added for sites $\Gamma_1, \Gamma_2, and \Gamma_3$. We assume that the ancestral state for each of those sites is known to be zero. Each node is labeled with a binary sequence of length four. The first position is for site 3 in M, and the next three positions are for sites $\Gamma_1, \Gamma_2, \Gamma_3$ in $\mathcal{G}$. This tree explains the genotypes in $\mathcal{G}$ at the first three sites in $\mathcal{G}$. Note that in order to place each mutation only once in the tree, the position for sites Γ_1 and Γ_2 are forced, and since the ancestral sequence is 0000, the position for site Γ_3 is also forced. From this labeled tree, we infer the phasing and haplotypes shown in table 12.2.

	Site 3 in M	Γ_1	Γ_2	Γ_3
r_1	0	0	0	1
r_1'	0	1	0	1
r_2	0	1	0	1
r_2'	1	0	1	0
r_3	1	0	1	0
r_3'	1	0	0	0

Table 12.2   Inferred paired haplotypes for individuals r_1, r_2, and r_3, at site 3 in M, and at the three sites Γ_1, Γ_2, and Γ_3 in $\mathcal{G}$.

12.5.2   Back to the Messy Reality

In the previous section we assumed that the historically correct ARG $\mathcal{N}$ for M was known, and therefore the marginal tree T_c was also known. These assumptions were for the purpose of explaining the *idealized logic* of using ARGs in this HI problem, following the spirit of the actual method in [246]. But of course, we don't know the historically correct ARG $\mathcal{N}$. To address this issue, a more formal probabilistic approach is developed in [246]. That approach requires generating a number of different *plausible* ARGs for M, using the program *Margarita* (discussed in section 9.1.7). Then, the HI problem is "essentially" solved, as described earlier, for each of the plausible ARGs. A maximum likelihood method is developed to integrate the information from the different HI solutions, to obtain the final phasing of $M \cup \mathcal{G}$. Those details are outside the scope of the book; the interested reader is referred to [246] for a full discussion. However, in section 13.4, we will discuss the use of *Margarita* in association mapping and a related probabilistic framework that is developed for that use.

12.6   Haplotyping to Minimize Lower Bounds

In this section we discuss haplotyping based on different criteria. The goal is to find criteria that allow efficient computation, and that are likely to guide the algorithm to find biologically valid HI solutions.

We have extensively discussed the problem and the value of computing lower bounds on $Rmin(M)$, where M is a binary matrix. In the context of this book, the sequences in M are thought to be SNP haplotypes for some set of sampled individuals. When haplotypes are not known, and only the set of genotypes, $\mathcal{G}$, are available, we would still like to have some estimate of the amount of recombination needed to explain $\mathcal{G}$. That problem can be formalized in a number of ways. In this section we will consider two approaches, and consider a third approach in the next section.

Definition Let $\mathcal{A}$ denote a recombination lower bound *method* that works on *binary* data M, and let $\mathcal{A}(M)$ be the lower bound given by method $\mathcal{A}$ when applied to M.

Definition Given a set of *genotypes* $\mathcal{G}$, we define $Min_{\mathcal{A}}(\mathcal{G})$ as the *Minimum* value of $\mathcal{A}(M)$ taken over *every* HI solution M for $\mathcal{G}$; that is, over all sets of haplotype pairs that explain $\mathcal{G}$. Similarly, we define $Max_{\mathcal{A}}(\mathcal{G})$ by changing "minimum" to "maximum" in the above definition.

The two quantities, $Min_{\mathcal{A}}(\mathcal{G})$ and $Max_{\mathcal{A}}(\mathcal{G})$, precisely define the *range* of values that method $\mathcal{A}$ would produce, over all possible HI solutions for $\mathcal{G}$. Note that $Min_{\mathcal{A}}(\mathcal{G}) \leq \mathcal{A}(M^*(\mathcal{G})) \leq Rmin(M^*(\mathcal{G}))$. However, it is not necessarily true that $Rmin(M^*(\mathcal{G})) \leq Max_{\mathcal{A}}(\mathcal{G})$; all we can say is that $\mathcal{A}(M^*(\mathcal{G})) \leq Max_{\mathcal{A}}(\mathcal{G})$.

In sections 12.6.2 we will show that $Min_{\mathcal{A}}(\mathcal{G})$ and $Max_{\mathcal{A}}(\mathcal{G})$ can be computed efficiently (i.e., in polynomial time), when $\mathcal{A}$ is the *HK bound*. In section 12.6.4, we will show that $Min_{\mathcal{A}}(\mathcal{G})$ can be computed efficiently, when $\mathcal{A}$ is the *connected-component* lower bound.

12.6.1 The Interpretation of $Min_{\mathcal{A}}(\mathcal{G})$ and $Max_{\mathcal{A}}(\mathcal{G})$

Min

It may seem odd to focus on the *minimum* value that method $\mathcal{A}$ can obtain, since we want to get as *high* an estimate as possible on the amount of recombination needed to generate $\mathcal{G}$. Certainly, if we had the set of haplotypes M, we would not want to compute the minimum possible lower bound on $Rmin(M)$ (which would trivially be zero). So why do we want to compute $Min_{\mathcal{A}}(\mathcal{G})$? The answer is that only the minimum value is *guaranteed* to be a true lower bound on $Rmin(M^*(\mathcal{G}))$. Any other values, such as $Max_{\mathcal{A}}(\mathcal{G})$, might be too large, i.e., larger than $Rmin(M^*(\mathcal{G}))$. But since $Min_{\mathcal{A}}(\mathcal{G}) \leq Rmin(M^*(\mathcal{G}))$, $Min_{\mathcal{A}}(\mathcal{G})$ is a valid *lower bound* on the amount of recombination in the historically correct set of haplotypes $M^*(\mathcal{G})$, and we can compute it even though we don't know $M^*(\mathcal{G})$.

Max

The biological motivation for $Max_{\mathcal{A}}(\mathcal{G})$ is less clear, although the problem of computing it is of natural interest from a purely combinatorial standpoint. Further, the *techniques* used to efficiently solve that problem may be of value in other more biologically motivated problems.

One (somewhat tortured) argument for the usefulness of $Max_A(\mathcal{G})$ is the following scenario. Suppose we are interested in the amount of recombination that must have occurred in the generation of the historically correct haplotypes $M^*(\mathcal{G})$ underlying the observed genotypes $\mathcal{G}$. Suppose also that the best available tool for computing a lower bound on $Rmin(M)$, given a set of haplotypes M, is a particular method $\mathcal{A}$. But since we only have the genotypes $\mathcal{G}$ and not the haplotypes $M^*(\mathcal{G})$, an obvious question is whether it would be worthwhile to determine more information about the haplotypes $M^*(\mathcal{G})$. With sufficient money and expertise, it is sometimes possible to experimentally determine $M^*(\mathcal{G})$ correctly. Of course, the point of finding the correct haplotypes $M^*(\mathcal{G})$ is to increase the resulting lower bound on $Rmin(M^*(\mathcal{G}))$, i.e., when $\mathcal{A}(M^*(\mathcal{G})) > Min_A(\mathcal{G})$. Note that $M^*(\mathcal{G})$ is a solution to the HI problem on $\mathcal{G}$.

Under this scenario, $Max_A(\mathcal{G})$ is the *highest* possible lower bound that can be obtained by using method $\mathcal{A}$, even if we knew $M^*(\mathcal{G})$. So computing $Max_A(\mathcal{G})$ could be helpful in *deciding* whether to expend the resources to determine $M^*(\mathcal{G})$ exactly. In particular, if the difference between $Max_A(G)$ and $Min_A(G)$ is small, then having the historically correct $M^*(\mathcal{G})$ may have little value. That is the best justification I know for $Max_A(G)$, other than the fact that we can efficiently compute it in some cases. I told you it was tortured.

12.6.2   The *HK Bound*: An Efficient Algorithm for $Min_{HK}(\mathcal{G})$

As discussed in chapter 5, the first and best-known lower bound on $Rmin(M)$ is the *HK bound* [183], although it is not the best-performing lower bound in theory or practice. When $\mathcal{A}$ is algorithm *HK bound*, we use the notation "$Min_{HK}(\mathcal{G})$" and "$Max_{HK}(\mathcal{G})$" in place of "$Min_A(\mathcal{G})$" and "$Max_A(\mathcal{G})$". $Min_{HK}(\mathcal{G})$ was studied in [434] where the following result was established:

Theorem 12.6.1 *$Min_{HK}(\mathcal{G})$ can be computed in polynomial time.*

Proof: We first discuss the *minimum* number of intervals that cover all the sites of $\mathcal{G}$, such that there is a PPH solution in each interval. We call such a set of intervals a *minimum PPH-cover* of $\mathcal{G}$, and claim that the leftmost interval in one minimum PPH-cover can be found as follows: Set c to 1, and find the largest column $c' \geq c$ such that there is a PPH solution in the matrix $\mathcal{G}$ restricted to the columns c through c'. Call this interval I. Interval I can be found by binary search, varying the value of d, and testing whether there is a PPH solution in the interval $[c, ..., d]$ of $\mathcal{G}$; c' is equal to the largest d allowing a PPH solution.

To see that interval I is in some minimum PPH-cover of $\mathcal{G}$, consider a minimum PPH-cover where the leftmost interval, I', is not I. By definition of I, I' must be strictly shorter than I. Therefore, we can replace I' with I and remove any interval, and/or a part of one interval, to the right of I' that overlaps interval I. The result is a minimum PPH-cover that contains I and has size no larger than the original PPH-cover. Hence, interval I is in some minimum PPH-cover.

Having found I, there is a minimum PPH-cover of $\mathcal{G}$ consisting of I, followed by a minimum PPH-cover of the interval $[c' + 1, ..., m]$ of $\mathcal{G}$. The first interval of a minimum PPH-cover of $[c' + 1, ..., m]$ can also be found by binary search. A minimum PPH-cover of $\mathcal{G}$ is therefore found by iterating the binary search until c' reaches m. This minimum PPH-cover can clearly be found in polynomial time. Let k be the number of intervals in that minimum PPH-cover, and call each of the intervals a *PPH-interval*.

We claim that $Min_{HK}(\mathcal{G}) = k - 1$. To prove that $Min_{HK}(\mathcal{G}) \leq k - 1$, consider the HI solution $M(\mathcal{G})$ to $\mathcal{G}$ composed of the PPH solutions in each of the k PPH-intervals. In that HI solution, there is no incompatible pair of sites strictly inside a single PPH-interval, so every incompatible pair of sites in $M(\mathcal{G})$ must be in two different PPH-intervals. Therefore, if we choose one point just after the end site (and before the next site) of each of the leftmost $k-1$ PPH-intervals, every incompatible pair of sites will straddle at least one of those $k-1$ points. Therefore, $HK(M(\mathcal{G})) \leq k - 1$, so $Min_{HK}(\mathcal{G}) \leq k - 1$.

To prove that $Min_{HK}(\mathcal{G})$ cannot be strictly less than $k - 1$, suppose not, and let M be an HI solution to $\mathcal{G}$ where $HK(M) = h < k - 1$. Let $\mathcal{P}$ be the h points that define $HK(M)$. Consider the partition of the sites of $\mathcal{G}$, before, between, and after the points in $\mathcal{P}$. That partition consists of exactly $h + 1 < k$ intervals. By the definition of the lower bound $HK(M)$, no pair of sites inside one of those $h + 1$ intervals can be incompatible in M. Therefore, M is an HI solution for $\mathcal{G}$ that defines $h + 1 < k$ intervals where there is a PPH solution inside each interval. But that contradicts the fact that k is the minimum number of intervals of $\mathcal{G}$ such that there is a PPH solution inside each interval. Therefore $Min_{HK}(\mathcal{G}) = k - 1$. ∎

We will next show that $Max_{HK}(\mathcal{G})$ can also be computed in polynomial time.

12.6.3   The *HK Bound*: An Efficient Algorithm for $Max_{HK}(\mathcal{G})$

Definition Given a set of genotypes $\mathcal{G}$, we say a pair of sites (c, d) in $\mathcal{G}$ is *potentially incompatible* if site-pair (c, d) can be phased so that the pair is incompatible. A site-pair that already has all four binary pairs, before phasing, is also considered *potentially incompatible*.

For example, suppose sites c and d in $\mathcal{G}$ are:

c	d
0	0
2	1
1	0
2	2

Pair (c, d) is potentially incompatible because row two will contribute the binary pair $(0, 1)$, in any phasing of (c, d), and we can phase sites (c, d) so that the 2s in row four become $(0, 0)$ and $(1, 1)$, contributing the needed pair $(1, 1)$. In that way, all four binary pairs will appear in the site-pair (c, d).

Definition Consider m integer points $1, 2, \ldots m$ on the real line, where m is the number of sites in $\mathcal{G}$. Let $\mathcal{I}_{\mathcal{G}}$ denote the set of closed intervals on the line, where closed interval $[c, d]$ is in $\mathcal{I}_{\mathcal{G}}$ if and only if site-pair (c, d) is potentially incompatible.

The set of intervals $\mathcal{I}_{\mathcal{G}}$ is a natural generalization of the set of intervals used for binary matrix M when computing $HK(M)$ (see section 5.2.2.2).

Definition Let $R^*(\mathcal{G})$ be the minimum-sized set, $R(\mathcal{G})$, of non-integer points, such that every interval in $\mathcal{I}_{\mathcal{G}}$ contains at least one point in $R(\mathcal{G})$.

Since the *interval coverage problem* (in section 5.2.2.2) was defined on an *arbitrary* set of closed intervals whose ends are on integer points on the line, we can use the same algorithm, algorithm *Interval-Scan*, to efficiently find $R^*(\mathcal{G})$.

Theorem 12.6.2 $Max_{HK}(\mathcal{G}) = |R^*(\mathcal{G})|$.

Proof: First, let $M(\mathcal{G})$ be the HI solution for $\mathcal{G}$ where the *HK bound* for $M(\mathcal{G})$ is equal to $Max_{HK}(\mathcal{G})$. Any interval I defined by an incompatible pair of sites in $M(\mathcal{G})$ is contained in $\mathcal{I}_{\mathcal{G}}$, and $R^*(\mathcal{G})$ contains a point in I. Hence the *HK bound* for $M(\mathcal{G})$ is at most $|R^*(\mathcal{G})|$, and so $Max_{HK}(\mathcal{G}) \leq |R^*(\mathcal{G})|$.

To show that $Max_{HK}(\mathcal{G}) \geq |R^*(\mathcal{G})|$, we only need to exhibit one HI solution $M(\mathcal{G})$ where the HI bound for $M(\mathcal{G})$ is at least $|R^*(\mathcal{G})|$. To find such an HI solution, let $\mathcal{I}$ be the set of $|R^*(\mathcal{G})|$ intervals selected from $\mathcal{I}_{\mathcal{G}}$ by algorithm *Interval-Scan*, and recall from corollary 5.2.1 that no pair of intervals in $\mathcal{I}$ can have a nontrivial intersection. That is, two intervals in $\mathcal{I}$ can share at most one integer point (at the end of one interval and the start of another). Hence if we can construct an HI solution $M(\mathcal{G})$ for $\mathcal{G}$, with the property that for each interval $[c, d]$ in $\mathcal{I}$, sites c and d are incompatible in $M(\mathcal{G})$, then $M(\mathcal{G})$ will be an HI solution of the desired type, that is, where the *HK bound* for $M(\mathcal{G})$ is at least $|R^*(\mathcal{G})|$.

Since no pair of intervals in $\mathcal{I}$ can have a nontrivial intersection, there is a well-defined ordering of the intervals, left to right, according to their left endpoints; we will process the intervals in that order. When processing a interval $[c, d]$ in $\mathcal{I}$, there are two cases: either point c is *not* the right endpoint of the previous interval in $\mathcal{I}$, or it is. If c is not the right endpoint of the previous interval in $\mathcal{I}$ (and this is certainly the case for the leftmost interval in the ordering), we phase the pair (c, d) in any way that makes them incompatible. This is always possible because, by construction, they are potentially incompatible. However, if c is the right endpoint of the previous interval $[c', d']$, then $c = d'$, and site d' has already been phased in the processing of the intervals. In that situation, we have to be more careful in how we phase site d so that it becomes incompatible with the already phased site c (which is d'). But that is simple to achieve. The only challenge is when sites c and d both have state 2 in a row f in $\mathcal{G}$, and to become incompatible, sites c, d must be phased to create a particular binary pair (say 0,1) in the two rows f and f'. That is always possible to do, by noting in which row, f or f', the phased site c contains a 0, and then phasing site d to place a 1 in that row. So, it is always possible to phase the endpoints in each interval in $\mathcal{I}$, so that each interval $[c, d]$ in $\mathcal{I}$ implies that (c, d) is an incompatible pair in $M(\mathcal{G})$. It then follows that the *HK bound* for $M(\mathcal{G})$ is at least $|R^*(\mathcal{G})|$, and so $Max_{HK}(\mathcal{G}) = |R^*(\mathcal{G})|$. ∎

12.6.4 The Connected-Component Bound: An Efficient Algorithm for $Min_{cc}(\mathcal{G})$

We now show how to compute $Min_A(\mathcal{G})$ efficiently when the lower bound $\mathcal{A}$ is the *connected-component* (cc) bound, discussed in section 7.1. In this case we use the notation "$Min_{cc}(\mathcal{G})$" in place of "$Min_A(\mathcal{G})$." Recall that for a binary matrix M, the *connected-component* lower bound for M, denoted $cc(M)$, is the number of nontrivial connected components in the incompatibility graph $G(M)$. So, $Min_{cc}(\mathcal{G}) = \min[cc(M(\mathcal{G})) : M(\mathcal{G})$ is an HI solution for $\mathcal{G}]$.

Definition Given $\mathcal{G}$ we define the *maximal incompatibility graph* for $\mathcal{G}$, denoted $MIG(\mathcal{G})$, as an undirected graph with one node for each site in $\mathcal{G}$, and an edge between a pair of nodes (c, d) if and only if the site-pair (c, d) is potentially incompatible.

Definition For any connected component C of $MIG(\mathcal{G})$, let $MIG(C)$ denote the subgraph of $MIG(\mathcal{G})$ induced by the nodes in C, and let $\mathcal{G}(C)$ denote $\mathcal{G}$ restricted to the sites in C. Clearly, $MIG(C)$ is $MIG(\mathcal{G}(C))$.

Definition Let $N_{\mathcal{G}}$ denote the number of connected components of $MIG(\mathcal{G})$ where $\mathcal{G}(C)$ does *not* have a PPH solution.

Note that there is a PPH solution for $\mathcal{G}(C)$, when C is a trivial connected component.

Theorem 12.6.3 $Min_{cc}(\mathcal{G}) = N_{\mathcal{G}}$.

Proof: We first show that the connected-component lower bound for any HI solution $M(\mathcal{G})$ for $\mathcal{G}$ is at least $N_{\mathcal{G}}$. Let $M(\mathcal{G})$ be an arbitrary HI solution for $\mathcal{G}$, and consider an arbitrary connected component C of $MIG(\mathcal{G})$, where $\mathcal{G}(C)$ does not have a PPH solution. There must be at least one incompatible pair of sites from C in $M(\mathcal{G})$, and so there must be at least one edge consisting of a pair of nodes from C, in the incompatibility graph $G(M(\mathcal{G}))$ for $M(\mathcal{G})$. Further, since $G(M(\mathcal{G}))$ is a subgraph of $MIG(\mathcal{G})$, every connected component of $G(M(\mathcal{G}))$ must be contained in a connected component of $MIG(\mathcal{G})$. Therefore, there must be at least one nontrivial connected component of $G(M(\mathcal{G}))$ contained in C. Since C was an arbitrary connected component of $MIG(\mathcal{G})$ where $\mathcal{G}(C)$ has no PPH solution, the number of nontrivial connected components of $G(M(\mathcal{G}))$ is at least $N_{\mathcal{G}}$. Hence, $Min_{cc}(\mathcal{G}) \geq N_{\mathcal{G}}$.

To finish the proof it suffices to find an HI solution $M(\mathcal{G})$ for $\mathcal{G}$ where the connected-component lower bound for $M(\mathcal{G})$ is $N_{\mathcal{G}}$. By the construction of $MIG(\mathcal{G})$, and the maximality of connected components, we can phase the sites in each connected component separately, without causing any pair of sites in different components to become incompatible.

To begin the construction of $M(\mathcal{G})$, for any connected component C of $MIG(\mathcal{G})$ where $\mathcal{G}(C)$ has a PPH solution, phase the sites in C to create a PPH solution for $\mathcal{G}(C)$. As a result, none of those sites will be incompatible with any other sites in $\mathcal{G}$. Next, we phase the sites of one of the $N_{\mathcal{G}}$ connected components, C, where $\mathcal{G}(C)$ does *not* have a PPH solution. In particular, we want to phase those sites so that the nodes of C form a connected component in the incompatibility graph for $M(\mathcal{G})$. To do this, find an arbitrary spanning tree T in $MIG(\mathcal{G})$, restricted to the sites in C; next choose a root of T and direct the edges away from the root. Then phase the site labeling the root of T, and a site labeling one child of the root, those two sites become incompatible. This is possible because the edge in T come from $MIG(\mathcal{G})$, so that the two sites are potentially incompatible. From that first phased pair, continue to phase the other sites in C by following the directed edges in T. In particular, phase any other site as soon as the site labeling its parent in T has been phased. As in the proof of theorem 12.6.2, and because each node has a unique parent, each site can be phased to be made incompatible with the site labeling its parent, no matter how that parent site was phased. The result is a phasing of the sites in C so that every pair of sites at the two endpoints of an edge in T become incompatible. Therefore, the sites of C will form a single connected component of the incompatibility graph, for the resulting HI solution $M(\mathcal{G})$. So, the *connected-component* lower bound for $M(\mathcal{G})$ is exactly $N_{\mathcal{G}}$, and hence $Min_{cc}(\mathcal{G}) \leq N_{\mathcal{G}}$. We earlier proved that $Min_{cc}(\mathcal{G}) \geq N_{\mathcal{G}}$, so $Min_{cc}(\mathcal{G}) = N_{\mathcal{G}}$.   ∎

12.7 Haplotyping to Minimize Recombinations

Our interest in computing $Min_\mathcal{A}(\mathcal{G})$ is in order to estimate the amount of recombination that occurred in the derivation of the historically correct (but unknown) haplotypes, $M^*(\mathcal{G})$, underlying the genotypes $\mathcal{G}$. For that goal, we want to compute a number that is as *large* as possible, while not exceeding $Rmin(M^*(\mathcal{G}))$. To be sure that we do not exceed $Rmin(M^*(\mathcal{G}))$, we use a lower bound method $\mathcal{A}$ to compute the *minimum* value of $\mathcal{A}(M(\mathcal{G}))$, over every HI solution $M(\mathcal{G})$ for $\mathcal{G}$. But computing the minimum value tends to reduce the resulting lower bound on $Rmin(M^*(\mathcal{G}))$. To increase the bound, and yet not exceed $Rmin(M^*(\mathcal{G}))$, we would like $\mathcal{A}$ to be a lower bound method that generally produces the *highest* lower bounds when applied to haplotype data.

12.7.1 Using the MinARG

In the previous section, we showed how to compute $min_\mathcal{A}(\mathcal{G})$ when $\mathcal{A}$ is either the *HK bound* or the *connected-component bound*. But we know from chapters 5 and 7 that the *HK* and *connected-component bounds* are generally weak in comparison to lower bounds produced by *RecMin* and *HapBound*. So, we would like to be able to efficiently compute, at least in practice, $min_\mathcal{A}(\mathcal{G})$ when $\mathcal{A}$ is the *HapBound* method, or *RecMin*, or other related bounds. Some effort has been made in that direction, and integer linear programs have been developed [447] for some of the lower bounds we have discussed, and for related lower bounds. However, the empirical running times of those approaches have been large, making the methods practical for only small-sized problem instances.

Since we are limited to small problem sizes, an alternative approach to increase the computed lower bound on $Rmin(M^*(\mathcal{G}))$, is to *avoid* using any lower bound method $\mathcal{A}$ for binary M, and instead use $Rmin(M)$ itself. The result will be the largest possible lower bound on $Rmin(M^*(\mathcal{G}))$.

Definition Given a set of *genotypes* $\mathcal{G}$, we define $Min_M(\mathcal{G})$ as the *minimum* value of $Rmin(M(\mathcal{G}))$, taken over *every* HI solution, $M(\mathcal{G})$, for $\mathcal{G}$.

OK, but how?

Clearly, $Min_M(\mathcal{G})$ will be larger than or equal to $Min_\mathcal{A}(\mathcal{G})$, for any lower bound method $\mathcal{A}$ defined on haplotypes. But it is also true that $Min_M(\mathcal{G}) \leq Rmin(M^*(\mathcal{G}))$. So, $Min_M(\mathcal{G})$ is a valid lower bound on the historically correct number of recombinations that occurred in the generation of $M*(\mathcal{G})$ (provided, as assumed, that the haplotypes $M^*(\mathcal{G})$ were historically derived on in a process describable by an ARG). $Min_M(\mathcal{G})$ is a great concept, but how can it be computed? Surprisingly, for small but meaningful data, practical computation of $Min_M(\mathcal{G})$ is possible.

The high-level approach

The approach is to build a search tree, using a fairly simple *branch-and-bound* strategy. Each search path will derive a set of haplotypes, denoted H. We will explain the search idea first without any bounding. To start, the algorithm branches on each choice of ancestral sequence, s_r. So at the start of the algorithm, each search path has derived the singleton set $H = \{s_r\}$. At a general step, on each of the search paths, the method will have derived a set of haplotypes H that can explain some subset $\mathcal{G}'$ of $\mathcal{G}$. That search path can then be expanded by adding another haplotype h to H, where h is either derived by a recombination of two haplotypes in H, or derived by a mutation, at a site c, from one of the haplotypes in H. Such a mutation is allowed if all of the haplotypes in H have the same state at site c, guaranteeing that no mutation at site c has yet occurred.

Since there may be (and generally will be) multiple choices for h, a search path is expanded by branching on all feasible choices for h. Note that if h can be generated by a mutation from a haplotype in H, then it cannot be generated by a recombination of haplotypes in H; conversely, if h can be generated by a recombination, it cannot be generated by a mutation. Hence, any expansion of a search path is determined solely by the new haplotype h. The algorithm builds a *search tree*, deriving a different set of haplotypes H, along each search path. A search path terminates when it has derived a set of haplotypes that explain $\mathcal{G}$. Since the search is exhaustive, the following is immediate:

Theorem 12.7.1 $Min_M(\mathcal{G})$ *is the minimum number of recombinations used over all of the terminated paths in the search tree.*

But it's exhausting!

Clearly, the exhaustive branching strategy described here will be effective only for very small data. However, the method can be made much more efficient in practice, using the well-known idea of *pruning* the search tree by computing lower bounds on the cost of expanding a path.

Suppose a search path $\mathcal{P}$ has derived a set of haplotypes H, and H explains a subset $\mathcal{G}' \subset \mathcal{G}$. In the pure branching strategy, the search would continue by branching from $\mathcal{P}$ in all the ways that derive a new haplotype h from H. However, suppose we already have an explanation for $\mathcal{G}$ using $R(\mathcal{G})$ recombinations, so we know that $Min_M(\mathcal{G}) \leq R(\mathcal{G})$. If we compute a lower bound on the number of recombinations needed to explain $H \cup (\mathcal{G} - \mathcal{G}')$, and the bound is no less than $R(\mathcal{G})$, then there is no need to expand from $\mathcal{P}$; we could never get a better solution in that way. Note that a haplotype is also a genotype, so we can use

$Min_{HK}(H \cup (\mathcal{G} - \mathcal{G}'))$ or $Min_{cc}(H \cup (\mathcal{G} - \mathcal{G}'))$ to obtain a lower bound on the number of recombinations needed to explain $H \cup (\mathcal{G} - \mathcal{G}')$.

Another approach (suggested by Yufeng Wu [438]) to getting a good lower bound is to first solve the *pure-parsimony* haplotyping problem for input $H \cup (\mathcal{G} - \mathcal{G}')$. The pure-parsimony problem was discussed earlier in this chapter (page 412). Suppose that the solution uses $h(G)$ haplotypes, which by definition are distinct. Then, similar to the *haplotype bound*, $n(G) - m - 1$ is a valid lower bound on the number of recombination nodes in any ARG that generates haplotypes that solve the HI problem for $H \cup (\mathcal{G} - \mathcal{G}')$.

The basic branch-and-bound approach discussed here was implemented in program *GenoMinARG* and tested on small, but real biological data [447]. Some of the tests are described in the next section. In the actual program, additional heuristics are added to make the bounding more effective, and to decide which feasible branches should be explored first. The goal in the latter case is to compute low values of $R(\mathcal{G})$ quickly, to increase the amount of bounding. We refer the reader to [447].

12.7.2 Empirical Tests and Applications

Three questions were explored empirically in [447]: How well does program *GenoMinARG* perform in locating *recombination hotspots*, given genotype (not haplotype) data? How accurately does it solve the HI problem? And, in estimating the amount of recombination necessary to derive a set of genotypes $\mathcal{G}$, how much accuracy is gained if we know the historically correct haplotypes that generated $\mathcal{G}$?

12.7.2.1 Finding Hotspots Given Genotypes, Not Haplotypes

Recall the discussion of *recombination hotspots* and *haplotype blocks* in section 12.3.1.2.

Interest in recombination hotspots is driven by fundamental questions in molecular biology and genetics, but the location of hotspots and the determination of recombination rates also has practical importance in improving the accuracy of methods such as association mapping. Thus, there has been intense research on hotspot detection and on characterization of recombination rates in genomes. For example, see [15, 16, 19, 111, 204, 205, 206, 207, 208, 229, 252, 300, 344, 365, 382, 423].

Tests using haplotypes
In [18, 19], Bafna and Bansal used a computational approach to locate recombination hotspots, and tested it by comparing its results to laboratory-derived hotspot data obtained from sperm-crossover analysis [207, 208]. The computational approach was based on computing recombination *lower bounds* on *haplotype* sequences, to find regions where

the lower bounds were large compared to other regions. Although lower bounds can be much lower than the true number of recombinations, the lower bounds seemed to track the laboratory-derived number of recombinations fairly well. That is, regions where the lower bounds were high were actually regions of recombination hotspots found in [207, 208]; and regions where the lower bounds were low were generally haplotype blocks. The correspondence was perfect in the first set of data [207], finding all five hotspots with no false-positives; but in the second set of data [208], the method missed one hotspot out of seven. The missed hotspot was the one with the lowest actual recombination rate. The method also compared very well to other computational approaches for finding recombination hotspots.

Tests using genotypes

The results in [19] indicate that recombination lower bounds using *haplotype* data are useful in identifying recombination hotspots. However, real biological data (e.g., the human MHC [207] and MS32 data [208]) consists of genotypes, *not* haplotypes. To find recombination hotspots, the method in [19] first inferred haplotypes from genotypes (using the program *PHASE*), and then computed recombination lower bounds using the derived haplotypes. It is not clear what effect haplotyping errors have on recombination hotspot detection. In [447], a more direct approach was developed and tested on the *genotype* data from [207, 208]: program *GenoMinARG* was used to find regions in the genotype data where any ARG explaining the data would need a large number of recombinations, relative to other regions in the data.

Definition Given a set of genotypes $\mathcal{G}$, and an interval I, let $\mathcal{G}(I)$ denote the sequences in $\mathcal{G}$ restricted to interval I.

In [447], the length of I was limited to six SNPs. Overlapping intervals were used, where every pair of consecutive intervals shared three SNP sites. In each interval I, *GenoMinARG* computed $Min_M(\mathcal{G}(I))$, and these results were used to compute the *average* minimum number of recombinations needed per kilobase. Regions with a large average minimum number of recombinations were predicted to be recombination hotspots, and regions with a small average minimum number were predicted to be recombination deserts. For the MHC data from [207], which involved 277 SNPs and 50 genotypes over a 216 Kb range, the computation took 242 seconds on a 2.0 GHz machine, when intervals of length five SNPs were used. With interval lengths of six SNPs, the computation took about 48 minutes. In a few intervals, the computation of $Min_M(\mathcal{G}(I))$ was too slow, and in those intervals an efficiently computed lower bound based on genotype data was used instead. The exact results are in [447], but essentially the results matched those in [19, 207], identifying the same hotspots with no false positives. The approach was also tested on the MS32 data from

[208], and again the results matched those in [19], but both methods missed the weakest hotspot detected in [207]. These results suggest that the use of genotype data, instead of haplotype data, allows equally good detection of recombination hotspots and haplotype blocks in humans.

12.7.2.2 Haplotyping Using a MinARG

The MinARG constructed by program *GenoMinARG* derives haplotypes that explain a given set of input genotypes $\mathcal{G}$. That is, the MinARG produces an HI solution for $\mathcal{G}$. So, it is natural to study the accuracy of the HI solutions obtained in this way. In addition to the practical utility of assessing program *GenoMinARG*, a study of its accuracy allows a partial evaluation of the commonly stated hypothesis that the historically correct derivation of haplotypes in a population involves a relatively small number of recombinations, that is, nature is fairly parsimonious in the use of recombinations (at least observable recombinations). The accuracy of program *GenoMinARG* was examined in comparison to program *PHASE* [398], which is generally accepted as the most accurate phasing program for relatively short genotype sequences.

There are two common ways to assess the quality of an HI solution, the *switch error* [223, 254] and the *standard error*. The *switch error* is the minimum number of runs of contiguous sites in an HI solution that need to be exchanged between the computed haplotype pairs, in order to make the resulting haplotype pairs agree with the correct haplotype pairs, divided by the number of heterozygous sites in the data. The *standard error* is the percentage of haplotype pairs in an HI solution that do not agree completely with the corresponding correct pair. The switch error is generally considered a more informative way to assess error, although the standard error is easier to understand.

The tests reported in [447] compared the accuracy of the HI solutions from *GenoMinARG* and *PHASE*. Those results indicate that the two programs have similar accuracy, although *GenoMinARG* had somewhat lower switch error and somewhat higher standard error than *PHASE*.

12.7.2.3 Evaluating Recombination Accuracy

To assess how well the results of programs *GenoMinARG* and *PHASE* reflect the amounts of required recombination, the values $Min_M(\mathcal{G})$ and $Rmin(M^*(\mathcal{G}))$ were compared in [447] on simulated and laboratory-derived data.

The simulated data were generated by running program *ms* [182] to create a set of $2n$ haplotypes, under differing recombination rates. For each combination of parameter settings, one hundred genotype datasets were generated, each one formed by randomly pairing haplotypes to create a genotype dataset $\mathcal{G}$. For each dataset $\mathcal{G}$, $Min_M(\mathcal{G})$ was computed by using program *GenoMinARG*, and compared to $Rmin(M^*(\mathcal{G}))$, computed by using the MinARG building program *Beagle* (see section 9.1.5). Program *PHASE* was used to compute an HI solution, H_p, for each $\mathcal{G}$, and $Rmin(H_p)$ was then computed by *Beagle*. Of course, $Min_M(\mathcal{G}) \leq Rmin(M^*(\mathcal{G}))$ and $Min_M(\mathcal{G}) \leq Rmin(H_p)$ in each dataset, but the precise distribution of the values required an empirical investigation.

The main empirical result is that for a large fraction of the data simulated, $Min_M(\mathcal{G}) = Rmin(M^*(\mathcal{G}))$, and that the average difference between these two values was less than 1 for simulations with each parameter setting. Another interesting result is that $Rmin(H_p)$ tended to be between $Min_M(\mathcal{G})$ and $Rmin(M^*(\mathcal{G}))$. The experimental results are consistent with the suggestion that having the correct haplotypes does not greatly improve the ability to estimate the amount of recombination that occurred in the derivation of the haplotypes. Having only the (much cheaper) genotype data seems almost as good.

13   Tree and ARG-Based Association Mapping

Associating yourself around the right people, brings you to the right places.
— kendra rose dion

Association mapping is a widely used, population-based approach to try to efficiently locate genes and mutations influencing genetic traits of interest (diseases or important commercial traits). Already there have been thousands of association studies, and it is expected that the utility of association studies will increase as the cost of genomic sequencing continues to decline, allowing much larger sample sizes. The central roles of recombination and of ARGs in the *logic* of association mapping was discussed in section 4.4, illustrated by the simplest case of locating a mutation causing a *simple-Mendelian* trait.

Association mapping was first developed for applications where there was already a *candidate region* conjectured to contain a gene or mutation that contributes to a genetic trait of interest. A locus with such a contributing gene or mutation is called a *causal locus* or *causal site*. In those applications, association mapping is used to verify or refute the conjecture, or to more precisely locate a causal site. For example, see [101].

Genome-wide association studies (GWAS)

When *no* candidate region is conjectured, a more ambitious form of association mapping, called a *genome-wide association study* (GWAS) begins without one, scanning the *entire* genome for a site or locus that may contribute to the trait of interest. Today, GWAS is the dominant way that association mapping is done, and almost two thousand GWAS studies have been reported [234]. Already, enough GWAS and other association studies have been conducted that meta-analysis is now possible [303, 319]. In meta-analysis many association mapping studies are combined to form the statistical equivalent of a single large study, so that results that were not statistically significant in any of the individual studies might be significant in the meta-analysis.

Most GWAS methods simply do medium-range association mapping in a large series of overlapping regions that cover the genome, using a medium-range method that is fast enough for such repeated use. Of course, multiple regions can be examined in parallel using a machine with multiple processors. Besides issues of speed, and hence practicality, the most significant methodological differences between candidate-region association mapping, and GWAS efforts, concern statistical issues, because the problems of *multiple testing* and *false positives* become much more serious in GWAS [187].

13.1   Successes and Controversies of Association Mapping

It is generally agreed that association mapping has led to significant genetic and genomic discoveries, and that its full potential has not yet been realized. See [9] for an excellent discussion of the logical foundations of gene mapping from 1913 to 2008, with an emphasis on mapping in humans that became feasible in the 1970s. But, despite the discoveries from association mapping, there is continued controversy.

> It is difficult to discuss GWAS without sounding megalomaniacal. [227][1]

The ultimate utility of association mapping, particularly GWAS, is not settled [213], although it has very strong proponents [9, 187, 417, 462]. Writing on the contributions of association mapping only to 2007:

> This first wave of genome-wide association studies is producing an impressive list of unexpected associations between genes or chromosomal regions and a broad range of diseases. There have been few, if any, similar bursts of discovery in the history of medical research. [187]

In [9], the authors simply state:

> GWASs work.
>
> Scores of publications have reported the localization of common SNPs associated with a wide range of common diseases and clinical conditions.
>
> Most associations do not involve previous candidate genes. In some cases, GWAS results immediately suggest new biological hypotheses.

---

1 I don't fully understand the point of this statement, but it is irresistible.

Missing heritability?

One of the controversial aspects of association mapping is that for many genetic traits, the mutations that have been found only explain a *small* percentage of the genetic contribution to the incidence of the trait (i.e. *heritability* of the trait). This is the "missing heritability problem" [213, 266, 462]. For example, *human height* is clearly a heritable trait, influenced by genetics, since there is strong correlation between the heights of parents and their offspring. The heritability of height is estimated to be above 80 percent. That is, more than 80 percent of the observed variance in height can be explained by genetics (even without knowing the particular genes involved). However, the causal sites and mutations that have been found by GWAS, individually explain about 5 percent of the genetic contribution to height. As another example, genetic factors are believed to explain about a quarter of the risk for type 2 diabetes,[2] but so far, the sites and mutations found by association mapping only explain around 10 percent of the genetic contribution to the risk for type 2 diabetes. Hence overall, the known sites explain under 3 percent of the *relative risk* for type 2 diabetes.

The missing heritability problem has dampened some of the initial enthusiasm for GWAS, but some researchers have argued that the problem is not significant [462]. They argue that the heritability of a trait is partly due to interactions between genes—*epistasis*—that *multiplies* the genetic influence of the genes involved. However, epistasis is difficult to measure, so that when the genetic contributions of *individual* sites, found by GWAS, are *added* rather than multiplied, the result is well below the estimated heritability of the trait. Supporters of this view argue that this explains the perceived missing heritability, and that most of the significant sites are found by GWAS. A recent detailed study of *yeast* supports that view [36]. A different argument that the amount of missing heritability has been overstated is found in [452]. They argue that "most of the heritability is not missing but has not previously been detected because the individual effects are too small to pass stringent significance tests." By considering the SNPs as a whole, rather than summing the individual effects that are large enough to pass the significance tests, they estimate that known SNPs explain more than 45 percent of the heritability of height in humans.

Proponents of GWAS further point out that even if each identified mutation only explains a small percentage of the genetic basis of a trait, identifying the gene(s) where the mutations occur can lead to significant insight into the *biology* of the trait. This is a point that is strongly made in [270, 271, 417, 462]. In fact, GWAS generally identifies genes and sites

---

2 This estimate is obtained from studies on identical twins that are outside of the scope of this book.

that were not previously known to influence the traits of interest, and this can lead to a deeper understanding of the biology of the trait. For example:

> The identification of so many loci has uncovered new gene functions in megakaryopoiesis and platelet formation. That is, new biology has resulted directly from the identification of SNPs that are associated with variation in platelet phenotypes. [417]

The common disease, common variant controversy

A major part of the expectation that association mapping would be more informative, was the "common disease, common variant" hypothesis. That hypothesis is that the genetic contribution to common diseases (such as hypertension) would be due to a small number of genes and a small number of *common* alleles, each with a large effect. That hypothesis appears now to be incorrect.

What is now more accepted is that the genetic contribution to common, complex diseases is either due to a large set of very rare mutations, each with a large effect, or is due to many different genes and different mutations, each with a small effect. The first of these two hypotheses is called the "rare allele model," and the second is called the "infinitesimal model." But there is still considerable "divergence of opinion over whether most of the variance is hidden as numerous rare variants of large effect or as common variants of very small effect" [125]. Evidence for the rare allele model is that many rare mutations do seem to contribute to the total genetic effect [197, 213, 309, 359, 407]. The large number of rare alleles is thought to be due to the rapid, recent expansion of the human population (within the last five thousand years), and the slow rate that deleterious alleles are removed by natural selection (i.e., *purifying selection*). The consequence is that many mutations seen in the human population are recent and of low frequency [2, 119]. Despite the different explanations for missing heritability, it is expected that genome-wide association studies will be more successful when based on full DNA sequence data rather than SNPs or exomes (the part of the genome that codes for protein), and when tens, or hundreds of thousands of individuals are sampled [1, 2, 213, 381, 428]. Of course, as in much of this area, there is still some uncertainty about that future [221].

13.2   Back to Methods and ARGs

Now we return to more technical issues, central to the focus of this book. We start by (partially) repeating quotations given in section 4.4:

> All the information about association is contained in the genealogy. [461]

If the true ARG were known, it would provide the optimal amount of information for mapping ... [286]

Few explicit ARGs

Despite the fact that trees and ARGs are at the heart of the *logic* of association mapping, there have only been a few attempts to *explicitly* use them for association mapping. There were some early attempts to use ARGs for association mapping, and to determine parameters of recombination under full coalescent likelihood or Bayesian frameworks [110, 245], but those approaches required the generation of a huge number of ARGs,[3] and have generally been viewed as practical for only small-sized data. Instead, until recently most inference methods that follow likelihood or Bayesian frameworks needed to make simplifying assumptions for computational practicality, and did not explicitly use trees or ARGs. The following statement characterized the situation through 2005:

> Though one might ideally wish to perform inference using the ancestral recombination graph (Nordburg [315]) this turns out to be extremely challenging computationally (e.g., Fearnhead and Donnelly [110]; Larribe et al. [245]). Instead, most of the existing methods make progress by simplifying the full model in various ways to make the problem more computationally tractable. [461]

However, in the last several years, new methods have been developed that do statistical inference in population genetics by explicitly building and sampling ARGs or features of ARGs [330, 331, 348, 377]. The key computational improvement, compared to the earlier methods, comes from the idea of building an ARG for a set of sequences by adding one sequence at a time. At each site c, a new sequence adds an edge that joins the marginal tree for c constructed from the sequences already considered. This approach was first developed in [330, 331] and was more recently built on and implemented in the program ARGweaver [348]. It is reported in [348] that ARGweaver can analyze human data of length two Mb in 36 hours using a single processor.

For some examples of simpler statistical models used for inferring patterns of recombination without explicitly building ARGs, see [300, 322, 455]. Also, some methods avoid formal statistical inference and rely instead on simpler, efficiently computed statistics or patterns that are *derived* from an unknown tree or ARG. These methods also do not explicitly construct ARGs.

---

3 We will not detail the workings of those methods, but simply comment (for readers who are familiar with the MCMC approach) that they collect statistics using the Markov-chain-Monte-Carlo approach, generating a long series of ARGs by successive probabilistically driven modifications.

We do ARGs here!

In this chapter we discuss association mapping methods that explicitly construct ARGs and/or trees, but in ways that are more practical than in the earlier association mapping efforts based on explicit ARG construction. The methods we discuss here are contained in the literature [91, 268, 286, 439, 440, 461]. Related methods that we will not discuss appear in the papers [32, 51, 75, 163, 222, 242, 323, 374, 405, 459].

All of the methods we discuss in detail are efficient, and are reported to be effective, at least for *medium-range* association mapping when a candidate region in the genome has first been identified. Some of the methods we discuss have also been used for GWAS, but some have not. However, even if an ARG or tree-based method is not usable for a full GWAS, it may still be valuable. As noted in [187]

> ... the *sine qua non* for belief in any specific result from a genome-wide association study is ... the consistency and strength of the association across one or more large-scale replication studies.
>
> The unprecedented number of comparisons being made in genome-wide association studies using "SNP chips" has led to the recognition that no initially identified association can be relied on until it has been replicated in one or more studies of adequate size.

Thus, an initial GWAS identifies candidate regions that *must* then be examined more closely using larger samples. These subsequent studies can use association mapping methods that may only be practical for regions considerably smaller than a whole genome. Hence, medium-range association methods still have importance in the GWAS era.

Beyond their practical value, methods that have only been used for medium-range association mapping are of interest here, since our focus is on the underlying *ideas* and *logic* of methods. Our goal is not to evaluate and recommend particular existing programs, but to develop and explain ideas: Ideas can grow and have future lives that current implementations might not have.

13.3   The Basic Logical Foundation

All of the methods that we will discuss share (at some level, and with some variations) the *same* logical foundation, consisting of three parts.

The first part of the foundation

The first part is the fact that an ARG deriving a set of extant SNP sequences M (usually under the infinite-sites model), can be viewed as the result of merging a sequence of trees. More specifically, the ARG defines a genealogical tree at each SNP site, and due to recombination, the genealogical trees change as one scans along the chromosome. This view was developed in section 9.2, where we discussed *tree-scanning* and the merge of a set of trees. The genealogical trees defined by an ARG are precisely the *sequence-labeled marginal trees* of the ARG; theorem 9.2.1 (page 332) shows how the marginal trees relate to the ARG. The reader is encouraged to reread or scan sections 9.2, 9.2.1, 9.2.2, and 9.2.3 before continuing.

In our discussion, it will not matter whether we focus on sequence-labeled marginal trees or just on marginal trees, so we will only use the term "marginal tree" in this discussion. Recall that the marginal tree at a SNP site is well defined, and that a tree T may be the marginal tree at all the sites in an *interval I_T* of consecutive SNP sites. This particularly occurs due to LD (linkage disequilibrium).

The second part of the foundation

Recall that in the context of association mapping, the individuals who possess a trait of interest are called the *cases*, and the individuals who don't possess the trait are called the *controls*.

The second, and most critical, part of the foundation is the following: Suppose that a causative mutation (one of perhaps several mutations that contribute to the trait of interest) occurs between SNP sites $c - 1$ and c, and that $\mathcal{N}^*$ is the historically correct ARG. Then, one or both of the marginal trees of $\mathcal{N}^*$ at sites $c - 1$ and c, should have an edge e, whose removal induces a significant *clustering* or *over-representation* of cases, compared to controls, at leaves beneath edge e. For Mendelian traits, there should be *one* such edge, and for non-Mendelian traits, there may be several such edges.

Conversely, if the marginal tree T at a SNP site c has an edge that defines a significant clustering of the cases, compared to controls, then this is evidence supporting the hypothesis that a causative mutation occurs "near" SNP site c on the chromosome. More specifically, such a clustering on a marginal tree T at site c supports the hypothesis that a causative mutation occurs in the interval of sites, I_T (containing both SNP and non-SNP sites), where T is the marginal tree. Interval I_T contains SNP site c, but its exact left and right endpoints are unknown, because we only know the marginal trees at SNP sites, and recombination breakpoints can physically occur between two consecutive SNP sites. If I_T contains SNP sites starting with c' and ending with c'', then I_T begins somewhere between SNP sites $c' - 1$ and c', and ends somewhere between SNP sites c'' and $c'' + 1$.

In addition, if there is an edge e that roughly separates the cases from the controls in a marginal tree T, this supports the hypothesis that a mutation contributing to the trait occurs during the *time* represented by edge e.

The third part of the foundation

The third part of the foundation is the assumption (implicit or explicit) that the historically correct ARG $\mathcal{N}^*$, for a set of extant sequences M, contains a "relatively small" number of detectable recombination nodes. ARG $\mathcal{N}^*$ might not be a MinARG for M, but the number of recombination nodes in $\mathcal{N}^*$ should not be "much larger" than $Rmin(M)$. Similarly, in the tree-scan of the historically correct ARG $\mathcal{N}^*$, the number of SNP sites where the marginal tree changes (i.e., differs from the marginal tree at the previous SNP site) is "close to" $Rmin(M)$.

Universal foundations?

The first two parts of the foundation are in the logic of all association mapping methods that we know of. It is the third part that is not common to *all* association mapping methods, but is part of all the methods discussed here. The following statement illustrates the first two parts of the foundation:

> If there is a disease-predisposing mutation at a particular chromosomal location, it would have occurred on some internal branch of the marginal tree at that location. So, one way to find disease associations is to scan across the marginal trees, looking for those branches that discriminate well between cases and controls – that is, that have a large number of cases beneath them and significantly fewer controls. Such a clustering of cases underneath a branch suggests that a causative mutation arose on that branch. [286]

What this statement does not explicitly say is that finding such a clustering under an edge (branch) in a marginal tree T, suggests that the mutation occurs in the chromosomal interval I_T, as discussed above. The statement also does not mention the third part of the foundation, but we will see that the method in [286] does implicitly rely on it. See figure 13.1 for an example of the logical foundation.

13.3.1   A First Attempt Use of the Foundations

The logical foundations of association mapping suggest several simple approaches to identify sites where a mutation contributes to the genetic trait of interest. These simple approaches have deficiencies that are generally corrected by adding *heuristic* ideas and *statistical* tests, and by heavily tuning the methods using real or simulated data. The end result sometimes obscures the foundational ideas, so it is useful to describe a "first attempt" method, and then discuss the deficiencies. The following is one natural first attempt:

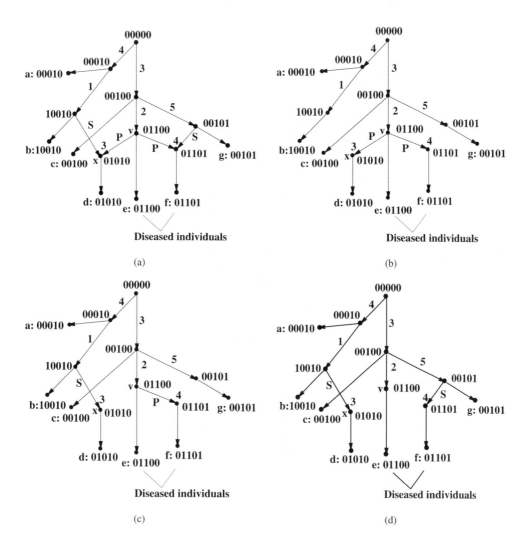

Figure 13.1 Panel (a) shows the ARG from figure 4.11 in section 4.4. Panels (b), (c), and (d) respectively show the marginal trees for SNP sites to the left of site 3, at SNP site 3, and at SNP site 4 and to its right. Assuming that the cases are individuals *e* and *f*, the edge labeled with mutation 2 in the marginal tree in panel (b) (for site 3) gives a perfect separation of the diseased individuals (the cases) from the non-diseased individuals (the controls). If the trait is assumed to be a simple-Mendelian trait, this tree supports the hypothesis that the mutation occurs on the chromosome *between* SNP sites 2 and 4, and occurs during the *time* represented by the edge labeled with mutation 2. This is the same conclusion reached earlier, in section 4.4. In this example, the deduced interval only contains a single SNP site, but that is not true in general.

A first attempt association mapping method

1. Given a set of SNP sequences M, from individuals divided into cases and controls, construct an ARG $\mathcal{N}$ for M that is "plausibly similar" to the historically correct, but unknown, ARG for M. For example, construct a *MinARG* for M; or use an ARG where the number of recombination nodes is "close" to some lower bound on $Rmin(M)$; or use an ARG constructed by programs *SHRUB* (section 9.1.4) or *KwARG*, or *Margarita* (section 9.1.7).

2. Find the set of distinct marginal trees $\mathcal{T}$ of $\mathcal{N}$, at the sites of M.

3. For any edge e in a tree $T \in \mathcal{T}$, let the *support of e* be a measure of how much the removal of edge e from T separates the leaves labeled by cases, from the leaves labeled by controls. Then, for each tree T in $\mathcal{T}$, find an edge e in T with the best support, and set the *support for T* equal to the support for e. Finally, associate the support for T with each SNP site in the interval I_T, the interval where T is the marginal tree.

4. Let T be the marginal tree in $\mathcal{T}$ with the largest support, i.e., containing the edge e with the largest support over all edges in all marginal trees in $\mathcal{T}$. If the support for T is "good enough," report the conclusion that a causative mutation likely occurs in interval I_T. Also report the conclusion that the mutation likely occurred during the time represented by edge e, found in tree T in step (3).

This *first attempt* method agrees fairly well with the high-level description of several methods. For example, the abstract in [286] states:

> There are two stages to our analysis. First, we infer plausible ARGs using a heuristic algorithm ... Second, we test the genealogical trees at each locus for a clustering of the disease cases beneath a branch, suggesting that a causative mutation occurred on that branch.

However, we will see that the full method developed in [286] contains additional ideas that are needed to overcome deficiencies in the *first attempt* method.

13.3.2 Deficiencies of the First Attempt

The first attempt method has three main deficiencies. The first deficiency is that it is not clear how to construct an ARG that is "plausibly similar" to the historically correct one. One suggestion is to use a MinARG of M, but it is not known how much the historically correct ARG agrees with a MinARG. Further, even if a MinARG is a good surrogate for the historically correct ARG, or if the correct ARG is itself a MinARG, computing MinARGs is computationally difficult, and there may be many different MinARGs for M. So, we don't know which ARG should be computed in step (1) of the first attempt method.

The second deficiency is that unless the trait of interest is a simple-Mendelian trait, no marginal tree is expected to have a single edge e where all of the cases, and none of the controls, are below e. How then should we define and recognize an informative "clustering of cases" below an edge? We need a more sophisticated way to select good edges in marginal trees, that is, a more sophisticated way to accomplish step (3). The third deficiency of the first attempt method is in step (4). How should we determine whether the support for a tree or a SNP site is a "good enough"? More generally, the second and third deficiencies point to the need for more sophisticated ways to determine how well a marginal tree *fits* the data.

Handling the deficiencies

We will next discuss several association mapping methods that roughly follow the first attempt method, paying attention to how they address the three stated deficiencies.

13.4   Association Mapping Using Program *Margarita*

In [286], each ARG is constructed by the program *Margarita* (discussed in section 9.1.7) which makes random choices during its execution. Hence, for the same data M, *Margarita* can return a different ARG each time it is run. When used for association mapping, *Margarita* is run one hundred times, generally returning one hundred different ARGs for M. In each of those ARGs, the marginal trees are extracted, giving one hundred marginal trees for every SNP site. At each SNP site c, the one hundred marginal trees are examined, and in each marginal tree T, each edge is scored for how well it separates the cases from the controls (the score is detailed below). Then the highest of those edge scores is associated with T. Finally, the scores associated with the one hundred marginal trees at c are averaged, to give an overall *association score* for site c. The association score is a measure of the support for the hypothesis that a causative mutation occurred near site c.

13.4.1   Assessing Support and Significance

The score given to an edge e in a marginal tree T for site c is the χ^2 *test-of-independence* statistic, using a two-by-two table. One axis of the table partitions the individuals into cases and controls, and the other axis partitions the individuals into those whose associated leaves are below e, and those whose associated leaves are not below e, in T. The χ^2 statistic computed for e tests the null hypothesis that the trait of interest, and the property of being below e in T, are independent. Hence, it also provides a score for the hypothesis that a causative mutation occurred on edge e in T, and a score for the hypothesis that the mutation occurred near site c in the chromosome. The higher the statistic, the greater the support. To combine the scores for site c, the one hundred χ^2 test statistics are averaged, to give the *association score* for site c.

Next, a *permutation test* is used to assess the statistical significance of the association score for c. Multiple times (10,000 times in [286]), the "case" and "control" labels are randomly permuted to create a random partition of the individuals into hypothetical cases and controls, without changing the number of each type. An association score for c is then computed as detailed above, for each permutation of the case-control labels. This gives an *empirical null-distribution* of association scores for site c. A *P-value* for c is defined as the proportion of permutations where the computed association score for c is above the association score obtained using the correct partition of the individuals into cases and controls. A small P-value for a site c, below some threshold of significance, supports the hypothesis that a causal mutation occurred near c.

Ensembles address the first deficiency

From these details, we see that the association mapping method in [286] addresses the first deficiency of the first attempt approach by computing an *ensemble* of plausible ARGs for M, and averaging the analysis over this ensemble. However, what justification is there that the inferred ARGs capture enough of the true history to be of use in association mapping? There is no theoretical assurance that these ARGs sufficiently reflect the historically correct ARG for M, and there is even no formal definition of what a "plausible" ARG is. The strongest argument is empirical: The "justification is that the inferred ARGs work for disease mapping" [286]. The method was extensively tested, and shown to work well on both simulated and real association data. The interested reader is referred to [286] for details.

But why do the plausible ARGs work?

> Well, it may be all right in practice, but it will never work in theory.
> — Warren Buffett

We can perhaps give some "hand-waving" explanation for why the approach in *Margarita* works. The fact that the method "works," and the fact that *Margarita* uses construction rules that implicitly try to *limit* the number of recombination nodes, supports the basic assumption that the historically correct ARG, $\mathcal{N}^*$, for M, also has a small number of recombination nodes, relative to the set of all ARGs for M. It further suggests that certain *topological* features appear frequently in the ARGs set of ARGs for M that have a relatively small number of recombination nodes. Perhaps those topological features are helpful, or necessary, in order to have a small number of recombination nodes. Perhaps the constraint of having a small number of recombination nodes forces, or encourages, those topological features. If so, then since the historically correct ARG $\mathcal{N}^*$ has a small number of recombination nodes, the plausible ARGs computed by *Margarita* tend to share some of the topology of $\mathcal{N}^*$. That sharing is sufficiently strong to allow *Margarita*, and similar ARG methods, to "work." It's hand-waving, but it's the best explanation we know. If correct, a method that uses MinARGs in place of the ARGs found by *Margarita* should also "work." We will shortly describe such a method.

The second and third deficiencies of the first attempt are handled in *Margarita* by the use of χ^2 statistics, and by the use of permutation tests to compute P-values. Additional statistical tests are discussed in [286]. Since our main focus is on the algorithms and the use of ARGs, we refer the interested reader to that paper.

13.5   TMARG: Association Mapping with MinARGs

In the previous section we stated that the reported success of association mapping, using *Margarita*, suggests that an association mapping method which *explicitly* constructs MinARGs (instead of "plausible ARGs"), would be of value. In this section we describe such an approach, developed in [439, 440], and encoded in the program *TMARG*. At the high level, this method follows the first attempt method and is similar to *Margarita*, but with several important differences. First, rather than constructing "plausible" ARGs for M, *true* MinARGs for M are constructed. Further, for some data, the MinARGs constructed are *uniformly sampled* from the space of all *time-ordered* MinARGs for M. This provides a more principled, less heuristic, and more representative way to construct an *ensemble* of ARGs, and makes the overall semantics of the method much clearer. Second, given a marginal tree T, the method assesses the support for the hypothesis that a causative mutation occurs in interval I_T, by using an explicit *likelihood model*, based on a *disease model*, rather than with the use of the χ^2-test. This again makes the semantics clearer, and likely makes the method more effective. We will explain these points below.

13.5.1   Uniform Sampling of Time-Ordered MinARGs

The key to understanding uniform sampling of MinARGs starts with the concepts of *self-derivability* and *unary-self-derivability*, introduced and used in sections 5.2.7.3 and 9.3. We begin by developing some deeper results about self-derivability and unary-self-derivability.

Lemma 13.5.1 *If M has two identical sites c and c', and M is self-derived on an ARG $\mathcal{N}$, then sites c and c' must label the same edge of $\mathcal{N}$.*

Proof: Suppose that identical sites c and c' label different edges e and e'. Since $\mathcal{N}$ is acyclic, one of those two edges is not on any path from the other edge. Without loss of generality, assume that e is not on any directed path from e'. Since M is self-derived on $\mathcal{N}$, every node label, including the ancestral sequence s_r, is in M, and since c and c' are identical, they must have the same state in s_r. It follows that the state of c must be different from the state of c' in the sequence s labeling the node at the head of edge e. But this is impossible, since $s \in M$, and in every sequence in M the state of c is identical to the state of c'. Hence, we conclude that c and c' label the same edge of $\mathcal{N}$. ∎

Note that lemma 13.5.1 does not say that all of the sites on any site-labeled edge must be identical, only that all identical sites label the same edge.

Corollary 13.5.1 *If M is unary-self-derivable, then all the sites in M are distinct.*

Now we can prove theorem 5.2.11 (page 155), restated below, that was first stated in section 5.2.7.4.

Theorem 5.2.11 *If M is unary-self-derived on an ARG $\mathcal{N}$, then $\mathcal{N}$ contains exactly $H(M)$ recombination nodes. Hence, $\mathcal{N}$ is a MinARG for M. Moreover, when M is unary-self-derivable, every MinARG for M must unary-self-derive M.*

Proof: Without loss of generality, assume that no two nodes in $\mathcal{N}$ have the same label, except for a leaf and its parent, if they are connected by an unlabeled edge. Recall that $D_r(M)$ and $D_c(M)$ are the number of distinct rows and columns of M, respectively, and that M has n rows and m columns. By corollary 13.5.1 $D_c(M) = m$, and so there are exactly m edges in $\mathcal{N}$ that are labeled with a single site of M. Therefore, exactly $m + 1$ distinct sequences label non-recombination nodes in $\mathcal{N}$. But then, exactly $D_r(M) - m - 1 = D_r(M) - D_c(M) - 1 = H(M)$ sequences of M label recombination nodes. No two recombination nodes have the same label, so $\mathcal{N}$ has exactly $H(M)$ recombination nodes, and hence $\mathcal{N}$ is a MinARG for M.

To prove the second part of the theorem, note that we established above that when M is unary-self-derivable, $Rmin(M) = H(M)$, i.e., that the *haplotype bound* is tight. Hence any MinARG $\mathcal{N}$ for M will have exactly $H(M)$ recombination nodes, labeled with *at most* $H(M)$ distinct sequences in M. By corollary 13.5.1, all sites in M are distinct, and hence M has exactly $D_c(M)$ sites. So, there are at most $D_c(M)$ site-labeled tree edges in $\mathcal{N}$. If fewer than $H(M)$ sequences from M label recombination nodes, or fewer than $D_c(M)$ sequences from M label tree or leaf edges, then strictly fewer than $H(M) + D_c(M) = (D_r(M) - D_c(M) - 1) + D_c(M) = D_r(M) - 1$ sequences from M label non-root nodes of $\mathcal{N}$. The root can be labeled with a different sequence from M, but still, $\mathcal{N}$ would generate strictly fewer than $D_r(M)$ sequences from M. Therefore, *exactly* $H(M)$ distinct sequences from M must label the recombination nodes, and exactly $D_c(M)$ different distinct sequences from M must label the heads of tree or leaf edges. But there are only $D_c(M)$ sites in M, so any labeled edge must be labeled with *only one* site, and each sequence labeling the head of a tree edge must be in M. Hence $\mathcal{N}$ unary-self-derives M. ■

13.5.2 Uniform Sampling When M Is Unary-Self-Derivable

Recall that in a *time-ordered* ARG (defined on page 315 in section 9.1.5), there is a total order on the nodes in the ARG, or equivalently, a total order on the events (mutation, recombination, and coalescence) that occur in the ARG. Recall also that theorem 9.1.8 (page 317) shows that every time-ordered ARG that derives a set of sequences M, and uses the fewest recombination nodes over all time-ordered ARGs for M, is a MinARG for M. Conversely, every MinARG can be converted to a time-ordered MinARG. So, sampling from the space of time-ordered MinARGs samples from the space of all MinARGs. However, it does not follow that a *uniform* sampling of time-ordered MinARGs is a uniform sampling of all MinARGs.

Counting MinARGs
We will develop a method that uniformly samples from the set of *time-ordered* MinARGs for M. The first step in that development is to give a method to *count* the number of *time-ordered* MinARGs that *unary-self-derive* M. In this counting, two MinARGs both contribute to the count if their topologies differ, or if the total-order of events in the MinARGs differ. However, two MinARGs that have the same topologies and the same total orders of events, but have differing specified crossover indexes at recombination nodes, are only counted once—that is, do not both contribute to the count.[4]

The counting method is a modification of the dynamic programming approach in algorithm *SD* (page 357), used to find the minimum number of recombination nodes in any ARG that self-derives a set of sequences.

---

4 It would be simple to modify the method so that all of the possible crossover indexes would contribute to the count. We leave that to the reader.

Algorithm Count-USD (M)

Set $c(K)$ to one for every singleton subset K of M.

for each subset $K \subseteq M$, where $\mid K \mid > 1$, (in order of the size of K) **do**
  Set $c(K)$ to 0
  for each sequence $s \in K$ **do**

  Set counter $\delta(K, s)$ to zero.
  $\{\delta(K, s)$ will count the number of ways that s can be created from $K - \{s\}.\}$

  for each pair of sequence s', s'' in $K - \{s\}$ **do**
    Test if s can be created by a recombination of s' and s''.
    If so, increment $\delta(K, s)$ by one.
  endfor

  for each sequence $\tilde{s} \in K - \{s\}$ **do**
    Test if s differs from $\tilde{s}$ at precisely one site c,
    such that all of the sequences in $K - \{s\}$ have the same state at site c.
    $\{$i.e., test if s can be derived from $\tilde{s}$ by a single (unused) mutation.$\}$
    If so, set $\delta(K, s) = 1$.
  endfor

  Set $c(K) = c(K) + c(K - \{s\}) \times \delta(K, s)$.

  endfor
endfor
Output $c(M)$.

Figure 13.2   Algorithm *Count-USD* counts the number of time-ordered MinARGs that unary-self-derive M. It returns $c(M)$, which is proved to be $C(M)$.

Definition For a set of sequences K, $C(K)$ denotes the number of time-ordered ARGs that unary-self-derive K.

Given M, algorithm *Count-USD*, shown in figure 13.2, computes the programming variable $c(K)$, for each subset of sequences, $K \subseteq M$. At the termination of the algorithm, $K = M$, and we will show that $c(M) = C(M)$. In fact, $c(K) = C(K)$, for every subset $K \subseteq M$. This is true even when M is not unary-self-derivable, in which case, $c(M) = C(M) = 0$.

Theorem 13.5.1 *Algorithm Count-USD correctly counts the number of time-ordered MinARGs that unary-self-derive M.*

Proof: It is straightforward to convert algorithm *Count-USD* from an algorithm that computes $c(M)$ to an algorithm that *constructs* $c(M)$ distinct, time-ordered ARGs for M. The construction algorithm can be organized as a branching *search-tree*, where each path on the tree records a sequence of choices for s, and $\tilde{s}$ or (s', s''); and constructs the time-ordered ARG dictated by those choices. A leaf v in the search tree represents one ARG constructed so far. If the ARG at v does not generate M, the construction algorithm extends the path from v in (possibly) multiple ways, by adding a new edge from v for every next combination of choices for s, $\tilde{s}$ or (s', s'') that are possible in algorithm *Count-USD*. When a sequence $s \in K$ is chosen, and a successful $\tilde{s}$ or (s', s'') is found, a node labeled s is added to the ARG at v, creating a time-ordered ARG for K. Sequence s will be the last sequence (in the time-ordering) generated by that ARG. The $c(M)$ time-ordered ARGs constructed for M in this way are all different, and each one unary-self-derives M. By theorem 5.2.11, all the constructed ARGs for M are MinARGs for M. Hence, the computed value $c(M)$ is less than or equal to $C(M)$, the algorithm *Count-USD* does not overcount.

For the converse, we show that no time-ordered MinARG that unary-self-derives M is missed. First, suppose that s differs from some $\tilde{s} \in K - \{s\}$ at exactly one site, c, and all sequences in $K - \{s\}$ have the same state at c. Then s cannot be created by a recombination of two sequences in $K - \{s\}$, since none of those sequences has the correct state at c. Further, since $\tilde{s}$ is identical to s at every site except c, there cannot be another sequence $s' \neq \tilde{s}$ in $K - \{s\}$, which differs from s at one site $c' \neq c$, where all sequences in $K - \{s\}$ have the same state. Hence, at most, the sequence in $K - \{s\}$ satisfies the conditions for $\tilde{s}$. Second, suppose that s can be created by a recombination of two sequences (s', s'') in $K - \{s\}$. Then, s is identical to s' at all the sites it contributes to s, and s is identical to s'' at the sites it contributes. So there is no site c where s differs from every sequence in $K - \{s\}$, hence there is no sequence in $K - \{s\}$ that satisfies the conditions for $\tilde{s}$.

What we have shown is that s cannot both be created by a recombination of a pair of sequences (s', s'') in $K - \{s\}$, and also be created by a single mutation from $\tilde{s} \in K - \{s\}$. Further, if s can be created by a single mutation from some $\tilde{s} \in K - s$, then there is only one such $\tilde{s}$. So, algorithm *Count-USD* correctly considers all cases in extending $K - s$ to K. ∎

A crude worst-case time for algorithm *Count-USD* is $O(n^2 m^2 2^n)$.

13.5.2.1   *From Counting to Uniform Sampling*

Algorithm *Count-USD* can be extended from an algorithm that counts MinARGs, to one that uses the count to *uniformly sample* (at random) from the set of MinARGs that unary-self-derive M. The algorithm, *Sample-USD*, is shown in figure 13.3. The sampling algorithm constructs a MinARG from the leaves upward. The key new idea is that at each point where the sampling algorithm can make a choice, it will use the counting information to determine how many MinARGs are possible for each choice, and form probabilities for each choice, based on those counts. It then makes the choice randomly based on those probabilities.

As shown in the proof of correctness of algorithm *Count-USD*, the two conditions in the "if" statements are mutually exclusive. It follows that the algorithm can create *any* ARG that unary-self-derives M. Similarly, any ARG created by the algorithm does unary-self-derive M, and so (by theorem 5.2.11), is a MinARG for M. The reverse order of the sequences in L gives the total order in which sequences are generated by the time-ordered ARG $\mathcal{N}$.

The algorithm can be implemented more efficiently if the information that is associated with each sequence s is precomputed during algorithm *Count-USD*. That is, as algorithm *Count-USD* is run, when a sequence s is found that can be generated from $K' = K - \{s\}$ by a mutation of a sequence $\tilde{s}$, associate that information with s. When s can be generated by a recombination of two sequences (s', s'') in K', associate that information with s. This avoids the need to find this information again in the latter part of algorithm *Sample-USD*.

Theorem 13.5.2 *Algorithm Sample-USD uniformly samples time-ordered MinARGs that unary-self-derive* M.

Proof: To prove the theorem, it is helpful to think of a directed graph G with 2^n nodes, where each node represents one distinct subset of the rows of M. We use v_K to name the node representing subset K. For every pair of nodes $(v_K, v_{K'})$ in G, there is a directed edge from v_K to $v_{K'}$, if and only if, $|K| = |K'| + 1$, and the unique sequence $s = K - K'$ can be generated from K'. The generation of s from K' is either by a single mutation of a sequence $\tilde{s} \in K'$ at a site c, where all of the sequences in K' have the same state at c; or it is by a recombination between some pair of sequences in K'. As shown in the proof of theorem 13.5.1, s cannot be generated from K', both by a mutation and by a recombination. In the case that s is generated by a mutation of $\tilde{s}$, we mark the single directed edge (K, K') with s'. In the case of a recombination, there may be several pairs of sequences in K' that can recombine to generate s. For *each* distinct pair of sequences (s', s'') in K' that can recombine to create s, we create in G a copy of the directed edge $(v_K, v_{K'})$, and mark it

---

Algorithm Sample-USD (M)

Run algorithm *Count-USD* to compute $C(K)$ and $\delta(K, s)$ for each subset $K \subseteq M$, and $s \in K$.

Set K to M, and set stack L to empty.

while (K is not empty) **do**

    for (each sequence $s \in K$) **do**
      compute $p(s) = \frac{\delta(K,s) \times C(K-\{s\})}{C(K)}$.
    endfor

    Randomly select a sequence $s \in K$, where each sequence s is selected with probability $p(s)$.

    if s can be generated from $K' = K - \{s\}$ by a mutation in a sequence $\tilde{s}$ **then**
      associate $\tilde{s}$ with s.
    endif

    if s can be generated by a recombination of sequences in $K' = K - \{s\}$ **then**
      uniformly select such a pair (s', s'') at random, and associate the pair with s.
    endif

    Push the selected sequence s on stack L, along with its associated information.
    Set K to $K' = K - \{s\}$.
endwhile

Popping the stack L, create an ARG $\mathcal{N}$ backward in time, by creating mutations or recombinations in reverse of the way they were added to L.
{The precise details of how $\mathcal{N}$ is built from L are straightforward and are left to the reader.}

Figure 13.3 Algorithm *Sample-USD* uniformly samples a time-ordered MinARG that unary-self-derives M.

with (s', s''). Note that there can be an edge, or edges, from K to a node K', and also an edge or edges from K to another node K''. The edge(s) from K to K' are associated with the sequence $K - K'$, and the edge(s) from K to K'' are associated with the sequence $K'' - K$. Note also that graph G can easily be constructed in parallel with the execution of algorithm *Count-USD*, so G encodes the information obtained by algorithm *Count-USD*.

The key observation is that a directed path P in G from node v_M to node $v_\emptyset$ specifies the construction of a time-ordered ARG $\mathcal{N}$ that unary-self-derives M. As in algorithm *Sample-USD*, the construction is bottom up (i.e., backward in time), with each traversed edge $(v_K, v_{K'})$ in P specifying either a mutation or a recombination event that should be added to the growing ARG in order to generate sequence $s = K - K'$, from K'.

More generally, there is a one-to-one correspondence between the directed paths from node v_M to node $v_\emptyset$ in G, and the time-ordered unary-self-derived ARGs for M. Let $\mathcal{P}$ be the set of all paths in G from v_M to $v_\emptyset$. Then, to randomly (uniformly) select a time-ordered ARG that unary self-derives M, one *could* generate all of the paths in $\mathcal{P}$, and select one at random. Of course, generating $\mathcal{P}$ would be very inefficient, so we next show that a *random* traversal of G has the same result, and that algorithm *Sample-USD* essentially finds such a random traversal.

Consider a directed edge $e = (v_M, v_{K'})$ for a subset K' of $m - 1$ rows of M. The proportion of all paths in $\mathcal{P}$ that contain edge e is $\frac{C(K')}{C(M)}$. So, in order to uniformly sample paths from $\mathcal{P}$, we should use edge e as the first edge on a randomly sampled path, with a frequency of $\frac{C(K')}{C(M)}$. That is what algorithm *Sample-USD* achieves when it randomly picks the first edge on a path from v_M to $V_\emptyset$.

In more detail, if an edge e out of v_M is labeled with a single sequence $\tilde{s}$, meaning that $s = M - K'$ can be derived from $\tilde{s}$ by one mutation, then $\delta(M, s) = 1$, and $p(s) = \frac{\delta(M,s) \times C(M - \{s\})}{C(M)} = \frac{C(M-\{s\})}{C(M)} = \frac{C(K')}{C(M)}$. So, the algorithm chooses edge e with the correct probability in this case. Alternatively, if e is labeled with a pair of sequences (s', s''), then $s = M - K'$ can be generated by a recombination between s' and s''. In that case, $\delta(M, s)$ is the number of different ways that s can be generated by a recombination of two sequences in K'; so edge e in particular should be selected with probability $\frac{p(s)}{\delta(M,s)} = \frac{C(M-\{s\})}{C(M)} = \frac{C(K')}{C(M)}$. In algorithm *Sample-USD*, edge e is (effectively) selected if and only if s is chosen in the random selection of an edge, and then the particular pair (s', s'') is chosen in the random selection of pairs that can recombine to create s. So, algorithm *Sample-USD* selects e with probability $\frac{p(s)}{\delta(M,s)}$, and the algorithm chooses e with the correct probability in this case as well.

Now consider all paths in $\mathcal{P}$ that go through a node v_K, and consider an edge e from v_K to $v_{K'}$. Among all paths in $\mathcal{P}$ that go through v_K, the fraction $\frac{C(K')}{C(K)}$ contain the edge e. As above (when $K = M$), what algorithm *Sample-USD* does when it reaches node v_K results in selecting edge e with a rough frequency of $\frac{C(K')}{C(K)}$. Inductively then, we conclude that the algorithm selects a path uniformly from $\mathcal{P}$, and hence uniformly samples time-ordered MinARGs that unary-self-derive M. ∎

Another way to look at algorithm *Count-USD* and algorithm *Sample-USD* is to consider the latter algorithm as finding a particular *traceback* path in the dynamic programming table established by the former algorithm. Note that although algorithm *Count-USD* takes exponential time, the remainder of algorithm *Sample-USD* runs in polynomial time.

13.5.2.2   Uniform Sampling of MinARGs When *M* Is Not Unary-Self-Derivable

The uniform sampling method presented works only for the case that M is unary-self-derivable, which may not always be the case. So, we need to show how the uniform sampling method can be extended to the more general case. The simplest extension is when only one Steiner sequence is needed, in addition to the sequences in M. In that case, the Steiner sequence s must either differ from some sequence $s' \in M$ by one mutation, or be generated from a pair of sequences (s', s'') in M by a recombination. Hence there are only $O(n^2)$ candidate sequences for s. We evaluate a candidate sequence s by determining if $M \cup \{s\}$ is unary-self-derivable, and put s in a set K, if and only if, $M \cup \{s\}$ is unary-self-derivable. Then we can uniformly sample from the MinARGs that generate M, by first uniformly picking a sequence $s \in K$, and then uniformly sampling from the MinARGs that unary-self-derive $M \cup \{s\}$.

If two Steiner sequences must be added to M, then either each is obtained from M by a mutation or a recombination, or one Steiner sequence is obtained from M, and then the second Steiner sequence is obtained from the first Steiner sequence by a mutation; or the second Steiner sequence is obtained by recombination of the first Steiner sequence, and a sequence in M. Again, there are only a polynomial number of candidate pairs that have to be evaluated, and again we can uniformly sample from the MinARGs that generate M using two Steiner sequences. This approach can be extended to any number of required Steiner sequences, but the computational time grows very rapidly with the number of Steiner sequences required. Hence, in [440], additional heuristic ideas are developed to make this approach more practical, and other ideas are developed to uniformly sample a set of ARGs which might not be MinARGs, but generally have a small number of recombinations, compared to a MinARG. We will not develop these ideas here, but refer the interested reader to [440].

13.5.3   Assessing the Support for a Marginal Tree

Having discussed the way that TMARG generates MinARGs, and how it differs from the way that *Margarita* generates "plausible" ARGs, we move to the second major difference between TMARG and *Margarita*, that is, in the way that TMARG assesses the *support* for any particular marginal tree.

We continue to assume that we are only looking for a single mutation that influences the trait of interest.[5] The basic task is to evaluate how well a marginal tree T "fits" the observed case/control data, that is, to determine the support for the hypothesis that a causative mutation occurs in interval I_T, or more simply, that a causative mutation occurs near a site c in I_T. The general approach in TMARG builds on the probabilistic framework developed by Zöllner and Pritchard in [461].

In the extreme case of a simple-Mendelian trait, this reduces to determining if there is a single edge in T whose removal perfectly separates the cases from the controls. But generally, we do not expect to find such a perfectly separating single edge. In those cases, the task of defining and determining the support for a tree is more involved, and usually depends on an explicit *disease* model. To explain this, we first introduce the most important part of a disease model, which is the *penetrance model* of a mutation (allele) contributing to the trait of interest.

13.5.3.1 *Penetrance and Disease Models*

Definition If it is known, or hypothesized, that a trait of interest can be caused, or influenced, by some (perhaps unknown) mutation (allele), then the *penetrance* of the mutation is the probability that an individual with the mutation also has the trait. Formally, we use D to denote the event of having a particular trait (**D**isease), and use μ to denote the event of having the **mu**-tation (μ-tation). Then the *penetrance of μ* is the probability of D given μ, written $p[D \mid \mu]$.

The case of $p[D \mid \mu] = 1$ is called *complete penetrance*, and the case of $p[D \mid \mu] < 1$ is called *incomplete penetrance*. Note that the only information about the mutation μ in the definition of penetrance is that it exists—that the trait is influenced by some (usually unknown) mutation.

Definition When a particular site c is specified, we use μ_c to denote the event that mutation μ influencing the trait occurs at site c. Then the *penetrance of mutation μ at site c* is $p[D \mid \mu_c]$. This is just a formal definition; in most cases, the value of $p[D \mid \mu_c]$ is unknown.

---

5 It is simple to extend this to mutations at multiple sites, if each individual mutation in the set has the same effect, or if the multiple mutations are close enough that they are in the same marginal tree.

Penetrance model

To fully detail a *penetrance model* for the trait, we also need to specify the probability of having the trait *without* having a mutation, or without having a mutation at a specified site c. If no site is specified, we write this as $p[D \mid \bar{\mu}]$, and if a site c is specified, we write this as $p[D \mid \bar{\mu_c}]$.

Definition A *penetrance model* for a trait is specified by the probabilities $p[D \mid \mu_c]$ and $p[D \mid \bar{\mu_c}]$, for each site c.

As mentioned, we often do not know the probabilities for specific sites. In that case, if we know $p[D \mid m]$ and $p[D \mid \bar{m}]$ then we set $p[D \mid m_c]$ to $p[D \mid m]$ and $p[D \mid \bar{m_c}]$ to $p[D \mid \bar{m}]$ at each site c where the mutation might occur.

A full disease model

The penetrance model is the key component of a disease model, but a full disease model may include additional detail, for example, specifying how many and which mutations are allowed or preferred; or specifying more precisely how traits are influenced by mutations; or specifying the probability that the trait can be detected, if it is present. We will discuss some of those extensions later, but the key concepts underlying support can be understood with a disease model that assumes only *one causal* mutation; that the trait can be *perfectly detected* when it exists; and assumes the *basic penetrance model* defined above. We make all of those assumptions in the following discussion.

13.5.4   The Disease Model Affects the Support for an Edge and a Marginal Tree

When a trait is simple-Mendelian, the support for an edge e in a marginal tree T is simple to determine: the removal of edge e has to *perfectly* separate the cases from the controls. But in general, support is not so clear and is strongly affected by the disease model.

Consider the ARG in figure 13.1 (page 441), which we first examined in section 4.4 (page 121). Assume again that the cases are individuals e and f. Then, the best-supported marginal tree (the tree that best fits the data) depends on the disease model for the trait. A simple-Mendelian trait corresponds to the disease model:

$$\{p[D \mid \mu_c] = 1; p[D \mid \bar{\mu_c}] = 0\}$$

at the causal site c, and

$$\{p[D \mid \mu_c] = 0; p[D \mid \bar{\mu_c}] = 0\}$$

at each noncausal site c. In that case, the marginal tree in panel (c) is the best-supported tree, and the best supported edge is the one labeled with site 2. The conclusion is that the mutation is definitely between sites 2 and 4. This is the same result obtained earlier in section 4.4. But when the disease model is not simple-Mendelian, for example, when

$$\{p[D \mid \mu] = 2/3; p[D \mid \overline{\mu}] = 0\},$$

the best-supported tree is in panel (b), and the *most likely* interval containing the causal mutation is to the left of site 3, although the associated edge is again the edge labeled 2. When the disease model is

$$\{p[D \mid \mu] = 1/2; p[D \mid \overline{\mu}] = 0\}$$

then the best-supported tree is in panel (d), so the mutation is most likely to the right of site 3, and the associated edge is the one labeled 3.

For another example, suppose that the cases are now $\{b, d, f\}$, as shown in figure 13.4. Assume that the disease model is

$$\{p[D \mid \mu] = 1; p[D \mid \overline{\mu}] = 1/4\}.$$

Then the two trees in panels (c) and (d) are equally well supported, and are better supported than is the tree in panel (b). In that case, we can conclude only that the most likely location of the mutation is to the right of site 2 and is on the edge labeled 1.

Note that when the trait is not simple-Mendelian, i.e., when either

$$p[D \mid \mu] < 1$$

or

$$p[D \mid \overline{\mu}] > 0,$$

we cannot know for certain which is the best supported tree or be certain of the edge where the mutation occurs, or the interval in the chromosome where the mutation occurs. We can only make probabilistic statements, usually in a likelihood or Bayesian framework.

Formalizing support

We now formalize the concept of *support* for an edge in a tree, and support for a tree.

Definition Given a marginal tree T with leaves partitioned into cases and controls, and given a disease model, $\mathcal{D}$, the *support for an edge e in T* is the probability of obtaining the observed cases and controls, assuming the mutation occurs on edge e, and that the disease model is $\mathcal{D}$.

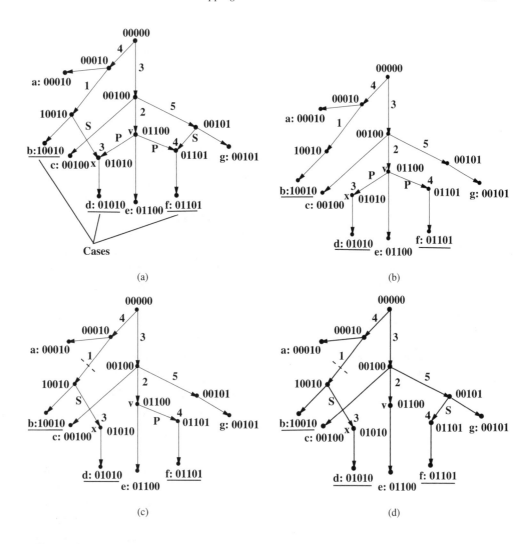

(a)

(b)

(c)

(d)

Figure 13.4 Assume now that the cases, shown underlined, are $\{b, d, f\}$, and that the disease model is $\{p[D \mid m] = 1; p[D \mid \overline{m}] = 1/4\}$. The edge labeled 1 in the trees in panels (c) and (d) fits the disease model well, while no edge in the tree in panel (b) does. The conclusion is that the most likely location of the mutation is to the right of site 2, and is on the edge labeled 1.

Definition Given a marginal tree T with leaves labeled cases and controls, and given a disease model, $\mathcal{D}$, the *max-support for T* is the *maximum* support of any of the edges in T. The *total-support for T* is the *sum* of the support of each of the edges in T.

What to do with unknown penetrance

With these definitions, computing the support of an edge in a tree T is a simple matter, as is computing the max or total-support of T, assuming we know specified values for $p[D \mid \mu]$ and $p[D \mid \overline{\mu}]$. The problem, however, is that those values are often *not* known. In that case, what is typical in likelihood and Bayesian approaches is to consider, and average over, a set of possible penetrance values in a reasonable range of values.[6] For example, the set of possible values for $p[D \mid \mu]$ could be given by some number of evenly spaced points in the interval $[0, 1]$, as could the set of possible values for $p[D \mid \overline{\mu}]$. In [440] and [461], the set of values for $p[D \mid \mu]$, $p[D \mid \overline{\mu}]$ are given by all the points in a 20 by 20 grid of equally spaced points with range $[0, 1]$. For any given marginal tree T, the support (max or total) of T, is then computed at each of these four hundred grid points. The *final*, reported support for T could be the *maximum*, or *median* or *mean* support obtained from those four hundred support values.

To compute the significance of the final support for a tree T, permutation tests can be performed, as described in section 13.4. However, a fully-deterministic, polynomial-time alternative to permutation tests is described and tested in [440]. Since the focus of this book is combinatorial, rather than statistical, we direct the interested reader to that paper.

Why should this work?

It is reasonable to think that the use of a *set* of (mostly incorrect) penetrance values would lead to *poor* computed support values, even when the correct marginal tree is examined. So it is reasonable to ask why this approach should locate the correct marginal tree. The standard answer is that the computed support for the correct marginal tree is much less affected by incorrect penetrance values than is an incorrect marginal tree. Moreover, the max-support for the correct tree, probably obtained when the correct penetrance values are used as one of the four-hundred choices, will be much larger than the max-support of an incorrect tree, even when the correct penetrance values are used. Is the standard answer convincing? It's hand-waving, but it may be the best we have.

---

6 Technically, this is called *integrating* over those values.

13.5.4.1 Extensions to the Disease Model

There are many extensions that have been considered in order to improve the biological fidelity of the models and methods. One extension is to include a term for the probability of a mutation on each edge, and to have different probabilities for different edges. For example, edges closer to the leaves might have a higher probability of a mutation than edges further from the leaves. That would reflect the belief that certain traits of interest originated relatively recently. Another example, when edges have lengths (as they do in the full coalescent model), is to make the probability of a mutation on an edge a function of the edge length; the longer the edge (modeling a longer passage of time), the higher the probability of a mutation. A second extension is to include the possibility of multiple mutations on any marginal tree. This models *allelic heterogeneity*. In this case, the change to the definition of support is immediate, but the computation of support becomes more intense, since $2^k - 1$ combinations of mutations must be considered on a tree with k edges. Further, we have to specify how multiple mutations influence the trait. The simplest assumption is that every mutation has the same effect, and multiple mutations on edges leading to a leaf do not increase the probability of having the trait. But other richer models are possible. Another extension is to include the probability of detecting the trait in an individual, given that the individual has the trait; a related extension is include a term for the population frequency of the trait, and a term for the population frequency of the mutation.

13.6 Association Methods Using Only Trees

The general *first attempt*, and the specific methods leading to *Margarita* and *TMARG*, all build a series of ARGs and extract the marginal trees from those ARGs at all, or a chosen subset, of the SNP sites. Then each marginal tree is examined to find any significant clustering of cases below some edges(s) in the tree. The methods differ in the way that they build and sample ARGs, and in the way that they assess the fit of the case/control data to a tree, and the significance of that fit.

Alternative, but related, approaches have also been studied where the step of building an ARG is bypassed. Those methods go *directly* to a set of trees, which might not be marginal trees of any ARG, and then assess the fit of the data to the trees. In this section, we discuss two such *direct-to-the-tree* association mapping methods.

13.6.1   Blossoc: BLOck aSSOCiation

This method, and its implementation in program *Blossoc*, developed by Mailmund et al. [268], are based on looking for intervals of sites in M that contain no incompatibilities, assuming the infinite-sites mutation model. That is, they look for intervals in which there is a *perfect phylogeny*. As in previously discussed methods, they then look for significant clustering of cases in one or more subtrees of the perfect phylogeny:

> We treat the prefect phylogeny as a decision tree and measure how well it explains the case/control classification: If the tree explains the classification well, there is an association between the tree topology and the disease, if the tree does not explain the classification, the tree and the disease status are considered independent. [268]

> If we can split the sequences into regions where no recombination has occurred, we can infer the trees of these regions and test for significant clustering of affected individuals. The regions containing the significant clusterings are our candidates for containing the disease affecting locus. [268]

In the second quotation, the "affected individuals" are the cases, and the "trees of these regions" are the perfect phylogenies for the sites in those regions. The name of program, *Blossoc*, highlights the centrality of regions where no recombination has occurred: "*Blossoc*" comes from "*BLOck*" and "*aSSOCiation*," where a *block* refers to a *haplotype block* (introduced in section 12.3.1.2), that is, an interval which has experienced little or no recombination.

Of course, it is not possible to know for sure whether a region has experienced recombinations or not, so instead, maximal intervals of sites (regions) are found that have no incompatibilities. A perfect phylogeny can be built for each of those intervals. Two such maximal intervals might overlap, so there is still a question of which intervals to use. A sensible approach might be to use all of them, simply looking for the perfect phylogeny that best fits the case/control data. But that is not done in *Blossoc*. Instead, starting from each SNP site c, a largest set of pairwise-compatible sites are found that contain and enclose c. Then a perfect phylogeny is constructed for that set of sites.

> If the markers used to build the tree are independent of the affected/unaffected status, we would expect that the affected individuals are distributed identically in all the clusters of the tree; if on the other hand, the tree is similar to the true genealogy of the disease locus, the affected individuals are expected to be overrepresented in one or more of the subtrees. The level of significant clusterings gives us a score for the tree, and this score is assigned to the locus for which the tree was built.

Handling some incompatibilities

In cases when the length of a maximal interval of pairwise compatible sites around a site c is small, *Blossoc* does allow some incompatible sites to be included. *Blossoc* adds sites incrementally in order of their distance from c. Suppose, in such a process, that all the sites added so far are pairwise compatible, and so fit a directed perfect phylogeny T. If a site c' is then encountered which is incompatible with one or more of the sites in T, there can be no single edge in T that separates the leaves in T with character c', from the leaves without c'. Still, *Blossoc* may add c' to the interval. *Blossoc* sets a minimum length interval requirement, and when the length of the longest perfect phylogeny interval enclosing c is less than the minimum, incompatible sites can be added until the minimum length is reached.

When an incompatible site is added, *Blossoc* finds the smallest set of edges in the existing T whose removal creates subtrees where no subtree has both leaves with c' and leaves without c'. It then labels those edges with site c'. The resulting tree for the interval will not be a perfect phylogeny, but it can still be used to cluster cases and controls, and be scored for how well it fits the data.

Scoring and assessing the fit

The specific details of how clusters in the tree are defined, scored, and assessed in *Blossoc* differ from the specific details in *Margarita*, *TMARG*, and *LATAG* (to be discussed next). However, the approach in *Blossoc* is similar *in spirit* to the approaches already discussed in detail. Moreover, once trees are obtained, one could conceivably apply any of the suggested clustering, scoring, and assessment methods. This is not to suggest that those details do not matter in the effectiveness and efficiency of different association mapping methods—they are very important. But the focus in this book is on the conceptual, algorithmic, biological, and bio-logical issues in obtaining the trees. So, we direct the interested reader to [268] for the details of clustering, scoring, and assessment in *Blossoc*.

13.6.2   Related Methods Based on Perfect Phylogenies

We mention here two additional association methods based on perfect phylogeny. The association mapping method in [75] exploits the existence of haplotype blocks, by assuming that the true derivation of the haplotypes in the block fits a *perfect phylogeny*: "We will assume the blocks are consistent with a 'perfect phylogeny,' where there is no recombination within the block and SNPs are in high linkage disequilibrium (LD) (highly correlated or a high degree of non-random association) with each other." Constraining the solution space to that of perfect phylogenies allows for faster computation, while not reducing the accuracy of the associations in the simulated and real data examined in [75]. The paper also notes that when only genotype data is given, the PPH problem can be solved to find regions where underlying haplotypes could have been derived on a perfect phylogeny, allowing their method to be applied to the haplotypes in those regions.

Program *CAMP* (coalescent-based association mapping) [222] is another program that is "strongly based on the theory of *perfect phylogeny* of SNPs and haplotypes." As in *Blossoc*, in regions without recombination, a perfect phylogeny is built from the observed SNPs; then, unobserved mutations that are consistent with the perfect phylogeny are inferred and tested for association with the phenotype of interest. Some recombination can be handled by a modification of the method that represents an "approximation of the perfect-phylogeny model."

13.6.3   LATAG: Local Approximation of the ARG

Zöllner and Pritchard [461] developed a coalescent-based association and fine-mapping method that does not construct full ARGs but does explicitly construct *trees* that they call *local approximations to the ARG*. Their method and program (called *LATAG*) actually predate both of the papers [286, 440] and the programs *Margarita* and *TMARG*, already discussed. In fact, the likelihood framework and disease model used in [440] is largely derived (but not identical to) the framework and model formulated in [461]. The critical difference, in the context of this book, is that explicit ARGs are generated in [440] but not in [461].

Zöllner and Pritchard stress the importance of making inferences based on a proper probabilistic population-genetic model. However, they also state:

> Our approach, based on a local approximation to the ARG, strikes a compromise between modeling the population genetic processes that produce the data and the need for a model that is computationally tractable for large datasets.

The method

The approach in [461] is to focus on one site at a time, and construct and assess a series of n-leaf trees that could explain the data at the site. Because no ARGs are built, the trees are not necessarily marginal trees of ARGs for the full input M, but we can think of each tree as a conjectured marginal tree for some unknown ARG for M. The trees are actually constructed by a Markov-chain-Monte-Carlo (MCMC) computation, which builds a series of trees, creating each tree from its predecessor by varying the topology and other parameters of the tree.

Since this book is focused on combinatorial structure of ARGs and trees, we will not attempt to describe in detail either the general MCMC approach, or the specific details of its use in *LATAG*. But, to give a sense of the approach, and why it is so time-consuming, we note that trees are constructed based on a very detailed coalescent model with parameters for tree topology; mutation rate θ; recombination rate ρ; a vector of recombination events that change sequence information at nodes; times between coalescent events; edge lengths; and a (partial) sequence label at each node. Recombinations and recombination rates are part of the model, even though no ARGs are built, because the partial sequences, deduced for internal nodes of a tree, are influenced by hypothesized recombination events. The MCMC computation uses the model and its variables to create the series of trees, as explained in [461]:

> We start with an initial value for each of these variables, chosen either at random from the prior or using some heuristic guess. Then at each step of the algorithm, we propose a change of one or more parameters. Each step include the "local update of internal nodes" for all nodes and one or more of the topology rearrangements. The updates for θ and ρ are performed less often. Each proposal is accepted according to the *Metropolis-Hastings* ratio.

The MCMC computation creates a huge number of trees, from which a small sample is taken to measure the fit of the trees to the observed data at a site c. Those values are combined to determine the support for the hypothesis that a causal mutation occurs near c. The general assessment approach is similar in *spirit* to the two methods already discussed, but of course differs in many details, and we point the interested reader to [461].

13.7 Whole-Genome Association Mapping with Unphased Data

The discussion so far has assumed that haplotype data is available for use by association mapping methods. But as discussed in chapter 12, directly-observed haplotype data is rare, and most SNP data originates in the form of *unphased, genotype* data. Some association mapping methods have the ability to make inferences using unphased data, but most methods assume that the *Haplotype Inference (HI)* problem has first been correctly solved for genotype data; then the predicted haplotypes are used for association mapping.

This *two-stage* approach may work well in practice, but it has deficiencies. The first is the time needed to solve the HI problem. It is reported in [91] that for genome-wide association mapping, *FastPhase* (which is a widely used haplotyping program) required three days of computation to produce its solution. The second, and potentially most serious problem, is that information is lost in replacing collected genotypes with deduced haplotypes: Errors in haplotyping can cause errors in mapping. Association mapping methods that make inferences based directly on the original genotype data do not have the loss-of-information problem, and they gain computational efficiency by avoiding the HI problem. Further, the two-stage approach (sometimes in a cryptic way) can still be implemented in those methods, in which case they should perform at least as well as an explicit two-stage approach. So, one might expect that genotype-based methods would be more accurate than haplotype-based methods (and faster when one adds in the time to find an HI solution), although they may be conceptually more difficult. Some research has investigated the effectiveness of genotype-based methods compared to haplotype-based methods, in a variety of inference problems, both when the true haplotypes are known, and when they have only been inferred [70, 226, 291, 447].

Unphased genotypes

In this section we discuss an association mapping method that is explicitly designed to make inferences using *unphased*, genotype data [91]. The method combines ideas from the *Perfect-Phylogeny Haplotyping* method (section 12.3) and the association mapping method *Blossoc* [268] discussed in section 13.6.1. However, it is not a two-stage method: It does not first solve the PPH problem and then feed haplotypes to *Blossoc*. Unfortunately, the resulting program that handles genotype data was also named *Blossoc*, since it is an extension of the original *Blossoc* program developed in [268].

The new version of *Blossoc* [91] takes in unphased genotype data and successively focuses on each SNP site c, finding the longest *balanced* interval around c where the genotype data allows a PPH solution. A "balanced interval" around c is one that roughly has the same number of SNP sites on either side of c. The interval is found by successively solving instances of the PPH problem on increasingly long, balanced intervals around c. If the longest balanced interval around c that allows a PPH solution is "long enough" (a parameter in the method), then the PPH tree T is used to assess the support for the hypothesis that a causative mutation occurs near SNP site c, that is, in the balanced interval around c, where T is found. Since the PPH solution might not be unique, a sampling of the PPH trees can be obtained, and an average value of support can be computed.

When the balanced interval around c that allows a PPH solution is not long enough, the new *Blossoc* method uses a haplotyping program in that interval, and then applies the logic in the original *Blossoc* method to the obtained haplotypes.

The specific way that the new version of *Blossoc* assesses support for the hypothesis that a causative mutation occurs near c, is different from the way it was done in the original version of *Blossoc*. However, it is again based on looking for non-random clusterings of cases in the tree, and so it is in the same spirit as the other assessment methods already mentioned. The details of those assessment methods are not the focus of this book, and so we direct the interested reader to [91].

13.8 Computational Efficiency and Mapping Accuracy

We have not said much about the accuracies of the mapping methods discussed, or discussed in detail their computational efficiency. Most papers compare the accuracy and efficiency of their method to competing prior methods, and report (not surprisingly) that their method looks attractive on some particular data, or ranges of problem parameters. But the accuracy results do not generally seem comprehensive or definitive, and we will simply conclude here that the tests show the discussed methods are effective for the kind of mapping they were designed for. There is, however, much greater variation in the computational *times* needed for the methods, and this impacts which method can be used for GWAS. Here again, we can only make broad-brush statements. Generally, methods based on detailed statistical models, which require statistical-based computational solutions, such as the Markov-chain-Monte-Carlo method, will be the slowest, and might not be practical for GWAS applications. Conversely, methods that replace detailed statistical models with simpler combinatorial structures and objective functions, such as finding the best case/control separations on a marginal tree, will be the fastest, and have been used for GWAS applications. A final issue is the time needed to solve the HI problem, if the mapping method requires haplotype input, but only genotype data is available. For GWAS applications, the time for haplotyping can be considerable. With some haplotyping methods, phasing requires days on genome-scale data. So, methods that directly use genotype data will be much faster, although questions about the relative accuracy of the two approaches (phasing vs. not-phasing) are still open.

13.9 A Related Problem: SNP Discovery Using ARGs

So far, we have assumed that SNP data is always available, and have not discussed how that data is obtained. This issue is largely outside of the scope of the book, and we do not intend a general discussion of SNP discovery here. However, we will discuss a method developed in [246], that uses program *Margarita* to build ARGs, and exploits shared haplotypes to help identify SNP sites. This method is conceptually related to association mapping, and in particular, to the method discussed in section 13.4, association mapping with program *Margarita*.

13.9.1   The SNP versus Error Problem

Ideally, when a region of the genome is sequenced in a sample of individuals, and two distinct nucleotides (above some frequency threshold) are seen at a particular site in the sample, we would like to conclude that the site is a true SNP site. That is, that there is real variation in the population at that site. But how can we be sure that the differences seen at that site are not sequencing errors? Older, more expensive sequencing machines have fairly low error rates, and can sequence DNA strings of length 500 to 1500 bases in a single *read*. Such long reads allow more accurate *sequence assembly*, the last major step in *whole-genome shotgun sequencing*. In sequence assembly, short overlapping strings of sequenced DNA (the reads) must be assembled to form long deduced DNA sequences (full chromosomes), each with hundreds of millions of base pairs.

In contrast, current, faster, and cheaper sequencing machines have significantly higher error rates, and have much shorter reads, between 30 and 150 bases, which leads to additional errors during sequence assembly. Compounding the problem is the desire to sequence many individuals in order to detect less common variants, and to make association mapping more effective. But with budget constraints, sequencing a large number of individuals is only practical with *low coverage* sequencing (defined below), which again makes it harder to avoid or correct errors during sequence assembly. The paper [246] addresses the problem of distinguishing *sequencing errors* from true *SNPs*, when the sample size is large, and the sequencing coverage is very low. *Sequence coverage* is roughly the expected number of times an individual base is contained in one of the reads, that is, sequenced strings. As coverage falls, errors in the sequence assembly increase. Very high quality shotgun sequencing often has coverage levels between 40 and 100. In contrast, the low-coverage pilot of the *Thousand Genomes Project* [1], sequenced around 180 individuals with sequence coverage under *four*. So the quality of those 180 sequences is much lower than in high-quality sequencing projects.

13.9.2   The Idealized Case and Logic

As introduced in section 12.5, the experimental setup considered in [246] is that a set of SNP haplotypes, M, are initially obtained at a *sparse* set of SNP sites, for a sample of individuals (as in the *HapMap 3* Project). Later, *low-coverage sequencing* is done, to obtain full-genome sequences of the same individuals (as in the *1000 Genomes Project*). The sequencing identifies an additional set, G, of *candidate* SNP sites not in M: polymorphic sites in the sequences, with variation above some minimum threshold, are considered candidate SNP sites. The problem is to decide whether a candidate SNP site in G is a true SNP, or just a site of a sequencing error.

To explain the method, assume first, as in section 12.5.1, that we have the historically correct ARG $\mathcal{N}^*$ for the known haplotypes, M. Let c be a site in M, and T_c be the marginal tree in $\mathcal{N}^*$ at site c, and let $\tilde{c}$ be the first site in $\mathcal{G}$ to the right of c. Let $\widetilde{\mathcal{N}}^*$ be the historically correct (but unknown) ARG for the sites in $M \cup \mathcal{G}$, and let $T_{\tilde{c}}$ be the (unknown) marginal tree of $\widetilde{\mathcal{N}}^*$ at $\tilde{c}$.

The key insight

The key insight behind the *linkage disequilibrium analysis* (LDA) method in [246], is:

> If $\tilde{c}$ is a true SNP site, so that the variation at $\tilde{c}$ is due to one historical mutation (under the infinite-sites model) at site $\tilde{c}$, then there should be a single edge in $T_{\tilde{c}}$ whose removal separates the leaves with state 1 at $\tilde{c}$, from the leaves with state 0 at $\tilde{c}$.

> Conversely, such a separation in $T_{\tilde{c}}$ is very unlikely to happen by chance. So, if the variation at $\tilde{c}$ is due to *random* sequencing error, then the states at $\tilde{c}$ will be widely and randomly distributed among the haplotypes at the leaves of $T_{\tilde{c}}$.

This insight is strongly related to the second part of the *foundation for association mapping*, discussed in section 13.3:

> If a causative mutation occurs between two SNP sites in M, and if $\mathcal{N}^*$ is the historically correct ARG for M, then one or both of the marginal trees of $\mathcal{N}^*$ at those sites should have an edge e which has significant clustering of cases, compared to controls, at leaves beneath edge e.

In fact, if we consider one particular state at $\tilde{c}$ to be an *observable trait* (i.e., a phenotype), the haplotypes with that state at $\tilde{c}$ can be considered *cases*, and the other haplotypes can be considered *controls*. Then, the key insight behind the LDA method is simply that there is one *causal mutation* (i.e., a true SNP) at $\tilde{c}$, if and only if, there is a single edge in $T_{\tilde{c}}$ that roughly separates the cases from the controls.

So, the idealized LDA method to determine whether site $\tilde{c}$ is a true SNP, is similar to methods that use ARGs to determine if a causal site influencing a genetic trait is in a specified genomic region. The idealized LDA method determines how well the (assumed known) marginal tree $T_{\tilde{c}}$ separates the cases from the controls. If it strongly separates them, then $\tilde{c}$ is declared a true SNP-site, and otherwise, $\tilde{c}$ is not declared a true SNP-site.

13.9.3 Back to Reality, Again

There are two main problems with the idealized version of the LDA method. First, we don't know the marginal tree $T_{\tilde{c}}$. But as discussed in section 12.5, $T_{\tilde{c}}$ should be similar to the marginal tree T_c, so we may be able to use T_c as a substitute for $T_{\tilde{c}}$. Moreover, the level of separation of the cases from the controls in T_c can partly indicate the similarity of the two marginal trees. If there is a strong separation of the cases from the controls in T_c, we can be fairly confident that (a) T_c and $T_{\tilde{c}}$ are highly similar and (b) $\tilde{c}$ is a true SNP site. If there isn't a strong separation, then one or both of these conclusions is unlikely to be correct. In that case, the LDA program does not report $\tilde{c}$ as a true SNP site. Hence, the program may overlook true SNP sites, but is unlikely to mistake sequencing errors for true SNPs. That is, it is more likely to make a *false-negative* mistake than a *false-positive* mistake. The dissimilarity between T_c and $T_{\tilde{c}}$ is due to the number of recombination events between c and $\tilde{c}$, and where they occur in $\widetilde{\mathcal{N}}^*$. So a suggestive approach to improve the LDA method is to generate all of the trees that are one or two rooted-SPR operations away from T_c, and evaluate how well any of these separates the cases from the controls. That extension was not implemented in the LDA program.

The second problem with the idealized version of the LDA method is that we don't know the true ARG $\mathcal{N}^*$ for M. That problem is again handled by a standard probabilistic (maximum-likelihood or Bayesian) framework. The LDA program uses *Margarita* to build twenty ARGs for M. Then, for each ARG, and for each candidate SNP site $\tilde{c}$ in $\mathcal{G}$, it uses the marginal trees from the ARG, at the two flanking sites of $\tilde{c}$ (essentially in the way described in the idealized version of the LDA method), to assess the likelihood that $\tilde{c}$ is a true SNP site. How the results from the twenty ARGs are integrated into a single measure of support is outside the scope of this discussion, but the details follow well-developed probabilistic methodologies, related to the discussions in sections 13.4.1 and 13.5.3. The reader is directed to [246] for details.

14 Extensions and Connections

In this chapter, we discuss several topics that extend and connect to material in earlier chapters. In contrast to those earlier chapters, the material presented here is more introductory, and the citations are meant to be representative, orienting the reader to the field.

We start by returning to *perfect phylogeny*, introducing the perfect-phylogeny problem when *more than two states* are allowed for a character. Then we discuss a recombination model, the *mosaic model*, that is appropriate for shorter time frames than was assumed so far. Next, at the other time extreme, we discuss phylogenetic network problems that arise in studying the evolution of *species* rather than populations. Despite the biological ways that these models differ from the population genetics model used for the MinARG problem, there are mathematical commonalities that can be exploited. So, one of the main points in that material, is to show how solutions of species-motivated problems relate to solutions of population-motivated problems. As one example, we explore in detail a structural theorem that was developed in the context of phylogenetics, but that can be exploited to compute the minimum number of rSPR operations needed to convert one marginal tree into another, which we introduced when we discussed *tree-scanning* for building a MinARG. Finally, we briefly discuss several other types of phylogenetic networks that have been actively studied.

14.1 Extensions of Perfect Phylogeny to *Nonbinary* Data

So far in this book, we have assumed that the data is binary, and to date, this is the nearly universal assumption in studies of phylogenetic networks. The binary assumption is biologically valid in many contexts, for example in modeling SNP data in populations or modeling the presence or absence of complex traits in species. However, despite the current and anticipated centrality of binary data, other important *multistate* (nonbinary) variation data are now being systematically collected in populations, and the frequencies of these variations are much greater than was assumed in the past [176, 210, 249, 352, 375, 400, 437]. Some of these variations have functional consequences and may be under selective pressure [317, 334, 376]. Also, the multistate model is appropriate for problems outside of biology; for example, see [306].

Nonbinary data consists of sequences, one from each sampled individual in a population, where the value at a single site is not restricted to 0 or 1, but can be a larger *integer* (the allele is *multiallelic* rather than *biallelic*). The meaning of an integer at a site varies by the type of variation it represents. In some cases it is an actual count (of something), and has an ordinal meaning. But in most cases, it only specifies a partition of the taxa, and hence it does not matter which particular integers are used. The need to analyze multiallelic data has led to a generalization of the (binary character) perfect-phylogeny model, called the *multistate perfect-phylogeny* model.

In this section, we introduce the multistate perfect-phylogeny problem and discuss the *three-state* perfect-phylogeny problem in detail, since it is closely related to the binary case.

14.1.1   Introduction to k-State Perfect Phylogeny

In the **k-state perfect-phylogeny problem**, the input is an n by m matrix M whose values are *integers* from 1 to k. Each row of M again represents a single taxon; each column of M represents a single character; and each value in cell (f, c) is the state of character c that taxon f possesses.

A *rooted k-state perfect phylogeny* for M is a generalization of a perfect phylogeny (for binary data). In the binary case, each character changes (mutates) exactly once from its ancestral state to its derived state. The natural generalization to k states is to allow a character to mutate $k - 1$ times, but to insist that for any character c, and any state i of character c, there is at most one mutation in the tree that changes the state of character c *to* state i.

Although true evolutionary history is always directed (indicating the flow of time), the historically correct ancestral sequence and the true root may not be known. Consequently, the literature on multistate perfect phylogeny has usually addressed the *unrooted* version, and we will do that here. We now state more formal definitions.

Definition For any character c, and any state i of c, the set of taxa in M that possess state i for character c is denoted $O_c(i)$; the set of taxa that do not possess state i is denoted by $\overline{O_c(i)}$.

Sets $O_c(i)$ and $\overline{O_c(i)}$ are the natural generalizations of sets O_c and $\overline{O_c}$ (defined on page 46) for binary characters. Note that $(O_c(i), \overline{O_c(i)})$ defines a *split* of the taxa even though character c may have more than two states.

Definition For a character c in M, let k_c be the number of distinct states that character c possesses, as specified in M. If M has m columns (characters), let Vm be the set of all m-length vectors, where entry c in a vector is an integer from $[1, ..., k_c]$.

Definition Given M as above, a *taxa-labeled* tree T for M is an undirected tree with n leaves, where each leaf is labeled by a distinct taxon in M, and each internal node of T is labeled by a vector from Vm (which need not be a vector in M), specifying a state of each of the m characters. We define $T_c(i)$ as the subgraph of T, induced by the nodes in T that are labeled with state i for character c.

Definition A taxa-labeled tree T for M is a *multistate perfect phylogeny* for M, if and only if, for every character c, and every state i of c, the subgraph $T_c(i)$ is a *connected subtree* of T. An example is shown in figure 14.1.

This definition of a *multistate* perfect phylogeny is the natural generalization of the second definition given for a (rooted, binary) perfect phylogeny in chapter 2 (page 39). The requirement in the definition of a perfect phylogeny that each subgraph $T_c(i)$ be a subtree is called the *convexity* requirement, introduced earlier in the case of binary characters. Note that for any character c and states $i \neq j$, the convexity requirement implies that subtrees $T_c(i)$ and $T_c(j)$ of a perfect phylogeny T must be *node disjoint*.

Another view of convexity

For another way to view convexity in T, arbitrarily designate a node in T as the root and direct all the edges in T away from it, and consider this directed tree as giving a history of character mutations. The convexity requirement is then equivalent to saying that for any character/state pair (c, i), there is at most one edge in T where the state of character c mutates to i, and there is no mutation into the ancestral state of any character. See figure 14.2.

Definition If k is the maximum number of states k_c for any character c in M, then a perfect phylogeny for M is called a *k-state perfect phylogeny* for M.

Definition The *k-state perfect-phylogeny problem* is to determine, for input M, whether there is a k-state perfect phylogeny for M, and to construct one if there is one.

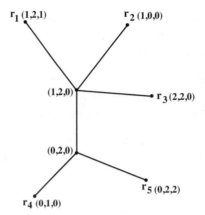

Figure 14.1   A three-state perfect phylogeny with $n = 5, m = 3$, adapted from the example in [114]. The input M is given in table 14.1. The subtree $T_3(0)$ contains the leaves labeled r_2, r_3, r_4, and the two interior nodes.

	c_1	c_2	c_3
r_1	1	2	1
r_2	1	0	0
r_3	2	2	0
r_4	0	1	0
r_5	0	2	2

Table 14.1   Input matrix M for the three-state perfect-phylogeny problem shown in figure 14.1.

14.1.2   Does k-State Reduce to Two-State?

A natural first conjecture is that the k-state perfect-phylogeny problem can be efficiently *reduced* to the binary perfect-phylogeny problem. One approach is to write the k states in binary notation and then solve the resulting binary perfect-phylogeny problem. For example, for three states, state 0 becomes 00, state 1 becomes 01, and state 2 becomes 10. This reduction does not work, as shown in table 14.2 and figure 14.3. However, it is true that *if* the resulting binary problem has a perfect phylogeny, then the original k-state problem has a k-state perfect phylogeny. We leave this to the reader to verify.

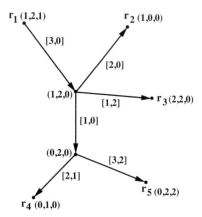

Figure 14.2   The three-state perfect phylogeny from figure 14.1, after rooting the tree at node r_1, and directing all edges away from the root. The pair of numbers labeling an edge represent the mutation that occurs on that edge. The first number indicates a character, and the second number indicates the state of that character at the head of the edge. For any character c and any state i, there is at most one edge where state c mutates *into* state i, and there is no mutation into a character-states present at the root node. This view of multistate perfect phylogeny is a natural generalization of the binary case, where for each character, the ancestral state mutates to the derived state exactly once.

It is possible to reduce any instance M of a k-state perfect-phylogeny problem to an instance M' of a *two-state* perfect-phylogeny problem, but only if we allow some *missing data* in M': M has a k-state perfect phylogeny if and only if binary values can be *imputed* for the missing data in M', with the result that the full data has a binary perfect phylogeny [399]. However, the binary imputation problem is NP-complete. An integer-programming approach to that problem has been developed and tested, and works well in practice on the tested instances [148]. Moreover, if even one row of the binary input matrix M has no missing data, then the binary imputation problem can be solved by an elegant polynomial-time algorithm [332]. This has been exploited to obtain a practical, but worst-case exponential-time algorithm to impute the values of missing data in binary perfect-phylogeny problems [368].

14.1.3   The Complexity of the k-State Problem

If none of the parameters k, n, or m is fixed (so k can grow with n), then the k-state perfect-phylogeny problem is NP-complete [38, 395]. In contrast, if k is any *fixed* integer, independent of n, then the problem can be solved in time that is polynomial in n and m. In

	c_1	c_2	$c_{1,1}$	$c_{1,2}$	$c_{2,1}$	$c_{2,2}$
r_1	0	1	0	0	0	1
r_2	1	0	0	1	0	0
r_3	2	2	1	0	1	0
r_4	2	1	1	0	0	1

Table 14.2 Example showing that the suggested reduction of a three-state problem instance to a binary problem instance does not work. The columns c_1 and c_2 describe a three-state perfect-phylogeny problem that has the three-state perfect phylogeny shown in figure 14.3. The states of character c_1 are written in binary notation in columns $c_{1,1}$ and $c_{1,2}$. Similarly, the states of character c_2 are written in binary in columns $c_{2,1}$ and $c_{2,2}$. So, columns $c_{1,1}, c_{1,2}, c_{2,1}, c_{2,2}$ define the two-state problem derived from the three state problem defined in columns c_1 and c_2. The binary problem does not have a perfect phylogeny because characters $c_{1,1}$ and $c_{2,2}$ are incompatible, that is, have all four gametes.

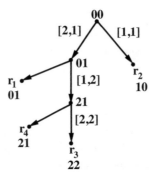

Figure 14.3 A three-state perfect phylogeny for the data from table 14.2; this is a counterexample to the correctness of the natural reduction of three states to two states.

fact, for $k = 2$, we showed in theorem 2.1.3 (page 45), that the problem can be solved in *linear* time. A polynomial-time solution for $k = 3$ was first shown in [97]; a polynomial-time solution for $k = 3$ or 4 was shown in [215]; and a polynomial bound for any *fixed* k was shown in [3]. The latter result was improved in [216] to a time bound of $O(2^{2k}nm^2)$. An excellent survey of most of these results appears in [114].

	c_1	c_2	c_3
r_1	2	1	0
r_2	1	2	1
r_3	2	1	2
r_4	0	0	2
r_5	0	1	2

Table 14.3 Input matrix M for a three-state perfect-phylogeny problem.

	$C_1(0)$	$C_1(1)$	$C_1(2)$	$C_2(0)$	$C_2(1)$	$C_2(2)$	$C_3(0)$	$C_3(1)$	$C_3(2)$
r_1	0	0	1	0	1	0	1	0	0
r_2	0	1	0	0	0	1	0	1	0
r_3	0	0	1	0	1	0	0	0	1
r_4	1	0	0	1	0	0	0	0	1
r_5	1	0	0	0	1	0	0	0	1

Table 14.4 Matrix $\vec{M}$ resulting from expanding the matrix M shown in table 14.3.

14.1.4 Three-State Perfect-Phylogeny Problem

The polynomial-time algorithm for $k = 3$ developed in the paper by Dress and Steel [97] is relatively simple in comparison to the other methods, and is related to, but does not reduce to, the solution to the binary case. Hence, we will discuss the case of $k = 3$ in detail.

14.1.4.1 The Dress-Steel Solution

First, create another matrix $\vec{M}$ derived from the input matrix M, with three characters $C_c(0), C_c(1), C_c(2)$ for each character c in M. All the taxa that have state i for character c in M are given state 1 for character $C_c(i)$ in $\vec{M}$, and the other taxa are given state 0 for $C_c(i)$. So, the original input matrix M is recoded as a *binary* matrix $\vec{M}$, with three *expanded* binary characters for each three-state character in M. Clearly, $\vec{M}$ can be constructed from M in polynomial time. An example of three-state data is given in table 14.3. Table 14.4 shows the expanded binary matrix $\vec{M}$. Note that each expanded character defines a split of the taxa.

Key structural result

The key structural result in [97], interpreted in terms of $\vec{M}$ is:

Theorem 14.1.1 *Given matrix M with $k = 3$, there is a three-state perfect phylogeny for M if and only if there is a set of (binary) characters $\mathcal{K}$ in $\vec{M}$, which are pairwise compatible, where for each character c in M, $\mathcal{K}$ contains at least two of the characters $C_c(1), C_c(2), C_c(3)$.*

Proof: Suppose there is a three-state perfect phylogeny T for M. For any character c of M, the subtrees $T_c(0), T_c(1)$ and $T_c(2)$ are node disjoint and contain all the nodes of T. For any character c, contract to a single node, all of the nodes of the subtree $T_c(i)$ of T. The resulting graph must be a path P_c with three nodes; we label each node v in P_c with the distinct state (0, 1, or 2) possessed by the nodes that contract to v. For example, in the perfect phylogeny T shown in figure 14.1, if we contract each of the subtrees $T_1(0), T_1(1), T_1(2)$ to a single node, we get a path P_3 with end nodes labeled 0 and 2, and interior node labeled 1.

   In general, we use i and j to denote the state-labels of the two nodes at the leaves of P_c. Since P_c is a path with two edges, there is an edge e in P_c whose removal would separate the leaf labeled i from the interior node, and the leaf labeled j. That is, both the interior node and the leaf labeled j are on the same side of edge e. Edge e is an uncontracted edge from T, and so edge e separates all the taxa with state i for character c, from all the taxa with the other two states, and hence defines the split $(O_c(i), \overline{O_c(i)})$. Similarly, there is also an edge in T that defines the split $(O_c(j), \overline{O_c(j)})$. Then, for character c, select characters $C_c(i)$ and $C_c(j)$ to be in $\mathcal{K}$. Repeating this for each character c in M, we select a set $\mathcal{K}$ of characters of $\vec{M}$ that contains exactly two expanded characters for each character c in M. Further, since each selected split is defined by an edge in T, and every pair of splits defined by edges in T are compatible (by theorem 2.4.2), the characters in $\mathcal{K}$ are pairwise compatible, and the necessary direction of theorem 14.1.1 is proved.

   Conversely, suppose there is a set of characters $\mathcal{K}$ in $\vec{M}$ satisfying the conditions of theorem 14.1.1. Let Z denote the set of taxa in M. By construction, each character in $\vec{M}$ defines a split of the taxa Z, and so $\mathcal{K}$ defines a set of pairwise compatible splits of the taxa. For a taxon f in M, the "trivial split" for f is the bipartition $\{f, Z - f\}$, which is clearly compatible with any other split. We augment the splits defined by $\mathcal{K}$ with these n trivial splits, and call the resulting set of splits $\mathcal{K}'$. By the *splits-equivalent theorem*, there is some tree T' with n leaves, each labeled with a distinct taxon in Z, and containing edges that define the splits in $\mathcal{K}'$. We can assume that each edge in T' actually defines one of the splits in $\mathcal{K}'$, by contracting any edge that does not define such a split. Also, we can assume that no internal node of T' has degree two, since otherwise two neighboring edges define the same split, in which case one edge can be contracted. We now show how to map the taxa to leaves of T, and how to label the interior nodes in T', so that T becomes a perfect phylogeny for M.

Because of the trivial splits in $\mathcal{K}'$, each taxon in Z labels a leaf of T', satisfying one requirement for a perfect phylogeny for M. We next need to show how to label the interior nodes of T', so that for every character c and every state i for c, $T'_c(i)$ is a connected subtree of T'. For a character c in M, suppose, without loss of generality, that characters $C_c(0)$ and $C_c(1)$ are in $\mathcal{K}'$, and let $e(0)$ and $e(1)$ be the edges in T' that define the splits $(O_c(0), \overline{O_c(0)})$, and $(O_c(1), \overline{O_c(1)})$. Removal of $e(0)$ from T' creates two connected subtrees, one which contains all and only the taxa in $O_c(0)$ labeling its leaves. Set the state of character c to 0, at each node in that subtree. This defines subtree $T'_c(0)$. Define T'' as the tree T' after the removal of all nodes and edges in $T'_c(0)$. Clearly, T'' contains all the leaves labeled by taxa in $O_c(1)$. T'' also contains edge $e(1)$; otherwise $e(1)$ would be an edge in $T'_c(0)$, and since all interior nodes have degree three or more, there would be a leaf labeled 0 on both sides of $e(1)$, contradicting the assumption that $e(1)$ defines the split $(O_c(1), \overline{O_c(1)})$. So, removal of $e(1)$ from T'' defines two connected subtrees of T', one that contains all and only the taxa in $O_c(1)$. We label the nodes in that subtree with state 1 for character c, defining $T'_c(1)$. Removing $T'_c(1)$ from T'' leaves a connected subtree of T' that must contain all and only the leaves labeled by taxa in $O_c(2)$. Label the nodes in that subtree with state 2 for character c, creating $T'_c(2)$. These three subtrees are node disjoint and show that character c obeys the convexity requirement. Since the argument holds for any c, we conclude that T' (with interior nodes labeled as above) is a three-state perfect phylogeny for M.   ∎

Note that second part of the proof (the sufficient direction) implicitly describes a polynomial-time algorithm to construct a three-state perfect phylogeny given a set of characters $\mathcal{K}$ in $\vec{M}$ satisfying the conditions of theorem 14.1.1. Thus, the remaining issue is how to select such a set $\mathcal{K}$ in polynomial time.

Polynomial-time selection of $\mathcal{K}$

In order to find a perfect phylogeny for M, theorem 14.1.1 requires that we select a set $\mathcal{K}$ of at least two characters from $C_c(0), C_c(1), C_c(2)$, for each character c in M, such that the characters in $\mathcal{K}$ are pairwise compatible in $\vec{M}$. This may seem at first to be a computationally difficult task since there are four possible choices for each character c, leading to a time of $\Omega(4^m)$ if all choices are explicitly considered. How can we make the selections efficiently?

Dress and Steel developed a polynomial-time algorithm for the selection problem [97], which is also described in [152]. However, a more direct approach, developed in [152], is to observe that this selection problem can be formulated as the *satisfiability* problem, where clauses only contain *two* literals. This is the classic *2-SAT* problem, which is well known [86, 123] to have a polynomial-time solution. In fact, the specialized algorithm given in [97] for the selection problem, follows the same logic as the more general *2-SAT* algorithm.

Reduction to 2-SAT

Given $\vec{M}$ we want to create, in polynomial time, a 2-SAT formula $\mathcal{F}$ that is satisfiable if and only if we can select a set of characters $\mathcal{K}$ in $\vec{M}$ that obeys the conditions described in theorem 14.1.1. To model the condition that $\mathcal{K}$ cannot contain two incompatible characters in $\vec{M}$, suppose $c(i)$ and $c'(i')$ are incompatible. The clause

$$(\neg c(i) \vee \neg c'(i')),$$

where the symbol "$\neg$" indicates *Boolean negation*, imposes the condition that $\mathcal{K}$ cannot contain both characters. Formula $\mathcal{F}$ should therefore contain one such clause for each pair of incompatible characters in $\vec{M}$. To model the condition that $\mathcal{K}$ must contain at least two characters from $C_c(0), C_c(1), C_c(2)$, create the following three clauses:

$$(C_c(0) \vee C_c(1))$$

$$(C_c(0) \vee C_c(2))$$

$$(C_c(1) \vee C_c(2))$$

Formula $\mathcal{F}$ should therefore contain such a set of three clauses for each character c in M. Clearly, $\mathcal{F}$ can be constructed in polynomial time from $\vec{M}$, and every clause in $\mathcal{F}$ has only two literals. We leave it to the reader to fully prove that $\mathcal{F}$ is satisfiable if and only if a proper set of characters $\mathcal{K}$ can be selected.

The polynomial-time construction of $\vec{M}$ from M, combined with theorem 14.1.1, combined with the reduction to 2-SAT, and the polynomial-time algorithm for 2-SAT, establishes:

Theorem 14.1.2 *The three-state perfect-phylogeny problem can be solved in polynomial time.*

14.1.5   Generalizations of the Four-Gametes and Splits-Equivalence Theorems

The four-gametes theorem (page 51) and, equivalently, the splits-equivalence theorem (page 55) contain two separable mathematical facts about the existence of a perfect phylogeny for a binary matrix M. One fact is that there is a perfect phylogeny for M if and only if there is a (root-unknown) perfect phylogeny for each *pair* of sites in M. A second fact is that there is a perfect phylogeny for a pair of sites in M if and only if the rows of M do not contain all four binary combinations (gametes) $\{0,0;\ 0,1,\ 1,0;\ 1,1\}$ at that pair of sites. One, or the other, or both, of these facts might be generalizable to multistate perfect-phylogeny problems. For the case of *three* states, a natural generalization has been found [238, 239]. The generalization of the first fact is:

Theorem 14.1.3 *Let M be a matrix with up to three states per site. There is a three-state perfect phylogeny for M if and only if there is a three-state perfect phylogeny for each subset of three sites in M.*

Much earlier, Walter Fitch [115] established that this is the "tightest" possible generalization for three states:

Theorem 14.1.4 *There is a matrix M, where every pair of sites in M has a three-state perfect phylogeny, but M does not have a three-state perfect phylogeny.*

The second fact has also been generalized to the case of three states [238, 239]:

Theorem 14.1.5 *A subset $\mathcal{K}$ of three sites in M has a three-state perfect phylogeny, if and only if $M(\mathcal{K})$ does not contain one of four specific patterns of data.*

One of those patterns is used in the construction given in [115]. We will not detail those four patterns here; the reader is referred to [238, 239, 378] for details. Theorem 14.1.5 was later simplified in [378].

For $k > 3$ the following theorem was stated in 1983:

Theorem 14.1.6 *For any fixed k, there is a matrix M where every subset of $k - 1$ sites has a k-state perfect phylogeny, but M does not have a k-state perfect phylogeny.*

This was stated in [285], and examples were given for $k = 3, 4, 5$. The result was more fully formalized and a full proof was given in [238, 239].

The connection to chordal graph theory
The extension of the four-gamete theorem to the case of three states was established by exploiting a deep connection between the multistate perfect-phylogeny problem and the

representation of chordal graphs. That connection was established by Peter Buneman [55], and has been further exploited in the study of multistate perfect phylogeny, for example in [143, 153, 154, 378].

What is the full generalization?

Given the generalizations of the four-gametes theorem to the case of three states, and theorem 14.1.6, it is natural to speculate on what the proper generalization is for the case of any $k > 3$ states. Surprisingly, unlike the case when k is two or three, for $k > 3$, it is *not* true that there is a k-state perfect phylogeny if and only if every subset of k sites has a k-state perfect phylogeny. This was first shown in [155], where it was established that there is a four-state problem instance where every subset of four sites has a four-state perfect phylogeny, but the entire instance does not have one.

More generally, it was shown later in [379], that for every $k \geq 2$, there is a k-state problem instance M with $(\lfloor k/2 \rfloor \times \lceil k/2 \rceil) + 1$ sites, such that M *does not* have a k-state perfect phylogeny, but every proper subset of sites of M does have a k-state perfect phylogeny. This bound agrees exactly with the known results for k from 2 to 4. This lower bound shows that in general, in order to determine if there is a k-state perfect phylogeny by looking at all subsets of sites of some size q; q must grow at least *quadratically* as a function of k. This bound has not been shown to be tight, except for k equal to 2 or 3, so the question of the proper generalization of the four-gametes theorem for $k > 3$ is still open. However, there is a claim that for four states, subsets of size five suffice [237].

14.2   The Mosaic Model of Recombination

In this section we introduce another (somewhat idealized) recombination model, called the *mosaic model*, that is appropriate for studying very recent recombination between individuals of different strains or species, that coinhabit some location (perhaps a corner of a hospital, a mosquito, your throat, etc.). The basic assumptions of the model are that extant, observed sequences are descendants of a *small number of founders* who coinhabited a location. Each extant sequence is either one of the founders, or consists of concatenated segments of the founder sequences, created by recombinations. Further, we assume that there has been no mutation (or no mutation that survived to be observed) since the formation of the community of coinhabitants. This is justified by the assumption that the community is relatively recent.

The classic example is of different strains of bacteria, or different species of bacteria, that co-locate, and exchange DNA. The original bacteria that started the community are the *founders*, and the derived types of bacteria, created by recombination between founders, are the *mosaic strains*. We can observe the bacterial strains that exist today, which include the mosaic strains, and might include some of the founders, but we often do not know which are the founders, even if they are among the observed strains. For a related founder reconstruction problem and its application to thirty-one phage genomes, see [404].

Deducing founders

Another fascinating example that has been given considerable attention (for example, see [419]) concerns the history of laboratory mice. Laboratory mice were bred and crossed to establish the current common laboratory strains, but the history of the crosses was not sufficiently recorded, or the identity of the founding strains was not sufficiently established, and the DNA sequence of the founders was certainly not known at the start of the breeding. So the genetic makeup of current laboratory mice is unclear. The general problem now is to *reconstruct* a likely set of founder sequences from which all of the observed strains can be derived, by recombination, thus explaining the mosaic structure of the current laboratory mice.

14.2.1 The Formal Problems

There are two related problems: the *recombination-deduction* problem, and the *founder-deduction* problem.

In the recombination-deduction problem, we are given the founder sequences F and the observed sequences M, and we want to explain how the founder sequences might have recombined to create the observed sequences. The assumption is that the recombinations were so recent that mutation can be ignored. The objective function is to *minimize* the number of recombination events to create M from F. This will be stated more precisely below.

In the founder-deduction problem, we are given the observed sequences M and a number k_F, which is the *target* number of founder sequences, but we do not know the founder sequences. The goal is to deduce k_F founder sequences, F, to *minimize* the solution to the recombination-deduction problem for M and F.

The mosaic model and the founder-deduction problem were introduced by Esko Ukkonen in [411], along with several other optimization problems and solutions defined on the mosaic model. A dynamic programming solution for the founder-deduction problem was developed there, with worst-case exponential time. More practical methods, capable of handling larger problem instances, were later developed in [350] and [448].

14.2.1.1   Recombination Deduction

Recall from section 9.1.3.3 (page 304) that given a set of binary sequences F, each of length m, and an m-length binary sequence s not in F, $Rmin(F; s)$ denotes the *minimum* number of *single-crossover* recombinations needed to create s from the sequences in F, without any mutations.

The *recombination-deduction problem* (RDP) is formally stated as:

> Given a set of founder sequences F and a set of observed sequences M, determine how to create each sequence $s \in M$ by single-crossover recombinations of sequences in F, to minimize
>
> $$\sum_{s \in M} Rmin(F; s)$$
>
> which is denoted $Rmin(F; M)$.

Note that a solution to the recombination-deduction problem partitions each sequence $s \in M$ into the minimum number of intervals $\mathcal{L}_s$, so that in each interval $I \in \mathcal{L}_s$, sequence s is identical to some sequence in F. A related problem is to partition the sequence in F into the minimum number of intervals $\mathcal{L}_F$, so that for each interval $I \in \mathcal{L}_F$, each sequence $s \in M$ is identical to some sequence of F. We will not address this second problem.

No history

In the recombination-deduction problem, we do not attempt to construct a full *history* of the recombinations, where one derived sequence may lead to another as in a genealogical network. Such a history of recombinations may use fewer than $Rmin(F; M)$ recombinations, if the same recombination is used to derive two sequences in M. However, the context of this problem is that the mosaic sequences were likely derived separately, and even when $Rmin(F, M)$ is larger than the historically correct number of recombinations, the objective function is sufficient to deduce much of the *pattern* of historical crossovers.

The recombination-deduction problem has arisen frequently in the biology and computational biology literature, independent of the founder-deduction problem. Typically, some variation of dynamic programming has been developed for its solution.[1] An example of the recombination-deduction problem in the biological literature is discussed in [451], where the DNA of fourteen strains of MF1 mice are expressed as mosaic sequences derived from

---

1 Sometimes when this problem is discussed in the context of probabilistic models, such as HMMs, the exposition may not explicitly state that dynamic programming is used.

four founders. That mosaic expression is shown in the beautiful, color-coded figure 2 in [451]. What stands out in that figure is that even without representing an explicit *history* of the recombinations, the patterns and similarities in the expressions (that is, how each of the fourteen strains is expressed as a mosaic of the founders) validate the mosaic model, and suggest that the result does reflect a historical process.

The purpose of expressing the MF1 sequences as mosaics is to amplify the effectiveness of association mapping, correlating observed traits with shared haplotype segments, instead of single markers. This approach led to the identification of a locus that modulates anxiety in mice: "This success was due to the unexpected finding that MF1 mice can be treated as mosaics of standard inbred strains and analyzed accordingly using probabilistic ancestral reconstruction" [451].

The solution

Recall from section 9.1.3.3 that algorithm *Min-Crossover* efficiently finds a set of $Rmin(F; s)$ crossovers to create s from F, when s can be derived from F. Hence, the solution to the recombination-deduction problem is to run algorithm *Min-Crossover*(F, s) separately for each $s \in M$, and then add and combine the separate solutions.

14.2.1.2   The Founder-Deduction Problem

We discuss here the solution developed in [448], which was implemented for binary sequences. However, the method easily generalizes (conceptually) to any finite alphabet, although the method becomes less practical as the size of the alphabet increases. It is useful to consider the set of founder sequences arranged in a matrix F.

Given the observed sequences M, and the target k_F, the founder-deduction problem can be solved, in principle, by enumerating all 2^{mk_F} combinations of k_F binary sequences of length m, considering each combination as a *candidate* set of founder sequences F. Then, the solution to the founder-deduction problem is found by solving the recombination-deduction problem for each candidate F, selecting the candidate F^* where $Rmin(F^*; M)$ is minimum over all the computed $Rmin(F; M)$ values. Of course, this approach quickly becomes impractical as m and k_F increase. The solution presented here, in worst case, could also require exhaustive enumeration, but it uses additional ideas that in practice dramatically reduce the enumeration, and hence reduce the running time.

First key idea

The first idea is to modify algorithm *Min-Crossover*(F, s), which is given all of F, to become algorithm *On-Line-Min-Crossover*, where F is revealed, and the algorithm must decide what to do, *one successive site* at a time.

***Algorithm** On-line-Min-Crossover*

In the case when all of F is given as input, each iteration in algorithm *Min-Crossover* finds the *longest* match between s and any sequence in F, starting from a given site c. It then assigns c to one position past the end of the match. But when F is revealed only one site at a time, the algorithm cannot look ahead to determine the longest match from site c. However, as sites are revealed, from c to the last examined site i, algorithm *On-line-Min-Crossover* can keep a *set*, $\Phi_s(i)$, of all the sequences in F that match s from site c to site i. When site $i + 1$ is revealed, character $i + 1$ of each sequence in $\Phi_s(i)$ is compared with character $s(i + 1)$, determining $\Phi_s(i + 1) \subseteq \Phi_s(i)$. If $\Phi_s(i + 1) \neq \emptyset$, then c remains the same and the algorithm moves on to position $i + 2$. But if $\Phi_s(i + 1) = \emptyset$, one new crossover is counted between sites i and $i + 1$; c is set to $i + 1$; $\Phi_s(i + 1)$ is assigned the set of sequences in F that have character $s(i + 1)$ at site $i + 1$; and then the algorithm moves to iteration $i + 2$. Notice that when $\Phi_s(i + 1) = \emptyset$, the algorithm determines that there will be a crossover between sites i and $i + 1$, but unless the new, assigned $\Phi_s(i + 1)$ contains only one sequence, it cannot determine (yet) which sequence in $\Phi_s(i + 1)$ is involved in the crossover.

What has been described is how to solve the recombination-deduction problem for a *single* sequence s, when F is revealed one site at a time. But algorithm *On-Line-Min-Crossover* can be simply extended to solve the recombination-deduction problem for *every* sequence $s \in M$, as each site i of F is revealed. The algorithm simply keeps a separate $\Phi_s(i)$ for each $s \in M$, and dovetails the computation for each s. We will assume that algorithm *On-line-Min-Crossover* has been modified as described.

Definition Let $F(1, i)$, $s(1, i)$ and $M(1, i)$ denote the portions of F, s and M, respectively, from site 1 to site i.

In summary,

Lemma 14.2.1 *Algorithm On-line-Min-Crossover can efficiently determine* $\sum_{s \in M} Rmin(F(1, i); s(1, i))$, *denoted* $Rmin(F(1, i); M(1, i))$, *as i is successively incremented from 1 to m.*

Reset

Later, we will need the following concept:

Definition When $\Phi_s(i + 1)$ is assigned the set of sequence in F that have character $s(i + 1)$ at site $i + 1$, this is called *resetting* Φ for s.

Suppose we modify algorithm *On-Line-Min-Crossover* so that at arbitrary points in its execution, Φ is reset for *each* $s \in M$ (in addition to resetting Φ for s when $\Phi_s(i + 1) = \emptyset$), and each time this resetting occurs, we add n to the total number of crossovers. Then, we will still produce a partition of each sequence $s \in M$ into intervals $\mathcal{L}_s$, so that in each interval $I \in \mathcal{L}_s$, sequence s is identical to some sequence in F. However, the total number of crossovers counted may be greater than $Rmin(F; M)$.

The second key idea

Given lemma 14.2.1, in principle, we can solve the founder-deduction problem one site at a time by building a search tree τ one level at a time, from level 1 to level m. When a node v at level i is expanded, it branches 2^{k_F} ways, enumerating every possible binary sequence for column $i + 1$. So the path to node v at level i represents one candidate founders matrix, that we call $F_v(1, i)$. Specifically what is recorded at node v is the ith column in candidate $F_v(1, i)$; the value $Rmin(F_v(1, i); M(1, i))$; and the set $\Phi_s(i)$, for each $s \in M$. (The root of τ, at level zero, could also branch to a node for each binary sequence of length 2^{k_F}, but by symmetry, the number of branches from the root can be reduced to $k_F - 1$. We leave this detail to the reader.)

When level m of τ is completed, the leaf v with the smallest value

$$Rmin(F_v(1, m); M(1, m))$$

determines the solution to the founder-deduction problem. The optimal founder sequences F are determined by traversing the path from the v back to the root, collecting the successive columns recorded at the nodes of the path.

As described, this solution using τ is no better than the original idea of enumerating and evaluating all 2^{mk_F} candidate founder sequences. However, the search tree can be *pruned*, using several ideas that dramatically speed up the search.

14.2.1.3 Pruning the Search Tree

The standard approach is to implement some variant of *branch-and-bound*. For each node at level i, we can use the same lower bound $L(i)$ on the number of crossovers needed in a solution to the founder-deduction problem for input M restricted to sites $i + 1$ through m. If, for a node v at level i, $Rmin(F_v(1, m); M(1, m)) + L(i)$ is larger or equal to the number of crossovers used for some set of founders F, that can create M, then there is no need to expand node v in the search tree. A simple, efficiently computed lower bound, $L(i)$, was developed in [448], and a much better lower bound was developed in [443]. However, an alternative to branch-and-bound was developed in [448]; we will discuss that in the next section. A branch-and-bound approach using the better lower bound was implemented and tested in [443], where extensive simulation results are discussed. In short, the empirical results show that the search-tree approach is very practical for a number of founders up to seven (and more recently, ten), with up to forty mosaic sequences of length forty. In that range of parameters, the solutions were found in an average of under six minutes (in 2010, and faster since then). See [443] for details.

An alternative pruning approach

Instead of using a *branch-and-bound* approach, ideas were developed in [448] to identify cases when the best solution from a node v in τ is no better than the best solution from a different node, u, in τ. In that case, node v can be pruned. The simplest version of this idea is given in:

Lemma 14.2.2 *If $Rmin(F_v(1, i); M(1, i)) - Rmin(F_u(1, i); M(1, i)) \geq n$ for two nodes v and u in τ at level i, where n is the number of sequences in M, then node v can be pruned. There is no need to expand τ from v.*

Proof: Consider expanding v to a new node w, by adding a k_F-length binary vector, b, to $F_v(1, i)$ at site $i + 1$. Next, consider expanding node u to a new node z, by adding the same vector b to $F_u(1, i)$, and then *resetting* $\Phi_s(i + 1)$ for each $s \in M$ at node z. This adds n to the number of crossovers, $Rmin(F_v(1, i); M(1, i))$ at z, but the effect is that the set $\Phi_s(i + 1)$ at z will be a superset (possibly equal) of the set $\Phi_s(i + 1)$ at w. Therefore, any path from w to a leaf of τ at level m is also a feasible path from z to a leaf of τ. This implies that if $Rmin(F_v(1, i); M(1, i)) - Rmin(F_u(1, i); M(1, i)) \geq n$, then the best solution in τ that goes through v will be no better than the best solution that goes through u, and hence the search tree need not expand v. $\blacksquare$

The pruning idea in lemma 14.2.2 is the simplest, and least effective, variant of the pruning ideas in [448]. In fact, the reader should be able to see simple improvements to the above idea, allowing more extensive pruning. But the simple idea illustrates the basic approach and the spirit of the pruning ideas.

The search-tree with pruning approach was implemented in a computer program called *RecBlock*, which has been empirically demonstrated to be practical for a number of founders up to five, with up to thirty mosaic sequences of length thirty. The branch-and-bound approach described here was later added as an option for program *RecBlock*.

14.2.2 Additional Mosaic Problems and Results

In [448], and implicitly in [411], the founder-deduction problem was shown to be solvable in $O(nm)$ time for *two* founders. A related mosaic problem was also shown to be solvable in $O(nm)$ time in [411]. The search-tree solution to the founder-deduction problem was generalized in [448] to the case that M consists of *genotype* data rather than *haplotype* data. Genotype data was similarly considered in [349] and [350]. Related variants of the mosaic model and founder-deduction problem were considered in [457, 458], and complexity results were obtained in [35].

14.3 Reticulation Networks

The focus of this book is primarily on ARGs, motivated by issues in population genetics. However, many of the structural and algorithmic results apply, or can be easily extended, to different types of phylogenetic networks that model different biological information and phenomena. We call these networks *reticulation* networks, although parts of the literature call them "hybridization networks," and other parts just call them "phylogenetic networks" [195, 305, 308, 372]. A precise definition will be given shortly. In this section, we give an introduction to reticulation networks and discuss some of their relationships to genealogical networks and ARGs.

Definition A directed acyclic graph (DAG) $\mathcal{D}$ with a single node r of in-degree zero, and no parallel edges, is called a *rooted DAG*; the node r is called the *root* of $\mathcal{D}$. Any node in a rooted DAG with in-degree more than one is called a *reticulation* node.

Note that a reticulation node can have in-degree greater than two. A reticulation node roughly represents the biological event where two or more entities (genes, individuals, communities, populations, species, breeds, strains, lines, etc.) coexist and interact in a common environment, with the result that one or more new *hybrid* entities emerge from the older entities. For example, different bacterial strains that coinhabit an environment frequently exchange DNA, creating new bacterial strains. Meiotic recombination also fits into this model, but the reticulation model is more flexible, and allows events other than recombination or events that are modeled by recombination.

Definition Given a set of taxa Z, a *reticulation* network for Z is a rooted DAG $\mathcal{D}$, where each leaf is labeled by a distinct taxon in Z, and each taxon in Z labels one leaf of $\mathcal{D}$.

Note that in contrast to an ARG, in a reticulation network, only the *leaves* of $\mathcal{D}$ are labeled; any internal node is unlabeled. Further, edges are unlabeled.

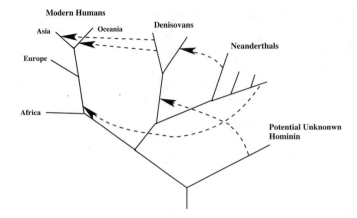

Figure 14.4   Inferred gene flow events from a potential hominin, and from Neanderthals, and from Denisovans, to modern humans. The figure is adapted from a more detailed figure in [343] that also shows the inferred magnitudes of the gene flows.

Definition A reticulation network where each reticulation node has in-degree two is called a *binary reticulation network*.

Reticulation networks represent evolution

We have defined a reticulation network, but of course we are interested in reticulation networks that represent some evolutionary information about a set of taxa. For example, figure 14.4 shows a recently inferred history [343] of gene flow events from a potential hominin, from Neanderthals, and from Denisovans, to modern humans. Here we begin to define how reticulation networks represent the evolutionary relationships of interest.

Definition If, for each reticulation node x in a rooted DAG $\mathcal{D}$, all but one of the edges into x are removed, then the result becomes a rooted *tree* $T_\mathcal{D}$ containing all of the nodes in $\mathcal{D}$. We say that $T_\mathcal{D}$ is a *directed spanning tree* of $\mathcal{D}$.

Note $T_\mathcal{D}$ will typically contain internal nodes with in-degree and out-degree one. It is also possible that an internal node v of $\mathcal{D}$ becomes a leaf when $T_\mathcal{D}$ is created. In that case, v is an *unlabeled* leaf.

Definition A rooted leaf-labeled tree T' is a *refinement* of a rooted tree T, if T' is obtained from T by successively removing any unlabeled leaf and its incident edge, and then contracting any (necessarily internal) nodes in the resulting tree that have in-degree and out-degree one.

The successive removal of unlabeled leaves must be done first, since a removal may create an additional unlabeled leaf, or an additional node of in-degree and out-degree one. For example, see figure 14.5.

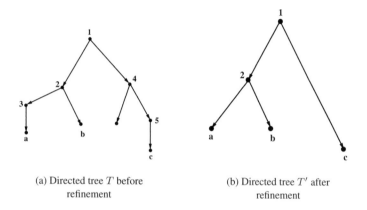

(a) Directed tree T before refinement

(b) Directed tree T' after refinement

Figure 14.5 Panel (a) shows a directed tree T rooted at node 1, with three labeled leaves, one unlabeled leaf, and two (internal) nodes, 3 and 5, that have in-degree and out-degree of one. The removal of the unlabeled leaf creates another node, 4, with in and out-degrees of one. Then the contraction of those three nodes creates the tree T' shown in panel (b). T' is the refinement of T. Note that T displays T'.

Definition Let T be a rooted tree where each leaf is labeled by a distinct taxon in a set $Z_T \subseteq Z$. T is *displayed* in a reticulation network $\mathcal{D}$ with leaf labels Z, if there is some directed spanning tree $\widehat{T}$ of $\mathcal{D}$, and a node $v_{\widehat{T}}$ in $\widehat{T}$, such that T is identical (including leaf labels) to the *refinement* of the subtree of $\widehat{T}$ rooted at node $v_{\widehat{T}}$. We also say that $\widehat{T}$ *displays* T.

Note that in this definition, a subtree rooted at v contains *all* the nodes and edges reachable from $v_{\widehat{T}}$ in $\widehat{T}$. See figure 14.5.

Sometimes, the definition of *display* is weakened so that T only needs to be identical to the refinement of *some* subtree of $\widehat{T}$ rooted at v. That is, in addition to edge contractions, any node or edge reachable from v in $\widehat{T}$ may be removed. Most results concerning trees displayed in a network hold under both of these definitions. We use the first definition in this book because it is simpler and we believe it to be more biologically informative. For example, see figure 14.6. Also, tree T displays T' in figure 14.5.

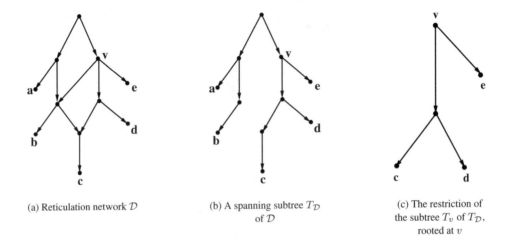

(a) Reticulation network $\mathcal{D}$

(b) A spanning subtree $T_{\mathcal{D}}$ of $\mathcal{D}$

(c) The restriction of the subtree T_v of $T_{\mathcal{D}}$, rooted at v

Figure 14.6   Example of tree displayed in a reticulation network $\mathcal{D}$. The tree shown in panel (c) is displayed in the reticulation network $\mathcal{D}$, shown in panel (a).

14.3.1   The Minimum-Reticulation and Minimum-Hybridization Problems

With the preceding definitions, we can define the major object of interest in this section: a rooted DAG that *displays* a given set of trees.

Definition Given a set $\mathcal{T}$ of rooted, leaf-labeled trees $\{T_1, T_2, ..., T_k\}$, where Z is the union of all the leaf labels in $\mathcal{T}$, a reticulation network $\mathcal{D}$, with leaf labels Z, is called a *reticulation network for* $\mathcal{T}$, if $\mathcal{D}$ displays each tree in $\mathcal{T}$.

Definition The number of reticulation nodes in a reticulation network $\mathcal{D}$ is denoted $\mathcal{R}(\mathcal{D})$.

Definition The *minimum-reticulation problem*: Given a set $\mathcal{T}$ of rooted, leaf-labeled trees $\{T_1, T_2, ..., T_k\}$, construct a reticulation network $\mathcal{D}^*$ for $\mathcal{T}$, that *minimizes* $\mathcal{R}(\mathcal{D})$ over all reticulation networks for $\mathcal{T}$. That minimum number of reticulations is denoted $\mathcal{R}min(\mathcal{T})$.

The rough biological meaning of the minimum-reticulation problem is that each tree in $\mathcal{T}$ is a known, deduced phylogeny for a subset of taxa based on some character(s). The characters in different trees might be incompatible. However, it is believed that the taxa jointly evolved on a reticulation network, where reticulation events are relatively rare, so that the historically correct network does not have many unnecessary reticulations. Thus, reticulation network $\mathcal{D}^*$ is an evolutionary hypothesis (or explanation) for the joint origin of the observed trees. By minimizing the number of reticulation nodes, it is expected that the biological fidelity of the deduced network will be increased.

The *minimum-reticulation problem* is NP-hard [45], but some insightful structural results about the problem are known, and the best-studied of these will be discussed in this chapter.

14.3.1.1   The Hybridization-Minimization Problem

A single biological reticulation event (such as species hybridization) generally involves only *two* taxa. Therefore, a reticulation node with in-degree greater than two usually indicates a lack of known resolution. That is, it represents a lack of certainly about which pairs of taxa were involved in the reticulation events, or the relative order that those events occurred. However, there are situations, such as in a community of many different bacterial strains, where new taxa are created from more than two older taxa. In such cases, reticulation nodes of in-degree more than two could reflect biological reality. So, rather than just counting the number of reticulation nodes in $\mathcal{D}$, the following alternative scoring scheme has often been used to quantify the amount of reticulation in a network $\mathcal{D}$:

$$\mathcal{H}(\mathcal{D}) = \sum_{v \in \mathcal{D}} d^-(v) - 1$$

where $d^-(v)$ is the *in-degree* of node v. Since each node in $\mathcal{D}$, other than the root, has at least one parent, and has exactly one parent if $\mathcal{D}$ is a tree, $\mathcal{H}(\mathcal{D})$ quantifies the number of *excess* parents there are in $\mathcal{D}$, over what would occur if $\mathcal{D}$ were a tree.

Definition $\mathcal{H}(\mathcal{D})$ is called the *hybridization number* of $\mathcal{D}$.

Definition We define the *hybridization-minimization problem* (HMP) as the problem of finding a reticulation network $\mathcal{D}^*$ for $\mathcal{T}$ that minimizes $\mathcal{H}(\mathcal{D})$ over all reticulation networks for $\mathcal{T}$. We let $\mathcal{H}(\mathcal{T})$ denote the number of reticulation nodes in $\mathcal{D}^*$.

An alternative view

An alternative way to view $\mathcal{H}(\mathcal{D})$ is to relate reticulation networks that allow nodes of in-degree *more than* two to reticulation networks where the in-degree is *at most* two, that is, to *binary* reticulation networks. Any reticulation node with in-degree $d > 2$ can be converted to exactly $d - 1$ reticulation nodes, each with in-degree two, so that the resulting network $\mathcal{D}'$ also displays all of the trees in $\mathcal{T}$. The essential idea is given by example in figure 14.7.

Theorem 14.3.1 *Given a set of directed trees $\mathcal{T}$, there is a reticulation network $\mathcal{D}$ that displays all of the trees in $\mathcal{T}$ with hybridization number $\mathcal{H}(\mathcal{D})$, if and only if there is a binary reticulation network that displays all of the trees in $\mathcal{T}$, with $\mathcal{H}(\mathcal{D})$ reticulation nodes.*

Definition The problem of minimizing the number of reticulation nodes in a binary reticulation network that displays a given set of trees, $\mathcal{T}$, is called the *minimum-binary-reticulation problem* (MBRP).

Corollary 14.3.1 *The hybridization-minimization problem is equivalent to the minimum-binary-reticulation problem. So, $\mathcal{H}(\mathcal{T})$ is the minimum number of reticulation nodes in any binary reticulation network for $\mathcal{T}$.*

The statement of the *minimum-binary-reticulation problem*, based on the *number* of reticulation nodes, is more consistent with the other problems in this book than is the statement of the minimum-hybridization problem. However, corollary 14.3.1 allows us to reference these two problems interchangeably.

14.4 Minimizing Binary Reticulations

Although the *minimum-reticulation* and the *minimum-binary-reticulation* problems are NP-hard [45, 372], there are elegant combinatorial results for special cases. In this section, we discuss the most insightful result, for *two* rooted, binary trees T and T'. See [305, 372] for a more complete review of related structural results. See [336] for recent algorithmic progress on computing the hybridization number of two trees.

In order to define the objects of interest in a unified way, and to facilitate the exposition, we add a new root, r^+, and a directed edge (r^+, r) from r^+ to the original root, r, in both T and T'. We also create a new leaf, r^*, and a new directed edge r^+ to r^*, in both T and T'. See figure 14.8a. Hence the leaf-set is augmented with r^*. The purpose of this will be clarified below. Recall that in a rooted *binary tree*, every non-leaf node has out-degree two.

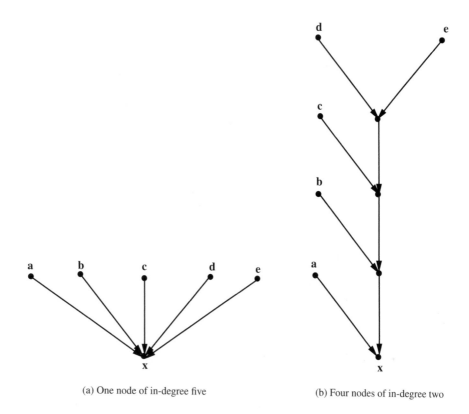

(a) One node of in-degree five   (b) Four nodes of in-degree two

Figure 14.7 Conversion of one reticulation node of in-degree five, to four reticulation nodes of in-degree two each. Any tree that can be displayed in a reticulation network containing the subgraph shown in (a) can be displayed in a reticulation network where the subgraph shown in (b) replaces the subgraph into x, shown in (a).

Definition A rooted, binary tree which is modified by adding a new root r^+, and a new leaf r^*, and directed edges (r^+, r) and (r^+, r^*), is called a *planted tree*. Note that a planted tree is only defined when it is derived from a rooted, *binary* tree.

Definition Given a planted tree T with leaf-set Z, and given a subset $Z' \subset Z$, we define $T(Z')$ as the *smallest* subtree of T containing the leaf-set Z'. $T'(Z')$ is similarly defined. See figure 14.8.

Definition Given two planted trees T and T' with the *same* leaf-set Z, $\mathcal{H}(T, T')$ denotes the *minimum hybridization number*, over all reticulation networks that display T and T'.

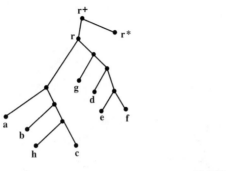

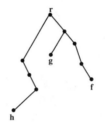

(a) Planted binary tree T

(b) $T(Z')$, the smallest subtree of
T connecting the set of leaves
$Z' = \{f, g, h\}$

(c) Refinement
of $T(Z')$

Figure 14.8   (a) A planted tree T with leaf-set Z.
(b) The smallest subtree $T(Z')$ of T connecting $Z' = \{f, g, h\}$.
(c) the refinement of $T(Z')$. When $|Z'| = 3$, the refinement of $T(Z')$ is called a *rooted-triple*, as will be
explained in section 14.5.1.

14.4.1   Agreement Forests

Definition Let T and T' be two planted binary trees with the *same* leaf-set Z. An *agreement forest* $\mathcal{F}$ for T, T' (on leaf-set Z) is a set of rooted, leaf-labeled trees $\{T_1, T_2, ..., T_k\}$,
where each tree $T_i \in \mathcal{F}$ has leaf-set $Z_i \subseteq Z$, $Z_i \neq \emptyset$, such that the following conditions
are satisfied:

Conditions for an agreement forest

(1) The leaf-sets $\{Z_i : T_i \in \mathcal{F}\}$ partition Z. That is, each leaf label in Z is in exactly one
tree in $\mathcal{F}$.

(2) Each tree $T_i \in \mathcal{F}$ is identical to the refinement of tree $T(Z_i)$, and also identical to the
refinement of $T'(Z_i)$.

(3) For each pair of trees (T_i, T_j) in $\mathcal{F}$, subtrees $T(Z_i)$ and $T(Z_j)$ share no nodes.
Similarly, $T'(Z_i)$ and $T'(Z_j)$ share no nodes.

See figure 14.9 for an example of an agreement forest. The purpose of adding the new
root r^+ and the new leaf r^* to T and T' is to force $T(Z_i)$ and $T'(Z_i)$ to contain the original
root r, for some $T_i \in \mathcal{F}$.

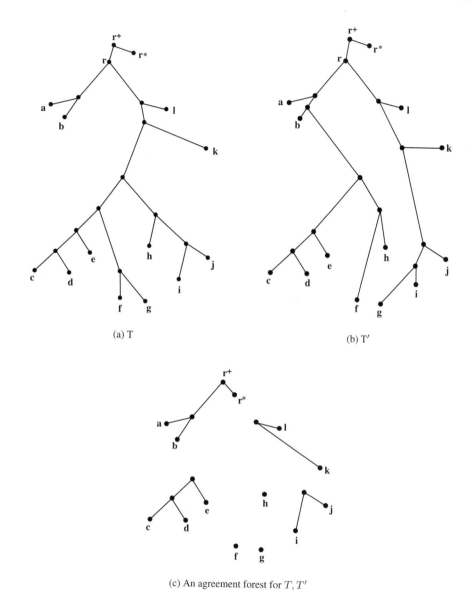

(a) T

(b) T'

(c) An agreement forest for T, T'

Figure 14.9 Panels (a) and (b) show two planted trees T and T' with the same leaf-set. Panel (c) shows an agreement forest for T, T' containing seven trees. Each of those trees are displayed in both T and T'. The direction of the edges is not shown; all edges are directed away from the root of their respective trees.

Definition An agreement forest for two trees T, T' is called a *maximum agreement forest (Maf)* for T, T' if it has the *fewest* number of trees over all agreement forests for T, T'. We use $m(T, T')$ for the number of trees in a maximum agreement forest for T, T'.

The concept of an agreement forest was introduced in the context of phylogenetic networks in [167], but was modified in later discussions to include the root node [44]. Recall the *rooted subtree-prune-and-regraft (rSPR)* operation, defined in section 9.2.3.1 (page 336).

Definition The minimum number of rSPR operations needed to transform the rooted tree T to the rooted tree T' is denoted $rSPR(T, T')$.

The following theorem, proved in [26], first established the central role of a maximum agreement forest in the study of reticulation networks.

Theorem 14.4.1 *For any two planted trees T, T', $rSPR(T, T') = m(T, T') \leq \mathcal{H}(T, T')$.*

We will not prove theorem 14.4.1, but we will exploit the first part of it in section 14.5 to obtain a practical approach to computing $rSPR(T, T')$. Next, we refine the concept of an agreement forest, to prove an equality for $\mathcal{H}(T, T')$.

14.4.1.1   Acyclic Agreement

Definition The tree $T_i \in \mathcal{F}$ is said to *dominate* the tree $T_j \in \mathcal{F}$ if and only if the root of $T(Z_i)$ is an ancestor of the root of $T(Z_j)$, or the root of $T'(Z_i)$ is an ancestor of the root of $T'(Z_j)$.

Definition An agreement forest $\mathcal{F} = \{T_1, ..., T_k\}$ for T, T' is called *acyclic* if the dominance relation has no cycles. That is, if the trees in $\mathcal{F}$ can be *ordered* so that every tree T_i appears in the order before every tree that T_i dominates. Such an ordering is called an *acyclic ordering*, or a *topological ordering* of $\mathcal{F}$.

The standard way to determine whether an agreement forest is acyclic is to build a directed graph H representing the dominance relation, where each node represents a tree in $\mathcal{F}$, and there is a directed edge (T_i, T_j) in H if and only if T_i dominates T_j. Then the agreement forest is acyclic if and only if H is a DAG. See figure 14.10.

Definition A *maximum acyclic agreement forest (Maaf)* for T, T', denoted *Maaf*(T, T'), is an acyclic agreement forest for T, T' with the fewest number of trees over all acyclic agreement forests for T, T'. We use $m_a(T, T')$ to denote the size of a Maaf for T, T'.

14.4.2 Main Result of This Section

We can now state and prove the main theorem of this section, the most insightful result that is known for the *minimum-binary-reticulation problem*.

Theorem 14.4.2 *Let T and T' be two planted trees with the same leaf-set Z. Then* $\mathcal{R}min(\{T,T'\}) = \mathcal{H}(T,T') = m_a(T,T') - 1$.

Proof: No reticulation network that displays two trees, T and T', needs to have a reticulation node with in-degree more than two. Hence, $\mathcal{R}min(\{T,T'\}) = \mathcal{H}(T,T')$.

To prove that $\mathcal{H}(T,T') = m_a(T,T') - 1$, we first show that $\mathcal{H}(T,T') \geq m_a(T,T') - 1$, by induction on $\mathcal{H}(T,T')$. To start, suppose T and T' are a pair of trees where $\mathcal{H}(T,T') = 0$. We need to prove in that case, that $m_a(T,T') = 1$. Both T and T' are *displayed* in some *tree* T'' with leaf-set Z. T'' does not need to have any nodes with in and out-degree of one, nor any unlabeled leaves, so we assume that it does not have either. Then, since T'' does not have any reticulation edges, the refinement of a subtree T''_v of T'' is precisely T''_v. So, tree T must be identical to some subtree T''_v of T'', rooted at a node v in T''. If v is not the root of T'', then since the root has out-degree greater than one, there would be a path from the root of T'' to some leaf f that does not go through v. Therefore, f would not be in Z, contradicting the assumption that T'' and T have the same leaf-set. Hence, v must be the root of T'', so T and T'' must be identical. The same conclusion holds for T', so T and T' must be identical, and hence $m_a(T,T') = 1$, proving the basis of the induction.

Next, assume that the theorem is true for every pair of trees (T,T') where $\mathcal{H}(T,T') \leq k - 1$ for some $k \geq 1$, and let (T,T') be a pair of trees where $\mathcal{H}(T,T') = k$. We will show that there is an acyclic agreement forest for T, T' containing at most $k + 1$ trees, implying that $\mathcal{H}(T,T') \geq m_a(T,T') - 1$. Let $\mathcal{D}$ be a reticulation network that displays T and T' using k reticulation nodes, and let x be a reticulation node in $\mathcal{D}$ such that no other reticulation node is reachable from x. Therefore, the subnetwork of $\mathcal{D}$ reachable from x is a *tree* T_x, with leaf-set $Z_x \subseteq Z$. We assume that T_x has no nodes with in-degree and out-degree of one, and no unlabeled leaves, as none are needed. Since tree T is displayed in $\mathcal{D}$, and $\mathcal{D}$ and T have the same leaf-set Z, there must be a node v in T such the subtree T_v of T, rooted at v, is identical to T_x. Similarly, there must be a node v' in T' such that the subtree $T'_{v'}$ of T', rooted at v', is identical to T_x. Therefore T_v and $T'_{v'}$ are identical.

Now consider the reticulation network $\widehat{\mathcal{D}}$ obtained from $\mathcal{D}$ be removing node x, subtree T_x, and the two reticulation edges entering x. Similarly, let $\widehat{T}$ (respectively $\widehat{T}'$) be the tree obtained from T (T') by removing node v (v'), subtree T_v ($T'_{v'}$), and the tree edge into v (v'). Clearly, $\widehat{\mathcal{D}}$ displays $\widehat{T}$ and $\widehat{T}'$, and has only $k - 1$ reticulation nodes, so $\mathcal{H}(\widehat{T},\widehat{T}') \leq k - 1$. Then, by the induction hypothesis, $\mathcal{H}(\widehat{T},\widehat{T}') = m_a(\widehat{T},\widehat{T}') - 1$, so $m_a(\widehat{T},\widehat{T}') \leq k$.

Let $Maaf(\widehat{T}, \widehat{T}')$ be the maximum acyclic agreement forest for $\widehat{T}, \widehat{T}'$, with leaf-set $Z - Z_x$. It follows that $Maaf(\widehat{T}, \widehat{T}') \cup T_x$, is an agreement forest for T, T', with leaf set Z. The number of trees in this agreement forest is at most $k + 1$.

Since T_v (respectively $T'_{v'}$) is a subtree of T (T') containing all nodes reachable from v (v'), no root node of a tree in $Maaf(\widehat{T}, \widehat{T}')$ can be a descendant of v in T, or a descendant of v' in T'. So, adding T_x to the end of the acyclic order for the trees in $Maaf(\widehat{T}, \widehat{T}')$, we obtain an acyclic order for $Maaf(\widehat{T}, \widehat{T}') \cup T_x$. Thus, $Maaf(\widehat{T}, \widehat{T}') \cup T_x$ is an *acyclic agreement forest* for T, T', of size $k + 1$. Hence, $m_a(T, T') \le k + 1$, so $m_a(T, T') - 1 \le k = \mathcal{H}(T, T')$, and the induction is complete.

The converse To prove the converse, that $\mathcal{H}(T, T') \le m_a(T, T') - 1$, let $Maaf(T, T')$ be a maximum acyclic agreement forest for T, T', and let $T_1, ..., T_k$ be an acyclic ordering of the subtrees. By definition, $Maaf(T, T')$ has $m_a(T, T')$ trees. We will construct a reticulation network $\mathcal{D}$ that displays T and T', using $m_a(T, T') - 1$ reticulation nodes, proving that $\mathcal{H}(T, T') \le m_a(T, T') - 1$. The construction will successively build DAGs $\mathcal{D}_1, \mathcal{D}_2, ..., \mathcal{D}_k$, where each DAG $\mathcal{D}_i$ displays $T(\bigcup_{j=1}^{j=i} Z_j)$ using $i - 1$ reticulation nodes. Hence, the final DAG $\mathcal{D}$ will be a reticulation network that displays all of T and T', using $k - 1$ reticulation nodes. The construction depends critically on the acyclic ordering of the trees in $Maaf(T, T')$.

To start, recall that tree T_1 in $Maaf(T, T')$ is a refinement of $T(Z_1)$ and also a refinement of $T'(Z_1)$. So, by setting $\mathcal{D}_1$ to T_1, $\mathcal{D}_1$ displays both $T(Z_1)$ and $T'(Z_1)$. In order to prepare for the construction of $\mathcal{D}_2$, we next show that $T(Z_1)$, $T'(Z_1)$, and T_1 all contain nodes r^+ and r^*.

If r^+ is not in $T(Z_1)$, then it is not in T_1, since T_1 is a refinement of $T(Z_1)$, and a refinement of a subtree of T will have the same root as that subtree. Similarly, if r^+ is not in T_1, then it is not in $T'(Z_1)$. So r^+ is either the root of $T(Z_1)$, $T'(Z_1)$, and T_1, or it is in none of those trees. Further, if r^+ is not in $T(Z_1)$, it cannot be in $T(Z_i)$ for any T_i in $Maaf(T, T')$, since if it were, T_i would dominate T_1, contradicting the acyclic ordering of the trees in $Maaf(T, T')$. But if r^+ is not in $T(Z_i)$ or $T'(Z_i)$, for any T_i in $Maaf(T, T')$, then $Z_j = \{r^*\}$ for some j, and hence $T_j = \{r^*\}$ would be an undominated tree that dominates no other tree in $Maaf(T, T')$. Then there would be at least two undominated trees in $Maaf(T, T')$. In that case, we can construct a smaller acyclic agreement forest for T, T' as follows: for any $T_i \ne T_j$, which is an undominated tree in $Maaf(T, T')$, add r^*, (r^+, r^*), (r^+, v) to T_i, where v is the root of T_i. That new path in T_i from r^* to v is a refinement of the path in $T(Z_i \cup r^*)$ from r^* to v, which does not intersect any node in $T(Z_h)$, for any $h \ne i, j$, by the third condition of an agreement forest, and the fact that T_i is undominated. Similarly, the new path does not intersect any node in $T'(Z_h)$, for $h \ne i, j$. Hence, adding the path r^*, r^+, r, v to T_i (adding r^* to Z_h) creates an acyclic agreement forest with one fewer tree than $Maaf(T, T')$, which is a contradiction. So, Z_1 must contain r^* and r^+, which means that r^+ is the root node of T_1, $T(Z_1)$, and $T'(Z_1)$.

Now we can describe the construction of $\mathcal{D}_2$. Tree T_2 appears next in the acyclic ordering of the trees in $Maaf(T, T')$, and T and T' each have one root node, that is, r^+, so the root of $T(Z_1)$ is an ancestor of the root of $T(Z_2)$ in both T and T'. Further, since T_1 and T_2 are consecutively ordered in the acyclic ordering, the root of $T(Z_2)$ (respectively $T'(Z_2)$) must be connected to some node v in $T(Z_1)$ (respectively $v' \in T'(Z_1)$) by a path in T (or T') that doesn't contain any edge in $T(Z_h)$ (or $T'(Z_h)$), for $k > 2$. See figure 14.11. Then, to construct $\mathcal{D}_2$, if nodes v and v' are in $\mathcal{D}_1$, we connect the root node x of T_2 to node v in $\mathcal{D}_1$ by a single edge, and also connect node x to node v' in $\mathcal{D}_1$ by a single edge, creating the network $\mathcal{D}_2$, with the single reticulation node x. If v (or v') is not in $\mathcal{D}_1$, then we can easily find the edge in $\mathcal{D}_1$ where v (or v') can be added to correspond to its location in T (T'), and then add edges to x as before. The removal of the edge (v', x) from $\mathcal{D}_2$ creates a tree that is a refinement of $T(Z_1 \cup Z_2)$, and the removal of (v, x) from $\mathcal{D}_2$ creates a refinement of $T'(Z_1 \cup Z_2)$. Hence, $\mathcal{D}_2$ displays $T(Z_1 \cup Z_2)$ and $T'(Z_1 \cup Z_2)$. See figure 14.11 and figure 14.12a.

In general, suppose we have a reticulation network $\mathcal{D}_i$ that displays all the trees $T_1, ..., T_i$, using $i - 1$ reticulation nodes. The root of $T(Z_{i+1})$ must be connected to some node v in $T(Z_j)$, for some $j \le i$, by a path in T that doesn't contain any edge in $T(Z_h)$ for $h > i$. Similarly, the root of $T'(Z_{i+1})$ must be connected to some node v' in $T'(Z_{j'})$, for $j' \le i$, by a path in T' that doesn't contain any edge in $T(Z_{h'})$ for $h' \ge i$. As before, we identify or create nodes v, v' in $\mathcal{D}_i$, and connect those nodes to the root of T_{i+1}, making that root a reticulation node. This creates $\mathcal{D}_{i+1}$, a reticulation network that displays $T(Z_1 \cup ... \cup Z_{i+1})$ and $T'(Z_1 \cup ... \cup Z_{i+1})$, using i reticulation nodes. After all of the trees in $Maaf(T, T')$ have been connected, additional edges can be added so that each reticulation node has out-degree one. The resulting reticulation network $\mathcal{D}$ displays T and T', and has $k - 1 = m_a(T, T') - 1$ reticulation nodes. See figure 14.12. Hence, $\mathcal{H}(T, T') \le m_a(T, T') - 1$. ∎

An alternate view of a Maaf

Notice that in figure 14.10, if we remove the edges in $T - T'$ from T, we get four subtrees whose refinements are the four trees in $Maaf(T, T')$. Similarly, if we remove the edges in $T' - T$ from T', we again get four subtrees whose refinements are the four trees in $Maaf(T, T')$. This is not accidental. In fact, one way to create $Maaf(T, T')$ for two rooted binary trees T, T', and hence to compute $\mathcal{H}(T, T')$, is to find the *minimum* number of edges to delete from T and T' respectively, so that the refinements of the resulting subtrees form the same acyclic agreement forest for T, T'. In section 14.5, we will develop and prove theorem 14.5.2, which is a variant of this fact. Theorem 14.4.2 and this optimization version of it have been exploited in several methods that compute $\mathcal{H}(T, T')$. The most practical of those methods appears in [450].

An alternative proof

Earlier, we proved that $m_a(T, T') - 1 \leq \mathcal{H}(T, T')$ by a formal induction. That induction suggests a more constructive proof, presented in [26]. The heart of that proof is the claim that if $\mathcal{D}$ is a reticulation network displaying T and T' using $\mathcal{H}(T, T')$ reticulation nodes, then the removal of all reticulation edges from $\mathcal{D}$ creates an *acyclic agreement forest* for T, T' with $\mathcal{H}(T, T') + 1$ trees. Hence, $m_a(T, T') \leq \mathcal{H}(T, T') + 1$. See figures 14.10 and 14.13. This is visually intuitive, and attractive as a high level argument, but when all the details are included to prove the claim rigorously, the overall exposition is similar in difficulty to the induction proof provided here.

Extension to multiple trees

It is possible to generalize part of theorem 14.4.2 from two trees to any set of trees. Before stating the generalization, we have to explain what an agreement forest, and an acyclic agreement forest, are for a set of trees $\mathcal{T}$. The definition of an agreement forest for $\mathcal{T}$ is easily obtained from the definition of an agreement forest for two trees (on page 494): replace every reference to the pair (T, T'), with a reference to the set $\mathcal{T}$; and replace each reference to an individual tree T or T', with a generic reference to any tree in $\mathcal{T}$. The generalization of the definition of an *acyclic* agreement forest for two trees, to a definition for a set of trees, $\mathcal{T}$, is obtained in the same way. Then $m_a(\mathcal{T})$ denotes the size of a *smallest* acyclic agreement forest for $\mathcal{T}$. The following generalization of theorem 14.4.2 was proved by Simone Linz in [255]:

Theorem 14.4.3 $\mathcal{R}min(\mathcal{T}) = m_a(\mathcal{T}) - 1$.

No similar generalization of theorem 14.4.2 is known for $\mathcal{H}(\mathcal{T})$, instead of $\mathcal{R}min(\mathcal{T})$.

A proof of theorem 14.4.3 can be obtained by a straightforward modification of the proof of theorem 14.4.2. In particular, in the first part of the proof, whenever T_v and $T'_{v'}$ are referred to, extend the reference to one such subtree in each of the k trees in $\mathcal{T}$. In the second part of the proof, whenever two trees are attached to a partial reticulation network, attach $|\mathcal{T}|$ trees in the analogous way.

Algorithms and software

Several algorithms and programs have been developed for reticulation problems with more than two trees. The paper [446] develops a bottom-up, coalescent-style approach to solve the *exact* minimum-binary-reticulation problem (equivalently, the minimum-hybridization

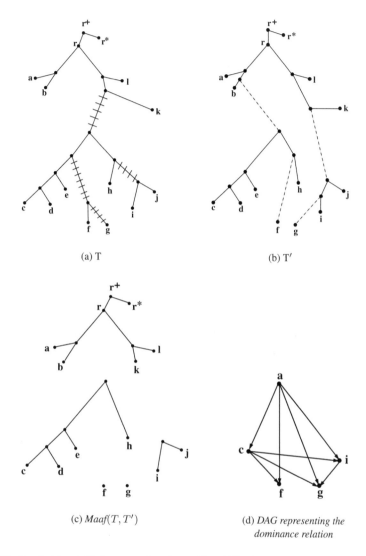

(a) T

(b) T′

(c) *Maaf*(T, T′)

(d) *DAG representing the dominance relation*

Figure 14.10 Panels (a) and (b) show two planted trees T and T' with the same leaf-set. The four edges in $T - T'$ are shown with perpendicular lines crossing the edges. The four edges in $T' - T$ are shown as dashed edges. Panel (c) shows a *maximum acyclic agreement forest (Maaf)* for T, T', with only five trees. Note that the removal of the edges in $T - T'$ from T creates five subtrees. After the refinement of each, the resulting set of trees forms *Maaf*(T, T'). The same Maaf is created by removing the edges in $T' - T$ from T', and refining the five resulting subtrees. Panel (d) shows the dominance relation defined on the trees in *Maaf*(T, T'), where each tree is represented by the lexicographically least leaf label in the tree. One topological ordering of the trees is (a, c, i, f, g), inducing an acyclic ordering on the trees in the Maaf.

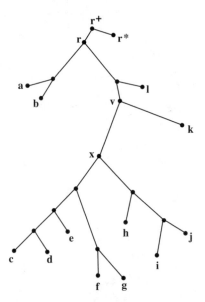

Figure 14.11 $Z_1 = \{a, b, l, k, r^*\}$, and $Z_2 = \{a, d, e, h\}$. The root node of subtree $T(Z_2)$ in T is labeled x. Node x connects to $T(Z_1)$ by a path (one edge in this case) that intersects $T(Z_1)$ on the edge into leaf k. Node v is added to that edge, creating two edges.

problem), for any number of trees. The input trees guide the bottom-up construction, making the method practical for problem instances beyond what would be practical by more brute-force methods. Three earlier results on multiple-tree hybridization problems appear in [66, 67, 444].

14.5   Computing $rSPR(T, T')$ with Integer Programming

In this section, we will develop an *integer linear programming (ILP)* formulation that computes $rSPR(T, T')$ for two leaf-labeled, rooted binary trees T, T', with the same leaf-set Z. The approach developed here is a modification of the approach due to Yufeng Wu [442]. See [171] for a discussion of earlier methods to compute the rSPR-distance, and some empirical testing of those methods. See [8, 43, 308, 386, 429] for additional results on SPR-distance.

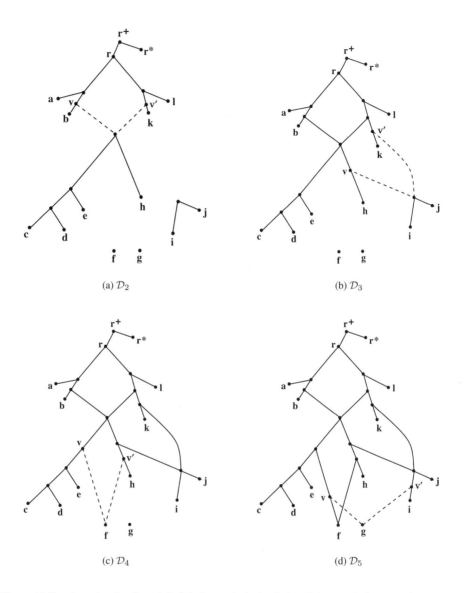

Figure 14.12 Assuming that (a, c, i, f, g) is the topological ordering of the trees in figure 14.10c, the construction of reticulation network $\mathcal{D}$ (detailed in the proof of theorem 14.4.2) takes four steps. At step i, nodes x, $v \in T$ and $v' \in T'$ are found, and subtree T_{i+1} is added to $\mathcal{D}_i$, creating $\mathcal{D}_{i+1}$. At each step, the added reticulation edges are drawn with dashes.

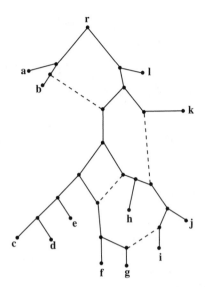

Figure 14.13   Binary reticulation network $\mathcal{D}$ displaying T and T' from figure 14.10, built from $Maaf(T, T')$. The dashed reticulation edges come from T', and the solid reticulation edges come from T. Removal of all the reticulation edges creates five subtrees, and after the refinement of each one, the trees in $Maaf(T, T')$ are created.

In order to describe the ILP formulation, we first have to prove a fundamental result about rooted binary trees, and second, we must reformulate the way we think about agreement forests.

14.5.1   Rooted-Triples of a Rooted Binary Tree

Recall that for any subset of leaves $Z' \subseteq Z$ of a tree T, $T(Z')$ is the smallest subtree of T that connects all of the leaves of Z'.

Definition Given a rooted, leaf-labeled binary tree T, and given three distinct leaves $\{f, g, h\}$ of T, the *rooted-triple* for $\{f, g, h\}$, denoted $Tr(f, g, h)$, is the refinement of subtree $T(\{f, g, h\})$ of T. See figure 14.8 (page 494), which shows a tree T, and the rooted-triple $Tr(f, g, h)$.

Because T is a rooted binary tree, $Tr(f, g, h)$ is also a rooted binary tree, for any three leaves $\{f, g, h\}$. Hence $Tr(f, g, h)$ must have a topology where two of the three leaves have a parent in common; and that parent and the other leaf have a parent in common. Thus any rooted-triple uniquely identifies the two leaves of T that are *siblings* in $Tr(f, g, h)$. For example, in figure 14.8, the siblings are $\{f, g\}$.

Definition Tr and $T'r$ denote the set of rooted-triples of trees $T = (V, E)$ and $T' = (V', E')$, respectively.

Theorem 14.5.1 *The set of rooted-triples, Tr, of a rooted, leaf-labeled binary tree T with three or more leaves, uniquely defines T.*

Proof: The proof is by induction on the number of leaves of T. The result is clearly true for three leaves. Assume it is true for trees with up to q leaves, and let T be a rooted, leaf-labeled tree with $q + 1$ leaves. Let $\{u, v\}$ be the two children of the root, r, and let T_u, T_v be the subtrees of T, rooted at u and v, respectively. Let $\sigma(f, g : h)$ be an *indicator variable* that takes the value one if leaves f and g are siblings in $Tr(f, g, h)$, and otherwise takes the value of zero.

Suppose, without loss of generality, that f is in T_u. It is easy to see that $\Sigma_{h \neq f, g}\, \sigma(f, g : h) = 0$, if and only if g is in T_v. Hence, the set of rooted-triples determines whether f and g are in the same *cluster* (rooted split) in T, defined by one of the two edges in T incident with root r. The relation of being together in one of those two clusters is *forced* by the rooted-triples, and is transitive. So the rooted-triples precisely determine the partition of the leaves into the two clusters defined by the two edges incident with r. By the induction hypothesis, the rooted-triples in one cluster uniquely define T_u, and the rooted-triples in the other cluster uniquely define T_v. Tree T is reconstructed from T_u and T_v by adding the directed edges (r, u) and (r, v). Since the first partition is forced, and T_u and T_v are unique, it follows that T is the unique rooted binary tree defined by its rooted-triples. ∎

The theorem can technically be extended to all rooted, leaf-labeled binary trees. If T has only one or two leaves, then T has no rooted-triples, but T is uniquely defined by the number of leaves it has. So the general theorem is that T is uniquely defined by the number of leaves it has, together with its rooted-triples. Thus, two leaf-labeled, rooted binary trees are identical if and only if they have the same set of rooted-triples, and the same number of leaves. The above proof also gives an implicit algorithm to reconstruct a rooted binary tree from its rooted-triples. Theorem 14.5.1 is generally credited to [5].

Theorem 14.5.1, and the fact that the rooted-triples of a rooted tree T are the same as the rooted-triples of the refinement of T, together imply:

Corollary 14.5.1 *If T is a rooted, leaf-labeled tree where each non-leaf node has out-degree at most two, then the rooted-triples of T uniquely define the refinement of T.*

Corollary 14.5.1 will be central to the development of the ILP for rSPR(T, T').

14.5.2   Rethinking Agreement Forests

Recall that $m(T, T')$ is the number of trees in a maximum agreement forest for T, T'. By theorem 14.4.1, we can compute $m(T, T')$ to get $rSPR(T, T')$. To that end, we first develop a different way to think about an *agreement forest* $\mathcal{F} = \{T_1, T_2, ..., T_k\}$, where each tree T_i has leaf-set $Z_i \subset Z$.

Lemma 14.5.1 *There exists a set of exactly $k - 1$ edges in T, denoted $\widehat{E}$, whose removal results in k subtrees of T, denoted Σ_T, such that each tree $T(Z_i)$, for $i \in [1, ..., k]$, is fully contained in a distinct subtree of Σ_T. The symmetric result holds for T'; the removed edges from T' are denoted $\widehat{E}'$.*

Proof: Let Δ_T be the set of edges that are in no tree $T(Z_i)$, for $i \in [1, ..., k]$. For every pair $i, j \in [1, ..., k]$, $T(Z_i)$ and $T(Z_j)$ have no node in common, so there must be an edge (possibly many choices for an edge) e in Δ_T whose removal from T separates all of the leaves in Z_i from all of the leaves in Z_j, and does not separate any leaves in the leaf-set of a tree in $\mathcal{F}$. After removing edge e from T, if any resulting subtree of T has leaves from two subsets, Z_p and Z_q, of leaves from trees in $\mathcal{F}$, then again there must be an edge in Δ_T whose removal separates the leaves of Z_p from the leaves of Z_q, but does not separate any leaves in the leaf-set of a tree in $\mathcal{F}$. Since there are k trees in $\mathcal{F}$, exactly $k - 1$ edges of Δ_T must be removed to separate all of the k leaf-subsets defined by $\mathcal{F}$. At that point, there are k remaining subtrees of T, and these define Σ_T. Clearly, each $T(Z_i)$ is completely contained in one distinct tree of Σ_T. Since T and T' are arbitrary designations, the same result holds for T'.   ∎

The containment of a subtree $T(Z_i)$ in a tree in Σ_T can be proper, for example if each Z_i contains a single leaf. Note that $k - 1$ is the minimum size of any set of edges whose removal from T partitions the leaves in Z into k subsets, e.g., as is done in $\mathcal{F}$.

Figure 14.10 (page 501) illustrates lemma 14.5.1. In figure 14.10a, one choice for $\widehat{E}$ consists of the edges shown with perpendicular crossing lines. One case is of particular interest. The singleton $\{f\}$ is a tree in $\mathcal{F}$, and the corresponding tree $\vec{T}$ in Σ_T consists of leaf f and its parent node. The single node $\{f\}$, which is a tree in $\mathcal{F}$, is not the refinement of $\vec{T}$, but is fully contained in $\vec{T}$, as required by lemma 14.5.1.

Definition For any three leaves, $\{f, g, h\}$ in T, if any edge $e \in T(\{f, g, h\})$ is removed from T, we say that the rooted-triple $Tr(f, g, h)$ (which is the refinement of $T(\{f, g, h\})$) is *broken*.

So, the process of removing the edges $\widehat{E}$ from T, also breaks some rooted-triples in Tr. Clearly, the rooted-triple $Tr(f, g, h)$ is *not* broken by the removal of $\widehat{E}$ if and only if all three nodes f, g, and h, are together in a subtree $T(Z_i)$; if and only if $\{f, g, h\}$ are together in Z_i, for some $i \in [1, ..., k]$.

Lemma 14.5.2 *For every Z_i, $i \in [1, ..., k]$, the set of rooted-triples in $T(Z_i)$ and $T'(Z_i)$ are the same.*

Proof: Suppose $T(Z_i)$ is a tree with some internal node(s) of in and out-degree of one. The refinement of $T(Z_i)$ contains exactly the same rooted-triples as in $T(Z_i)$. Hence $T_i \in \mathcal{F}$ has the same rooted-triples as $T(Z_i)$. By the same argument, T_i has the same rooted-triples as $T'(Z_i)$. So, $T(Z_i)$ and $T'(Z_i)$ have the same rooted-triples. ∎

Note that the case when $|Z_i| < 3$ is included in lemma 14.5.2. In that case, neither $T(Z_i)$ nor $T'(Z_i)$ contains any triples, so the lemma holds vacuously.

Definition Let $\widetilde{Tr}$ and $\widetilde{T'r}$ be the sets of *unbroken* rooted-triples of Tr and $T'r$, after the removal of edges $\widehat{E}$ and $\widehat{E}'$ from T and T', respectively.

Lemma 14.5.3 *An agreement forest, $\mathcal{F}$ for T, T' induces two equal-size sets of edges, $\widehat{E}$ and $\widehat{E}'$, whose removal from T and T', respectively, leaves forests of subtrees of T and T', that identically partition the leaves of Z. Further, $\widetilde{Tr}$ is identical to $\widetilde{T'r}$.*

Proof: Lemma 14.5.1 established that $|\widehat{E}| = k - 1 = |\widehat{E}'|$, when $\mathcal{F}$ has k trees. The two sets of subtrees partition Z in the same way because each partitions Z into the subsets $\{Z_i, ..., Z_k\}$ defined by $\mathcal{F}$. The fact that $\widetilde{Tr}$ and $\widetilde{T'r}$ are identical is established in lemma 14.5.2. ∎

Conversely:

Lemma 14.5.4 *Consider the removal of any $k-1$ edges, $\widehat{E}$ and $\widehat{E}'$, from trees $T = (V, E)$ and $T' = (V, E')$ respectively, and let $\vec{\mathcal{F}}$ and $\vec{\mathcal{F}}'$ denote the two resulting forests of k trees each. Suppose that the trees of $\vec{\mathcal{F}}$ and $\vec{\mathcal{F}}'$ partition Z in the same way; and that the sets of unbroken rooted-triples $\widetilde{Tr} \subset Tr$, and $\widetilde{T'r} \subset T'r$, are identical. Then there is an agreement forest $\mathcal{F}$ for T, T' such that the edge sets $\widehat{E}$ and $\widehat{E}'$ satisfy the conditions of lemma 14.5.1.*

Proof: We construct an agreement forest $\mathcal{F}$ from $\vec{\mathcal{F}}$ (or $\vec{\mathcal{F}}'$). Let $\vec{Z} = \{\vec{Z}_1, ..., \vec{Z}_k\}$ be the partition of Z defined by the trees of $\vec{\mathcal{F}}$ (or $\vec{\mathcal{F}}'$). We claim that the desired agreement forest $\mathcal{F}$ consists of the refinements of the k trees $\{T(\vec{Z}_1), ..., T(\vec{Z}_k)\}$. Consider any tree $T(\vec{Z}_i)$

for some $i \in [1, ..., k]$. By the fact that the trees of $\vec{\mathcal{F}}$ and $\vec{\mathcal{F}}'$ partition Z in the same way, there will be a tree in $\vec{\mathcal{F}}'$ whose leaf-set from Z is exactly $\vec{Z}_i$.

Suppose that $|\vec{Z}_i| \leq 2$. There is only one possible refinement of a tree with two leaves, and a tree with one leaf is its own refinement. So, the refinements of $T(\vec{Z}_i)$ and $T'(\vec{Z}_i)$ are identical in this case. Now suppose that $\vec{Z}_i$ has three or more leaves. By corollary 14.5.1, the refinements of $T(\vec{Z}_i)$ and $T'(\vec{Z}_i)$ will be identical if $T(\vec{Z}_i)$ and $T'(\vec{Z}_i)$ have the same set of rooted-triples. All of the rooted-triples of $T(\vec{Z}_i)$ and $T'(\vec{Z}_i)$ are unbroken, because they are each contained in a (*connected*) tree in $\vec{\mathcal{F}}$ and $\vec{\mathcal{F}}'$, respectively. Then, since the sets of *unbroken* rooted-triples, $\widetilde{T}r$ and $\widetilde{T}'r$, are identical, the rooted-triples of $T(\vec{Z}_i)$ and $T'(\vec{Z}_i)$ are identical. So, the refinement of $T(\vec{Z}_i)$ is identical to the refinement of $T'(\vec{Z}_i)$.

By construction, no pair of trees in $\vec{\mathcal{F}}$ shares a node, and no pair of trees in $\vec{\mathcal{F}}'$ shares a node, so none of the k trees $\{T(\vec{Z}_1), ..., T(\vec{Z}_k)\}$, shares a node. Hence the refinements of the k trees $\{T(\vec{Z}_1), ..., T(\vec{Z}_k)\}$ form the desired agreement forest $\mathcal{F}$.   ∎

In summary:

Theorem 14.5.2 *There is an agreement forest $\mathcal{F}$ for T, T' of size k if and only if there are two sets of $k - 1$ edges each, $\widehat{E} \subseteq E$ and $\widehat{E}' \subseteq E'$, such that:*

    (a) The forests $\vec{\mathcal{F}}$ and $\vec{\mathcal{F}}'$ of k trees each, created by removing $\widehat{E}$ and $\widehat{E}'$ from T and T', respectively, identically partition the leaves in Z; and

    (b) The unbroken rooted-triples $\widetilde{T}r \subset Tr$ and $\widetilde{T}'r \subset T'r$ are identical.

Hence, $m(T, T')$ equals the smallest such k.

Modifications

We note that condition (a) is equivalent to:

    (a') For each *pair* of leaves (f, g) in Z, f and g are connected by a path consisting of edges in $E - \widehat{E}$ in $\vec{\mathcal{F}}$, if and only if f and g are connected by a path consisting of edges in $E' - \widehat{E}'$ in $\vec{\mathcal{F}}'$.

Condition (a') is closer to what will be needed for the ILP formulation for $m(T, T')$.

We will see additional improvements in the next section, after the basic ILP formulation is introduced.

14.5.3   An ILP Formulation to Compute $m(T, T')$

Now we formulate an *integer linear program* to compute $m(T, T')$, and hence $rSPR(T, T')$. The idea is to encode the specifications of theorem 14.5.2 (using condition (a') in place of (a), so that an optimal solution to the ILP gives the smallest value of k that satisfies the conditions of theorem 14.5.2.

The variables
For each edge e in T, the ILP formulation has a binary, integer programming variable, D_e, that will be set to one (in a solution) to indicate that edge e should be removed; otherwise it is set to zero. Symmetrically, the formulation has a binary, integer programming variable, $D_{e'}$, for each edge e' in T'.

Enforcing equal-size edge sets
The following equality enforces the condition that $|\widehat{E}| = |\widehat{E'}|$:

$$(*) \quad \sum_{e \in E} D_e = \sum_{e' \in E'} D_{e'}$$

Enforcing identical partitions
Let $P(f, g)$ be the path in T between two leaves f and g. Leaves f and g will be in the same tree in $\vec{\mathcal{F}}$ if and only if no edge on $P(f, g)$ is removed. That is, if and only if

$$\sum_{\text{edge } e \in P(f,g)} D_e = 0$$

Similarly, let $P'(f, g)$ be the path in T' between the leaves f and g. They will be in the same tree in $\vec{\mathcal{F}'}$ if and only if no edge on $P'(f, g)$ is removed; which occurs if and only if

$$\sum_{\text{edge } e' \in P'(f,g)} D_{e'} = 0$$

Therefore, condition (a') of theorem 14.5.2 can be expressed by:

For each pair of leaves $(f, g) \in S$,

$$\sum_{\text{edge } e \in P(f,g)} D_e \geq 1$$

if and only if

$$\sum_{\text{edge } e' \in P'(f,g)} D_{e'} \geq 1,$$

which can be expressed in the ILP formulation by the pair of integer, linear inequalities:

$$(i) \quad \sum_{\text{edge } e \in P(f,g)} D_e \ \leq \ |P(f,g)| \ \times \sum_{\text{edge } e' \in P'(f,g)} D_{e'}$$

and

$$(ii) \quad \sum_{\text{edge } e' \in P'(f,g)} D_{e'} \ \leq \ |P'(f,g)| \ \times \sum_{\text{edge } e \in P(f,g)} D_e$$

where $|P(f,g)|$ and $|P'(f,g)|$ are the number of edges on the two paths, respectively. Inequality (i) ensures that if any edge in $P(f,g)$ is removed, then some edge on $P'(f,g)$ is also removed; inequality (ii) ensures the converse.

Enforcing identical unbroken rooted-triples

Creating inequalities to enforce condition (b), that the removed edges leave identical sets $\widetilde{T}r$ and $\widetilde{T}'r$, is less direct. For each triple of leaves $\{f, g, h\}$ in Z, we examine the rooted-triples $Tr(f,g,h)$ and $T'r(f,g,h)$ in T and T'. If those rooted-triples are not identical (that is, they have different sibling pairs), we call $\{f, g, h\}$ a *conflicted triple*.

Condition (b) of theorem 14.5.2 requires that for each conflicted triple $\{f, g, h\}$, both $Tr(f,g,h)$ and $T'r(f,g,h)$ must be broken, by removing at least one edge in $T(\{f, g, h\})$, and at least one edge in $T'(\{f, g, h\})$. This can be specified in the ILP formulation with the following inequalities:

For each conflicted triple $\{f, g, h\}$,

$$(iii) \quad \sum_{e \in T(\{f,g,h\})} D_e \geq 1$$

and

$$(iv) \quad \sum_{e' \in T'(\{f,g,h\})} D_{e'} \geq 1$$

The objective

Finally, the objective function is:

$$\text{Minimize} \sum_{e \in E} D_e$$

An optimal solution to this ILP formulation will give values to the variables so that $\sum_{e \in E} D_e = m(T, T') = rSPR(T, T')$.

14.5.3.1 Improvements

The formulation described so far correctly implements the conditions of theorem 14.5.2, but equality $(*)$ can be eliminated—it is "mathematically redundant" (implied by the other inequalities). We leave the proof of this to the reader. However, sometimes having some redundant (in)equalities helps the ILP-solver find the optimal solution faster.

A more significant improvement is that all of the inequalities of type (iv) are also redundant and can be removed from the ILP formulation. When all the inequalities of types (i), (ii), and (iii) are satisfied, all of the type (iv) inequalities will *automatically* be satisfied.

To see this claim, note that in any ILP solution, inequalities (i) and (ii) ensure that the leaves of Z are identically partitioned in $\vec{\mathcal{F}}$ and $\vec{\mathcal{F}}'$. Now consider two trees $\vec{T}$ and $\vec{T}'$ in $\vec{\mathcal{F}}$ and $\vec{\mathcal{F}}'$, respectively, that have the same leaf-set. The inequalities (iii) ensure that there is no conflicted triple consisting of leaves that are all in $\vec{T}$. Since any three leaves $\{f, g, h\}$ in $\vec{T}'$ are also in $\vec{T}$ (and conversely, but that is unneeded), and there are no conflicted triples in $\vec{T}$, it follows by the definition of a conflicted triple that $T'r(f, g, h) = Tr(f, g, h)$. So, $\vec{T}'$ has no conflicted triple. In general, no tree in $\vec{\mathcal{F}}'$ has a conflicted triple. Further, as noted before, if $T'r(f, g, h)$ is an unbroken rooted-triple, then leaves $\{f, g, h\}$ *must* be together in the same tree in $\vec{\mathcal{F}}'$. Hence, all of the conflicted triples in Tr' will be broken when all the inequalities $(i), (ii), (iii)$ are satisfied. Therefore, all the inequalities of type (iv) will also be satisfied—they are redundant.

The final ILP formulation

In summary, the ILP formulation to compute $rSPR(T, T')$ is shown in figure 14.14.

Size

For two binary trees with n leaves, the formulation has $4n - 4$ variables and $2\binom{n}{2} + \binom{n}{3} + 1 = \theta(n^3)$ inequalities. Hence the size of the formulation is polynomially-bounded in the size of the problem instance.

$$\text{Minimize} \sum_{e \in E} D_e$$

st:

$$(*) \quad \sum_{e \in E} D_e = \sum_{e' \in E'} D_{e'}$$

For each pair of leaves $(f, g) \in S$,

$$(i) \quad \sum_{\text{edge } e \in P(f,g)} D_e \leq |P(f,g)| \times \sum_{\text{edge } e' \in P'(f,g)} D_{e'}$$

$$(ii) \quad \sum_{\text{edge } e' \in P'(f,g)} D_{e'} \leq |P'(f,g)| \times \sum_{\text{edge } e \in P(f,g)} D_e$$

For each conflicted triple $\{f, g, h\}$,

$$(iii) \quad \sum_{e \in T(\{f,g,h\})} D_e \geq 1$$

All variables are binary.

Figure 14.14   The final ILP formulation to compute $rSPR(T, T')$.

Empirical results

Two ILP formulations are discussed in [442], both a bit different from the one presented here, but following similar logic. Empirical results are presented that show that the ILP solves quickly for moderate sized problems (trees with up to 100 nodes, and $rSPR(T, T') \leq 10$), faster and more accurately than previous methods, which were heuristic and did not guarantee the exact solution.

14.6   Displaying Clusters: A Weaker but More Achievable Goal

Although the *minimum-reticulation* and the *minimum-binary-reticulation* problems have intuitive phylogenetic meaning, exact solutions are practically obtainable for only restricted versions or for small instances of the problems. Further, it is often the case that the currently established topologies of the specified set of trees $\mathcal{T}$ are not completely reliable. It is more strongly accepted that the *clusters* (*rooted splits or clades*) contained in the trees in $\mathcal{T}$ are more reliable than are the actual topologies. Recall that a cluster (defined on page 58) in a rooted tree T is the set of taxa that label the leaves reachable from some edge in T.[2] This leads to the problem of finding a reticulation network $\mathcal{D}$ that explains the *clusters* contained in the trees in $\mathcal{T}$, using the fewest number of reticulation nodes, even if $\mathcal{D}$ does not actually display the *trees* in $\mathcal{T}$. We now begin to formalize this goal.

Definition A cluster c in a rooted tree T is *displayed* in a reticulation network $\mathcal{D}$ if there is *some* directed spanning tree $T_{\mathcal{D}}$ of $\mathcal{D}$, such that c is a cluster contained in $T_{\mathcal{D}}$. Note that $T_{\mathcal{D}}$ need not be T. We also say that $\mathcal{D}$ *displays* c.

Definition The set of clusters in a set of rooted leaf-labeled trees $\mathcal{T}$ are displayed in a reticulation network $\mathcal{D}$, if each cluster in the set is displayed in $\mathcal{D}$.

Note that each cluster in a tree T in $\mathcal{T}$ may be displayed in a different tree directed spanning tree $T_{\mathcal{D}}$, and any of those trees may be different from T.

Recall that theorem 2.4.4 (page 58) says that the set of clusters of rooted tree T *uniquely* determines T. So it is natural to conjecture that if a reticulation network $\mathcal{D}$ displays *all* of the clusters in a set of trees $\mathcal{T}$, then it will also display the trees in $\mathcal{T}$. However, this is not generally true.[3] Certainly, if tree T is displayed in $\mathcal{D}$, then $\mathcal{D}$ displays the clusters in T, but the converse is not true in general.

Definition The *soft minimum-reticulation (SMR) problem*: Given a set of clusters $\mathcal{K}$ from a set of rooted trees $\mathcal{T}$, find a reticulation network $\mathcal{D}$ that displays the clusters in $\mathcal{K}$, minimizing the number of reticulation nodes over all reticulation networks that displays those clusters. (Recall that there is no bound on the in-degree of a reticulation node.) Such a network is called a *minimum-reticulation network* (MRN) for $\mathcal{K}$. The number of reticulation nodes in an MRN for $\mathcal{K}$ is denoted $RminRN(\mathcal{K})$.

The use of the word "soft" in the "soft minimum-reticulation problem" is borrowed from the concept of "soft-wired cluster" and "networks that represent a set of clusters in the soft-wired sense" defined in [195].

---

2 Since a cluster is a set of taxa, it is essentially a character, and we will use c (the symbol usually used to denote a character or site) to denote a cluster.

3 One special case where it is true is when $\mathcal{D}$ is a *cluster network* (discussed in [195]) and all the trees in $\mathcal{T}$ are all binary. But cluster networks are highly structured, and a reticulation network is not necessarily a cluster network.

14.6.1   The Close Relationship of the MinARG and SMR Problems

The problem of finding a MinARG is actually closely related to the *soft minimum-reticulation problem*. In this section, we formally explain that relationship.

14.6.1.1   The Forward Direction

To start, let $\mathcal{K}$ be the set of clusters given by an instance of the SMR problem. We will represent $\mathcal{K}$ with a binary matrix, $M_{\mathcal{K}}$, where each row represents a taxon, and each column represents a cluster. We assume that each cluster is identified by the column that represents it. In detail, cell $M_{\mathcal{K}}(f, c)$ has value 1 if cluster c contains taxon f, and otherwise $M_{\mathcal{K}}(f, c)$ has value 0. Since $\mathcal{K}$ is a *set*, the particular assignment of clusters to columns is arbitrary, and any permutation of the columns of $M_{\mathcal{K}}$ is also a valid representation of $\mathcal{K}$. But for any *specific* representation, $M_{\mathcal{K}}$, we have:

Theorem 14.6.1 *Any ARG $\mathcal{N}$ for $M_{\mathcal{K}}$, with all-zero ancestral sequence, defines a reticulation network $\mathcal{D}$ that displays the clusters in $\mathcal{K}$. The number of recombination nodes in $\mathcal{N}$ equals the number of reticulation nodes in $\mathcal{D}$.*

Proof: For any cluster $c \in \mathcal{K}$, let T_c be the *marginal tree* in ARG $\mathcal{N}$ for site c. Clearly, the edge e in $\mathcal{N}$ that is labeled with c must be in T_c, and the leaves below e in T_c exactly define the taxa in cluster c. Therefore T_c is a tree that is displayed in $\mathcal{N}$ and contains cluster c. Hence, $\mathcal{N}$ displays all the clusters in $\mathcal{K}$. To formally create the reticulation network $\mathcal{D}$ for $\mathcal{K}$, remove the edge and node labels from $\mathcal{N}$, and label the leaves of $\mathcal{D}$ with the taxa they represent.   ∎

Corollary 14.6.1 *Let matrix $M_{\mathcal{K}}$ be a fixed representation of the set of clusters $\mathcal{K}$. Then $RminRN(\mathcal{K}) \leq Rmin_0(M_{\mathcal{K}})$.*

If $M_{\mathcal{K}}'$ is a permutation of the columns of $M_{\mathcal{K}}$, then it is possible that $Rmin_0(M_{\mathcal{K}}') < Rmin_0(M_{\mathcal{K}})$, in which case $RminRN(\mathcal{K}) < Rmin_0(M_{\mathcal{K}})$.

ARG results extend to reticulation networks displaying clusters

theorem 14.6.1 implies that many results about the construction of ARGs can be adapted to apply to reticulation networks that display clusters. For example, the full-decomposition theorem (theorem 7.2.1) can be adapted. In fact, since the set of conflicting pairs of columns in a binary matrix M is not changed by permuting the columns, and hence the set of connected components in the conflict graph is not changed by permuting the columns, we obtain:

Theorem 14.6.2 *Given a set of clusters $\mathcal{K}$, there is a fully decomposed reticulation network $\mathcal{D}$ that displays $\mathcal{K}$, and the backbone tree of $\mathcal{D}$ is unique.*

Theorem 14.6.2 was first stated and proved in [193]. See also [192].

14.6.1.2 The Converse Direction

Let $\mathcal{D}$ be a reticulation network that displays a set of clusters $\mathcal{K}$ represented by a matrix $M_{\mathcal{K}}$. By theorem 3.2.1, we know there is an ARG that derives $M_{\mathcal{K}}$, but is there always an ARG for $M_{\mathcal{K}}$ (with all-zero ancestral sequence) where the number of recombination nodes is the same as the number of reticulation nodes in $\mathcal{D}$? The answer depends on the in-degrees of the reticulation nodes in $\mathcal{D}$, and the type of recombination allowed in an ARG.

Theorem 14.6.3 *Let $\mathcal{D}$ be a* binary *reticulation network displaying the clusters in $\mathcal{K}$, and let $M_{\mathcal{K}}$ be any binary matrix representing the clusters in $\mathcal{K}$. Then there is an ARG $\mathcal{N}$ for $M_{\mathcal{K}}$ where multiple-crossover recombination is allowed, such that the number of recombination nodes in $\mathcal{N}$ equals the number of reticulation nodes in $\mathcal{D}$.*

Proof: The proof is constructive, creating $\mathcal{N}$ from $\mathcal{D}$. The outline of the argument is: Start by making $\mathcal{N}$ a copy of $\mathcal{D}$; assign each site in $M_{\mathcal{K}}$ to an edge in $\mathcal{N}$; convert each reticulation node in $\mathcal{D}$ to a recombination node in $\mathcal{N}$; and label each node in $\mathcal{N}$ with a binary sequence. The result is that $\mathcal{N}$ is an ARG for $M_{\mathcal{K}}$, where multiple-crossover recombination is allowed, and the number of recombination nodes in $\mathcal{N}$ is the same as the number of reticulation nodes in $\mathcal{D}$.

The construction in detail

Since $\mathcal{D}$ displays all the clusters in $\mathcal{K}$, for each cluster $c \in \mathcal{K}$, there is a directed spanning tree $T_{\mathcal{D}}$ in $\mathcal{D}$ that contains c. Thus, there is a *lowest* edge e_c in $T_{\mathcal{D}}$ such that the taxa in c are precisely the taxa labeling the leaves reachable from e_c. We label edge e_c in $\mathcal{N}$ with c. Let $T(c)$ be the refinement of the subtree of $T_{\mathcal{D}}$ consisting of the nodes reachable from the head of edge e_c. For each $c \in \mathcal{K}$, we find and label e_c, and determine $T(c)$.

Next, we determine for each reticulation node $x \in \mathcal{D}$, how to convert x to a recombination node in $\mathcal{N}$. For each $c \in \mathcal{K}$, and for each reticulation node x in $\mathcal{D}$ that is also a node in tree $T(c)$, let p_x denote the parent of x in $T(c)$. For the required recombination event at node x in $\mathcal{N}$, we *specify* that the sequence we will (later) create to label node x *must* obtain its value for site c from the sequence (not yet determined) we will create to label node p_x. Finally, we create the binary sequences labeling the nodes in $\mathcal{N}$, by first setting the ancestral sequence to the all-zero sequence, and then propagating sequences down through

$\mathcal{N}$, changing the value at a site c from 0 to 1 when traversing edge e_c; and obeying any of the specifications determined earlier for a recombination node. If, at any recombination node x in $\mathcal{N}$, there is no parental specification for a site c, then x is not in $T(c)$, so neither parent of x is reachable from edge e_c in $\mathcal{N}$. Hence, the propagation of sequences in $\mathcal{N}$ will assign state 0 to site c for both parents of x, and therefore x will also have state 0 at site c. It is easy to verify that the resulting ARG $\mathcal{N}$ generates $M_{\mathcal{K}}$.   ∎

Theorem 14.6.3 requires that $\mathcal{D}$ be a binary reticulation network, and the theorem does not extend to all reticulation networks with in-degrees larger than two. For more on the relationship of ARGs and reticulation networks, see [416].

MinARG results extend to binary reticulation networks

Theorem 14.6.3 implies that many results proved for MinARGs where multiple-crossover recombination is allowed also hold for the problem of finding a binary reticulation network with the fewest reticulation nodes. For example, theorem 7.1.1 holds, establishing that the number of nontrivial connected components in the conflict graph derived from $M_{\mathcal{K}}$ is a lower bound on the number of needed reticulation nodes in any binary reticulation network for $\mathcal{K}$.

14.6.2   A Static Definition of the History Bound

A final reflection of the close relationship of the MinARG and SMR problems comes by connecting the *history bound* on $Rmin_0(M)$ to $RminRN(\mathcal{K})$.

Given a binary matrix M, let $\mathcal{K}$ be the set of clusters obtained by interpreting each column c of M as a cluster c in $\mathcal{K}$ (with 0 the ancestral state, and 1 the derived state). The following theorem (which we will not prove here) states a *static definition* for the *history bound*, settling a problem that had been open for some time.

Theorem 14.6.4 *The history bound on* $Rmin_0(M)$, *also denoted* $MLB(M)$, *equals* $RminRN(\mathcal{K})$.

This was stated by several people [220], and partly established in [219]. It was explicitly proved in [279]. The theorem can be extended to the case that no ancestral sequence is known.

Theorem 14.6.4 is one connection between the MinARG and SMR problems, but it also highlights a difference. The theorem states that the solution to the SMR problem for $\mathcal{K}$ equals the *history bound* on the same matrix, called M in that context. Then since the *history bound* can be computed in *exponential* time, the SMR problem can be solved in exponential time. This is in contrast to the MinARG problem, where only *super-exponential* time methods are known (see Section 9.3.1).

14.6.3 Clusters from Rooted Trees

Theorem 14.6.3 was motivated by clusters derived from rooted trees, but it actually applies to any arbitrary clusters. However, when the clusters do come from a set of trees $\mathcal{T}$, the minimum number of reticulation nodes needed in a reticulation network that displays the clusters will be at most $\mathcal{H}(\mathcal{T})$, so theorem 14.6.3 implies:

Corollary 14.6.2 $\mathcal{H}(\mathcal{T}) \geq R^m min_0(M_{\mathcal{K}})$.

Also, when we restrict set $\mathcal{T}$ to contain only *two* trees, we can get a sharper result.

Definition Let $M(T, T')$ be a binary matrix that represents the set of clusters, $\mathcal{K}$, defined by the edges in *two* rooted trees T and T'. Assume that in $M(T, T')$, the columns representing clusters from T are to the left of the columns representing clusters from T'.

Corollary 14.6.3 *Let $\mathcal{D}$ be a binary reticulation network that displays the clusters $\mathcal{K}$, defined by the edges in the rooted trees T and T'. Then there is an ARG $\mathcal{N}$ for $M(T, T')$ that only uses single-crossover recombination, and where the number of recombination nodes in $\mathcal{N}$ equals the number of reticulation nodes in $\mathcal{D}$.*

Proof: The proof is identical to the proof of theorem 14.6.3, but we observe that when we convert a reticulation node x in $\mathcal{D}$ to a recombination node in $\mathcal{N}$, only a *single-crossover* is needed. The reason is that all of the columns in $M(T, T')$ representing clusters from T are together to the left of all of the columns representing clusters from T'. So all of the columns from T will choose the same parent of x, and all of the columns from T' will choose the other parent. Hence, only one crossover is required at x. ∎

Generalizing, we have:

Definition Let $\mathcal{T} = \{T_1, ..., T_k\}$ be a set of k rooted trees. Let $M(\mathcal{T})$ be a binary matrix that represents the clusters defined by the edges in $\mathcal{T}$, where the columns representing clusters from any tree $T_i \in \mathcal{T}$ are grouped together.

Theorem 14.6.5 *Let $\mathcal{D}$ be a binary reticulation network that displays the clusters defined by the k trees in $\mathcal{T}$. Then there is an ARG $\mathcal{N}$ that generates $M(\mathcal{T})$, with the all-zero ancestral sequence, where the number of recombination nodes in $\mathcal{N}$ equals the number* *of reticulation nodes in* $\mathcal{D}$, *and where each recombination event allows at most* $k - 1$ *crossovers.*

We leave the proof to the reader.

14.6.4   Back to Minimum Binary Reticulation

Now we turn our attention back to the *minimum-binary-reticulation problem* for two directed trees T and T'. That is, we want a reticulation network that displays the actual *trees* T and T', not just the clusters defined by those trees.

Theorem 14.6.6 *Let* $\mathcal{D}$ *be a binary reticulation network that displays two rooted (not necessarily binary) trees* T *and* T', *with hybridization number* $\mathcal{H}(\mathcal{D}) = \mathcal{H}(T, T')$. *Then there is an ARG* $\mathcal{N}$ *that generates* $M(T, T')$, *with the all-zero ancestral sequence, using* $\mathcal{H}(T, T')$ *single-crossover recombinations.*

Proof: The proof follows the same general ideas as the proofs of theorem 14.6.3 and corollary 14.6.3, with some modifications. Again, we start by setting $\mathcal{N}$ equal to a copy of $\mathcal{D}$. $\mathcal{D}$ displays T and T', and each column c in $M(T, T')$ represents a cluster in T or T', and hence tree T_c referenced in the proof of theorem 14.6.3, is either tree T or T'.

In more detail, for any cluster c represented by a column c in $M(T, T')$, there is an originating edge e_c in T or T'. By assumption, both T and T' are displayed in $\mathcal{D}$. So, for any column c in $M(T, T')$ that was generated by an edge e_c in T, we can identify the corresponding edge e_c in the subgraph of $\mathcal{D}$ that displays T. We label edge e_c with site c. We do the similar labeling for any column in $M(T, T')$ that was generated by an edge in T'.

Since $\mathcal{D}$ is only required to display the two trees, T and T', and $\mathcal{H}(\mathcal{D}) = \mathcal{H}(T, T')$, every reticulation node x in $\mathcal{D}$ will have in-degree exactly two; one of the edges into x will be in the subgraph that displays T, and the other will be in the subgraph that displays T'. Then, to convert a reticulation node in $\mathcal{D}$ to a recombination node in $\mathcal{N}$, we use the proof of corollary 14.6.3 to establish that only single-crossover recombination is needed.   ∎

Theorem 14.6.6 was first proved in [45]. It can be generalized for a set of k trees. As before, let $\mathcal{T} = \{T_1, ..., T_k\}$ be a set of k rooted (not necessarily binary) trees, and $M(\mathcal{T})$ be a binary matrix that represents the clusters defined by $\mathcal{T}$, where the columns representing clusters from any tree $T_i \in \mathcal{T}$ are grouped together.

Theorem 14.6.7 *Let* $\mathcal{D}$ *be a reticulation network that displays the* k *trees in* $\mathcal{T}$, *where* $\mathcal{D}$ *has the smallest hybridization number,* $\mathcal{H}(\mathcal{D})$, *of any reticulation network that displays* $\mathcal{T}$.

By theorem 14.3.1, we can assume that $\mathcal{D}$ is a binary reticulation network. Then there is an ARG $\mathcal{N}$ that generates $M(\mathcal{T})$, with the all-zero ancestral sequence, using $\mathcal{H}(\mathcal{D}) = \mathcal{HT}$ recombinations nodes, where each recombination event allows at most $k - 1$ crossovers.

We again leave the proof to the reader.

14.6.4.1   The Converse Direction

Theorem 14.6.6 shows that $Rmin_0(M(T, T') \leq \mathcal{H}(T, T')$. Now we discuss a converse result, established in [426]. In contrast to the first direction, this result only applies to two rooted *binary* trees.

Theorem 14.6.8 *Given two rooted binary trees T and T', with n labeled leaves, let $\mathcal{N}$ be a MinARG for $M(T, T')$, with the all-zero ancestral sequence, using only single-crossover recombination. Then there is a reticulation network $\mathcal{D}$ that displays T and T', using only $Rmin_0(M(T, T')$ reticulation nodes.*

Proof: Let $M(T)$ be the matrix $M(T, T')$, restricted to the sites corresponding to clusters in T; similarly, let $M(T')$ be $M(T, T')$ restricted to the sites corresponding to clusters in T'.

The proof of the theorem is by induction on the number of recombination nodes in $\mathcal{N}$. First, consider the case of no recombinations, so $\mathcal{N}$ is a binary tree (binary, by the assumption that every tree node in an ARG has out-degree two). Restricting $\mathcal{N}$ to the edges labeled with sites in $M(T)$ produces a perfect phylogeny for $M(T)$. But if we label each edge e in T with the site in $M(T)$ that represents the cluster defined by edge e, then T is also a perfect phylogeny for $M(T)$. (This is a bit circular, since that site in $M(T)$ was derived from e in the first place.) Now, by corollary 2.1.1, the perfect phylogeny for $M(T)$ is unique, so T must be identical to $\mathcal{N}$. By the same reasoning, T' must be identical to $\mathcal{N}$. Hence, after removing edge labels from $\mathcal{N}$, the three trees $\mathcal{N}$, T, and T' must be identical, leaf-labeled rooted trees. So $\mathcal{N}$ displays both T and T', and this establishes the theorem in the case when $\mathcal{N}$ contains no recombinations.

Suppose that the theorem holds when $\mathcal{N}$ contains $k > 0$ recombinations, and consider the case when $\mathcal{N}$ contains $k + 1$ recombinations. Let x be a recombination node in $\mathcal{N}$ such that no other recombination node is reachable from x. Since x is a recombination node, there is only one edge out of x, say to a node v, which might be a leaf. So v is the root of directed binary tree $\mathcal{N}_v$, which might only consist of node v.

Let Z_v be the set of taxa that label the leaves of $\mathcal{N}_v$, and let tree $T(Z_v)$ be the smallest subtree of T that connects all of the leaves in Z_v. As before, label each edge e in $T(Z_v)$ with the site in $M(T, T')$ that represents the cluster defined by edge e, and note that every edge in $T(Z_v)$ gets labeled. Since T is a binary tree, and $T(Z_v)$ is minimal, the root node u of $T(Z_v)$ must have out-degree two. We want to establish that $T(Z_v)$ is a binary tree, which is equivalent to the claim that the subtree of T rooted at u, denoted T_u, is exactly $T(Z_v)$. If not, there is a taxon f labeling a leaf of T_u, where f is not in Z_v. Then, the path from f to u must first intersect $T(Z_v)$ at some node $w \neq f$. Node w cannot be the root u of $T(Z_v)$, since then u would have out-degree greater than two. So there is an edge e in $T(Z_v)$ into w. Edge e is labeled with a character, c_e, from $M(T, T')$, so f must have state 1 for character c_e in $M(T, T')$. But tree $\mathcal{N}_v$ contains the only edge labeled c_e in $\mathcal{N}$, and $\mathcal{N}$ generates $M(T, T')$, so f must be a leaf in $\mathcal{N}$ reachable from node v, contradicting the assumption that $f \notin Z_v$. Hence, we conclude that $T(Z_v)$ is a binary tree.

With similar reasoning, we can establish that the set of sites labeling edges in $T(Z_v)$, is the set of sites from $M(T)$ labeling edges in $\mathcal{N}_v$. Let $M_T(Z_v)$ be the matrix $M(T)$ induced by the taxa in Z_v and the sites from $M(T)$ that label edges in $\mathcal{N}_v$. Hence, $T(Z_v)$ is a perfect phylogeny for $M_T(Z_v)$. But if we remove all the edge-labels in $\mathcal{N}_v$ that correspond to sites in $M(T')$, we also obtain a perfect phylogeny for $M_T(Z_v)$. Hence $T(Z_v)$ and $\mathcal{N}_v$ are identical trees. Similarly, the smallest subtree, $T'(Z_v)$, of T' that connects the leaves in Z_v, is identical to $\mathcal{N}_v$ and to $T(Z_v)$.

To finish the proof of the theorem, we need to construct a reticulation network with $k+1$ reticulation nodes that displays T and T'. To do that, remove node x and the subtree T_x rooted at x, from $\mathcal{N}$. Also, remove subtree $T(Z_v)$ from T, but leave its root node u; and remove subtree $T'(Z_v)$ from T', but leave its root u'. By the induction hypothesis, the resulting ARG, $\mathcal{N} - T_x$, can be used to create a reticulation network $\mathcal{D}$ with k reticulation nodes that displays the trees $T - T(Z_v) \cup \{u\}$ and $T' - T'(Z_v) \cup \{u'\}$. Next, direct the edges in $\mathcal{D}$ incident with u and u', into u and u', respectively, and then superimpose nodes u and u', and call the new node v^*. Finally, superimpose the root of subtree T_v and node v^*. The result is a reticulation network with $k + 1$ reticulation nodes that displays T and T'.   ∎

Theorems 14.6.6 and 14.6.8 establish the following result, first stated in [45]:

Theorem 14.6.9 *For two rooted binary trees T, T', $\mathcal{H}(T, T') = Rmin_0(M(T, T'))$.*

Corollary 14.6.4 *For two rooted binary trees T, T', $\mathcal{R}min(\{T, T'\}) = \mathcal{H}(T, T') = Rmin_0(M(T, T')) = MLB(M)$.*

Proof: The first equality is from corollary 14.6.3, and the last equality is from theorem 14.6.4.   ∎

Hence, the *history bound* on $Rmin_0(M(T, T'))$ is tight when M represents *two* rooted binary trees. That result was obtained directly, without the use of theorem 14.6.4, in [219].

Time constraints

All of the results in this chapter ignore the constraint of *simultaneity*. When two taxa are involved in a reticulation event, they must exist at some common point in time. In the case of species hybridization, that time window may be very large, but it is still constraining. Incorporation of temporal constraints into hybridization networks has been considered in [186, 257, 256], with *tree-child* networks [60], where every internal node must be the tail of a tree-edge. That implies that from every node in the network, there must be a path to some leaf that does not go through a reticulation node. That is similar to, but stronger than, the *rec-visibility* condition discussed in section 10.1.3.1.

14.7 Other Phylogenetic Networks

This book emphasizes specific types of phylogenetic networks, but the world of phylogenetic networks is broader than the coverage here. So, I would be remiss if I did not at least point the reader to some of the literature on other major types of phylogenetic networks and results. Some major types of phylogenetic networks that were not discussed in this book are *median networks* (reduced and quasi) [22, 23]; *Buneman graphs, split networks, split tree, split decomposition networks* [24, 54, 95, 98, 190]; *neighbor nets* [53]; *level-k networks* [179, 415]; *normal networks* [179]; and many more (authors of those, please forgive me). There is also a body of literature on comparing networks and computing metric distances between them. For example, see [57, 58, 60, 304]. And again, the books by Huson et al. [195] and Morrison [294] discuss a broad range of phylogenetic networks.

Appendix A

A Short Introduction to Integer Linear Programming

Throughout this book, we have presented a range of *integer linear programming* (ILP) *formulations* (also called *models*) for computational problems related to ARGs, recombination and phylogenetic networks. We have also presented some empirical results obtained from *solving* the formulations on real and simulated data. Integer linear programming (more commonly called *integer programming*) is a versatile and often effective computational approach that is widely used in engineering, business, urban planning, transportation scheduling, and more. However, it is almost unknown to biologists, although computer scientists and mathematicians are increasingly using integer programming in computational biology. Highly engineered, proprietary computer code is available to *solve* ILP formulations, and the (partial) effectiveness of these codes has encouraged the broader adoption of integer programming. Alternative *open-source* code is also available to solve ILP formulations, and it, too, is highly effective in many applications. In this appendix, we give a brief introduction to integer linear programming. We start with the foundational topic of *linear programming*.

Linear Programming (LP) and Its Use

It is helpful to divide the discussion of linear programming (and integer linear programming) into three parts: the *concrete formulation* of a linear program (or model), given all the data required to specify a problem instance; the *solution* of a concrete formulation; and the *abstract formulation* of a linear program.

A concrete formulation
Consider the following contrived problem. Suppose you have a laboratory making and selling two kinds of chemicals, and you have two lab assistants working for you. One gram of chemical *A* requires three hours of work by Assistant 1, and another six hours of

work by Assistant 2. Similarly, one gram of chemical B requires four hours of work from Assistant 1, and another two hours of work from Assistant 2. The amount that can be made of either chemical is completely divisible, as is the time that either assistant can work on either chemical. That is, any fractional amount of the two chemicals can be made by the appropriate proportional allocation of time. For example, to make 2.6 grams of chemical A requires 7.8 hours from Assistant 1 and 15.6 hours from Assistant 2.

Chemical A can be sold for $12 per gram, and chemical B can be sold for $7 per gram. Fractional amounts can be made and sold. Suppose that at least 1.3 grams of chemical B must be made (due to some prior commitment), but any amount of chemical A is allowed. The main constraint on how much can be made is that Assistant 1 has only thirteen hours available to work, and Assistant 2 has only seventeen hours available.

The problem

How much of each of the two chemicals should be made in order to *maximize* the total value of what is produced? More generally, *how* can we figure out how much of each chemical to make?

The answer to the second question is to *formulate* a linear program *model* that describes (or expresses) the details of the first question. That formulation, with all the details of the problem instance, is called a *concrete* LP formulation. Then to answer the first question, we *solve* the concrete LP formulation. We will discuss formulation and solution next.

Creating an LP formulation for a problem instance

To create an LP formulation (model) for a problem instance, we begin by creating linear programming *variables* to express the unknown values that we want to ultimately determine: variable X_A denotes the amount of chemical A to be made; and variable X_B denotes the amount of chemical B to be made.

The next step in formulating an LP model for a problem instance is to write *linear inequalities* (which could be equalities) that express the *constraints* on the values that can be assigned to the variables. In this example, the two main constraints are the amounts of time that the two assistants have available to work:

$$3X_A \;+\; 4X_B \le 13$$
$$6X_A \;+\; 2X_B \le 17$$

The first inequality expresses the time constraint of Assistant 1, and the second inequality expresses the time constraint of Assistant 2. Note that left side of each inequality is a *linear combination* of some of the LP variables, that the right side is a constant, and that the two sides are connected by an arithmetic relation.[1] This is the general form for any constraint in a linear program.

We also need inequalities to express constraints on the amounts of each chemical to be made:

$$X_A \geq 0$$
$$X_B \geq 1.3$$

Feasible solutions Given the inequalities in the formulation, some assignments of values for variables X_A and X_B are feasible, i.e., the assignment satisfies all of the constraints. Other assignments and some assignments are infeasible, because some constraint is violated. For example, $X_A = 2$ and $X_B = 1.4$ is feasible, but $X_A = 2.5, X_B = 1.3$ is infeasible, because it violates the second time constraint, although it satisfies the first one.

Definition Any assignment of values to the variables that satisfies all of the constraints is called a *feasible solution*. If there is no feasible solution, then the formulation is called an *infeasible formulation*.

Objective function
So far, no part of the concrete formulation includes the money received for sale of the chemicals, nor is the goal of making the most valuable mix of chemicals expressed in the formulation. The dollar value of what is produced is expressed by the *linear* combination:

$$12X_A + 7X_B,$$

since one gram of chemical A is sold for \$12 and one gram of chemical B is sold for \$7. The goal is to maximize the value of what is produced, and this is expressed in the following *linear objective function*:

$$\text{Maximize } 12X_A + 7X_B$$

The full formulation Summarizing the discussion, the full formulation of the problem is shown in figure A.1.

---

1 The relation is "less-than-or-equal" in the two above constraints, but "greater-than-or-equal" or "equal to" are also allowed.

$$\text{Maximize } 12X_A \quad + \quad 7X_B$$

subject to the constraints:

$$3X_A \quad + \quad 4X_B \le 21$$

$$6X_A \quad + \quad 2X_B \le 17$$

$$X_A \quad \ge \quad 0$$

$$X_B \quad \ge \quad 1.3$$

Figure A.1   The full concrete formulation of the chemical assignment problem. This is called a "concrete" formulation because it contains all of the information in this particular problem *instance*. Usually, the phrase "subject to the constraints" is abbreviated to "st." Note that the objective function is a linear function of a subset (possibly the whole set) of the LP variables, and that each of the constraints is a linear inequality, defined on some of the LP variables. The last two inequalities are also called *bounds* because each one provides a bound (upper or lower) on a single LP variable.

A concrete LP formulation has all the information required to allow a solution to the problem instance. The formulation can then be input to an *LP-solver* (in the proper format). The solution to a concrete LP formulation is usually referred to as an *optimal* solution. What the LP-solver determines and returns is a *feasible solution* to the concrete LP formulation that *optimizes* the objective function. Notice that the phrasing allows for the possibility that there is more than one optimal solution, which is often the case.

In the example of the concrete LP formulation in figure A.1, an optimal solution has value 46.5 and is achieved by setting (subject to roundoff) the value of X_A to 1.44444, and X_B to 4.166667.

Refinements to the model

Once we know the optimal values for X_A and X_B, we can figure out how much work each assistant should do for each chemical. However, it is convenient to create LP variables for those amounts and have them explicitly determined by the LP-solver. To do this, we let variables $X_{1,A}$ and $X_{1,B}$ denote the amounts of time Assistant 1 work on chemicals A and B respectively, and let variables $X_{2,A}, X_{2,B}$ denote the amounts of time Assistant 2 works on chemicals A and B respectively. Using those new variables, we augment the concrete

$$\text{Maximize } 12X_A \ + \ 7X_B$$

subject to the constraints:

$$3X_A \ + \ 4X_B \le 13$$

$$6X_A \ + \ 2X_B \le 21$$

$$X_{1,A} \ = \ 3X_A$$

$$X_{1,B} \ = \ 4X_B$$

$$X_{2,A} \ = \ 6X_A$$

$$X_{2,B} \ = \ 2X_B$$

$$X_A \ \ge \ 0$$

$$X_B \ \ge \ 1.3$$

Figure A.2   The concrete LP formulation now includes variables for the individual amounts of time worked by the two assistants on the two chemicals. Notice that the added constraints are linear *equalities*.

LP formulation with the equalities:

$$X_{1,A} \ = \ 3X_A$$

$$X_{1,B} \ = \ 4X_A$$

$$X_{2,A} \ = \ 6X_A$$

$$X_{2,B} \ = \ 2X_A$$

The new full concrete formulation is shown in figure A.2.

Now when we solve the concrete LP formulation, we explicitly learn that the optimal assigned times in the prior optimal solution are: $X_{1,A} = 4.333333$, $X_{1,B} = 16.666667$, $X_{2,A} = 8.666667$, $X_{2,B} = 8.333333$.

Summary

This small example illustrates the three parts of every linear programming formulation: an objective function (either to *maximize* or *minimize*) a *linear* function of a (sub)set of the LP variables; a set of *linear* (in)equalities (constraints), each defined on a (sub)set of the LP variables; and a set of *bounds*, each defined on a single LP variable. Each bound is actually a constraint, and so could be considered as part of the constraints, but are historically separated from them.

Solving Concrete LP Formulations

In the preceding example, we showed an optimal solution for the concrete LP formulation but did not say how that solution was obtained. There are several *algorithms* (well-specified methods) that can take any concrete LP formulation and find an optimal solution; or determine that the formulation is infeasible; or that the solution value is unbounded (which usually indicates an error in the LP formulation).

Algorithms

The first and most famous LP *algorithm* is called the *simplex algorithm*, developed by George Dantzig during World War II. It is still the basis for many practical LP-solvers, although additional refinements have been made to the original method. Further, other algorithms have been developed that are based on very different ideas than the simplex algorithm. Some of those algorithms have theoretical properties that the simplex algorithm lacks. For example, some LP algorithms are *provably efficient* in a theoretical sense—they solve the problem in worst case time bounded by a polynomial function of the size of the LP formulation, which is a property that does not hold for the simplex algorithm, despite its efficiency in practice. However, for the purposes of this book, we do not need the details of any of these algorithms, or any of their theoretical properties. What is important in this book is the fact that highly engineered computer programs have been developed that implement LP algorithms, and these programs are very effective in practice.

LP-solvers

When the details of an LP-algorithm are written into an executable computer program, the program is called an *LP-solver*. An LP-solver takes in a concrete LP formulation (usually in some very user-unfriendly format), and returns values assigned to the LP variables in an optimal solution. It is beyond the scope of this book to review available LP-solvers, but we point out that we have had good experience with two widely used LP-solvers. The LP-solver that we have used the most is called *CPLEX*. It is a proprietary, commercially created LP-solver that has been extensively tuned and engineered over many years. Currently, *CPLEX* is owned by IBM, and it has recently made *academic* licenses available for free. In our experience, CPLEX is very fast and capable of handling huge LP formulations. The second LP-solver that we have experience with is the *GNU linear programming kit (GLPK)*, an open source, free, LP-solver available from the *GNU Project*. In our experience, *GLPK* is also very effective and capable of handling large LP formulations, although it may run slower than CPLEX—much slower in some extreme cases.

Integer Linear Programming (ILP)

Linear programming allows the LP variables to be given *fractional*, that is, non-integer, values. For example $X_{2,B}$ has value 8.33333 in the optimal solution found above. Integer linear programming simply refines linear programming by *requiring* that the variables in a formulation only be given *integer* values. However, all the *coefficients* and the right-hand-side *constants* in the formulation are still allowed to be fractional. As before, the integer values assigned to the variables are required to be a feasible solution to the LP formulation. Such a set of values is called an *integer feasible solution*—the best integer feasible solution is called *integer optimal*. Clearly, the optimal solution to an LP formulation (without the integrality requirement) will be no worse, and is often strictly better, than the integer optimal solution. For example, the integer optimal solution to the LP formulation for the chemical production problem has value 40, in contrast to the optimal LP solution, which has value 46.5.

An LP formulation where all the variables are required to take on integer values is called an *ILP* formulation. An ILP formulation where the variables are further constrained to only take on values of *zero* or *one* is called a *binary* formulation. Binary formulations are very commonly used; for example all of the ILP formulations developed in this book for ARG and network problems, are binary formulations.

When we convert the concrete LP formulation for the chemical production problem to ILP formulation, by requiring that all the variables take on positive integer values, the optimal ILP solution (of value 40) sets X_A to 1, X_B to 4, $X_{1,A}$ to 3, $X_{1,B}$ to 16, $X_{2,A}$ to 6, and $X_{2,B}$ to 8.

ILP-Solvers

The two LP-solvers discussed earlier, CPLEX and GLPK, are also ILP-solvers, returning an integer optimal solution, or determining that no integer feasible solution exists, or that the solution value is unbounded. In addition to the LP formulation, the user must specify which variables are restricted to integer values, or binary values. Additional constraints, such as requiring a variable to be set in a certain range of values, can be specified in the constraints.

At the high-level, the approaches that ILP-solvers use are quite different from LP-solvers, but the methods usually require creating and solving many concrete LP formulations. Thus, the time to solve an LP formulation is generally much less than the time needed for the same ILP formulation, that is, when the variables are required to have integer values. Further, unlike the case of linear programming, where theoretically efficient LP-solvers have been created, no such ILP-solver exists. In fact, the problem of solving

ILP formulations is known to be *NP-hard*, which suggests that no theoretically efficient ILP-solver is possible. Such issues are out of the scope of this book, and despite those theoretical results, highly tuned programs such as CPLEX are surprisingly effective on many ILP formulations, including the specific ones discussed in this book.

Expressibility of ILP formulations

The fact that the problem of solving ILP formulations is *NP-hard* suggests that no theoretically efficient ILP-solver can exist. However, it also establishes a more *positive* fact: Any problem in a huge class of natural problems can provably and efficiently (i.e., with a polynomially bounded number of variables and constraints) be expressed by an ILP formulation.

More technically, any problem in the class *NP* or any optimization variant of a problem in *NP*, can be *reduced* in *worst-case polynomial time* to some ILP formulation. Again, a full discussion of this is outside the scope of this book, but it establishes that despite being restricted to *linear* constraints and *linear* objective functions and *integer* values, ILP formulations are actually very expressive; they are able to represent "most" natural discrete optimization problems. Rarely does one encounter a discrete optimization problem that cannot be efficiently formulated (through a fast reduction resulting in a small ILP formulation) as an ILP problem. Sometimes the formulation is tricky to find, but in many cases, the formulation is quite simple, although that does not guarantee that its solution will be simple. Moreover, sometimes a larger, more detailed ILP formulation is simpler to design, and more efficiently solved by the ILP-solver. For example, in section 14.5.3, we saw that the equality constraint

$$\sum_{e \in E} D_e = \sum_{e' \in E'} D_{e'}$$

could be removed from the ILP formulation. However, inclusion of such "redundant" constraints often helps the ILP-solver find an optimal solution faster. Because the constraint is mathematically redundant (i.e., implied by the other constraints), the ILP-solver will find an optimal solution where the equality constraint is satisfied, even if it is not explicitly in the ILP formulation. But having it explicitly stated in the formulation can reduce the time needed to find an optimal solution.

Abstract ILP Formulations

The LP and ILP examples given above are called "concrete" formulations, meaning that they completely specify a problem instance and (suitably formatted) can be input to an LP or ILP-solver. In contrast, when we demonstrate how a general problem can be solved by integer programming, we describe the "abstract" ILP formulation. For example, consider the following general problem:

> Given n assistants, and m chemicals, where each assistant can work a given amount of time, and one unit of each chemical requires the input of some specified amount of time (possibly zero) from each assistant, and each unit of a chemical has a given sales price, determine the amount of each chemical to make to maximize the total value of the chemicals produced, and find a feasible allocation of assistant times to achieve that combination of chemicals. Certain specified chemicals can only be made in whole-unit quantities, while the others can be made in fractional-unit quantities.

An *abstract* ILP formulation for this general problem is as follows: Let C_j denote the price of a unit of chemical j, let T_i denote the amount of time Assistant i can work, and let $T_{i,j}$ denote the amount of time required from Assistant i to produce a unit of chemical j. Let $\mathcal{I}$ be the set of chemicals that can only be produced in whole-unit quantities. Then, create a variable X_j for each chemical j, denoting the amount of chemical j to produce; create a variable $X_{i,j}$ for each Assistant i and each chemical j, to denote how much time Assistant i will contribute to the creation of chemical j.

The **abstract objective function** is:

$$\text{Maximize} \sum_j C_j \times X_j.$$

The constraints consist of one inequality for each Assistant i:

$$\sum_j T_{i,j} \times X_j \leq T_i$$

and one equality for each Assistant i and each chemical j:

$$X_{i,j} = T_{i,j} \times X_j.$$

The bounds consist of:

$$\text{for each chemical } j, \ X_j \geq 0$$

and

$$\text{for each chemical } j \in \mathcal{I}, \ X_j \text{ must be an integer.}$$

An abstract ILP formulation describes the *logic* and the general form of the formulation. And, for any specific problem instance, the abstract formulation can be converted into a *concrete* ILP formulation. The ILP formulations described in this book, for ARG and network problems, are *abstract* ILP formulations.

Bibliography

[1] The 1000 Genomes Project Consortium. A map of human genome variation from population-scale sequencing. *Nature*, 467:1061–1073, 2010.

[2] The 1000 Genomes Project Consortium. An integrated map of genetic variation from 1,092 genomes. *Nature*, 491:56, 2012.

[3] R. Agarwala and D. Fernandez-Baca. A polynomial-time algorithm for the perfect phylogeny problem when the number of character states is fixed. *SIAM Journal on Computing*, 23:1216–1224, 1994.

[4] R. Agarwala, D. Fernandez-Baca, and Giora Slutzki. Fast algorithms for inferring evolutionary trees. *Journal of Computational Biology*, 2:397–408, 1995.

[5] A. Aho, Y. Sagiv, T. Szymanski, and D. Ullman. Inferring a tree from lowest common ancestors with an application to the optimization of relational expressions. *SIAM Journal on Computing*, 10:405–421, 1981.

[6] R. K. Ahuja, T. L. Magnanti, and J. B. Orlin. *Network Flows: Theory, Algorithms, and Applications*. Prentice Hall, 1993.

[7] B. Albrecht, C. Scornavacca, A. Cenci, and D. Huson. Fast computation of minimum hybridization networks. *Bioinformatics*, 28:191–197, 2012.

[8] B. Allen and M. Steel. Subtree transfer operations and their induced metrics on evolutionary trees. *Annals of Combinatorics (online)*, 5, 2001.

[9] D. Altshuler, M. Daly, and E. Lander. Genetic mapping in human disease. *Science*, 322:881–888, 2008.

[10] M. Arenas, M. Patricio, D. Posada, and G. Valiente. Characterization of phylogenetic networks with NetTest. *BMC Bioinformatics*, 11:268, 2010.

[11] M. Arenas. Computer programs and methodologies for the simulation of DNA sequence data with recombination. *Frontiers in Genetics*, 4, 2013.

[12] M. Arenas. The importance and application of the ancestral recombination graph. *Frontiers in Genetics*, 4, 2013.

[13] M. Arenas, G. Valiente, and D. Posada. Characterization of reticulate networks based on the coalescent with recombination. *Molecular Biology and Evolution*, 25:2517–2520, 2008.

[14] F. Ariey and D. Menard et al. A molecular marker of artemisinin-resistant *Plasmodium falciparum* malaria. *Nature*, 505:50–55, 2014.

[15] N. Arnheim, P. Calabrese, and M. Nordborg. Hot and cold spots of recombination in the human genome: The reason we should find them and how this can be achieved. *American Journal of Human Genetics*, 73:5–16, 2003.

[16] N. Arnheim, P. Calabrese, and I. Tiemann-Boege. Mammalian meiotic recombination hotspots. *Annual Review of Genetics*, 41:363–399, 2007.

[17] V. Bafna and V. Bansal. The number of recombination events in a sample history: conflict graph and lower bounds. *IEEE/ACM Transactions on Computational Biology and Bioinformatics*, 1:78–90, 2004.

[18] V. Bafna and V. Bansal. Improved recombination lower bounds for haplotype data. In *RECOMB, The Annual International Conference on Research in Computational Molecular Biology*. LNBI 3500, Springer, 2005.

[19] V. Bafna and V. Bansal. Inference about recombination from haplotype data: Lower bounds and recombination hotspots. *Journal of Computational Biology*, 13:501–521, 2006.

[20] V. Bafna, D. Gusfield, S. Hannenhalli, and S. Yooseph. A note on efficient computation of haplotypes via perfect phylogeny. *Journal of Computational Biology*, 11(5):858–866, 2004.

[21] V. Bafna, D. Gusfield, G. Lancia, and S. Yooseph. Haplotyping as perfect phylogeny: A direct approach. *J. Computational Biology*, 10:323–340, 2003.

[22] H. J. Bandelt, P. Foster, and A. Röhl. Median-joining networks for inferring intraspecific phylogenies. *Molecular Biology and Evolution*, 16:37–48, 1999.

[23] H. J. Bandelt, P. Foster, B. Sykes, and M. Richards. Mitochondrial portaits of human populations using median networks. *Genetics*, 141:743–753, 1995.

[24] H. J. Bandelt and A. W. Dress. Split decomposition: A new and useful approach to phylogenetic analysis of distance data. *Molecular Phylogenetics and Evolution*, 1(3):242–252, 1992.

[25] E. Bapteste, Y. Boucher, and W.F. Doolittle. The fate of phylogenetics in the face of lateral gene transfer. In J.H. Schulz, editor, *Genetic Recombination Research Progress*, pages 139–162. Nova Science Press, 2008.

[26] M. Baroni, S. Grunewald, V. Moulton, and C. Semple. Bounding the number of hybridisation events for a consistent evolutionary history. *Journal of Mathematical Biology*, 51:171–182, 2005.

[27] M. Baroni, C. Semple, and M. Steel. A framework for representing reticulate evolution. *Annals of Combinatorics*, 8:391–408, 2004.

[28] T. Barzuza, J.S. Beckman, R. Shamir, and I. Pe'er. Computational problems in perfect phylogeny haplotyping: XOR genotypes and TAG SNPs. In *CPM, Symposium on Combinatorial Pattern Matching*, 2004.

[29] T. Barzuza, J.S. Beckmann, and R. Shamir. Computational problems in perfect phylogeny haplotyping: Typing without calling the allele. *IEEE/ACM Transactions on Computational Biology and Bioinformatics*, 5(1):101–109, 2008.

[30] T. Barzuza, J.S. Beckmann, R. Shamir, and I. Pe'er. Typing without calling the allele: A strategy for inferring SNP haplotypes. *European Journal of Human Genetics*, 13:898–901, 2005.

[31] T. Bersaglier and P. C. Sabeti et al. Genetic signatures of strong recent positive selection at the lactase gene. *American Journal of Human Genetics*, 74:1111–1120, 2004.

[32] S. Besenbacher, T. Mailund, and M. Schierup. Local phylogeny mapping of quantitative traits: Higher accuracy and better ranking than single-marker association in genomewide scans. *Genetics*, 181:747–753, 2009.

[33] A. Bigham and M.D. Schriver et al. Identifying signatures of natural selection in Tibetan and Andean populations using dense genome scan data. *PLoS Genetics*, 6(9):e1001116, 09 2010.

[34] R. E. Bixby and D. K. Wagner. An almost linear-time algorithm for graph realization. *Mathematics of Operations Research*, 13:99–123, 1988.

[35] G. Blin and R. Rizzi et al. Minimum mosaic inference of a set of recombinants. *International Journal of Foundations of Computer Science*, 24:51–66, 2011.

[36] J. S. Bloom, I.M. Ehrenreich, W. T. Loo, T.L.V Lite, and L. Kruglyak. Finding the sources of missing heritability in a yeast cross. *Nature*, 494:234–237, 2013.

[37] E. Bloomquist and M.A. Suchard. Unifying vertical and nonvertical evolution: a stochastic ARG-based framework. *Systematic Biology*, 59:2741, 2010.

[38] H. Bodlaender, M. Fellows, and T. Warnow. Two strikes against perfect phylogeny. *Proceedings of the 19th International Colloquium on Automata, Languages and Programming*, pages 273–283, 1992.

[39] A. Bondy and U. S. R. Murty. *Graph Theory*. Springer, Graduate Texts in Mathematics, 2008.

[40] P. Bonizzoni. A linear-time algorithm for the perfect phylogeny haplotype problem. *Algorithmica*, 48:267–285, 2007.

[41] P. Bonizzoni, A.P. Carrieri, G. Della Vedova, R. Dondi, and T.M. Przytycka. When and how the perfect phylogeny model explains evolution. In N. Jonoska and M. Saito, editors, *Discrete and Topological Models in Molecular Biology*, Natural Computing Series, chapter 4. Springer, 2013.

[42] P. Bonizzoni, G. Della Vedova, R. Dondi, and J. Li. The haplotyping problem: Models and solutions. *Journal of Computer Science and Technology*, 18:675–688, 2003.

[43] M. Bordewich, C. McCartin, and C. Semple. A 3-approximation algorithm for the subtree distance between phylogenies. *Journal of Discrete Algorithms*, 6(3):458–471, 2008.

[44] M. Bordewich and C. Semple. On the computational complexity of the rooted subtree prune and regraft distance. *Annals of Combinatorics*, 8:409–423, 2004.

[45] M. Bordewich and C. Semple. Computing the minimum number of hybridization events for a consistent evolutionary history. *Discrete Applied Math*, 155:914–928, 2007.

[46] D. Botstein, R. L. White, M. Skolnick, and R. W. Davis. Construction of a genetic linkage map in man using restriction fragment length polymorphisms. *American Journal of Human Genetics*, 32:314–331, 1980.

[47] U. Brandes and S. Cornelsen. Phylogenetic graph models beyond trees. *Discrete Applied Math*, 157:2361–2369, 2009.

[48] D. Brown and I. Harrower. A new integer programming formulation for the pure parsimony problem in haplotype analysis. In *WABI, Workshop on Algorithms in Bioinformatics*, volume 3240, pages 254–265. LNCS, Springer, 2004.

[49] D. Brown and I. Harrower. Toward an algebraic understanding of haplotype inference by pure parsimony. In *Computational Systems Bioinformatics Conference*. LNCS, Springer-Verlag, 2006.

[50] D. Brown and I.M. Harrower. Integer Programming Approaches to Haplotype Inference by Pure Parsimony. *IEEE/ACM Transactions on Computational Biology and Bioinformatics*, 3(2):141–154, 2006.

[51] S.R. Browning and B.L. Browning. Rapid and accurate haplotype phasing and missing-data inference for whole-genome association studies by use of localized haplotype clustering. *American Journal of Human Genetics*, 81:1084–1097, 2007.

[52] S.R. Browning and B.L. Browning. Haplotype phasing: Existing methods and new developments. *Nature Reviews Genetics*, 12:703–714, 2011.

[53] D. Bryant and V. Moulton. Neighbornet: An agglomerative method for the construction of planar phylogenetic networks. In R. Guigo and D. Gusfield, editors, *WABI, Workshop on Algorithms in Bioinformatics*, volume 2452 of *Lecture Notes in Computer Science*, pages 375–391. Springer, 2002.

[54] P. Buneman. The recovery of trees from measures of dissimilarity. In D.G. Kendall and P. Tautu, editors, *Mathematics in the archaeological and historical sciences*, pages 387–385. Edinburgh University Press, 1971.

[55] P. Buneman. A characterization of rigid circuit graphs. *Discrete Math*, 9:205–212, 1974.

[56] C. Campbell, Z. Wang, and Y. Qian. Ancestral recombination histories for error detection in genome sequencing. Technical report, University of Oxford, Statistics Department, 2010.

[57] G. Cardona, M. Llabres, F. Rossello, and G. Valiente. A distance metric for a class of tree-sibling phylogenetic networks. *Bioinformatics*, 24(13):1481–1488, 2008.

[58] G. Cardona, M. Llabres, F. Rossello, and G. Valiente. Metrics for phylogenetic networks i: Generalizations of the Robinson-Foulds metric. *IEEE/ACM Transactions on Computational Biology and Bioinformatics*, 6(1):46–61, 2009.

[59] G. Cardona, M. Llabres, F. Rossello, and G. Valiente. Comparison of galled trees. *IEEE/ACM Transactions on Computational Biology and Bioinformatics*, 8(2):410–427, 2011.

[60] G. Cardona, F. Rossello, and G. Valiente. Comparison of tree-child phylogenetic networks. *IEEE/ACM Transactions on Computational Biology and Bioinformatics*, 6:552–569, 2009.

[61] N. Casali and F. Drobniewski et al. Evolution and transmission of drug-resistant tuberculosis in a russian population. *Nature Genetics*, 46:279–286, 2014.

[62] A. Chakravarti. It's raining SNP's, hallelujah? *Nature Genetics*, 19:216–217, 1998.

[63] P. Charbit, M. Habib, W. Limouzy, F. de Montgolfier, M. Raffinot, and M. Rao. A note on computing set overlap classes. *Information Processing Letters*, 108:186–191, 2008.

[64] B. Charlesworth. Effective population size and patterns of molecular evolution and variation. *Nature Reviews: Genetics*, 10:195–205, 2009.

[65] N. Charlton, I. Carbone, S. Tavantzis, and M. Cubeta. Phylogenetic relatedness of the M2 double-stranded RNA in *Rhizoctonia* fungi. *Mycologia*, 100:555–64, 2008.

[66] Z. Z. Chen and L. Wang. Algorithms for reticulate networks of multiple phylogenetic trees. *IEEE/ACM Transactions on Computational Biology and Bioinformatics*, 9(2):372–384, 2012.

[67] Z. Z. Chen and L. Wang. An ultrafast algorithm for reticulate networks. *Journal of Computational Biology*, 20:38–41, 2013.

[68] C. Chewapreecha and S. Bentley et al. Dense genomic sampling identifies highways of pneumococcal recombination. *Nature Genetics*, 46:305–309, 2014.

[69] R.H. Chung and D. Gusfield. Empirical exploration of perfect phylogeny haplotyping and haplotypers. In *Proceedings of the 9th International Conference on Computing and Combinatorics COCOON03*, volume 2697 of *LNCS*, pages 5–19, 2003.

[70] A. Clark. Association testing with phased haplotypes vs. unphased diplotypes. Talk given at USC RECOMB workshop on Computational Methods for SNPs and Haplotypes, January 27, 2007.

[71] A. Clark. Inference of haplotypes from PCR-amplified samples of diploid populations. *Molecular Biology and Evolution*, 7:111–122, 1990.

[72] A. Clark. Finding genes underlying risk of complex disease by linkage disequilibrium mapping. *Current Opinion in Genetics & Development*, 13:296–302, 2003.

[73] A. Clark, X. Wang, and T. Matise. Contrasting methods of quantifying fine structure of human recombination. *Annual Review of Genomics and Human Genetics*, 11:4564, 2010.

[74] A. Clark, K. Weiss, and D. Nickerson et al. Haplotype structure and population genetic inferences from nucleotide-sequence variation in human lipoprotein lipase. *American Journal of Human Genetics*, 63:595–612, 1998.

[75] T. G. Clark, M. De Iorio, and R. C. Griffiths. Bayesian logistic regression using a perfect phylogeny. *Biostatistics*, 8:32–52, 2007.

[76] F. Cohan. What are bacterial species? *Annual Review of Microbiology*, 56:45787, 2002.

[77] F. Cole, S. Keeney, and M. Jasin. Preaching about the converted: How meiotic gene conversion influences genomic diversity. *Annals of the New York Academy of Sciences*, 1267(1):95–102, 2012.

[78] J. Comeron, R. Ratnappan, and S. Bailin. The many landscapes of recombination in *drosophila melanogaster*. *PLoS Genetics*, 8:e1002905, 10 2012.

[79] G. Coop and M. Przeworski. An evolutionary view of human recombination. *Nature Reviews Genetics*, 8:23–34, 2007.

[80] T. Cormen, C. Leiserson, R. Rivest, and C. Stein. *Introduction to Algorithms, 3rd edition*. MIT Press, 2009.

[81] N. C. Crawford and T. C. Glenn et al. More than 1000 ultraconserved elements provide evidence that turtles are the sister group of archosaurs. *Biology Letters*, 8:783 – 786, 2012.

[82] T. Dagan and W. Martin. Getting a better picture of microbial evolution en route to a network of genomes. *Philosophical Transactions of the Royal Society B*, 364(1527):2187–2196, 2009.

[83] E. Dahlhaus. Parallel algorithms for hierarchial clustering and applications to split decomposition and parity graph recognition. *Journal of Algorithms*, 36:205–240, 2000.

[84] M. Daly, J. Rioux, S. Schaffner, T. Hudson, and E. Lander. Fine-structure haplotype map of 5q31: Implications for gene-based studies and genomic LD mapping. Abstract of talk presented at the American Associate of Human Genetics National meeting, October 14, 2001.

[85] M. Daly, J. Rioux, S. Schaffner, T. Hudson, and E. Lander. High-resolution haplotype structure in the human genome. *Nature Genetics*, 29:229–232, 2001.

[86] S. Dasgupta, C. H. Papadimitriou, and U. V. Vazirani. *Algorithms*. McGraw Hill, 2008.

[87] W. H. Day and D. Sankoff. Computational complexity of inferring phylogenies by compatibility. *Systematic Zoology*, 35:224–229, 1986.

[88] A. de Vries and G. te Meerman. A haplotype sharing method for determining the relative age of SNP alleles. *Human Heredity*, 69:52–59, 2010.

[89] Z. Ding, V. Filkov, and D. Gusfield. A linear-time algorithm for the perfect phylogeny haplotyping problem. In *RECOMB, The Annual International Conference on Research in Computational Molecular Biology*, pages 585–600. LNCI 3500, Springer, 2005.

[90] Z. Ding, V. Filkov, and D. Gusfield. A linear-time algorithm for the perfect phylogeny haplotyping problem. *Journal of Computational Biology*, 13(2):522–553, 2006.

[91] Z. Ding, T. Mailund, and Y. S. Song. Efficient whole-genome association mapping using local phylogenies for unphased genotype data. *Bioinformatics*, 24(19):2215–2221, 2008.

[92] L. Dollo. Le lois de l'évolution. *Bulletin de la Societé Belge de Géologie de Paléontologie et d'Hydrologie*, 7:164–167, 1893.

[93] W. F. Doolittle. Phylogenetic classification and the universal tree. *Science*, 284:2124–2129, 1999.

[94] W. F. Doolittle. Uprooting the tree of life. *Scientific American*, 282:90–95, 2000.

[95] J. Dopazo, A. W. M. Dress, and A. von Haeseler. Split decomposition: a new technique to analyse viral evolution. *Proceedings of the National Academy of Sciences (USA)*, 90:10320–10324, 1993.

[96] A. Dress, K. T. Huber, J. Koolen, V. Moulton, and A. Spillner. *Basic Phylogenetic Combinatorics*. Cambridge University Press, 2012.

[97] A. Dress and M. Steel. Convex tree realizations of partitions. *Applied Math Letters*, 5:3–6, 1993.

[98] A. W. M. Dress and D. Huson. Constructing splits graphs. *IEEE/ACM Transactions on Computational Biology and Bioinformatics*, 1(3):109–115, 2004.

[99] C. Drysdale and S. Ligget et al. Complex promoter and coding region $\beta2$-adrenergic receptor haplotypes alter receptor expression and predict *in vivo* responsiveness. *Proceedings of the National Academy of Sciences (USA)*, 97:10483–10488, 2000.

[100] A. Efros and E. Halperin. Haplotype reconstruction using perfect phylogeny and sequence data. *BMC Bioinformatics*, 13, 2012.

[101] I. M. Ehrenreich and M. D. Purugganan et al. Candidate gene association mapping of arabidopsis flowering time. *Genetics*, 183:325–335, 2009.

[102] N. El-Mabrouk. Deriving haplotypes through recombination and gene conversion. *Journal of Bioinformatics and Computational Biology*, 2:241–256, 2004.

[103] N. El-Mabrouk and D. Labuda. Haplotypes histories as pathways of recombinations. *Bioinformatics*, 20:1836–1841, 2004.

[104] J.A. Endler. *Natural Selection in the Wild*. Princeton University Press, 1986.

[105] E. Eskin, E. Halperin, and R. M. Karp. Large scale reconstruction of haplotypes from genotype data. In *RECOMB, The Annual International Conference on Research in Computational Molecular Biology*, pages 104–113, 2003.

[106] E. Eskin, E. Halperin, and R.M. Karp. Efficient reconstruction of haplotype structure via perfect phylogeny. *Journal of Bioinformatics and Computational Biology*, 1:1–20, 2003.

[107] G. Estabrook, C. Johnson, and F. McMorris. An idealized concept of the true cladistic character. *Mathematical Bioscience*, 23:263–272, 1975.

[108] G. Estabrook, C. Johnson, and F. McMorris. An algebraic analysis of cladistic characters. *Discrete Math*, 16:141–147, 1976.

[109] G. Estabrook, C. Johnson, and F. McMorris. A mathematical foundation for the analysis of cladistic character compatibility. *Mathematical Bioscience*, 29:181–187, 1976.

[110] P. Fearnhead and P. Donnelly. Estimating recombination rates from population genetic data. *Genetics*, 159:1299–1318, 2001.

[111] P. Fearnhead, R.M. Harding, J.A. Schneider, S. Myers, and P. Donnelly. Application of coalescent methods to reveal fine scale rate variation and recombination hotspots. *Genetics*, 167:2067–2081, 2004.

[112] J. Felsenstein. *Inferring Phylogenies*. Sinauer, 2004.

[113] J. Felsenstein. Trees of genes in populations. In O. Gascuel and M. Steel, editors, *Reconstructing Evolution: New Mathematical and Computational Advances*, pages 3–25. Oxford University Press, 2007.

[114] D. Fernandez-Baca. The perfect phylogeny problem. In D.Z. Du and X. Cheng, editors, *Steiner Trees in Industries*. Kluwer Academic Publishers, 2000.

[115] W. Fitch. Towards finding the tree of maximum parsimony. In G. F. Estabrook, editor, *Proceedings of the Eighth International Conference on Numerical Taxonomy*, pages 189–230. W. H. Freeman, 1975.

[116] K. Frazer, E. Eskin, and D. R. Cox et al. A sequence-based variation map of 8.27 million SNPs in inbred mouse strains. *Nature*, 448:831–843, 2007.

[117] R. H. French-Constant. The molecular genetics of insecticide resistance. *Genetics*, 194:807–815, 2013.

[118] L. Friss, R. Hudson, A. Bartoszewicz, J. Wall, T. Donfalk, and A. Di Rienzo. Gene conversion and differential population histories may explain the contrast between polymorphism and linkage disequilibrium levels. *American Journal of Human Genetics*, 69:831–843, 2001.

[119] W. Fu and J.M. Akey et al. Analysis of 6,515 exomes reveals the recent origin of most human protein-coding variants. *Nature*, 493:216–220, 2013.

[120] P. Gambette. http://www.lirmm.fr/~gambette/RePhylogeneticNetworks.php.

[121] P. Gambette. Who is who in phylogenetic networks: Articles, authors and programs. http://www.atgc-montpellier.fr/phylnet.

[122] P. Gambette and K. Huber. On encodings of phylogenetic networks of bounded level. *Mathematical Biology*, 65:157–180, 2012.

[123] M. Garey and D. Johnson. *Computers and intractability*. W.H. Freeman, 1979.

[124] F. Gavril and R. Tamari. An algorithm for constructing edge-trees from hypergraphs. *Networks*, 13:377–388, 1983.

[125] G. Gibson. Rare and common variants: Twenty arguments. *Nature Review Genetics*, 13:135–145, 2012.

[126] P. Gogarten and J. P. Townsand. Horizontal gene transfer, genome innovation and evolution. *Nature Reviews, Microbiology*, 9:679–687, 2005.

[127] A. Graça, I. Lynce, J. Marques-Silva, and A. L. Oliveira. Haplotype inference by pure parsimony: A survey. *Journal of Computational Biology*, 17:969–992, 2010.

[128] J. Gramm, T. Nierhoff, R. Sharan, and T. Tantau. On the complexity of haplotyping via perfect phylogeny. In *Second RECOMB Satellite Workshop on Computational Methods for SNPs and Haplotypes*, pages 20–21. LNBI, Springer, 2004.

[129] J. Gramm, T. Nierhoff, and T. Tantau. Perfect path phylogeny haplotyping with missing data is fixed-parameter tractable. In *First International Workshop on Parametrized and Exact Computation (IWPEC 2004)*. LNCS, Springer, 2004.

[130] R. C. Griffiths and P. Marjoram. Ancestral inference from samples of DNA sequences with recombination. *Journal of Computational Biology*, 3:479–502, 1996.

[131] R. C. Griffiths and P. Marjoram. An ancestral recombination graph. In P. Donnelly and S. Tavare, editors, *Progress in Population Genetics and Human Evolution*, pages 257–270. IMA Volumes in Mathematics and Its Applications, vol. 87, 1997.

[132] S. R. Grossman and P. C. Sabeti et al. Identifying recent adaptations in large-scale genomic data. *Cell*, 152(4):703–713, 2013.

[133] A. Gupta, J. Manuch, L. Stacho, and X. Zhao. Haplotype inferring via galled-tree networks is NP-complete. In *Computing and Combinatorics*, volume 5092, pages 287–298. LNCS, Springer, 2008.

[134] A. Gupta, J. Manuch, L. Stacho, and X. Zhao. Algorithm for haplotype inference via galled-tree networks with simple galls. *Journal of Computational Biology*, 19:439–454, 2012.

[135] A. Gupta, J. Manuch, X. Zhao, and L. Stacho. Characterization of the existence of galled-tree networks. *Journal of Bioinformatics and Computational Biology*, 4:1309–1328, 2006.

[136] D. Gusfield. Efficient algorithms for inferring evolutionary history. *Networks*, 21:19–28, 1991.

[137] D. Gusfield. *Algorithms on Strings, Trees and Sequences: Computer Science and Computational Biology*. Cambridge University Press, 1997.

[138] D. Gusfield. A practical algorithm for deducing haplotypes in diploid populations. In *Proceedings of 8th International Conference on Intelligent Systems in Molecular Biology*, pages 183–189. AAAI Press, 2000.

[139] D. Gusfield. Haplotyping as perfect phylogeny: Conceptual framework and efficient solutions. In *RECOMB, The Annual International Conference on Research in Computational Molecular Biology*, pages 166–175. ACM Press, 2002.

[140] D. Gusfield. Haplotype inference by pure parsimony. In R. Baeza-Yates, E. Chavez, and M. Chrochemore, editors, *Proceedings of the Annual Symposium on Combinatorial Pattern Matching*, volume 2676, pages 144–155. LNCS, Springer, 2003.

[141] D. Gusfield. On the decomposition optimality conjecture for phylogenetic networks. Technical report, UC Davis, Department of Computer Science, 2005.

[142] D. Gusfield. Optimal, efficient reconstruction of root-unknown phylogenetic networks with constrained and structured recombination. *Journal of Computer and System Sciences*, 70:381–398, 2005.

[143] D. Gusfield. The multi-state perfect phylogeny problem with missing and removable data: Solutions via integer-programming and chordal graph theory. *Journal of Computational Biology*, 17:383–399, 2010.

[144] D. Gusfield and V. Bansal. A fundamental decomposition theory for phylogenetic networks and incompatible characters. In *RECOMB, The Annual International Conference on Research in Computational Molecular Biology*, pages 217–232. LNBI 3500, Springer, 2005.

[145] D. Gusfield, V. Bansal, V. Bafna, and Y. S. Song. A decomposition theory for phylogenetic networks and incompatible characters. *Journal of Computational Biology*, 14:1247–1272, 2007.

[146] D. Gusfield, S. Eddhu, and C. Langley. The fine structure of galls in phylogenetic networks. *INFORMS Journal on Computing, Special Issue on Computational Biology*, 16:459–469, 2004.

[147] D. Gusfield, S. Eddhu, and C. Langley. Optimal, efficient reconstruction of phylogenetic networks with constrained recombination. *Journal of Bioinformatics and Computational Biology*, 2(1):173–213, 2004.

[148] D. Gusfield, Y. Frid, and D. Brown. Integer programming formulations and computations solving phylogenetic and population genetic problems with missing or genotypic data. In *Proceedings of 13th Annual International Conference on Combinatorics and Computing*, pages 51–64. LNCS 4598, Springer, 2007.

[149] D. Gusfield and D. Hickerson. A new lower bound on the number of needed recombination nodes in both unrooted and rooted phylogenetic networks. Report UCD-ECS-2004-06. Technical report, University of California, Davis, 2004.

[150] D. Gusfield, D. Hickerson, and S. Eddhu. An efficiently-computed lower bound on the number of recombinations in phylogenetic networks: Theory and empirical study. *Discrete Applied Math, Special Issue on Computational Biology, 2007*, 155:806–830, 2007.

[151] D. Gusfield and S. Orzack. Haplotype inference. In S. Aluru, editor, *Handbook of Computational Molecular Biology*, pages 18/1–18/25. Chapman and Hall/CRC, 2005.

[152] D. Gusfield and Y. Wu. The three-state perfect phylogeny problem reduces to 2-SAT. *Communication and Information Sciences*, 9:195–201, 2009.

[153] R. Gysel and D. Gusfield. Extensions and improvements to the chordal graph approach to the multistate perfect phylogeny problem. *IEEE/ACM Transactions on Computational Biology and Bioinformatics*, 8(4):912–917, 2011.

[154] R. Gysel, F. Lam, and D. Gusfield. Constructing perfect phylogenies and proper triangulations for three-state characters. *Algorithms in Molecular Biology*, 7:26, 2012.

[155] M. Habib and T-H To. On a conjecture about compatibility of multi-states characters. In T. Przytycka and M.F. Sagot, editors, *Algorithms in Bioinformatics*, volume 6833 of *Lecture Notes in Computer Science*, pages 116–127. Springer, 2011.

[156] B. Halldorsson, V. Bafna, N. Edwards, R. Lipert, S. Yooseph, and S. Istrail. Combinatorial problems arising in SNP and haplotype analysis. In C. Calude, M. Dinneen, and V. Vajnovski, editors, *Discrete Mathematics and Theoretical Computer Science. Proceedings of DMTCS 2003*, volume 2731 of *Lecture Notes in Computer Science*, pages 26–47. Springer, 2003.

[157] B. Halldorsson, V. Bafna, N. Edwards, R. Lipert, S. Yooseph, and S. Istrail. A survey of computational methods for determining haplotypes. In *Proceedings of the First RECOMB Satellite on Computational Methods for SNPs and Haplotype Inference*, volume 2983 of *Lecture Notes in Bioinformatics, LNBI*, pages 26–47. Springer, 2004.

[158] E. Halperin and E. Eskin. Haplotype reconstruction from genotype data using imperfect phylogeny. *Bioinformatics*, 20:1842–1849, 2004.

[159] E. Halperin and R. M. Karp. Perfect phylogeny and haplotype assignment. In *RECOMB, The Annual International Conference on Research in Computational Molecular Biology*, pages 10–19. ACM Press, 2004.

[160] W. Hao, V.G. Allen, F.B. Jamieson, D.E. Low, and D.C. Alexander. Phylogenetic incongruence in E. coli O104: Understanding the evolutionary relationships of emerging pathogens in the face of homologous recombination. *PLoS ONE*, 7:e33971, 2012.

[161] D. Hartl. *A Primer of Population Genetics, 2nd Edition*. Sinauer, 1988.

[162] J. He and A. Zelikovsky. Linear reduction for haplotype inference. In *WABI, Workshop on Algorithms in Bioinformatics*, volume 3240, pages 242–253. LNCS, Springer, 2004.

[163] Y. He, C. Li, C. Amos, M. Xiong, and H. Ling L. Jin. Accelerating haplotype-based genome-wide association study using perfect phylogeny and phase-known reference data. *PLoS One*, 6:e22097, 2011.

[164] A.M. Heimberg and R. Cowper-Sal lari et al. MicroRNAs reveal the interrelationships of hagfish, lampreys and gnathostomes and the nature of the ancestral vertebrate. *Proceedings of the National Academy of Sciences (USA)*, 107:19379–19383, 2010.

[165] J. Hein. Reconstructing evolution of sequences subject to recombination using parsimony. *Mathematical Bioscience*, 98:185–200, 1990.

[166] J. Hein. A heuristic method to reconstruct the history of sequences subject to recombination. *Journal of Molecular Evolution*, 36:396–405, 1993.

[167] J. Hein, T. Jiang, L. Wang, and K. Zhang. On the complexity of comparing evolutionary trees. *Discrete Applied Math*, 71:153–169, 1996.

[168] J. Hein, M. Schierup, and C. Wiuf. *Gene Genealogies, Variation and Evolution: A Primer in Coalescent Theory*. Oxford University Press, 2005.

[169] G. Hellenthal, G. Busby, G. Band, J. Wilson, C. Capelli, D. Falush, and S. Myers. A genetic atlas of human admixture history. *Science*, 343:747–751, 2014.

[170] L. Helmuth. Genome research: Map of the human genome 3.0. *Science*, 293(5530):583–585, 2001.

[171] T. Hill and H. Schiöth et al. SPRIT: Identifying horizontal gene transfer in rooted phylogenetic trees. *BMC Evolutionary Biology (online)*, 10(42), 2010.

[172] D. Hillis and C. Moritz (eds). *Molecular Systematics*. Sinauer Associates, 1990.

[173] D. Hillis, C. Moritz, and B. Mable (eds). *Molecular Systematics, 2nd edition*. Sinauer Associates, 1996.

[174] D. M. Hillis. SINEs of the perfect character. *Proceedings of the National Academy of Sciences (USA)*, 96:9979–9981, 1999.

[175] A. Hinch, D. Reich, and S. Myers et al. The landscape of recombination in African Americans. *Nature*, 476:170–175, 2011.

[176] D. Hinds, L. Stuve, G. Nilsen, E. Halperin, E. Eskin, D. Gallinger, K. Frazer, and D. Cox. Whole-genome patterns of common DNA variation in three human populations. *Science*, 307:1072–1079, 2005.

[177] A. Van't Holt, N. Edmonds, M. Kalikova, F. Marec, and I. Saccheri. Industrial melanism in British peppered moths has a singular and recent mutational origin. *Science*, 332:958–960, 2011.

[178] E. Hubbel. Personal communication, 2000.

[179] K. T. Huber, L. van Iersel, S. Kelk, and R. Suchecki. A practical algorithm for reconstructing level-1 phylogenetic networks. *TCBB*, 8(3):607–620, 2011.

[180] R. Hudson. Properties of a neutral allele model with intragenic recombination. *Theoretical Population Biology*, 23:183201, 1983.

[181] R. Hudson. Gene genealogies and the coalescent process. *Oxford Survey of Evolutionary Biology*, 7:1–44, 1990.

[182] R. Hudson. Generating samples under the Wright-Fisher neutral model of genetic variation. *Bioinformatics*, 18(2):337–338, 2002.

[183] R. Hudson and N. Kaplan. Statistical properties of the number of recombination events in the history of a sample of DNA sequences. *Genetics*, 111:147–164, 1985.

[184] R. Hudson, N. Kaplan, and C. H. Langley. The hitchhiking effect revisited. *Genetics*, 123:887–899, 1989.

[185] R. Hudson, A. G. Saez, and F. J. Ayala. DNA variation at the Sod locus of *Drosophila melanogaster*: An unfolding story of natural selection. *Proceedings of the National Academy of Sciences (USA)*, 94:7725–7729, 1997.

[186] P. J. Humphries, S. Linz, and C. Semple. On the complexity of computing the temporal hybridization number for two phylogenies. *Discrete Applied Mathematics*, 161:871–880, 2013.

[187] D. J. Hunter and P. Kraft. Drinking from the fire hose — Statistical issues in genomewide association studies. *New England Journal of Medicine*, 357:436–439, 2007.

[188] M. Hurles. How homologous recombination generates a mutable genome. *Human Genomics*, 2:1016–1017, 2005.

[189] D. Huson. Split networks and reticulate networks. In O. Gascuel and M. Steel, editors, *Reconstructing Evolution: New Mathematical and Computational Advances*, pages 247–276. Oxford University Press, 2007.

[190] D. Huson and D. Bryant. Application of phylogenetic networks in evolutionary studies. *Molelular Biology and Evolution*, 23:254–267, 2006.

[191] D. Huson and H. Kloepper. Computing recombination networks from binary sequences. *Bioinformatics, supplement 2*, 21:ii159–ii165, 2005.

[192] D. Huson and T. Klopper. Beyond galled trees — Decomposition and computation of galled networks. In *RECOMB, The Annual International Conference on Research in Computational Molecular Biology*, pages 211–225. LNBI 4453, Springer, 2007.

[193] D. Huson, T. Klopper, P. Lockhart, and M. Steel. Reconstruction of reticulate networks from gene trees. In *RECOMB, The Annual International Conference on Research in Computational Molecular Biology*, pages 233–249. LNBI 3500, Springer, 2005.

[194] D. Huson, R. Rupp, V. Berry, P. Gambette, and C. Paul. Computing galled networks from real data. *Bioinformatics*, 25(12):i85–i93, 2009.

[195] D. Huson, R. Rupp, and C. Scornavacca. *Phylogenetic Networks*. Cambridge University Press, 2010.

[196] D. Huson and C. Scornavacca. A survey of combinatorial methods for phylogenetic networks. *Genome Biology and Evolution*, 3:23–35, 2010.

[197] J. R. Huyghe and K. L. Mohlke et al. Exome array analysis identifies new loci and low-frequency variants influencing insulin processing and secretion. *Nature Genetics*, 45:197–201, 2013.

[198] International HapMap Consortium. The HapMap project. *Nature*, 426:789–796, 2003.

[199] International HapMap Consortium. A haplotype map of the human genome. *Nature*, 437:1299 1320, 2005.

[200] International HapMap 3 Consortium. Integrating common and rare genetic variation in diverse human populations. *Nature*, 467:52–58, 2010.

[201] P. Janvier. MicroRNAs revive old views about jawless vertebrate divergence and evolution. *Proceedings of the National Academy of Sciences (USA)*, 107:19137–19138, 2010.

[202] A. Javed and L. Parida. *Recombinomics*: Population genomics from a recombination perspective. In *Proceedings of the Third C* Conference on Computer Science and Software Engineering*, pages 129–137, 2010.

[203] A. Javed, M. Pybus, M. Melé, F. Utro, J. Bertranpetit, F. Calafell, and L. Parida. IRiS: Construction of ARG networks at genomic scales. *Bioinformatics*, 27(17):2448–2450, 2011.

[204] A. Jeffreys and R. Neumann. The rise and fall of a human recombination hot spot. *Nature Genetics*, 41:625–629, 2009.

[205] A. J. Jeffreys and C. A. May. Intense and highly localized gene conversion activity in human meiotic crossover hot spots. *Nature Genetics*, 36:151–156, 2004.

[206] A. J. Jeffreys and R. Neumann. Reciprocal crossover asymmetry and meiotic drive in a human recombination hot spot. *Nature Genetics*, 31:267–271, 2002.

[207] A.J. Jeffreys, L. Kauppi, and R. Neumann. Intensely punctated meiotic recombination in the class ii region of the major histocompatibility complex. *Nature Genetics*, 29:217–222, 2001.

[208] A.J. Jeffreys, R. Neumann, M. Panayi, S. Myers, and P. Donnelly. Human recombination hot spots hidden in regions of strong marker association. *Nature Genetics*, 37:601–606, 2005.

[209] P.A. Jenkins, Y.S. Song, and R.B. Brem. Genealogy-based methods for inference of historical recombination and gene flow and their application in *Saccharomyces cerevisiae*. *PLoS ONE*, 7:e46947, 2012.

[210] Z. Jiang, P. Pevzner, and E. Eichler et al. Ancestral reconstruction of segmental duplications reveals punctuated cores of human genome evolution. *Nature Genetics*, 39:1361–1368, 2007.

[211] L. Jin, P. Underhill, V. Doctor, R. Davis, P. Shen, L. Luca Cavalli-Sforza, and P. Oefner. Distribution of haplotypes from a chromosome 21 region distinguishes multiple prehistoric human migrations. *Proceedings of the National Academy of Sciences (USA)*, 96:3796–3800, 1999.

[212] H.R. Johnston and D. J. Cutler. Population demographic history can cause the appearance of recombination hotspots. *American Journal of Human Genetics*, 90:774–783, 2012.

[213] J. Kaiser. Genetic influences on diseases remain rare. *Science*, 338:1016–1017, 2012.

[214] K. Kalpakis and P. Namjoshi. Haplotype phasing using semidefinite programming. In *Proceedings of IEEE Conference on Bioinformatics and Bioengineering*, pages 145–152, 2005.

[215] S. Kannan and T. Warnow. Inferring evolutionary history from DNA sequences. *SIAM Journal on Computing*, 23:713–737, 1994.

[216] S. Kannan and T. Warnow. A fast algorithm for the computation and enumeration of perfect phylogenies when the number of character states is fixed. *SIAM Journal on Computing*, 26:1749–1763, 1997.

[217] E. K. Karlsson and K. Lindblad-Toh et al. Efficient mapping of Mendelian traits in dogs through genome-wide association. *Nature Genetics*, 39:1321–1328, 2007.

[218] J. D. Kececioglu and D. Gusfield. Reconstructing a history of recombinations from a set of sequences. *Discrete Applied Mathematics*, 88:239–260, 1998.

[219] S. Kelk, C. Scornavacca, and L. van Iersel. On the elusiveness of clusters. *IEEE/ACM Transactions on Computational Biology and Bioinformatics*, 9:517–534, 2012.

[220] S. Kelk, C. Scornavacca, L. van Iersel, and C. Whidden. Personal communication, 2012.

[221] H. L. Kim, L. Hie, and S. Schuster. Poor man's 1000 genome project: Recent human population expansion confounds the detection of disease alleles in 7,098 complete mitochondrial genomes. *Frontiers in Genetics*, 4(13), 2013.

[222] G. Kimmel, R. Karp, M. Jordan, and E. Halperin. Association mapping and significance estimation via the coalescent. *American Journal of Human Genetics*, 83:675–683, 2008.

[223] G. Kimmel and R. Shamir. GERBIL: Genotype resolution and block identification using likelihood. *Proceedings of the National Academy of Sciences (USA)*, 102:158–162, 2005.

[224] G. Kimmel and R. Shamir. The incomplete perfect phylogeny haplotype problem. *Journal of Bioinformatics and Computational Biology*, 3:359–384, 2005.

[225] J. Kingman. On the genealogy of large populations. *Journal of Applied Probability*, 19:2743, 1982.

[226] B. Kirkpatrick. Haplotypes versus genotypes on pedigrees. In *WABI, Workshop on Algorithms in Bioinformatics*, volume 6293, pages 136–147. LNCS, Springer, 2010.

[227] C. Klein, K. Lohmann, and A. Ziegler. The promise and limitations of genome-wide association studies. *The Journal of the American Medical Association*, 308(18):1867–1868, 2012.

[228] J. Kleinberg and E. Tardos. *Algorithm Design*. Addison-Wesley Longman, 2005.

[229] A. Kong and K. Stefansson et al. A high-resolution recombination map of the human genome. *Nature Genetics*, 31:241–247, 2002.

[230] M. Kreitman. Nucleotide polymorphism at the alcohol dehydrogenase locus of *Drosophila melanogaster*. *Nature*, 304:412–417, 1983.

[231] M.K. Kuhner and J. Felsenstein. Sampling among haplotype resolutions in a coalescent-based genealogy sampler. *Genetic Epidemiology*, 19:S15–S21, 2000.

[232] V. Kunin, L. Goldovsky, N. Darzentas, and C. A. Ouzounis. The net of life: Reconstructing the microbial phylogenetic network. *Genome Research*, 15:954–959, 2005.

[233] P.Y. Kwok. Genomics: Genetic association by whole-genome analysis? *Science*, 294:1669–1670, 2001.

[234] L.A. Hindorff LA, J. MacArthur, J. Morales, H.A. Junkins HA, P.N. Hall, A.K. Klemm, and T.A. Manolio. A catalog of published genome-wide association studies. Available at: www.genome.gov/gwastudies.

[235] M. Lajoie and N. El-Mabrouk. Recovering haplotype structure through recombination and gene conversion. *Bioinformatics*, 21:1173–1179, 2005.

[236] F. Lam. Personal communication, 2010.

[237] F. Lam. Personal communication, 2012.

[238] F. Lam, D. Gusfield, and S. Sridhar. Generalizing the four gamete condition and splits equiv-alence theorem: Perfect phylogeny on three state characters. In S.L. Salzberg and T. Warnow, editors, *WABI, Workshop on Algorithms in Bioinformatics*, volume 5724 of *Lecture Notes in Computer Science*, pages 206–219. Springer, 2009.

[239] F. Lam, D. Gusfield, and S. Sridhar. Generalizing the four gamete condition and splits equiv-alence theorem: Perfect phylogeny on three state characters. *SIAM Journal on Discrete Math*, 25:1144–1175, 2011.

[240] F. Lam, C. H. Langley, and Y. S. Song. On the genealogy of asexual diploids. *Journal of Computational Biology*, 18:415–428, 2011.

[241] F. Lam, R. Tarpine, and S. Istrail. The imperfect ancestral recombination graph reconstruction problem. Upper bounds for recombination and homoplasy. *Journal of Computational Biology*, 17:767–781, 2010.

[242] J. Lam, K. Roeder, and B. Devlin. Haplotype fine mapping by evolutionary trees. *American Journal of Human Genetics*, 66:659–673, 2000.

[243] G. Lancia, C. Pinotti, and R. Rizzi. Haplotyping populations by pure parsimony: Complexity, exact and approximation algorithms. *INFORMS Journal on Computing, Special Issue on Computational Biology*, 16:348–359, 2004.

[244] E. Lander. Mapping heredity. In E. Lander and M. S. Waterman, editors, *Calculating the Secrets of Life*. National Academy Press, 1995.

[245] F. Larribe, S. Lessard, and N. J. Schork. Gene mapping via the ancestral recombination graph. *Theoretical Population Biology*, 62:215–229, 2002.

[246] S. Q. Le and R. Durbin. SNP detection and genotyping from low-coverage sequencing data on multiple diploid samples. *Genome Research*, 21:952–960, 2011.

[247] J. F. Lefebvre and D. Labuda. Fraction of informative recombinations: A heuristic approach to analyze recombination rates. *Genetics*, 178:20692079, 2008.

[248] W. J. LeQuesne. A method of selection of characters in numerical taxonomy. *Systematic Zoology*, 18:201–205, 1969.

[249] S. Levi, G. Sutton, and J. C. Venter et al. The diploid genome sequence of an individual human. *PLoS Biology*, 5, 2007.

[250] N. Lewis-Rogers, K.A. Crandall, and D. Posada. Evolutionary analysis of genetic recombi-nation. In V. Parisi, V. de Fonzo, and F. Aluffi-Pentini, editors, *Dynamical Genetics*, pages 49–78. Research Signpost, 2004.

[251] J. Li and L. Cavalli-Sforza et al. Worldwide human relationships inferred from genome-wide patterns of variation. *Science*, 319:1100–1104, 2008.

[252] N. Li and M. Stephens. Modelling linkage disequilibrium, and identifying recombination hotspots using SNP data. *Genetics*, 165:2213–2233, 2003.

[253] J.Z. Lin, A. Brown, and M. T. Clegg. Heterogeneous geographic patterns of nucleotide sequence diversity between two alcohol dehydrogenase genes in wild barley (*Hordeum vulgare* subspecies *spontaneum*). *Proceedings of the National Academy of Sciences (USA)*, 98:531–536, 2001.

[254] S. Lin, D. Cutler, M. Zwick, and A. Chakravarti. Haplotype inference in random population samples. *American Journal of Human Genetics*, 71:1129–1137, 2002.

[255] S. Linz. *Reticulation in Evolution*. PhD thesis, University of Düsseldorf, 2008.

[256] S. Linz and C. Semple. Hybridization in nonbinary trees. *IEEE/ACM Transactions on Computational Biology and Bioinformatics*, 6:30–45, January 2009.

[257] S. Linz, C. Semple, and T. Stadler. Analyzing and reconstructing reticulation networks under timing constraints. *Journal of Mathematical Biology*, 61:715–737, 2010.

[258] W. Lipski and F. Preparata. Efficient algorithms for finding maximum matchings in convex bipartite graphs and related problems. *Acta Informatica*, 15:329–346, 1988.

[259] X. Liu and Y. X. Fu. Algorithms to estimate the lower bounds of recombination with or without recurrent mutations. *BMC Genomics*, 9:S24, 2008.

[260] Y. Liu and C. Q. Zhang. A linear solution for haplotype perfect phylogeny problem (extended abstract). In *Advances in Bioinformatics and Its Applications*, pages 173–184. World Scientific, 2004.

[261] L. Löfgren. Irredundant and redundant Boolean branch networks. *IRE Transactions on Circuit Theory*, CT-6:158–175, 1959.

[262] J. R. Lupski. Genome mosaicism — One human, multiple genomes. *Science*, 341:358–359, 2013.

[263] R. Lyngsø. Specifying scoring schemes in Kwarg. http://www.stats.ox.ac.uk/~lyngsoe/section26/kwarg_scoring.pdf. [Online; accessed December 30, 2010].

[264] R. Lyngsø. Tools for recombination analysis in the coalescent. http://www.stats.ox.ac.uk/~lyngsoe/section26/. [Online; accessed December 30, 2010].

[265] R. Lyngsø, Y.S. Song, and J. Hein. Minimum recombination histories by branch and bound. In *WABI, Workshop on Algorithms in Bioinformatics*, volume 3692, pages 239–250. LNCS, Springer, 2005.

[266] B. Maher. The case of the missing heritability. *Nature*, 451:18–21, 2008.

[267] L. Mahler, T. Ingram, J. Revell, and J. Losos. Exceptional convergence on the macroevolutionary landscape in island lizard radiations. *Science*, 341:292–295, 2013.

[268] T. Mailund, S. Besenbacher, and M. Schierup. Whole genome association mapping by incompatibilities and local perfect phylogenies. *BMC Bioinformatics*, 7:454, 2006.

[269] E. Mancera, R. Bourgon, A. Brozzi, W. Huber, and L. Steinmetz. High-resolution mapping of meiotic crossovers and non-crossovers in yeast. *Nature*, 454:479–485, 2008.

[270] T. A. Manolio, L. D. Brooks, and F. S. Collins. A HapMap harvest of insights into the genetics of common disease. *Journal of Clinical Investgation 2*, 118(5), 2008.

[271] T. A. Manolio and F. S. Collins. The HapMap and genome-wide association studies in diagnosis and therapy. *Annual Review of Medicine*, 60:443–456, 2009.

[272] J. Marchini, D. Cutler, N. Patterson, M. Stephens, E. Eskin, E. Halperin, S. Lin, Z.S. Qin, H.M. Munro, G.R. Abecasis, and P. Donnelly. A comparison of phasing algorithms for trios and unrelated individuals. *American Journal of Human Genetics*, 78:437–450, 2006.

[273] J. Marchini, P. Donnelly, and L. Cardon. Genome-wide strategies for detecting multiple loci that influence complex diseases. *Nature Genetics*, 37:413–417, 2005.

[274] P. Marjoram and R. Joyce. Practical implications of coalescent theory. In L.S. Heath and N. Ramakrishnan, editors, *Problem Solving Handbook in Computational Biology and Bioinformatics*, pages 63–84. Springer, 2011.

[275] P. Marjoram and J. D. Wall. Fast "coalescent simulation". *BMC Genetics*, 7, 2006.

[276] B. Martin. Endosymbiosis and lateral gene transfer: Biologists know that the tree of life is not a tree, but what are mathematicians doing about it? Lecture Abstract for Current Challenges and Problems in Phylogenetics, Isaac Newton Institute, Cambridge University, September 2–7, 2007.

[277] D. P. Martin, P. Lemey, and D. Posada. Analysing recombination in nucleotide sequences. *Molecular Ecology Resources*, 11:943–955, 2011.

[278] E.R. Martin, J.R. Gilbert, and E.H. Lai et al. Analysis of association at single nucleotide polymorphisms in the APOE region. *Genomics*, 63, 2000.

[279] J. Matsieva. A static formulation of the history bound. Master's Thesis, University of California, Davis, Computer Science, 2014.

[280] J. Maynard-Smith and J. Haigh. The hitch-hiking effect of a favourable gene. *Genetical Research*, 23:23–35, 2 1974.

[281] T. A. McKee and F.R. McMorris. *Topics in Intersection Graph Theory*. SIAM Monographs on Discrete Mathematics, 1999.

[282] F. McMorris. On the compatability of binary qualitative taxonomic haracters. *Bulletin of Mathematical Biology*, 39:133–138, 1977.

[283] G. McVean and N. J. Cardin. Approximating the coalescent with recombination. *Philosophical Transactions of the Royal Society B*, 360:1387–1393, 2005.

[284] G. McVicker, D. Gordon, C. Davis, and P. Green. Widespread genomic signatures of natural selection in Hominid evolution. *PLoS Genetics*, 5:1–16, 2009.

[285] C. A. Meacham. Theoretical and computational considerations of the compatibility of qualitative taxonomic characters. In J. Felsenstein, editor, *Numerical Taxonomy*, pages 304–314. Springer-Verlag Nato ASI series Vol. G1, 1983.

[286] M. Minichiello and R. Durbin. Mapping trait loci using inferred ancestral recombination graphs. *American Journal of Human Genetics*, 79:910–922, 2006.

[287] T. Mitchell-Olds, J. H. Willis, and D. B. Goldstein. Which evolutionary processes influence natural genetic variation for phynotypic traits? *Nature Reviews: Genetics*, 8(11):845–856, 2007.

[288] B. Moret, L. Nakhleh, T. Warnow, C.R. Linder, A. Tholse, A. Padolina, J. Sun, and R. Timme. Phylogenetic networks: Modeling, reconstructibility, and accuracy. *IEEE/ACM Transactions on Computational Biology and Bioinformatics*, pages 13–23, 2004.

[289] P.L. Morrell, D.M. Toleno, K.E. Lundy, and M.T. Clegg. Estimating the contribution of mutation, recombination and gene conversion in the generation of haplotypic diversity. *Genetics*, 173, 2006.

[290] A. P. Morris, J. C. Whittaker, and D. J. Balding. Fine-scale mapping of disease loci via shattered coalescent modeling of genealogies. *American Journal of Human Genetics*, 70:686–707, 2002.

[291] A.P. Morris, J.C. Whittaker, and D.J. Balding. Little loss of information due to unknown phase for fine-scale linkage-disequilibrium mapping with single-nucleotide-polymorphism genotype data. *American Journal of Human Genetics*, 74:945–953, 2004.

[292] D. Morrison. Networks in phylogenetic analysis: new tools for population biology. *International Journal for Parasitology*, 35:567–582, 2005.

[293] D. Morrison. Phylogenetic networks in systematic biology (and elsewhere). In *Research Advances in Systematic Biology*, pages 1–48, Trivandrum, India, 2009. Global Research Network.

[294] D. Morrison. *Introduction to Phylogenetic Networks*. RJR Productions, 2011.

[295] J. Mu, R. Myers, and X. Su et al. *Plasmodium falciparum* genome-wide scans for positive selection, recombination hot spots and resistance to antimalarial drugs. *Nature Genetics*, 42:268–271, 2010.

[296] S. Mukherjee. *The Emperor of All Maladies*. Simon and Schuster, 2010.

[297] M. Mutsuddi, D. Morriss, S. Waggoner, M. Daly, E. Scolnick, and P. Sklar. Analysis of high-resolution HapMap of DTNBP1 (Dysbindin) suggests no consistency between reported common variant associations and schizophrenia. *American Journal of Human Genetics*, 79:903–909, 2006.

[298] S. Myers. Personal communication, 2004.

[299] S. Myers. *The Detection of Recombination Events Using DNA Sequence Data*. PhD thesis, University of Oxford, Department of Statistics, 2003.

[300] S. Myers, L. Bottolo, C. Freeman, G. McVean, and P. Donnelly. A fine-scale map of recombination rates and hotspots across the human genome. *Science*, 310:321–324, 2005.

[301] S. Myers, R. Bowden, A. Tumian, R. Bontrop, C. Freeman, T. MacFie, G. McVean, and P. Donnelly. Drive against hotspot motifs in primates implicates the prdm9 gene in meiotic recombination. *Science*, 327(5967):876–879, 2010.

[302] S. Myers and R. C. Griffiths. Bounds on the minimum number of recombination events in a sample history. *Genetics*, 163:375–394, 2003.

[303] A. Nahajan and A. Morris et al. Genome-wide trans-ancestry meta-analysis provides insight into the genetic architecture of type 2 diabetes susceptibility. *Nature Genetics*, 46:234–244, 2014.

[304] L. Nakhleh. A metric on the space of reduced phylogenetic networks. *IEEE/ACM Transactions on Computational Biology and Bioinformatics*, 7(2), 2010.

[305] L. Nakhleh. Evolutionary phylogenetic networks: Models and issues. In L. S. Heath and N. Ramakrishnan, editors, *Problem Solving Handbook in Computational Biology and Bioinformatics*, pages 125–158. Springer, 2011.

[306] L. Nakhleh, D. Ringe, and T. Warnow. Perfect phylogenetic networks: A new methodology for reconstructing the evolutionary history of natural languages. *Language*, 81:382–420, 2005.

[307] L. Nakhleh, J. Sun, T. Warnow, C.R. Linder, B.M.E. Moret, and A. Tholse. Towards the development of computational tools for evaluating phylogenetic network reconstruction methods. In *Proceedings of the Pacific Symposium on Biocomputing (PSB)*, pages 315–326, 2003.

[308] L. Nakhleh, T. Warnow, C.R. Linder, and K. St. John. Reconstructing reticulate evolution in species — Theory and practice. *Journal of Computational Biology*, 12:796–811, 2005.

[309] M. Nelson and V. Mooser et al. An abundance of rare functional variants in 202 drug target genes sequenced in 14,002 people. *Science*, 337:100–104, 2013.

[310] C. T. Nguyen, N. B. Nguyen, W. K. Sung, and L. Zhang. Reconstructing recombination network from sequence data: The small parsimony problem. *IEEE/ACM Transactions on Computational Biology and Bioinformatics*, 4(3):394–402, 2007.

[311] D. Nickerson, S. Taylor, K. Weiss, and A. Clark et al. DNA sequence diversity in a 9.7-kb region of the human lipoprotein lipase gene. *Nature Genetics*, 19:233–240, 1998.

[312] R. Nielsen, I. Hellman, M Hubisz, C. Bustamante, and A. Clark. Recent and ongoing selection in the human genome. *Nature Reviews: Genetics*, 8:857–868, 2007.

[313] NIH. National Library of Medicine, Online Mendelian inheritance in man, http://www.ncbi.nlm.nih.gov/omim.

[314] M. Nikaido, A. P. Rooney, and N. Okada. Phylogenetic relationships among cetartiodactyls based on insertions of short and long interpersed elements: Hippopotamuses are the closest extant relatives of whales. *Proceedings of the National Academy of Sciences (USA)*, 96:10261–10266, 1999.

[315] M. Nordborg. Coalescent theory. In D. Balding, M. Bishop, and C. Cannings, editors, *Handbook of Statistical Genetics*, pages 179–212. Wiley, 2001.

[316] M. Nordborg and S. Tavare. Linkage disequilibrium: What history has to tell us. *Trends in Genetics*, 18:83–90, 2002.

[317] J. Novembre, J. K. Pritchard, and G. Coop. Adaptive drool in the gene pool. *Nature Genetics*, 39:1188–1190, 2007.

[318] K. O'Donnell, H. Krister, B. Tacke, and H. Casper. Gene fenealogies reveal global phylogeographic structure and reproductive isolation among lineages of *Fusarium graminearum*, the fungus causing wheat scab. *Proceedings of the National Academy of Sciences (USA)*, 97:7905–7910, 2000.

[319] Y. Okada and R. M. Plenge et al. Genetics of rheumatoid arthritis contributes to biology and drug discovery. *Nature*, 506:376–381, 2014.

[320] International Commission on Zoological Nomenclature. *INTERNATIONAL CODE OF ZOOLOGICAL NOMENCLATURE, Fourth Edition*. The International Trust for Zoological Nomenclature, 1999.

[321] S.H. Orzack, D. Gusfield, J. Olson, S. Nesbitt, L. Subrahmanyan, and Jr. V. P. Stanton. Analysis and exploration of the use of rule-based algorithms and consensus methods for the inferral of haplotypes. *Genetics*, 165:915–928, 2003.

[322] B. Padhukasahasram and B. Rannala. Bayesian population genomic inference of crossing over and gene conversion. *Genetics*, 189:607–619, 2011.

[323] F. Pan, L. McMillan, F. Pardo-Manuel De Villena, D. Threadgill, and W. Wang. TreeQA: Quantitative genome wide association mapping using local perfect phylogeny trees. In *Proceedings of the Pacific Symposium on Biocomputing*, pages 415–426. World Scientific Press, 2009.

[324] L. Parida. Ancestral recombinations graph: A reconstructability perspective using random-graphs framework. *Journal of Computational Biology*, 17:1345–1370, 2010.

[325] L. Parida. Combinatorics in recombinational population genomics. In *ISBRA, International Symposium on Bioinformatics Research and Applications*, pages 126–127. LNCS 6053, Springer, 2010.

[326] L. Parida. Non-reduncant representation of ancestral recombination graphs. In M. Anisimova, editor, *Evolutionary Genomics: Statistical and Computational Methods, Volume 2*, volume 856 of *Methods in Molecular Biology*, chapter 13, pages 315–332. Springer, 2012.

[327] L. Parida, A. Javed, M. Melé, F. Calafell, J. Bertranpetit, and the Genographic Consortium. Minimizing recombinations in consensus networks for phylogeographic studies. *BMC Bioinformatics*, 10(S-1), 2009.

[328] L. Parida, M. Melé, F. Calafell, J. Bertranpetit, and the Genographic Consortium. Estimating the ancestral recombinations graph (ARG) as compatible networks of SNP patterns. *Journal of Computational Biology*, 15:1133–1153, 2008.

[329] L. Parida, P. F. Palamara, and A. Javed. A minimal descriptor of an ancestral recombinations graph. *BMC Bioinformatics*, 12 (Suppl 1):S6, 2011.

[330] J. Paul and Y. S. Song. A principled approach to deriving approximate conditional sampling distributions in population genetics models with recombination. *Genetics*, 186:321–338, 2010.

[331] J. Paul, M. Steinrücken, and Y. S. Song. An accurate sequentially markov conditional sampling distribution for the coalescent with recombination. *Genetics*, 187:1115–1128, 2011.

[332] I. Pe'er, T. Pupko, R. Shamir, and R. Sharan. Incomplete directed perfect phylogeny. *SIAM Journal on Computing*, 33:590–607, 2004.

[333] E. Pennisi. Human evolution: More genomes from Denisova cave show mixing of early human groups. *Science*, 340:799, 2013.

[334] G. Perry and N. J. Dominy et al. Diet and evolution of human amylase gene copy number variation. *Nature Genetics*, 39:1256–1260, 2007.

[335] J. Pickrell, G. Coop, J. Novembre, and J. Pritchard et al. Signals of recent positive selection in a worldwide sample of human populations. *Genome Research*, 19:826–837, 2009.

[336] T. Piovessan and S.M. Kelk. A simple fixed parameter tractable algorithm for computing the hybridization number of two (not necessarilly binary) trees. *IEEE/ACM Transactions on Computational Biology and Bioinformatics*, 10:18–25, 2013.

[337] C. Plowe. Malaria: Resistance nailed. *Nature*, 505:30–31, 2014.

[338] D. Posada and K. Crandall. Intraspecific gene genealogies: Trees grafting into networks. *Trends in Ecology and Evolution*, 16:37–45, 2001.

[339] D. Posada, K. A. Crandall, and E. Holmes. Recombination in evolutionary genomics. *Annual Review of Genetics*, 36:75–97, 2002.

[340] G.D. Poznik and C. D. Bustamante et al. Sequencing Y chromosomes resolves discrepancy in time to common ancestor of males versus females. *Science*, 341:562–565, 2013.

[341] J. K. Pritchard, J. K. Pickrell, and G. Coop. The genetics of human adaptation: Hard sweeps, soft sweeps, and polygenic adaptation. *Current Biology*, 20:R208–R215, 2010.

[342] S. Proulx, D. Promislow, and P. Phillips. Network thinking in ecology and evolution. *Trends in Ecology and Evolution*, 20, 2005.

[343] K. Prüfer and S. Pääbo et al. The complete sequence of a Neanderthal from the Altai mountains. *Nature*, 505:43 – 49, 2014.

[344] M. Przeworski. Motivating hotspots. *Science*, 310:247–248, 2005.

[345] T. Przytycka, G. Davis, N. Song, and D. Durand. Graph theoretical insights into evolution of multidomain proteins. *Journal of Computational Biology*, 13:351–363, 2006.

[346] P. Puigbo, Y. Wolf, and E. V. Koonin. The tree and net components of prokaryote evolution. *Genome Biology and Evolution*, 2:745–756, 2010.

[347] J. Qi, A. Wijeratne, L. Tomsho, Y. Hu, S Schuster, and H. Ma. Characterization of meiotic crossovers and gene conversion by whole-genome sequencing in *Saccharomyces cerevisiae*. *BMC Genomics (online)*, 10:475, 2009.

[348] M. Rasmussen, M. Hubisz, I. Gronau, and A. Siepel. Genome-wide inference of ancestral recombination graphs. ArXiv 1306.5110v3 [q-bio.PE] December 3, 2013.

[349] P. Rastas, M. Koivisto, H. Mannila, and E. Ukkonen. A hidden Markov technique for haplotype reconstruction. In *WABI, Workshop on Algorithms in Bioinformatics*, volume 3692, pages 140–151. LNCS, Springer, 2005.

[350] P. Rastas and E. Ukkonen. Haplotype inference via hierarchical genotype parsing. In *WABI, Workshop on Algorithms in Bioinformatics*, volume 4645, pages 85–97. LNCS, Springer, 2007.

[351] D. A. Ray, J. Xing, A-H. Salem, and M. A. Batzer. SINEs of the *nearly* perfect character. *Systematic Biology*, 55:928–935, 2006.

[352] R. Redon and M. Hurles et al. Global variation in copy number in the human genome. *Nature*, 444:444–454, 2006.

[353] N. Risch and K. Merikangas. The future of genetic studies of complex human diseases. *Science*, 275:1516–1517, 1996.

[354] R. J. Robbins. Introduction to the republication of the 1913 paper by A. H. Sturtevant: The linear arrangement of six sex-linked factors in *Drosophila*, as shown by their mode of association. Republication by Electronic Scholarly Publishing, www.esp.org, 1998.

[355] D.F. Robinson and L. R. Foulds. Comparison of phylogenetic trees. *Mathematical Bioscience*, 53:131–147, 1981.

[356] I. B. Rogozin, Y. I. Wolf, V. N. Babenko, and E. V. Koonin. Dollo parsimony and the reconstruction of genome evolution. In V. A. Albert, editor, *Parsimony, Phylogeny, and Genomics*. Oxford University Press, 2006.

[357] A. Rokas and P. Holland. Rare genomic changes as a tool for phylogenetics. *Trends in Evolution and Ecology*, 15:454–459, 2000.

[358] F. Rossell and G. Valiente. All that glisters is not galled. *Mathematical Biosciences*, 221:54–59, 2009.

[359] E. Ruark and M. Rahman et al. Mosaic PPM1D mutations are associated with predisposition to breast and ovarian cancer. *Nature*, 493:406–410, 2013.

[360] P. Sabeti, D. Reich, and E. Lander et al. Detecting recent positive selection in the human genome from haplotype structure. *Nature*, 419:832–837, 2002.

[361] P. Sabeti, S. Schaffner, and E. Lander et al. Positive natural selection in the human lineage. *Science*, 312:1614–1620, 2006.

[362] P. Sabeti, P. Varilly, and P. Fry et al. Genome-wide detection and characterization of positive selection in human populations. *Nature*, 449:913–918, 2007.

[363] R. Salari and S. Batzoglou et al. Inference of tumor phylogenies with improved somatic mutation discovery. In *RECOMB, The Annual International Conference on Research in Computational Molecular Biology*, pages 246–263. LNBI 7821, Springer, 2013.

[364] M. Sanderson and L. Hufford. *Homoplasy: The Recurrence of Similarity in Evolution*. Academic Press, 1996.

[365] I. Sandovici, S Kassovska-Bratinova, J. Vaughn, R. Stewart, M. Leppert, and C. Sapienza. Human imprinted chromosomal regions are historical hot-spots of recombination. *PLoS Genetics*, 2:944–954, 2006.

[366] R. V. Satya and A. Mukherjee. An optimal algorithm for perfect phylogeny haplotyping. In *Proceedings of the CSB Bioinformatics Conference*. IEEE Press, 2005.

[367] R. V. Satya and A. Mukherjee. An optimal algorithm for perfect phylogeny haplotyping. *Journal of Computational Biology*, 13(4):897–928, 2006.

[368] R. V. Satya and A. Mukherjee. The undirected incomplete perfect phylogeny problem. *IEEE/ACM Transactions on Computational Biology and Bioinformatics*, 5:618–629, 2008.

[369] R.V. Satya, A. Mukherjee, G. Alexe, L. Parida, and G. Bhanot. Constructing near-perfect phylogenies with multiple homoplasy events. *Bioinformatics*, 22:e514–i522, 2006. Bioinformatics Supplement, Proceedings of ISMB.

[370] P. Scheet and M. Stephens. A fast and flexible statistical model for large-scale population genotype data: applications to inferring missing genotypes and haplotypic phase. *American Journal of Human Genetics*, 78:629–644, 2006.

[371] M. H. Schierup and J. Hein. Consequences of recombination on traditional phylogenetic analysis. *Genetics*, 156:879–891, 2000.

[372] C. Semple. Hybridization networks. In O. Gascuel and M. Steel, editors, *Reconstructing Evolution: New Mathematical and Computational Advances*, pages 277–309. Oxford University Press, 2007.

[373] C. Semple and M. Steel. *Phylogenetics*. Oxford University Press, 2003.

[374] P. Sevon, H. Toivonen, and V. Ollikainen. TreeDT: Tree pattern mining for gene mapping. *IEEE/ACM Transactions on Computational Biology and Bioinformatics*, 3:174–185, 2006.

[375] A. J. Sharp and E. E. Eichler et al. Segmental duplications and copy-number variation in the human genome. *Amercian Journal of Human Genetics*, 77:78–88, 2005.

[376] C. J. Shaw and J. R. Lupski. Implications of human genome architecture for rearrangement-based disorders: The genomic basis of disease. *Human Molecular Genetics*, 13:R57–R64, 2004.

[377] S. Sheehan, K. Harris, and Y. S. Song. Estimating variable effective population sizes from multiple genomes: A sequentially markov conditional sampling distribution approach. *Genetics*, 194:647–66, 2013.

[378] B. Shutters and D. Fernandez-Baca. A simple characterization of the minimal obstruction sets for three-state perfect phylogenies. *Applied Math Letters*, 25:1226–1229, 2012.

[379] B. Shutters, S. Vakati, and D. Fernandez-Baca. Improved lower bounds on the compatibility of quartets, triplets, and multi-state characters. In B. Raphael and J. Tang, editors, *Algorithms in Bioinformatics*, volume 7534 of *Lecture Notes in Computer Science*, pages 190–200. Springer, Berlin, 2012.

[380] A. Siepel. Phylogenomics of primates and their ancestral populations. *Genome Research*, 19:1929–1941, 2009.

[381] R. Sladek and P. Froguel et al. A genome-wide association study identifies novel risk loci for type 2 diabetes. *Nature*, 445:881–885, 2007.

[382] F. Smagulova and G. Petukhova et al. Genome-wide analysis reveals novel molecular features of mouse recombination hotspots. *Nature*, 472:375–378, 2011.

[383] P. Sneath, M. J. Sackin, and R. P. Ambler. Detecting evolutionary incompatibilities from protein sequences. *Systematic Zoology*, 24:311–332, 1975.

[384] J. Soares and M. Stefanes. Algorithms for maximum independent set in convex bipartite graphs. *Algorithmica*, 53:35–49, 2009.

[385] Y. S. Song. On the combinatorics of rooted binary phylogenetic trees. *Annals of Combinatorics*, 7:365–379, 2003.

[386] Y. S. Song. Properties of subtree-prune-and-regraft operations on totally-ordered phylogenetic trees. *Annals of Combinatorics*, 10:129–146, 2006.

[387] Y. S. Song, Z. Ding, D. Gusfield, C. H. Langley, and Y. Wu. Algorithms to distinguish the role of gene-conversion from single-crossover recombination in the derivation of SNP sequences in populations. *Journal of Computational Biology*, 14:1273–1286, 2007.

[388] Y. S. Song and J. Hein. Parsimonious reconstruction of sequence evolution and haplotype blocks: Finding the minimum number of recombination events. In *WABI, Workshop on Algorithms in Bioinformatics*, volume 2812, pages 287–302. LNCS, Springer, 2003.

[389] Y. S. Song and J. Hein. On the minimum number of recombination events in the evolutionary history of DNA sequences. *Journal of Mathematical Biology*, 48:160–186, 2004.

[390] Y. S. Song, Y. Wu, and D. Gusfield. Efficient computation of close lower and upper bounds on the minimum number of needed recombinations in the evolution of biological sequences. *Bioinformatics*, 21:i413–i422, 2005. Bioinformatics Suppl. 1, Proceedings of ISMB 2005.

[391] Y. S. Song, Y. Wu, and D. Gusfield. Efficient computation of close lower and upper bounds on the minimum number of needed recombinations in the evolution of biological sequences. *Bioinformatics*, 21:i413–i422, 2005. Bioinformatics Suppl. 1, Proceedings of ISMB 2005.

[392] Y. S. Song, Y. Wu, and D. Gusfield. Haplotyping with one homoplasy or recombination event. In *WABI, Workshop on Algorithms in Bioinformatics*, volume 3692, pages 152–164. LNCS, Springer, 2005.

[393] Y. S. Song. A concise necessary and sufficient condition for the existence of a galled-tree. *IEEE/ACM Transactions on Computational Biology and Bioinformatics*, 3:186–191, 2006.

[394] Y. S. Song and J. Hein. Constructing minimal ancestral recombination graphs. *Journal of Computational Biology*, 12:159–178, 2005.

[395] M. Steel. The complexity of reconstructing trees from qualitative characters and subtrees. *Journal of Classification*, 9:91–116, 1992.

[396] J. C. Stephens and J. F. Vovis et al. Haplotype variation and linkage disequilibrium in 313 human genes. *Science*, 293:489–493, 2001.

[397] J.C. Stephens. On the frequency of undetectable recombination events. *Genetics*, 112:923–926, 1986.

[398] M. Stephens, N. Smith, and P. Donnelly. A new statistical method for haplotype reconstruction from population data. *American Journal of Human Genetics*, 68:978–989, 2001.

[399] K. Stevens and D. Gusfield. Reducing multi-state to binary perfect phylogeny with applications to missing, removable, inserted, and deleted data. In V. Moulton and M. Singh, editors, *WABI, Workshop on Algorithms in Bioinformatics*, volume 6293 of *Lecture Notes in Computer Science*, pages 274–287. Springer, 2010.

[400] B. E. Stranger, A. C. Nica, and E. T. Dermitzakis. Populations genomics of human gene expression. *Nature Genetics*, 39:1217–1224, 2007.

[401] A. H. Sturtevant. The linear arrangement of six sex-linked factors in *Drosophila*, as shown by their mode of association. *Journal of Experimental Zoology*, 14:43–59, 1913.

[402] Y. Sun and J. Ambrose et al. Deep genome-wide measurement of meiotic gene conversion using tetrad analysis in *Arabidopsis thaliana*. *PLoS Genetics*, 8:e1002968, 10 2012.

[403] N. Sutter, C. D. Bustamante, and E. Ostrander et al. A single IGF1 allele is a major determinant of small size in dogs. *Science*, 316:112–115, 2007.

[404] K. Swenson, P. Guertin, H. Deschênes, and A. Bergeron. Reconstructing the modular recombination history of staphylococcus aureus phages. *BMC Bioinformatics*, 14(Suppl 15:S17), 2013.

[405] I. Tachmazidou, C. Verzilli, and M. De Iorio. Genetic association mapping via evolution-based clustering of haplotypes. *PLoS Genetics*, 3:e111, 07 2007.

[406] K. Tang, K. Thornton, and M. Stoneking. A new approach for using genome scans to detect recent positive selection in the human genome. *PLoS Biology*, 5:1587–1602, 2007.

[407] M. Tennesen and J. Akey et al. Evolution and functional impact of rare coding variation from deep sequencing of human exomes. *Science*, 337:64–69, 2013.

[408] K.M. Teshima, G. Coop, and M. Przeworski. How reliable are empirical genome scans for selective sweeps? *Genome Research*, 16:702–712, 2006.

[409] S. Tishkoff and F. Reed et al. Convergent adaptation of human lactase persistence in Africa and Europe. *Nature Genetics*, 39:31–40, 2007.

[410] W.T. Tutte. An algorithm for determining whether a given binary matroid is graphic. *Proceedings of the American Mathematical Society*, 11:905–917, 1960.

[411] E. Ukkonen. Finding founder sequences from a set of recombinants. In *WABI, Workshop on Algorithms in Bioinformatics*, volume 2452, pages 277–286. LNCS, Springer, 2002.

[412] F. Utro, O. Cornejo, D. Livingstone, J.C. Motamayor, and L. Parida. ARG-based genome-wide analysis of cacao cultivars. *BMC Bioinformatics*, 13: Suppl 19, 2012.

[413] L. van Iersel. *Algorithms, Haplotypes and Phylogenetic Networks*. PhD thesis, Technische Universiteit Eindhoven, the Netherlands, 2009.

[414] L. van Iersel. Different topological restrictions of rooted phylogenetic networks. Which make biological sense?, March 3, 2013. Available at: phylonetworks.blogspot.com.

[415] L. van Iersel, J. Keijsper, S. Kelk, L. Stougie, F. Hagen, and T. Boekhout. Constructing level-2 phylogenetic networks from triplets. *IEEE/ACM Transactions on Computational Biology and Bioinformatics*, 6(4):667–681, 2009.

[416] L. van Iersel and S. Kelk. When two trees go to war. *Journal of Theoretical Biology*, 269:245–255, 2011.

[417] P. M. Visscher, M. A. Brown, M. I. McCarthy, and J. Yang. Five years of GWAS discovery. *American Journal of Human Genetics*, 90:7–24, 2012.

[418] B.F. Voight, S. Kudaravalli, X. Wen, and J.K. Pritchard. A map of recent positive selection in the human genome. *PLoS Biology*, 4, 2006.

[419] C. Wade and M. Daly et al. The mosaic structure of variation in the laboratory mouse genome. *Nature*, 420:574–578, 2002.

[420] D.B. Wake, M.H. Wake, and C.D. Specht. Homoplasy: From detecting pattern to determining process and mechanism of evolution. *Science*, 331:1032–1035, 2011.

[421] J. Wakeley. *Coalescent Theory*. Roberts and Co., 2009.

[422] J. D. Wall. A comparison of estimators of the population recombination rate. *Molecular Biology and Evolution*, 17:156–163, 2000.

[423] J. D. Wall. Close look at gene conversion hot spots. *Nature Genetics*, 36:114–115, 2004.

[424] J. D. Wall and J. K. Pritchard. Haplotype blocks and linkage disequilibrium in the human genome. *Nature Reviews — Genetics*, 4:587–597, 2003.

[425] A. Walsh, R. Kortschak, M. Gardner, T. Bertozzi, and D. Adelson. Widespread horizontal transfer of retrotransposons. *Proceedings of the National Academy of Sciences (USA)*, 110(3):1012–1016, 2013.

[426] L. Wang, K. Zhang, and L. Zhang. Perfect phylogenetic networks with recombination. *Journal of Computational Biology*, 8:69–78, 2001.

[427] J. Watson, N. Hopkins, J. Roberts, J. Steitz, and A. Weiner. *Molecular Biology of the Gene (4th edition)*. Benjamin Cummings, 1987.

[428] Wellcome Trust Case Control Consortium. Genome-wide association study of 14,000 cases of seven common diseases and 3,000 shared controls. *Nature*, 447:661–678, 2007.

[429] C. Whidden, R. Beiko, and N. Zeh. Fixed-parameter algorithms for maximum agreement forests. *SIAM Journal on Computing*, 42:1431–1466, 2013.

[430] H. Whitney. Congruent graphs and the connectivity of graphs. *American Journal of Mathematics*, 54:150–168, 1932.

[431] H. Whitney. 2-isomorphic graphs. *American Journal of Mathematics*, 55:245–254, 1933.

[432] E. O. Wilson. A consistency test for phylogenies based on contemporaneous species. *Systematic Zoology*, 14:214–220, 1965.

[433] W. Winckler, S. Myers, and D. Altschuler et al. Comparison of fine-scale recombination rates in humans and chimpanzees. *Science*, 308:107–111, 2005.

[434] C. Wiuf. Inference of recombination and block structure using unphased data. *Genetics*, 166:537–545, 2004.

[435] C. Wiuf, T. Christensen, and J. Hein. A simulation study of the reliability of recombination detection methods. *Molecular Biology and Evolution*, 18:1929–1939, 2001.

[436] C. Wiuf and J. Hein. Recombination as a point process along sequences. *Theoretical Population Biology*, 55:1217–1228, 1999.

[437] K. Wong and R. deLeeuw et al. A comprehensive analysis of common copy-number variations in the human genome. *American Journal of Human Genetics*, 80:91–104, 2007.

[438] Y. Wu. Personal communication, 2011.

[439] Y. Wu. Association mapping of complex diseases with ancestral recombination graphs: Models and efficient algorithms. In T. Speed and H. Huang, editors, *RECOMB, The Annual International Conference on Research in Computational Molecular Biology*, volume 4453, pages 488–502. LNBI, Springer, 2007.

[440] Y. Wu. Association mapping of complex diseases with ancestral recombination graphs: Models and efficient algorithms. *Journal of Computational Biology*, 15:667–684, 2008.

[441] Y. Wu. An analytical upper bound on the minimum number of recombinations in the history of SNP sequences in populations. *Information Processing Letters*, 109:427–431, 2009.

[442] Y. Wu. A practical method for exact computation of subtree prune and regraft distance. *Bioinformatics*, 25(2):190–196, 2009.

[443] Y. Wu. Bounds on the minimum mosaic of population sequences under recombination. In *Proceedings of the Annual Symposium on Combinatorial Pattern Matching*, volume 6129, pages 152–163. LNCS, Springer, 2010.

[444] Y. Wu. Close lower and upper bounds for the minimum reticulate network of multiple phylogenetic trees. *Bioinformatics*, 26:140–148, 2010.

[445] Y. Wu. New methods for inference of local tree topologies with recombinant SNP sequences in populations. *IEEE/ACM Transactions on Computational Biology and Bioinformatics*, 8:182–193, 2011.

[446] Y. Wu. An algorithm for constructing parsimonious hybridization networks with multiple phylogenetic trees. In *RECOMB, The Annual International Conference on Research in Computational Molecular Biology*, volume 7821, pages 291–303. LNBI, Springer, 2013.

[447] Y. Wu and D. Gusfield. Efficient computation of minimum recombination with genotypes (not haplotypes). *Journal of Bioinformatics and Computational Biology*, pages 181–200, 2007.

[448] Y. Wu and D. Gusfield. Improved algorithms for inferring the minimum mosaic of a set of recombinants. In *Proceedings of the Annual Symposium on Combinatorial Pattern Matching*, volume 4580, pages 150–161. LNCS, Springer, 2007.

[449] Y. Wu and D. Gusfield. A new recombination lower bound and the minimum perfect phylogenetic forest problem. *Journal of Combinatorial Optimization*, 16:229–247, 2008.

[450] Y. Wu and J. Wang. Fast computation of the exact hybridization number of two phylogenetic trees. In M. Borodovsky, J.P. Gogarten, T.M. Przytycka, and S. Rajasekaran, editors, *ISBRA, International Symposium on Bioinformatics Research and Applications*, volume 6053, pages 203–214. Springer, 2010.

[451] B. Yalcin, J. Flint, and R. Mott et al. Genetic dissection of a behavioral quantitative trait locus shows that Rgs2 modulates anxiety in mice. *Nature Genetics*, 36:1197–1202, 2004.

[452] J. Yang and M. Visscher. Common SNPs explain a large proportion of the heritability for human height. *Nature Genetics*, 42:565–569, 2010.

[453] T. Yang, H-W Deng, and T. Niu. Critical assessment of coalescent simulators in modeling recombination hotspots in genomic sequences. *BMC Bioinformatics*, 15:3, 2014.

[454] M. Yeager and N. Orr et al. Genome-wide association study of prostate cancer identifies a second risk locus at 8q24. *Nature Genetics*, 39:645–649, 2007.

[455] J. Yin, M. Jordan, and Y.S. Song. Joint estimation of gene conversion rates and mean conversion tract lengths from population SNP data. *Bioinformatics*, 25:Pp. i231–i239, 2009.

[456] J. Zhang, W. Rowe, A. Clark, and K. Buetow. Genomewide distribution of high-frequency, completely mismatching SNP haplotype pairs observed to be common across human populations. *American Journal of Human Genetics*, 73:1073–1081, 2003.

[457] Q. Zhang, L. McMillan, and D. Threadgill et al. Genotype sequence segmentation: Handling constraints and noise. In *WABI, Workshop on Algorithms in Bioinformatics*, volume 5251, pages 271–283. LNCS, Springer, 2008.

[458] Q. Zhang, L. McMillan, and D. Threadgill et al. Inferring genome-wide mosaic structure. In *Proceedings of Pacific Symposium on Biocomputing*, pages 150–161. World Scientific Press, 2009.

[459] Z. Zhang, X. Zhang, and W. Wang. HTreeQA: Using semi-perfect phylogeny trees in quantitative trait loci study on genotype data. *G3*, 2:175–189, 2012.

[460] C. Zimmer. DNA doubletake. *New York Times*, September 16, 2013.

[461] S. Zöllner and J.K. Pritchard. Coalescent-based association mapping and fine mapping of complex trait loci. *Genetics*, 169:1071–1092, 2005.

[462] O. Zuk, E. Hechter, S.R. Sunyaev, and E. S. Lander. The mystery of missing heritability: Genetic interactions create phantom heritability. *Proceedings of the National Academy of Sciences (USA)*, 109:1193–1198, 2012.

Index